suit against roofing company, 335
network coverage of balloon race, 335
tax audit probability, 337
network's new sitcoms, 337
selling solar heating units, 337
demand for new calculus text, 337
audience share, local TV news, **346, 355**
cars on hand in showroom, 350, 359
foreign car share of market, 359
laundry detergent market, **361, 364, 365**
put on ''hold'' calling phone company, 387
store locations in three towns, **394**
branch banks in nearby towns, 397
competing television stations, 407
auto assembly line absentees, **423, 425, 427**
advertising expenditures, 431
tensile strength of cables, 431, 432
moving company rates, 432
weekly sales records of realtors, 433
lifetimes of lightbulbs, 433, 460
auto towing service requests, **435, 436, 439**
plane reservation no-shows, 448, 461
diameter of machine part, 460
poultry demand at market, 460
salaries of workers, 461
typing speed of secretaries, 462
sales of tape decks, 463
credit purchasing, 473, **481,** 486, 489
retirement fund, **482,** 486
depreciation fund, **483,** 486, 489
gifting with investment potential, **484**
municipality's airport bond issue, 486
yearly inheritance to niece, 486
government agency's funding, 487
deposit of inheritance, 489
power company's fuel payments, 489

Social Sciences, Politics, Law, Government, Education

private school mailing community calendar, 36
lawyers' fee plans, 36
urban renewal government grants, 114
diagram of students passing exams, 201
categories of government loan applicants, 201
passing welfare reform compromise bill, 202
convictions of prison inmates, 202
political opinion research poll results, 202
survey of marital satisfaction, 203

IRS auditing tax returns, 203
election combinations, 208
auto consumer rating service, 208
court cases assigned to judges, 208
student exam answer combinations, 209
screening medical school applicants, 218
Supreme Court decisions, 219
grade assignment combinations, **226**
assignment of foreign service officers, 229
composite sketch of crime suspect, 232
mathematics instructor assignments, 232
defendant's sentence probabilities, 247
juror in Baker County, 256
cabinet members to NATO committee, 257
army volunteers examined, **264,** 280
employment of spouses, 270
voters for Senator Foghorn, 270, 280
matchmaking, 270
critics' reviews of French film, 270
commuter observation of late trains, 270
sociological survey of voters, 279
U.S. life table probabilities of survival, 280
court argument—murderer's car, 281
high school substance abusers, 281
law graduates passing bar exam, 290
chances of conviction or acquittal, 290
grades in Russian literature, 294
students in same class, 295
medical school survival rates, 295
international naval maneuvers, 295
screening of prospective airline pilots, 303
reading of textbook by A students, 303
voters for mass transit project, **309**
selecting jurors from registered voters, 311
multiple choice examination, 311
D.A. getting witnesses to testify, 312
test of extra-sensory perception, 313
interest group hiring lobbyist, 334
ethnic voter support, 336
time in graduate program, 337
unemployment trend, 348, 359
rural migration to cities, 349, 359
class mobility of children, 350, 359
political parties of children, 350, 359
steps to doctorate, **365,** 371
diplomatic personnel assignments, 369
judges approving wiretap request, 370
choosing between two defense systems, 397
senatorial candidates on nuclear power, 397

Finite Mathematics

Finite Mathematics

Daniel Gallin

Scott, Foresman and Company
Glenview, Illinois

Dallas, Tex.
Oakland, N.J.
Palo Alto, Cal.
Tucker, Ga.
London, England

To Janet, Juliana, and Dahlia

Library of Congress Cataloging in Publication Data

Gallin, Daniel.
Finite mathematics.

   Includes bibliographical references and index.
   1. Mathematics—1961–   . I. Title
QA39.2.G34 1984   519   83-11640
ISBN 0-673-16048-3

ISBN 0-673-16048-3

 23456-RRC-8887868584

Preface

This text is designed for a course in finite mathematics for students in the management, life, and social sciences. Such a course, offered at the freshman or sophomore level, has become a standard feature of the undergraduate curriculum in American colleges and universities. In evolving over 25 years, the finite math course has moved toward applications in business, economics, and life science, with an emphasis on the idea of mathematical modeling. This book is very much in harmony with that trend.

Prerequisites   It is assumed that the reader has studied at least one year of high school algebra. In Chapter 1 straight lines and linear equations are reviewed, and a review of inequalities has been integrated into Chapter 2. Beyond these topics the only prerequisite is a knowledge of real-number arithmetic. The need for review has been weighed against the risk of boredom; extensive classroom experience has shown that the present approach is satisfactory.

Major Features

Exercises   Both routine drill problems and more difficult word problems are in the exercises at the end of each section. The exercises have all been carefully composed to yield "nice" answers with a minimum of computation. Each section averages twenty to thirty exercises. Review exercises are part of the Summary following each chapter.

Problems of greater-than-average challenge are subtly set off with the exercise number in color.

Chapter Summaries   Following each chapter the Chapter Summary includes review exercises and a listing of important terms and formulas keyed to the sections where they are found.

Highlights Key concepts are boxed for easy reference and emphasis. Boxed procedures and methods appear in black print; boxed definitions, formulas, and other material are in color. Important words and phrases are in boldface type.

Questions that naturally come to a student's mind and pitfalls that often appear are anticipated in this helpful text. **Remarks to Reader** and **Questions/ Answers** address issues from a student's point of view.

Course Plan Most courses in finite mathematics are built around three major clusters of topics: (1) linear equations and matrices, (2) linear programming, and (3) probability theory. Linear programming is introduced early in this text for a logical development of the major ideas. However, the instructor has considerable flexibility in choosing the order of topics. The diagram below indicates the **interdependence of chapters.**

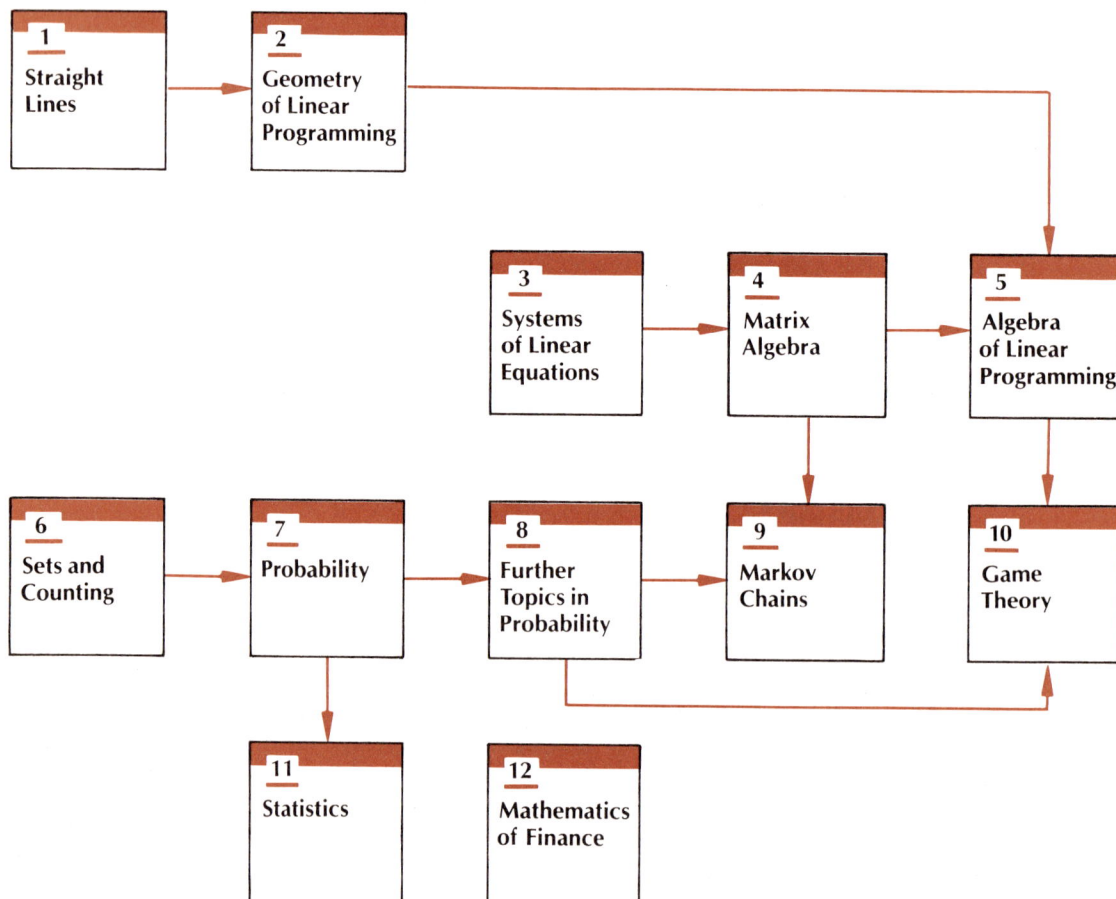

Linear mathematics and matrix algebra are given a thorough introduction in Chapters 2 through 5.

Permutations and combinations are treated in Chapter 6. Throughout Chapters 7 and 8 those examples and exercises that require these more difficult counting techniques are flagged so that the instructor can avoid them if a light treatment of probability is desired. The only exception to this rule is the section on Bernoulli trials (Section 8.2), where combinations are unavoidable. In Chapter 11 the discussion of binomial distributions (Sections 11.2 and 11.3) requires Bernoulli's Formula (Section 8.2).

Either Markov chains (Chapter 9) or game theory (Chapter 10) provides an excellent conclusion for the course, since they both illustrate the interdependence of matrix algebra and probability theory. Mathematics of finance (Chapter 12) can be covered at any point.

Section Length

Most of the sections in Chapters 1–10 can be covered satisfactorily in a 50-minute lecture. However, Gauss-Jordan elimination (Section 3.1), the simplex algorithm (5.1), and stochastic processes (7.5) may require two lectures. Some instructors may choose to spend more time on the Leontief input-output model (4.4), linear programming models (5.3), permutations and combinations (6.3), the meaning of probability (7.1), or laws of probability (7.3).

It is especially important to emphasize the pivot transformation in Section 3.1, since the same procedure enters into solving linear systems, inverting matrices, and solving linear programs by the simplex algorithm.

Applications

Special emphasis has been placed on applications, especially in business and economics, and also in the life and social sciences. Contrived or artificial examples have been avoided. The excitement in this course comes from seeing and understanding the application of a mathematical model to a real (or at least realistic) problem. An index of applications appears on the endsheets of the book.

Some examples are developed in steps throughout a section, interspersed with the general discussion. The application in Example 1, for instance, may reappear in Example 1, Continued, or Example 1, Revisited. The end of each Example is marked with a symbol: an open diamond $\diamond$ if the application will be used again in the section, a solid diamond $\blacklozenge$ if the application is finished.

The starred sections in the table of contents are all "optional" in the sense that they can be omitted without loss of continuity. However, the instructor is urged to take a close look at them; in particular, the applications of linear functions (Section 1.4), Leontief input-output model (4.4), linear programming models (5.3), and decision theory (8.4) are rather sophisticated mathematical models that even the average student can appreciate.

Calculators   It is anticipated that the student will have a hand calculator to work assigned exercises. However, the only topic for which a calculator is essential is mathematics of finance (Chapter 12). If this chapter is to be covered, students should be advised to obtain a calculator with an exponential function (y^x).

Computer Projects   A computer can be used very effectively as a pedagogical aid at various points in the text, for example, in carrying out the simplex algorithm for relatively large-scale programs (Section 5.3), and in Monte Carlo simulation (7.2 and following). At the end of these sections a number of computer projects are formulated which are directly related to the text discussion. Some of these projects are intended for students already familiar with BASIC. The others utilize the BASIC programs listed below, written specifically for this text, which require no computer background.

PIVOT   Performs a pivot transformation on a designated nonzero entry in a tableau or augmented matrix.

SIMPL   Computes the optimal solution to a linear program by the simplex method.

DICE   Simulates repeated rolls of two fair dice by generating a sequence of random integers i and j from 1 to 6. The program counts the frequency of each sum $s = i + j$.

SERIES   Simulates 2000 trials of a complete best-of-five series of games between two teams A and B, where team A has a prescribed probability p of winning each game.

BINOM   For a Bernoulli process with a given n, p, q, the program computes the theoretical probability of obtaining between k_1 and k_2 successes, inclusive.

TRIALS   Simulates a Bernoulli process with a given n, p, q. The program repeats the process a prescribed number of times and counts the frequency of k successes for $k = 0, 1, . . ., n$.

MARKOV   Computes the state vectors of a Markov chain at times $n = 0, 1, . . ., 25$.

RUIN   Simulates a series of games played for unit stakes between a player with A units and an opponent with B units. The series ends when either the player or his opponent is ruined.

GAME   Simulates 2000 plays of an $m \times n$ matrix game between players using prescribed strategies P and Q. The program computes the rowplayer's average net gain.

DATA   Computes the mean, median, and standard deviation for a given set of raw data and groups the data into prescribed class intervals.

STAT	Computes the mean and standard deviation for a given set of grouped data.
NORMAL	Computes the area $A(z)$ under a normal curve, to the left of z.

Instructor's Manual

The computer programs can be found in the Instructor's Manual which includes instructions on their use as well as sample output. The Manual contains answers to the even-numbered exercises and lecture guides with suggestions for presenting text material and pacing the course.

Acknowledgments

Since this project was conceived it has undergone many modifications and improvements. I am greatly indebted to Jack Pritchard and Menton Sveen of Scott, Foresman for their early and continuing enthusiasm and encouragement. Professor Gerald Bradley was responsible for a major restructuring of the work as well as many detailed improvements, and I am similarly indebted to Professors Garrett J. Etgen, Joseph F. Krebs, Alan Lambert, Anthony Peressini, Burton Rodin, and Arthur B. Simon for their very helpful reviews of successive drafts.

I wish to thank also my colleagues at U.S.F., Professors Allan Cruse, James Finch, Betty Gibson, Millianne Lehmann, William McClung, Richard Peddicord, and Robert Wolf, for many useful discussions and suggestions. Howard Sandler and Theresa Harned generously shared with me their expertise on sports and gambling, respectively, and Dr. Julius Greenberg provided an inexhaustible supply of information on medicine and physiology, as well as a good deal of personal encouragement and support. My thanks are also due to Mary Rose Kent, who helped me prepare the final version of the manuscript, and to Gordon Sheridan, for a number of helpful conversations.

I am deeply obligated to the approximately one thousand students who have taken my finite mathematics course. Their enthusiasm and interest inspired this book, and their comments on earlier drafts provided the most valuable reviews of all.

Daniel Gallin

To the Student

Technically speaking, all the mathematics in this book is "elementary." In other words, nothing here requires any special cleverness or ingenuity. A background of one year of high school algebra plus the ability to think clearly and read carefully are all you need to master the subject matter completely. If you get too relaxed, however, you will probably be in for a rude surprise: there are some subtle concepts in these pages, and a great many problems that cannot be solved simply by plugging numbers into a formula. Here are a few words of advice:

1. Read the text.   Don't make the mistake of heading straight for the homework problems armed only with your lecture notes. Also, bear in mind that a mathematics text is not intended for speed-reading. Keep a pencil handy, and make a note of any point you find unclear.

2. Ask questions.   If something confuses you, it probably confuses other students as well. Come to class prepared to ask about it, and clear it up before going on to new material. Don't ever be afraid of asking a "foolish" question.

3. Memorize definitions.   Terminology and symbolism have been held to a minimum in this text, but there is no way to do mathematics without a certain amount of technical jargon. Terms like "combination of objects," "independent event," and "union of sets" have precise, specific meanings. Learn them.

4. Practice.   Do a few additional exercises on your own, when necessary, to reinforce your understanding of a particular concept or technique.

5. Explain your work.   In working assigned exercises show all the details, so that someone reading your work does not have to guess what you had in mind. Remember that the method is far more important than the answer.

6. *Use the calculator wisely.* The hand calculator is a marvelous aid, but be aware of its limitations. An exact value is always preferable to an approximate (rounded-off) value, and exact values should be retained until the last stages of a problem, to avoid introducing roundoff error. For example, when we divide .125 by .875 on a calculator, we obtain a nonterminating decimal

$$\frac{.125}{.875} = .142857 \ldots$$

which is *approximately* equal to the rounded-off value .143 or .1429. To obtain the *exact* value we must convert the numerator and denominator to fractions.

$$\frac{.125}{.875} = \frac{125/1000}{875/1000} = \frac{125}{875} = \frac{1}{7}$$

Almost all the computational exercises in this text will yield "nice" answers in a reasonable amount of time, provided exact values are used. If a problem is taking too long, or unsightly numbers are cropping up (like 73/17), recheck your work. You have almost certainly made a numerical error.

Contents

Introduction   xvii

1 Straight Lines

1.1   **The Equation of a Line**   1
1.2   **Intersecting Lines**   8
1.3   **Linear Functions**   17
★1.4   **Applications of Linear Functions**   27
      Chapter 1 Summary   37

2 The Geometry of Linear Programming

2.1   **Formulation of Linear Programs**   40
2.2   **Linear Inequalities**   48
2.3   **Systems of Linear Inequalities**   57
2.4   **Geometric Solution of Linear Programs**   65
★2.5   **Geometric Solution in Higher Dimensions**   72
      Chapter 2 Summary   81

3 Systems of Linear Equations

3.1 **Gauss-Jordan Elimination**   83

3.2 **Analysis of Linear Systems**   93

★**3.3** **Applications of Linear Systems**   104

Chapter 3 Summary   113

4 Matrix Algebra

4.1 **Matrices and Vectors**   115

4.2 **Matrix Multiplication**   125

4.3 **Matrix Solution of Linear Systems**   135

★**4.4** **The Leontief Input-Output Model**   146

Chapter 4 Summary   158

5 The Algebra of Linear Programming

5.1 **The Simplex Algorithm**   160

5.2 **The Simplex Algorithm: Nonstandard Programs**   172

★**5.3** **Linear Programming Models**   181

Chapter 5 Summary   191

6 Sets and Counting

6.1 **Elements of Set Theory**   193

6.2 **Techniques of Counting**   203

6.3   **Permutations and Combinations**   209

★6.4   **The Binominal Theorem**   221

Chapter 6 Summary   231

7   Probability

7.1   **The Meaning of Probability**   233

7.2   **The Classical Formula**   248

7.3   **Basic Laws of Probability**   259

7.4   **Conditional Probability**   272

7.5   **Stochastic Processes**   281

Chapter 7 Summary   293

8   Further Topics in Probability

★8.1   **Bayes' Formula**   296

8.2   **Bernoulli Trials**   304

8.3   **Expected Value**   314

★8.4   **Decision Theory**   325

Chapter 8 Summary   336

9   Markov Chains

9.1   **Introduction to Markov Chains**   339

9.2   **Regular Markov Chains**   350

9.3   **Absorbing Markov Chains**   360

★9.4   **Gambler's Ruin**   371

★9.5   **Genetics**   377

Chapter 9 Summary   386

10 Game Theory

10.1   **Strictly Determined Games**   388
10.2   **Games of Strategy**   398
10.3   **Simplex Solution of $m \times n$ Games**   409
      Chapter 10 Summary   419

11 Statistics

11.1   **Descriptive Statistics**   421
11.2   **Probability Distributions**   435
11.3   **Normal Distributions**   449
      Chapter 11 Summary   462

12 Mathematics of Finance

12.1   **Simple and Compound Interest**   464
12.2   **Installment Loans and Annuities**   475
      Chapter 12 Summary   488

Appendix

Table 1   **Area Under a Normal Curve**   490
Table 2   **Random Binary Numbers**   492

Answers to Selected Exercises   493

Index   519

Introduction

For at least three centuries mathematics has been recognized as the official language of the natural sciences. Within the last few decades, however, mathematical methods have become increasingly important in business and the social sciences as well.

One reason for this development has been the growth of industry itself, both in size and complexity. Planners now find themselves facing a bewildering array of new problems: How should production levels and prices be determined to prevent shortages? How should limited resources be used to maximize profit? Where should new facilities be located to minimize production and shipping costs? Millions of dollars may ride on the answers to such questions, and it is therefore not surprising that decision makers have turned to mathematics in the search for precise, rational management techniques. Then too, the revolution in computer technology has made it physically and economically practical to apply mathematical methods to complex problems involving hundreds of different variables and vast quantities of empirical data.

The term "finite mathematics," first coined by John G. Kemeny in 1956, comprises those mathematical topics which fall outside of calculus—that is, which do not involve "continuous" variables or limiting processes. In this book we examine a number of these topics, all of which have widespread applications not only in business and economics but also in the physical and biological sciences. While our illustrative examples may involve only three variables rather than three hundred, the underlying principles are exactly the same. Every mathematical concept introduced in this book is actually used, every day, in solving practical problems.

Special attention has been focused throughout this text on **mathematical modeling,** or translating practical problems into mathematical terms. The ability to analyze a problem clearly, to isolate its important features and express them abstractly, is of far greater importance than the knowledge of any particular mathematical technique, for the simple reason that *a problem cannot be solved before it is fully understood*—no matter how much money or computer time is available. Indeed, the growing influence of mathematics in areas outside the physical sciences is not so much due to any purely mathematical discovery, but rather to the development of new mathematical models for real-world problems.

We begin our survey of finite mathematics with a discussion of straight lines and linear equations. Most of this material should be familiar to the reader, but it is worth reviewing since it will provide the basis for some of our later topics.

1

Straight Lines

The straight line is second only to the point in mathematical simplicity. Yet lines, and the algebraic equations that represent them, play a fundamental role in many practical applications. In this chapter we review the basic facts about linear equations and their graphs and give some illustrations of their uses.

1.1 The Equation of a Line

Although points, lines, and planes are geometric entities, algebra can be used to describe them and to derive many of their important properties. This was first discovered by the French philosopher and mathematician René Descartes (1596–1650), who introduced the idea of *rectangular coordinates,* described below, and the familiar method of graphing algebraic equations as curves in the plane.

The Real Number Line Rectangular coordinates in the plane are based on the notion of the **real number line.** To introduce a coordinate system on a given straight line: (i) designate a particular point 0 on the line as the **origin** of the system; (ii) select one of two directions on the line to be the **positive** direction; (iii) choose a **unit length** on the line for measuring distances. With such a coordinate system established, each real number x can be represented as a unique point on the line. To plot the value x, we start at the origin and move x units along the line, taking into account the sign (positive or negative) of x. As an illustration, several values are plotted as points on the number line of Figure 1. The positive direction is taken to the right, as indicated by the arrow. In this way every real number corresponds to exactly one point, and conversely every point corresponds to exactly one number, called its **coordinate.**

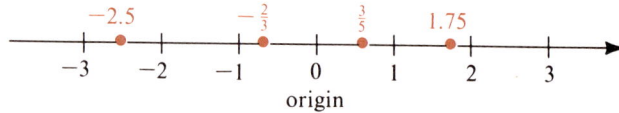

Figure 1

Rectangular Coordinates in the Plane To establish a coordinate system in the plane two perpendicular lines, or **coordinate axes,** are chosen. By convention these axes are horizontal and vertical, and are called the **x-axis** and **y-axis** respectively. We introduce a coordinate system on each axis in such a way that the lines share a common **origin** O at their point of intersection. The positive direction is usually taken to the right on the x-axis and upward on the y-axis. The unit length may or may not be the same on the two axes.

The location of a point P is now specified by an **ordered pair** (x, y) of real numbers. To plot the ordered pair (x, y), we start at the origin and move x units horizontally and then y units vertically, taking into account the signs of x and y. Figure 2 shows an xy-coordinate system in the plane, with several points plotted for illustration.

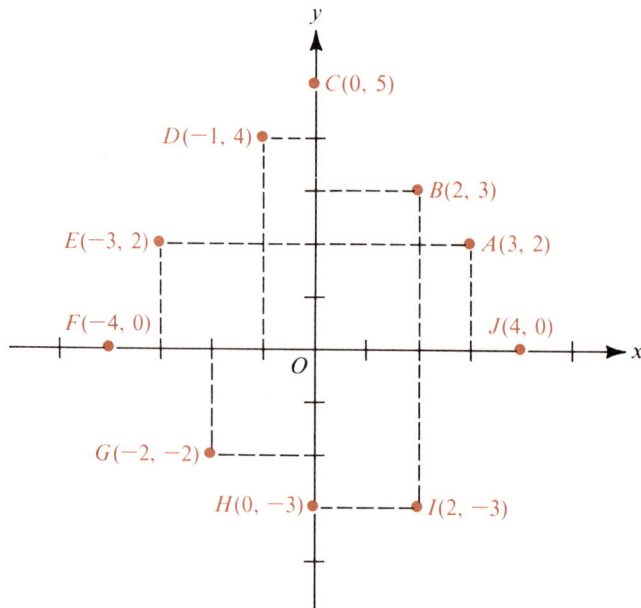

Figure 2

For example, the ordered pair (3, 2) corresponds to point A in the figure, obtained by starting at the origin and moving 3 units to the right, then 2 units upward. In the same way, every ordered pair (x, y) determines a unique point P in the plane, and conversely. The numbers x and y are called the **coordinates** of the point. For example, point D in Figure 2 has x-coordinate -1 and y-coordinate 4; we would therefore describe D as "the point $(-1, 4)$." Notice that points on the x-axis, such as $F(-4, 0)$ and $J(4, 0)$, have y-coordinate zero, while points on the y-axis, such as $C(0, 5)$ and $H(0, -3)$, have x-coordinate zero. The origin O has coordinates $(0, 0)$.

The coordinate axes divide the plane into four regions, called **quadrants.** They are conventionally numbered counterclockwise, starting from the upper right, as shown in Figure 3. The quadrant in which a point lies is determined by the signs (positive or negative) of its coordinates. Thus, a point (x, y) with both coordinates positive $(x > 0, y > 0)$ lies in Quadrant I, a point (x, y) with negative x-coordinate and positive y-coordinate lies in Quadrant II, etc.

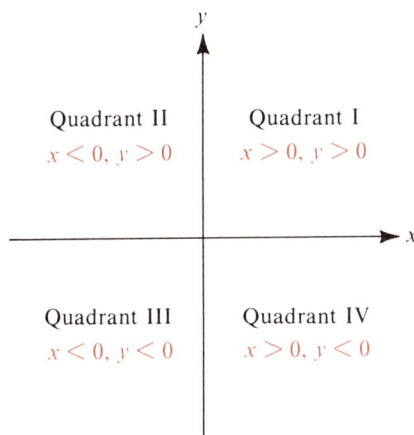

Figure 3

Linear Equations   An equation of the form

$$ax + by = c$$

where a, b, and c are constants and at least one of the coefficients a, b is nonzero, is called a **linear equation** in x and y. Such an equation has many different solutions (x, y). If we fix a particular xy-coordinate system in the plane, the points whose coordinates satisfy the equation make up the **graph** of the equation. The following basic fact justifies the term *linear* equation.

The graph of any linear equation $ax + by = c$ is a straight line, and conversely any straight line is the graph of such an equation.

For example, the equation

$$2x + 3y = 6$$

is of the form $ax + by = c$, with $a = 2$, $b = 3$, and $c = 6$. The ordered pair $(-3, 4)$ is a solution, since

$$2x + 3y = 2(-3) + 3(4) = -6 + 12 = 6$$

and the ordered pair $(2, 2/3)$ is also a solution, since

$$2x + 3y = 2(2) + 3\left(\frac{2}{3}\right) = 4 + 2 = 6.$$

Other solutions are $(1/2, 5/3)$ and $(6, -2)$, as the reader can check by substitution. These four points are plotted in Figure 4. As the picture suggests, the four points all lie along the same straight line. This line is the graph of the equation $2x + 3y = 6$.

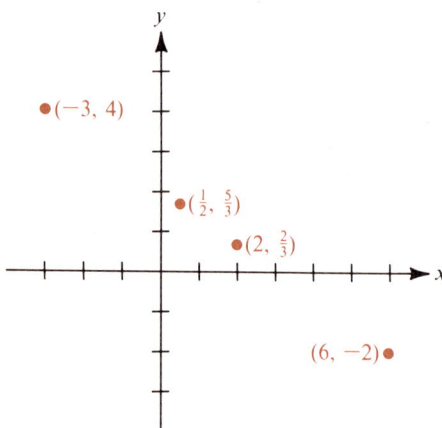

Figure 4

Graphing by Intercepts   The line determined by a given equation can be graphed very easily by finding its **intercepts,** or the points where it crosses the coordinate axes. This procedure is described below.

Graphing by Intercepts

To graph a linear equation $ax + by = c$:

1. Set $y = 0$ and solve for x. This determines the **x-intercept** of the line.
2. Set $x = 0$ and solve for y. This determines the **y-intercept** of the line.
3. Join these points to obtain the graph.

Example 1 Graph the equation $2x + 3y = 6$.

Solution We set $y = 0$ in the equation and solve for x.

$$2x + 3(0) = 6$$
$$2x = 6$$
$$x = 3$$

The point $(3, 0)$ thus lies on the graph of $2x + 3y = 6$, and it also lies on the x-axis (since its y-coordinate is zero). We have therefore found the x-intercept of the line, or the point where it crosses the x-axis. To find the y-intercept of the line, we set $x = 0$ and solve for y.

$$2(0) + 3y = 6$$
$$3y = 6$$
$$y = 2$$

The line therefore crosses the y-axis at $(0, 2)$. Since two points determine a single straight line, we have only to plot the intercepts and draw the line through them to obtain the graph in Figure 5.

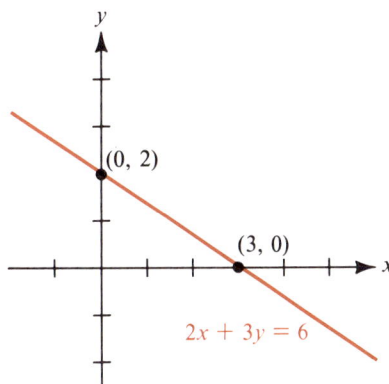

Figure 5

In the linear equation $ax + by = c$, one or two of the constants a, b, c may be zero. In this case the method of graphing by intercepts must be modified slightly.

Example 2 Graph the equation $4x - 5y = 0$.

Solution Here we have $ax + by = c$, where $a = 4$, $b = -5$, and $c = 0$. When we try to find the intercepts we encounter a slight problem: $y = 0$ yields $x = 0$, and vice-versa. This is because *the line goes through the origin* and thus has only one intercept, namely the point (0, 0). To solve the problem we need only find one additional point on the line. This can be accomplished by arbitrarily assigning a value to x in the equation, for instance, the value $x = 5$.

$$4(5) - 5y = 0$$
$$20 - 5y = 0$$
$$5y = 20$$
$$y = 4$$

We now have a second point (5, 4) on the line. Joining (0, 0) and (5, 4) we obtain the graph in Figure 6.

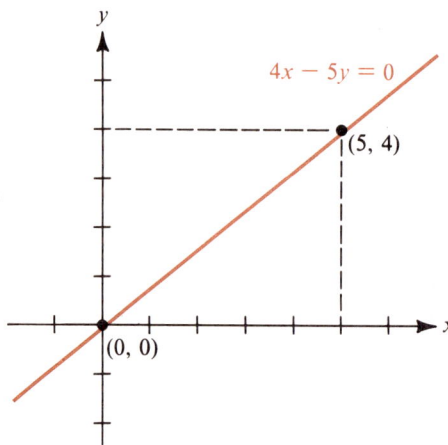

Figure 6

Example 3   Graph the equation $3x = 4$.

Solution   If we write this equation in the form $3x + 0y = 4$, we see that it is linear in x and y, with $a = 3$, $b = 0$, and $c = 4$. Therefore, it must have a straight line as its graph. Setting $y = 0$ we find that $x = 4/3$, so the x-intercept of the line is (4/3, 0). But setting $x = 0$ leads to an absurdity: $0 = 4$. This tells us that the equation $3x = 4$ *cannot be satisfied* by any point (0, y) on the y-axis. In other words, the line has no y-intercept. From this we conclude that the line must be parallel to the y-axis, or vertical. (See Figure 7a.)   ◆

Example 4   Graph the equation $5y = 10$.

Solution   In standard linear form the equation is $0x + 5y = 10$. The line described by this equation has y-intercept (0, 2) but no x-intercept. It is thus parallel to the x-axis, or horizontal. (See Figure 7b.)

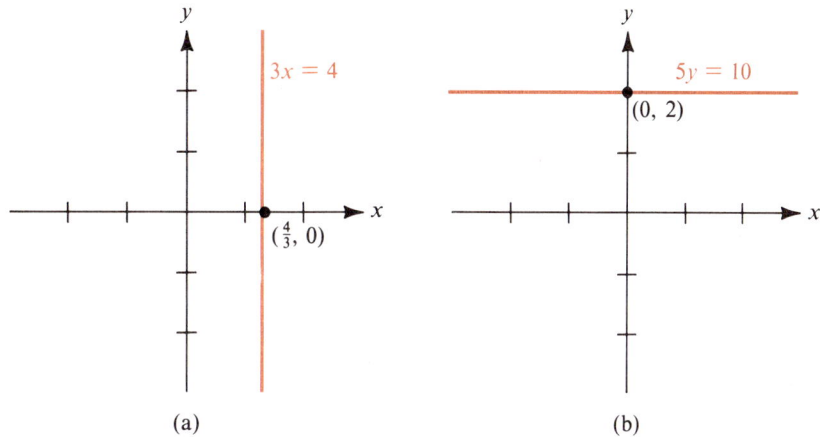

(a)   (b)

Figure 7   ◆

Notice that the equation $3x = 4$ can be written equivalently as $x = 4/3$, while the equation $5y = 10$ can be written as $y = 2$. In general:

The equation $x = k$ (*k* constant) represents a *vertical* line that crosses the *x*-axis at $(k, 0)$.

The equation $y = k$ (*k* constant) represents a *horizontal* line that crosses the *y*-axis at $(0, k)$.

In particular, letting $k = 0$ we see that the equation of the *y*-axis is $x = 0$, and the equation of the *x*-axis is $y = 0$.

The different forms of a linear equation and the various possibilities for its coefficients are summarized in the following table:

General linear equation $ax + by = c$	
$a \neq 0, b \neq 0, c \neq 0$	Typical case, two distinct intercepts
$a \neq 0, b \neq 0, c = 0$	Line through origin
$a \neq 0, b = 0, c \neq 0$	Vertical line, equation $x = k$
$a = 0, b \neq 0, c \neq 0$	Horizontal line, equation $y = k$
$a \neq 0, b = 0, c = 0$	*y*-axis, equation $x = 0$
$a = 0, b \neq 0, c = 0$	*x*-axis, equation $y = 0$

1.1 Exercises   1. Plot the following values as points on the real number line.

(a) 4   (c) $\frac{5}{2}$ (e) 1.4

(b) -1 (d) $-\frac{29}{12}$ (f) $\left(\frac{3}{4}\right)^2$

2. Plot the following values as points on the real number line.

   (a) 10   (c) 32   (e) -37.5

   (b) -25 (d) $\frac{278}{15}$ (f) $\left(-\frac{5}{2}\right)^3$

3. Plot the following ordered pairs as points in the xy-plane.

   (a) $(4, -2)$ (c) $(0, 0)$ (e) $(0, 3)$

   (b) $(-3, 5)$ (d) $\left(\frac{7}{2}, \frac{5}{2}\right)$ (f) $\left(-2, -\frac{5}{2}\right)$

4. Plot the following ordered pairs as points in the xy-plane.

   (a) $(0, 200)$ (c) $(125, 175)$ (e) $(240, 0)$

   (b) $(-300, 150)$ (d) $(380, -45)$ (f) $(315, 210)$

In Exercises 5–12, express the equation in linear form $(ax + by = c)$.

5. $2y + 5x - 1 = 3x + y$ 6. $x + \frac{1}{2}y + 2 = \frac{1}{3}x + y - 2$

7. $x = y$ 8. $y = -\frac{2}{3}x + 4$

9. $.2x + .3y = .05(x + y)$ 10. $.8(.5x + .7y + 1) = .4(x + .8y + .5)$

11. $-\frac{1}{5}(x + 2y + 1) + \frac{2}{5}(3x - 1) = 2y + \frac{27}{5}$

12. $\frac{2}{3}(x - 5y + 1) = \frac{1}{2}[2x + y - \frac{2}{3}(x - 2y - 11)]$

In Exercises 13–32, graph the linear equation.

13. $2x + 5y = 100$ 14. $4x - 4y = 1$

15. $3x - 8y = 24$ 16. $5x + 3y = 30$

17. $\frac{1}{3}x + \frac{4}{3}y = 0$ 18. $\frac{3}{5}x - \frac{2}{5}y = 12$

19. $x - y = 15$ 20. $x + y = 80$

21. $-3.75x + 1.25y = 45$ 22. $2x - y = 0$

23. $x + y = 0$ 24. $3x = 100$

25. $-2y = 5$ 26. $2y = 0$

27. $x = -1$ 28. $x - y = 0$

29. $8x + 3y = 5$ 30. $4x + 5y = -10$

31. $-3x + 4y = 8$ 32. $3y = 7$

1.2 Intersecting Lines

The geometric problem of finding the point where two given lines intersect corresponds to the algebraic problem of solving two linear equations in two unknowns. In this section we review the methods of substitution and elimination, the generalizations of which will play an important role in Chapter 3.

Systems of Equations   We begin with the fact, familiar from geometry, that two nonparallel lines intersect in a point. If we know the equations of the lines, we should be able to find the coordinates of the point.

Example 1   Find the point where the lines $2x - 3y = 9$ and $x + y = 2$ intersect.

Discussion   Since the point of intersection lies on *both* lines, its coordinates (x, y) must satisfy both equations. We must therefore solve the **system of two linear equations in two unknowns**

$$2x - 3y = 9$$
$$x + y = 2.$$

Two different methods of solution are illustrated.

Substitution Method

To solve a system of two linear equations in two unknowns:

1. Use one of the equations to solve for one variable in terms of the other.

2. Substitute this quantity into the remaining equation.

In our example we can solve for y in terms of x using the second equation.

$$x + y = 2$$
$$y = 2 - x \tag{1}$$

Substituting $2 - x$ for y in the first equation of the system, we obtain

$$2x - 3(2 - x) = 9$$
$$2x - 6 + 3x = 9$$
$$5x = 15$$
$$x = 3.$$

Finally, letting $x = 3$ in equation (1), we obtain y.

$$y = 2 - x = 2 - 3 = -1$$

The point of intersection is therefore $(3, -1)$. We can verify this solution by direct substitution in the original equations.

Check: $2x - 3y = 2(3) - 3(-1) = 6 + 3 = 9$
$x + y = 3 + (-1) = 2$

The situation is illustrated in Figure 1 on the next page.

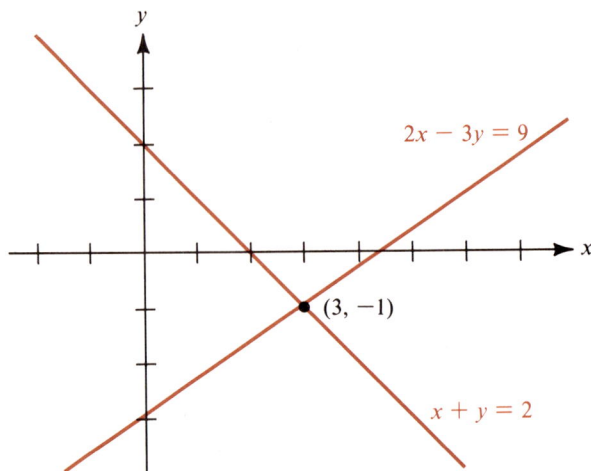

Figure 1

The second method of solution is called **elimination.**

Elimination Method

To solve a system of two linear equations in two unknowns:

1. Multiply one or both equations by appropriate constants so that one of the variables has coefficient c in one equation and $-c$ in the other.
2. Add the equations together to eliminate this variable.

To apply this method to the system of Example 1 we begin by writing down the given equations.

$$2x - 3y = 9$$
$$x + y = 2$$

Multiplying the second equation by 3, we obtain the equivalent system below.

$$2x - 3y = 9$$
$$3x + 3y = 6$$

If we now add these equations, the y-term is eliminated.

$$5x = 15$$
$$x = 3$$

Finally, we substitute $x = 3$ into either of the original equations—say, the second—and solve for y.

$$3 + y = 2$$
$$y = -1$$

We thus arrive at the solution $(3, -1)$, as before.

In the next example, the elimination technique is applied to the y-term rather than the x-term.

Example 2 Find the point where the lines $3x + 7y = 27$ and $5x + 4y = 22$ intersect.

Solution We must solve the following system of equations:

$$3x + 7y = 27$$
$$5x + 4y = 22$$

If we multiply the first equation by 5 and the second by -3, we obtain a system with the same solutions.

$$15x + 35y = 135$$
$$-15x - 12y = -66$$

When we add these equations the x-term is eliminated.

$$23y = 69$$
$$y = 3$$

We can now find x by substituting $y = 3$ into the first equation.

$$3x + 7(3) = 27$$
$$3x + 21 = 27$$
$$3x = 6$$
$$x = 2$$

The lines intersect at the point $(2, 3)$, as shown in Figure 2.

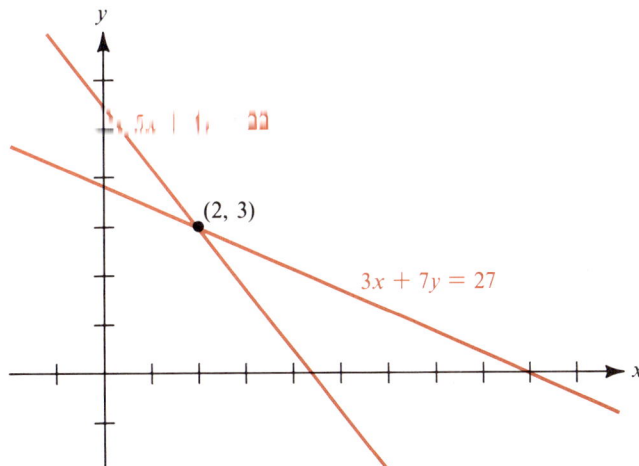

Figure 2

We would have arrived at the same result in Example 2 using the substitution method, of course. Here, as in most cases, however, elimination requires less computation.

When two lines fail to intersect in a unique point, as in the following examples, the elimination method breaks down.

Example 3   Find the point where the lines $2x - 4y = 7$ and $-3x + 6y = 8$ intersect.

Solution   We apply the elimination method to the system

$$2x - 4y = 7$$
$$-3x + 6y = 8.$$

Multiplying the first equation by 3 and the second by 2, we transform the system into

$$6x - 12y = 21$$
$$-6x + 12y = 16.$$

When we add these equations both the x-term and y-term drop out, leaving us with an absurdity:

$$0 = 37.$$

We conclude that the given equations form an **inconsistent system;** that is, they have no common solution (x, y). This means that the lines they represent are parallel, as in Figure 3.

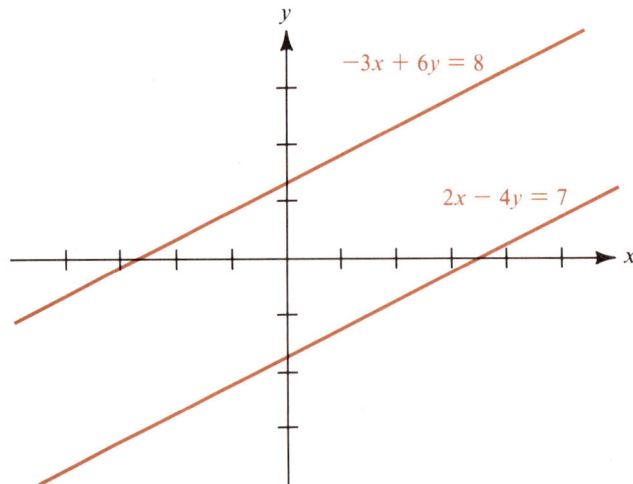

Figure 3

Example 4 Find the point where the lines $8x + 6y = 18$ and $20x + 15y = 45$ intersect.

Solution Here we must solve the following system:

$$8x + 6y = 18$$
$$20x + 15y = 45$$

We multiply the first equation by 5 and the second by -2.

$$40x + 30y = 90$$
$$-40x - 30y = -90$$

When we add these together we obtain

$$0 = 0$$

which is true but not very informative. The problem here is that the given system is a **redundant system:** the first equation implies the second, as we see by multiplying it by 5/2. Both equations thus have the straight line graph shown in Figure 4, and every point on this line is a common solution to the given equations.

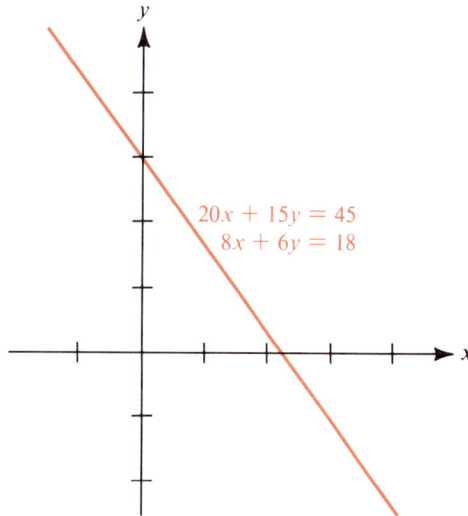

$20x + 15y = 45$
$8x + 6y = 18$

Figure 4

◆

The foregoing examples illustrate all the possible cases that may arise in solving two linear equations in two unknowns. The results are summarized on the next page.

Every system

$$a_1 x + b_1 y = c_1$$
$$a_2 x + b_2 y = c_2$$

of linear equations leads to exactly one of the following three cases:

Typical case: The system has a unique solution, and the corresponding lines intersect in one point.

Inconsistent case: The system has no solution, and the corresponding lines are parallel.

Redundant case: The system has infinitely many solutions, and the corresponding lines coincide.

In Chapter 3 we shall solve systems involving more than two equations in more than two unknowns, using a general technique based on the elimination method. As we shall see, the basic classification given above applies to these more complicated systems as well.

Applications Linear equations play a central role in many practical problems. Example 5 below illustrates the three basic steps involved in formulating a problem mathematically:

Step 1 Introduce variables to represent certain unknown quantities.

Step 2 Summarize the given information in a convenient form.

Step 3 Translate the conditions of the problem into mathematical statements.

Example 5   A furniture manufacturer makes two styles of chair, "Antique" and "Baroque," for sale to retail outlets. Each Antique chair requires two work-hours of construction and three work-hours of finishing, while each Baroque chair requires three work-hours of construction and five work-hours of finishing. The company has a total of 60 work-hours of construction labor and 95 work-hours of finishing labor available each day. Is it possible to use up exactly the available amount of labor?

Solution   Begin by introducing two unknowns (Step 1).

Let x = number of Antique chairs produced daily

y = number of Baroque chairs produced daily.

Next (Step 2), summarize the given data in the form of a table, indicating the total available amounts of labor at the bottom of the corresponding columns (wh = work-hours).

Style	Number of units	Labor per unit	
		Construction	*Finishing*
Antique	x	2 wh	3 wh
Baroque	y	3 wh	5 wh
Available		60 wh	95 wh

We now make the following simple but crucial observation: *If one Antique chair requires 2 work-hours of construction, then x Antique chairs will require 2x work-hours of construction.* Similarly, y Baroque chairs will require $3y$ work-hours of construction. Hence, in making x Antique and y Baroque chairs we use a total of $2x + 3y$ work-hours of construction labor. By the same kind of reasoning, we arrive at the expression $3x + 5y$ for the total amount of finishing labor used. The condition that exactly the available amount of labor be used up therefore translates into the following pair of equations:

$$2x + 3y = 60$$
$$3x + 5y = 95$$

We will solve this system by the elimination method. Multiply the first equation by 5 and the second by -3.

$$10x + 15y = 300$$
$$-9x - 15y = -285$$

Add these to eliminate the y-term.

$$x = 15$$

Finally, substitute $x = 15$ back into the first equation.

$$2(15) + 3y = 60$$
$$30 + 3y = 60$$
$$3y = 30$$
$$y = 10$$

We now have our answer: the company can use up all the available labor by producing 15 Antique chairs and 10 Baroque chairs each day.   ◆

Question: What if either x or y had come out negative in Example 5?

Answer: If we had obtained negative values for x or y we would have had to conclude that *the problem has no solution,* since the physical interpretation of x and y makes such values meaningless. For example, if the manufacturer had only 80 work-hours of finishing labor available instead of 95, the corresponding equations would be

$$2x + 3y = 60$$
$$3x + 5y = 80$$

which have the solution $x = 60$, $y = -20$. In this case it would not be possible to use up exactly the available supply of labor.

1.2 Exercises   *In Exercises 1–14, find the point (if any) where the lines intersect, and illustrate by means of a graph.*

1. $3x + 2y = 8$
   $x - y = 1$

2. $x + 2y = 7$
   $3x + y = 6$

3. $6x + 5y = 21$
   $3x - 4y = 30$

4. $2x - 5y = 14$
   $5x - 2y = 14$

5. $7x + 5y = 41$
   $4x - 3y = 0$

6. $5x + 4y = 0$
   $4x + 5y = 9$

7. $.9x + .8y = 72$
   $.1x + .2y = 10$

8. $1.6x - .8y = 8$
   $-1.2x + .6y = 0$

9. $x + y = 8$
   $x - y = 3$

10. $2x - 3y = 6$
    $3x - 5y = 15$

11. $-\frac{2}{3}x + \frac{1}{2}y = 15$
    $\frac{4}{5}x - \frac{3}{5}y = 12$

12. $\frac{2}{3}x + y = 8$
    $x + \frac{1}{2}y = 8$

13. $2x + 3y = 75$
    $4y = 20$

14. $-x + \frac{2}{5}y = -2$
    $\frac{5}{2}x - y = 5$

15. Each ounce of Food A contains 2 gm of carbohydrate and 2 gm of protein, while each ounce of Food B contains 4 gm of carbohydrate and 2 gm of protein. By combining these foods, is it possible to obtain exactly 60 gm of carbohydrate and 50 gm of protein?

16. The Exxess Oil Company operates two eastern refineries. The Albany facility produces 100 barrels (bbl) of high-grade oil and 300 bbl of medium-grade oil per day. The Boston facility produces 200 bbl of high-grade and 100 bbl of medium-grade per day. By operating each facility for a certain number of days, can the company produce exactly 12,000 bbl of high-grade oil and 24,000 bbl of medium-grade oil?

17. The House of Coffee sells two different blends of Brazilian and Colombian beans. Each pound of Blend A contains 1/4 lb of Brazilian and 3/4 lb of Colombian. Each pound of Blend B contains 2/3 lb of Brazilian and 1/3 lb of Colombian. The store would like to use up its entire supply of Brazilian and Colombian beans by producing a certain amount of each blend. Is this possible

   (a) if the store has 1200 lb of Brazilian and 1500 lb of Colombian?

   (b) if the store has 3000 lb of Brazilian and 1400 lb of Colombian?

18. World Travel Service has agreed to transport a group of 1800 football fans from Los Angeles to Chicago for one of the season's big games. The agency can charter two different types of aircraft for the trip: Type A has 20 first-class seats and 80 economy seats. Type B has 40 first-class seats and 60 economy seats. By using

a certain number of planes of each type, the agency hopes to accommodate the group's needs exactly, with no seats left empty on any plane.

(a) Is this possible, if 600 fans want first-class seats and 1200 want economy seats?

(b) Is this possible, if 650 fans want first-class seats and 1150 want economy seats?

19. A chemist has available two strengths of acid solution. Solution A is 10% acid (by volume), while Solution B is 20% acid. How many liters of each solution should the chemist mix together to obtain a total of 100 liters of 12% acid solution?

20. A marine biologist uses two different fish nutrients. Nutrient 1 contains 7 units of protein per gm, while Nutrient 2 contains 3 units of protein per gm. How many grams of each should be mixed together, if the biologist needs 300 gm of nutrient containing 4 units of protein per gm?

21. Arnold had twice as much as Berenice when they first arrived in Las Vegas. He lost $500 and she lost $100, and now the situation is reversed: she has twice as much as he has. How much did each of them arrive with?

22. Customs officials suspect Al the Weasel of smuggling jade rings and bracelets into the U.S., where he sells them for $150 and $250, respectively. Their investigation shows that his latest shipment sold here for a total of $50,000, and officials in Taiwan, where the items cost 40% and 50% less, respectively, have proof that he paid $28,000 there for the merchandise. How many rings and how many bracelets did the shipment contain?

23. During a certain year, Laura had a profit of $5600 from two investments, the first earning 5% and the second 12%. Had she put all her money into the second investment, her profit would have been $8400. How much did she put into each investment?

24. A woman holds 45 shares of stock A and 24 shares of stock B. On the morning of a certain day the value of her holdings was $4800. During the day the price of stock A doubled, while the price of stock B fell by 50%. At day's end her holdings were worth $7800. What was the initial price of each stock?

1.3 Linear Functions

Almost all practical problems involve **variables,** quantities that may assume various possible values. Often one of the quantities depends on the others: in mathematical language, one variable is a **function** of the others. For example, the position of a moving object is a function of time, and the profit of a corporation is a function of both sales and production costs. In this section we consider an important special case of the function concept directly related to our study of straight lines.

Slope of a Line It is useful to have a numerical measure of the "steepness" of a line. For this purpose we introduce the following notion:

Definition

Let L be a nonvertical line passing through the distinct points $P(x_1, y_1)$ and $Q(x_2, y_2)$. The **slope m** of L is given by

$$m = \frac{y_2 - y_1}{x_2 - x_1}. \tag{1}$$

The value of this quotient is the same no matter which points are used.

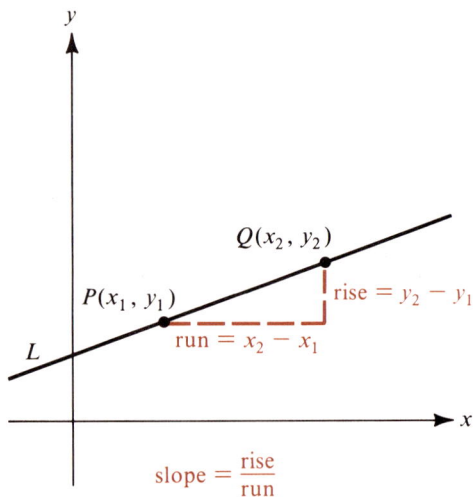

$$\text{slope} = \frac{\text{rise}}{\text{run}}$$

Figure 1

The meaning of the slope is illustrated in Figure 1. As we move from P to Q along the line L, the x-coordinate changes by an amount $x_2 - x_1$ and the y-coordinate changes by $y_2 - y_1$. The slope m thus represents the ratio of the change in y to the change in x, or the "rise over the run." That this value is indeed independent of the points used is seen from Figure 2, using similar triangles.

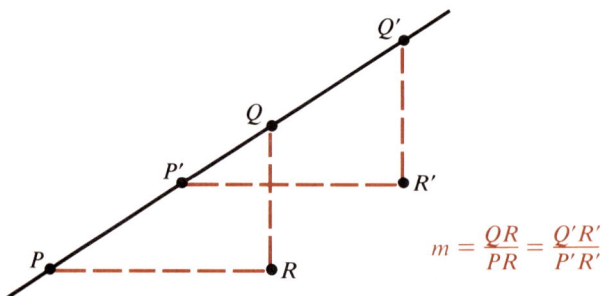

$$m = \frac{QR}{PR} = \frac{Q'R'}{P'R'}$$

Figure 2

Note that the slope of a vertical line is undefined. This is because points on a vertical line all have the same x-coordinate, and the quotient in (1) would therefore have denominator zero. In Figure 3 the fixed point $P(2, 1)$ has been joined to various other points Q. The slope of each line is computed using formula (1). For example, when Q is the point $(5, 2)$ the line PQ has slope

$$m = \frac{y_2 - y_1}{x_2 - x_1} = \frac{2 - 1}{5 - 2} = \frac{1}{3}.$$

Note that interchanging the points P and Q in the formula gives the same result:

$$m = \frac{y_1 - y_2}{x_1 - x_2} = \frac{1 - 2}{2 - 5} = \frac{-1}{-3} = \frac{1}{3}$$

Note also that lines with *positive* slope slant upward from left to right, while lines with *negative* slope slant downward. A horizontal line has slope zero.

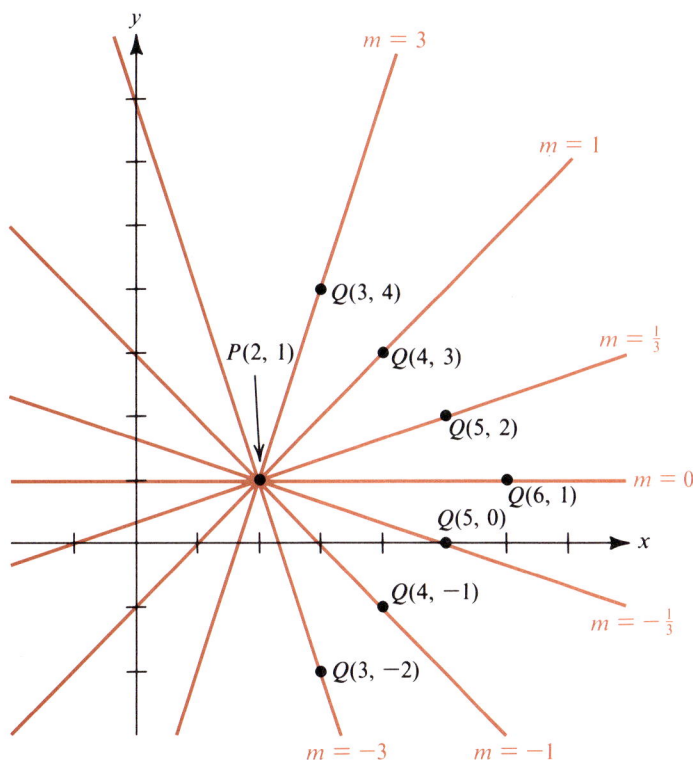

Figure 3

Slope-Intercept Form   A nonvertical line L can always be described by a linear equation $ax + by = c$, where $b \neq 0$. If we solve this equation for y in terms of x, we obtain

$$by = -ax + c$$

$$y = \left(-\frac{a}{b}\right)x + \left(\frac{c}{b}\right).$$

That is, the equation of the line can be written in the form

$$y = mx + k$$

where $m = -a/b$ and $k = c/b$. The significance of the numbers m and k is given by the following result:

Slope-Intercept Form

When the equation of a line L is written in the form

$$y = mx + k \qquad\qquad (2)$$

m is the slope and $(0, k)$ is the y-intercept of L. This is called the **slope-intercept form** of the linear equation.

To see this, let $P(x_1, y_1)$ and $Q(x_2, y_2)$ be two distinct points on L. Then the coordinates of P and Q both satisfy equation (2).

$$y_1 = mx_1 + k \qquad \text{and} \qquad y_2 = mx_2 + k$$

Subtracting the first equation from the second, we have

$$y_2 - y_1 = mx_2 - mx_1 = m(x_2 - x_1)$$

and therefore

$$m = \frac{y_2 - y_1}{x_2 - x_1} = \text{slope of } L.$$

Setting $x = 0$ in equation (2) yields $y = k$, so that $(0, k)$ is the y-intercept of the line.

Example 1   Find the slope and y-intercept of the line $-2x + 5y = 10$.

Solution   Solve the given equation for y in terms of x to obtain the slope-intercept form.

$$5y = 2x + 10$$

$$y = \frac{2}{5}x + 2$$

For a line in this form, the slope is the coefficient of x, so $m = 2/5$. The y-intercept is given by the constant term $k = 2$. To graph the line, let x take an arbitrary value in the above equation and calculate y. For instance,

$$x = 5 \quad \text{gives} \quad y = \frac{2}{5}(5) + 2 = 4.$$

Hence the line goes through the points $(0, 2)$ and $(5, 4)$, as in Figure 4.

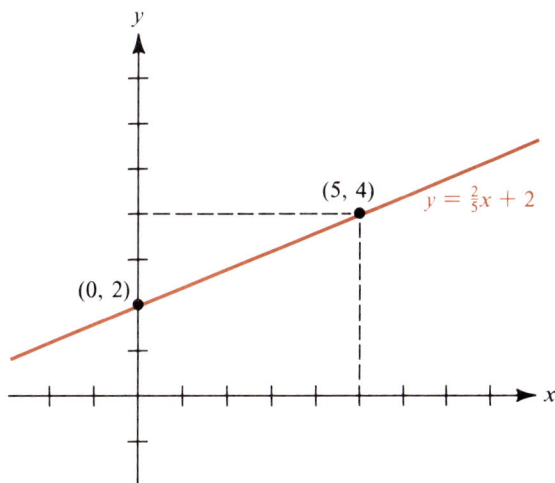

Figure 4

Point-Slope Form   If a point $P(x_1, y_1)$ and a number m are given, there will be one and only one line L through P with slope m. If $Q(x, y)$ is any other point on L, the slope of L is equal to the "rise over run" as we go from P to Q; that is,

$$m = \frac{y - y_1}{x - x_1}.$$

Multiplying by $x - x_1$ we obtain the equation

$$y - y_1 = m(x - x_1).$$

This equation also holds when Q is P, since both sides are zero in this case. In other words, this equation is satisfied by every point (x, y) on L. We have therefore derived the following result:

Point-Slope Form

The line through $P(x_1, y_1)$ with slope m has equation

$$y - y_1 = m(x - x_1). \tag{3}$$

This is called the **point-slope form** of the linear equation.

Example 2   Find the equation of the line through (4, 5) with slope 3/2.

Solution   By the point-slope formula (3), the equation is

$$y - 5 = \frac{3}{2}(x - 4)$$

which simplifies to

$$y = \frac{3}{2}x - 1$$

or, equivalently, $3x - 2y = 2$. (See Figure 5.)

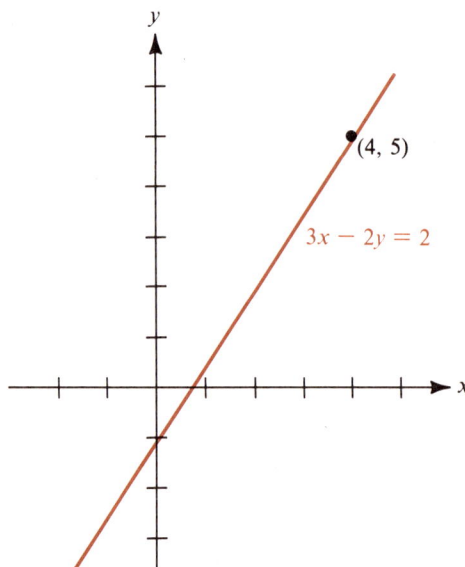

Figure 5   ◆

The point-slope formula can also be used to find the equation of a line through two given points, as in the next example.

Example 3   Find the equation of the line through $P(-1, 2)$ and $Q(3, -1)$.

Solution   First find the slope of the line using formula (1).

$$m = \frac{y_2 - y_1}{x_2 - x_1} = \frac{-1 - 2}{3 - (-1)} = -\frac{3}{4}$$

Since the line passes through $P(-1, 2)$ and has slope $-3/4$, its point-slope equation is

$$y - 2 = -\frac{3}{4}(x + 1)$$

which simplifies to $3x + 4y = 5$. (See Figure 6.)

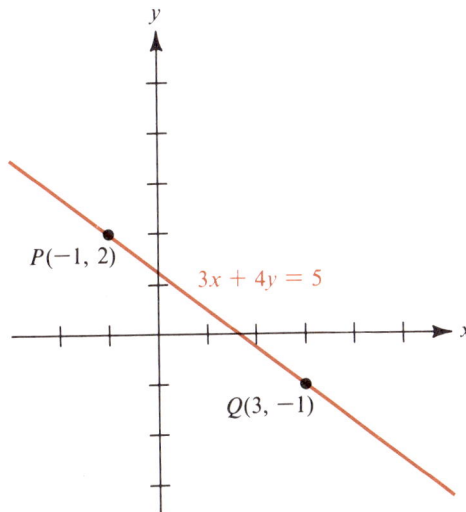

Figure 6

Linear Functions When two variables x and y satisfy an equation $ax + by = c$, they are said to be **linearly related.** If $b \neq 0$ this equation can be written equivalently in slope-intercept form as

$$y = mx + k.$$

The above equation expresses y as a **linear function** of x, with x the **independent** variable and y the **dependent** variable. The constant k represents the **initial value** of the function: it is the value of y when $x = 0$. The slope m represents the **rate of change** of the function: the change in y produced by a one-unit increase in x.

Example 4 A bank offers a checking plan in which the customer pays a monthly service charge of $1.75 plus an additional 5¢ for each check processed. Express the total monthly cost as a function of the number of checks written.

Solution If $x =$ number of checks written per month and $y =$ total monthly service cost (in dollars), then

$$y = .05x + 1.75.$$

This equation expresses y as a linear function of x, with $m = .05$ and $k = 1.75$. To graph this function, let x take an arbitrary value and solve for y. For example,

$$x = 10 \quad \text{gives} \quad y = .05(10) + 1.75 = 2.25.$$

Thus, the line goes through the points $(0, 1.75)$ and $(10, 2.25)$, as shown in Figure 7.

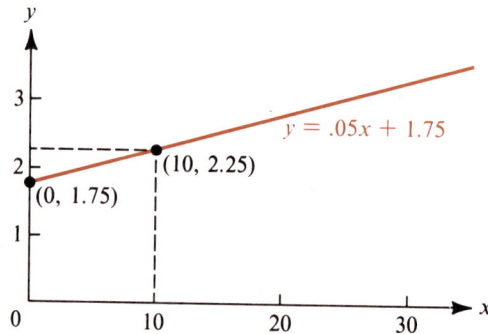

Figure 7

Since we are interested only in nonnegative values of x we graph only that part of the line which lies in the first quadrant. The fact that the line slants upward from left to right shows that the cost y increases as the number x of checks increases. In general, a linear function $y = mx + k$ is **increasing** if $m > 0$, **decreasing** if $m < 0$, and **constant** if $m = 0$. This is illustrated in Figure 8.

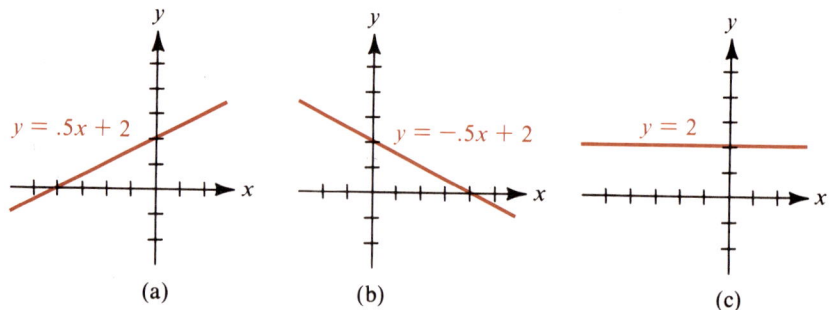

Figure 8

When two variables are linearly related, any two pairs of corresponding values can be used to determine the linear function, as shown by the following example.

Example 5   A middle-eastern country exports a constant amount of oil each year, so its oil reserves have been decreasing linearly with time. Industry experts estimate that the country's reserves were 350 million barrels in 1975 and 252 million barrels in 1982. If this trend continues, when will the oil run out?

Solution   Letting R = remaining oil reserves, in millions of barrels, and t = time (in years) since 1975, we have a linear relation $R = mt + k$. The given conditions say that $R = 350$ when $t = 0$ (in 1975), and $R = 252$ when $t = 7$ (in 1982). Thus, the points $(0, 350)$ and $(7, 252)$ lie on the graph of the function, as illustrated in Figure 9. By formula (1), the slope of this line is

$$m = \frac{252 - 350}{7 - 0} = \frac{-98}{7} = -14.$$

Since the line crosses the R-axis at $k = 350$, the slope-intercept equation is

$$R = -14t + 350.$$

Note that $m = -14$ represents the rate of change of the function, that is, the amount of oil pumped out each year. According to the above equation, the reserves will reach zero when

$$-14t + 350 = 0.$$

Solving this equation we obtain $14t = 350$, or $t = 25$. In other words, if the linear trend continues the oil will run out in the year 2000. (See Figure 9.)

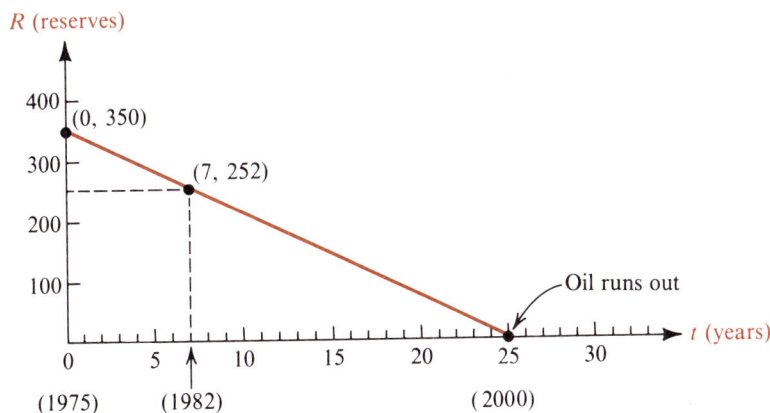

Figure 9

1.3 Exercises

In Exercises 1–10, find the slope and y-intercept of the line. Illustrate by means of a graph.

1. $2x + 5y = 10$

2. $x - y = 4$

3. $5.25x - 1.75y = 7$

4. $3x + 4y = 0$

5. $x = 200$

6. $5y + 2 = 3x$

7. $x = 5y + 4$

8. $\frac{3}{2}y = -15$

9. $x + y = 0$

10. $y = 0$

In Exercises 11–24, find the equation of the line satisfying the given conditions. Illustrate by means of a graph.

11. Slope $-3/4$, y-intercept $(0,4)$

12. Slope $1/2$, y-intercept $(0, -1)$

13. Slope 0, y-intercept $(0, 3)$

14. Slope undefined, x-intercept $(2, 0)$

15. Through $(-4, 2)$, slope $2/3$

16. Through $(5, -3)$, slope -2

17. Through $(-1, 1)$, slope undefined

18. Through $(0, 0)$, slope 1

19. Through $(2, -1)$ and $(-3, 4)$

20. Through $(-8, 6)$ and $(2, 4)$

21. Through $(3, -8)$ and $(4, 1)$

22. Through $(-3, 0)$ and $(-5, -4)$

23. Through $(45, 20)$ and $(0, 0)$

24. Through $(5, 2)$ and $(5, -4)$

In Exercises 25–28, use the following fact: two nonvertical lines are parallel if and only if they have the same slope. Find the equation of the line described, and illustrate by means of a graph.

25. Through $(2, -1)$, parallel to $3x + 4y = 8$

26. Through $(1, 0)$, parallel to $-x + y = 1$

27. Through $(2, 4)$, parallel to $2x - y = 5$

28. Through $(-1, 1)$, parallel to $y = 2$

In Exercises 29–32, use the following fact: two nonvertical lines are perpendicular if and only if their slopes are negative reciprocals of each other (that is, if the product of their slopes equals -1). Find the equation of the line described and illustrate by means of a graph.

29. Through $(1, 3)$, perpendicular to $x - y = 1$

30. Through $(2/5, 3/5)$, perpendicular to $4x - 5y = 1$

31. Through $(0, 0)$, perpendicular to $2x + 3y = 6$

32. Through $(5, 4)$, perpendicular to $x = 10$

33. Suppose a loan of $5000 is made at 9% simple interest per annum, to be repaid in one lump sum sometime in the future. The borrower must repay both the principal of $5000 and the interest charge of $(.09)(5000) = \$450$ for each year the loan is outstanding.

(a) How much will be owed after one year? Two years? Three years?

(b) Let S be the total amount owed after t years. Express S as a linear function of t.

(c) How long will it take for the debt to reach $7400?

34. A machine purchased for $10,000 is depreciated linearly over a 15-year period, at the end of which it has a scrap value of $2500. In other words, the value of the machine decreases linearly from $10,000 to $2500 over fifteen years.

(a) Let V be the value of the machine t years after its purchase. Express V as a linear function of t.

(b) What will the machine be worth after three years? Five years? Six years?

(c) How long will it take for the value to reach $4000?

35. The Celsius (or Centigrade) and Fahrenheit temperature scales are linearly related, with 0°C corresponding to 32°F (the freezing point of water) and 100°C corresponding to 212°F (the boiling point of water).

(a) Find the equation for Fahrenheit temperature F in terms of Celsius temperature C.

(b) If the temperature in a room is 20°C, what is it in Fahrenheit degrees?

36. The level of oxidant pollution in a certain area has been decreasing linearly since 1970, when an intensive pollution control program began. At that time the oxidant level in the air was measured at .15 parts per million (ppm); by 1980 the figure was .09 ppm.

(a) Let P be the oxidant level (in ppm) and let t be the time (in years) since 1970. Express P as a linear function of t.

(b) What will the oxidant level be in 1985? In 1987?

(c) Air with an oxidant level of .03 ppm or less is considered clean. If the present linear trend continues, when will this be achieved?

1.4 Applications of Linear Functions

It is often the case in a business operation that both the cost of production and the revenue from sales are linear functions of the output, or amount produced. In this section we analyze some problems of this type, as well as others in which the comparison of linear functions plays a major role.

Break-Even Analysis In any production process there are both **fixed costs** and **variable costs.** Fixed costs include such items as rent, equipment, depreciation, and interest, which remain the same no matter how much output is produced. Variable costs include such items as material and labor, which depend on the amount of output. The **total cost** of an operation is the sum of the fixed and variable costs, which will be a function of the number of units produced. The **revenue,** or sales income, will also depend on the amount produced. The point at which cost exactly equals revenue is called the **break-even point.**

Example 1   The Zeno Corporation has a division that makes color TV sets. The fixed costs of the operation are $15,000 per day, and the variable cost is $125 for each unit produced. The sets sell for $325 each. How many sets must be produced daily for the operation to break even?

Solution   If the firm produces x units per day, the total production cost C (in dollars) will be

$$C = 125x + 15,000$$

and the total revenue R (assuming all the units are sold) will be

$$R = 325x.$$

Note that both cost and revenue are linear functions of x. If they are graphed in a common xy-coordinate system, with y representing both cost and revenue, the **cost and revenue lines** shown in Figure 1 are obtained.

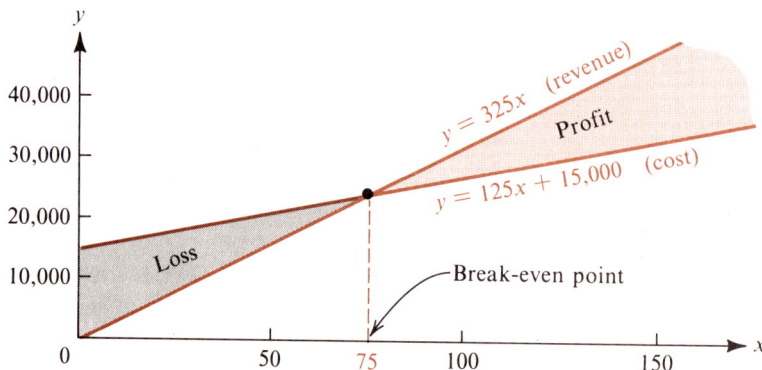

Figure 1

The break-even point is the value of x for which cost equals revenue.

$$C = R$$
$$125x + 15,000 = 325x$$
$$200x = 15,000$$
$$x = 75$$

This is the x-coordinate of the point where the cost and revenue lines intersect. For $x = 75$ we have $C = R = \$24,375$, so the operation shows neither a profit nor a loss. For values of $x < 75$ the cost line lies above the revenue line ($C > R$), so the operation is losing money. For values of $x > 75$ the cost line lies below the revenue line ($C < R$), so the operation shows a net profit.   ◇

In general, **profit** equals the difference between revenue and cost.

$$\text{profit} \quad P = R - C$$

This is another linear function of x. In the example above,

$$P = R - C$$
$$= 325x - (125x + 15,000)$$
$$= 200x - 15,000.$$

The graph of this function is the **profit line** shown in Figure 2. At the break-even point $x = 75$, the profit line crosses the x-axis ($P = 0$). For values of $x < 75$ the profit line lies below the x-axis, indicating a *negative* profit, that is, a net loss. For values of $x > 75$ the line lies above the x-axis, indicating a *positive* profit, or a net gain.

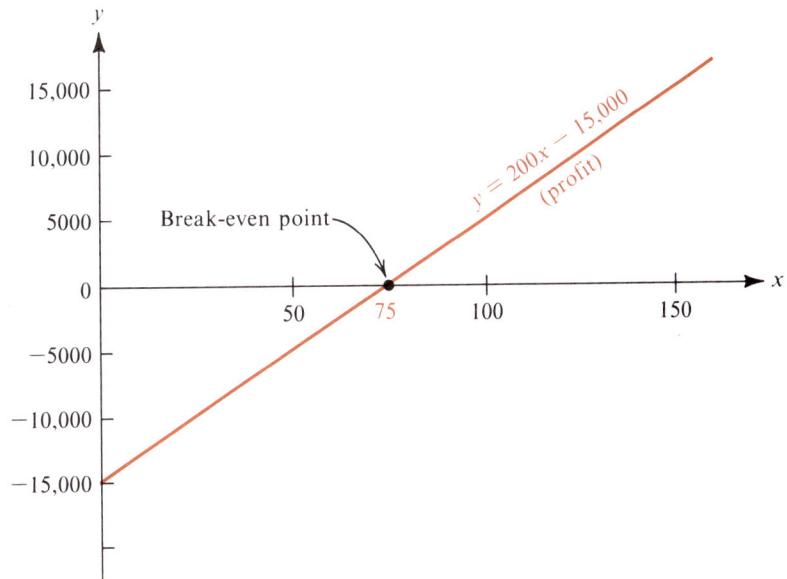

Figure 2

Using the profit function, we can also answer some other questions of interest.

Example 1, Continued   How many sets must be produced daily:
(a) to make a profit of $17,000 a day?
(b) to make a profit of $125 on each unit?
(c) to make a profit of 20% on sales?

Solution   In each case we begin with the profit function $P = 200x - 15,000$.
(a) Setting the profit equal to 17,000, we solve for x.

$$P = 17,000$$
$$200x - 15,000 = 17,000$$
$$200x = 32,000$$
$$x = 160$$

(b) The profit per unit is P/x. Hence, we must solve the equation

$$\frac{P}{x} = 125.$$

$$\frac{200x - 15,000}{x} = 125$$
$$200x - 15,000 = 125x$$
$$75x = 15,000$$
$$x = 200$$

(c) The condition that profit is 20% of sales translates into the equation $P = .20(R)$.

$$P = .20(R)$$
$$200x - 15,000 = .20(325x)$$
$$200x - 15,000 = 65x$$
$$135x = 15,000$$
$$x \doteq 111.1$$

(The symbol $\doteq$ indicates that the answer has been rounded off.) In practical terms, this means that the company must produce at least 112 units per day to achieve a profit margin of 20%.   ◆

Comparison of Profit Functions   In some applications we are presented with several alternative methods of production. By comparing their associated profit functions we can determine the conditions under which one method will be preferable to another.

Example 2   The firm of Ziferstein & Son manufactures a dental floss dispenser that sells for $1.55. Under the present production method the operation has fixed costs of $200 a day and a variable cost of 75¢ per unit. The junior owner wants to lease some new equipment that will save labor and bring the variable cost down to 35¢ per unit. The senior owner is opposed to the new plan, since payments for the equipment will increase the daily fixed costs by $250. The junior owner argues that the savings in variable costs will more than offset the increase in fixed costs. Is he right?

Discussion   Let x be the number of units produced per day. The cost functions for the two production methods are given by

$$C_1 = .75x + 200 \quad \text{(old method)}$$
$$C_2 = .35x + 450 \quad \text{(new method)}$$

while the revenue function is the same for both.

$$R = 1.55x$$

We calculate the profit function for each method.

$$P_1 = R - C_1 = .80x - 200 \quad \text{(old method)}$$
$$P_2 = R - C_2 = 1.20x - 450 \quad \text{(new method)}$$

The two profit lines are shown in Figure 3.

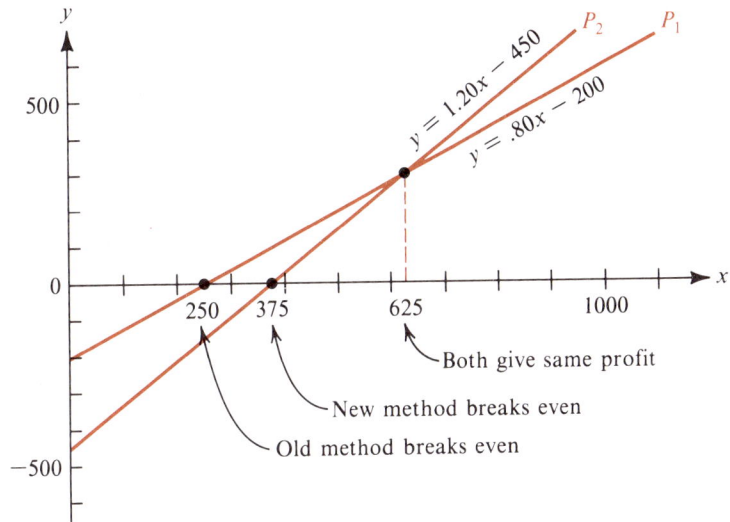

Figure 3

To find the break-even point under the old method, we set $P_1 = 0$.

$$.80x - 200 = 0$$
$$.80x = 200$$
$$x = 250$$

This is the point where the line P_1 crosses the x-axis in Figure 3. Similarly, the new method will break even when $P_2 = 0$.

$$1.20x - 450 = 0$$
$$1.20x = 450$$
$$x = 375$$

For values of x in the range $250 < x \leq 375$, therefore, only the old method will show a profit. We see from Figure 3 that the old method is more profitable (that is, $P_1 > P_2$) up to the point where the two profit lines cross, but beyond that point the new method is more profitable ($P_2 > P_1$). To find the intersection point we set $P_1 = P_2$ and solve for x.

$$.80x - 200 = 1.20x - 450$$
$$.40x = 250$$
$$x = 625$$

When $x = 625$ both methods yield the same profit: $P = \$300$. When $x < 625$ we have $P_1 > P_2$, and when $x > 625$ we have $P_2 > P_1$. Our answer to the problem, therefore, must be conditional: if it is possible to produce more than 625 units per day under the new method, and they can all be sold at the price stated, then the new method is preferable; otherwise the old method is preferable. ◆

Sometimes one alternative is definitely better than another, as the following example illustrates.

Example 3   Sylvia plans to run a private shuttle service between the Astor Hotel and the local airport. Van rental, insurance, and license costs total $3600 a year. Each round-trip run will yield $14 in revenues. Gasoline and maintenance costs average $2 per run. In addition, the city council intends to charge her either a flat franchise fee of $1200 a year or else a franchise tax of 50% of revenues received. Which plan will be more profitable for Sylvia?

Solution   Suppose she makes x round-trip runs per year. Under the franchise fee (Plan 1), the fixed yearly costs will be $3600 + $1200 = 4800, and the variable costs will be $2x$ dollars. The total annual cost is therefore given by

$$C_1 = 2x + 4800$$

while annual revenue is given by

$$R = 14x.$$

The profit function under Plan 1 is therefore

$$P_1 = R - C_1 = 12x - 4800.$$

Under the franchise tax (Plan 2), the fixed cost is reduced to $3600, but $.50(R) = .50(14x) = 7x$ dollars must be added to the variable cost, for a total annual cost of

$$C_2 = 9x + 3600.$$

Since the revenue function remains the same, the profit function under Plan 2 is given by

$$P_2 = R - C_2 = 5x - 3600.$$

The profit lines are shown in Figure 4.

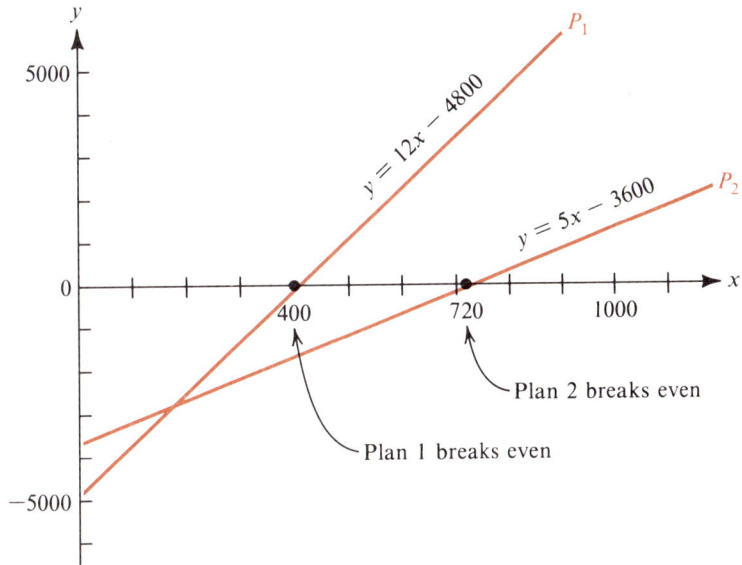

Figure 4

Under Plan 1, the operation will break even when $P_1 = 0$.

$$12x - 4800 = 0$$
$$12x = 4800$$
$$x = 400$$

Thus, Plan 1 breaks even when $x = 400$. Similarly, Plan 2 breaks even when $P_2 = 0$, or $x = 720$. From the graph, however, we see that Plan 2 never becomes more profitable than Plan 1, since $P_1 > P_2$ for all values of x beyond the first break-even point. Thus, Sylvia should favor the franchise fee instead of the franchise tax, and under this plan she will have to make 400 runs per year, or about 34 per month, to break even. ◆

In the following problem, the linear functions being compared represent cost rather than profit.

Example 4 A bank offers its customers three different checking plans. Under Plan 1, you pay a flat monthly service charge of $4.00, regardless of how many checks you write. Under Plan 2 you pay $1.75 a month, plus 5¢ for each check written. Under Plan 3 you pay 12¢ per check, with no fixed service charge. Under what conditions will Plan 2 be the least expensive of the three plans?

Solution   Let x be the number of checks written per month. The monthly cost functions for the three plans (in dollars) are as follows:

$$C_1 = 4$$
$$C_2 = .05x + 1.75$$
$$C_3 = .12x$$

These functions are graphed in Figure 5. The cost lines for Plans 2 and 3 cross when $C_2 = C_3$; that is, when

$$.05x + 1.75 = .12x$$
$$.07x = 1.75$$
$$x = 25.$$

Similarly, the cost lines for Plans 1 and 2 cross when $C_1 = C_2$:

$$4 = .05x + 1.75$$
$$.05x = 2.25$$
$$x = 45$$

From Figure 5 we see that the cost line for Plan 2 lies below the other two lines over the interval $25 \le x \le 45$. Hence, Plan 2 is the least expensive when the customer writes between 25 and 45 checks per month.

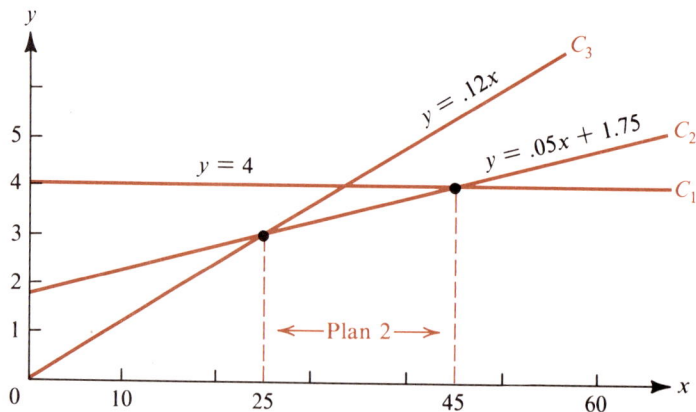

Figure 5

Remark to Reader   It is important in these problems to choose the right scale on each axis so your graph will show all the relevant intersection points. If your first attempt is unsuccessful, try a larger unit of measurement. As in the examples above, you may want to use different units of measurement on the x- and y-axes.

1.4 Exercises

1. A manufacturing operation has cost and revenue functions

$$C = 1.25x + 8400$$
$$R = 2x$$

where x is the number of units of output. How many units must be produced in order for the operation to break even? Draw a graph of the profit function.

2. A power company has fixed costs of $35,000 a day, with a variable cost of 2¢ for each kilowatt-hour (kwh) of electricity produced. The company charges its customers 4¢ per kwh. How many kwh must be sold daily in order for the company to break even? Draw a graph of the profit function.

3. An electronics firm has a plant that produces a single type of silicon chip, which the firm sells for $1.40. If the plant's fixed costs are $1200 a day, and the chips cost 60¢ each to produce, how many must be made daily in order for the plant to break even? Draw a graph of the profit function.

4. An agricultural corporation finds that its asparagus processing division has fixed costs of $14,000 a year, with a variable cost of 15¢ for each pound processed. If the asparagus is sold to a distributor at 35¢/lb, how much must be processed per year in order to break even? Draw a graph of the profit function.

5. In Exercise 1, how many units must be produced

   (a) to make a profit of $6000?

   (b) to make a profit of 25¢ per unit?

   (c) to make a profit of 25% on sales?

6. In Exercise 2, how many kwh must be sold daily

   (a) to make a profit of $25,000 a day?

   (b) to make a profit of 1¢ per kwh?

   (c) to make a profit of 15% on sales?

7. In Exercise 3, how many silicon chips must be produced daily

   (a) to make a profit of $3000 a day?

   (b) to make a profit of 50¢ on each chip?

   (c) to make a profit of 40% on sales?

8. In Exercise 4, how many pounds of asparagus must be processed per year

   (a) to make an annual profit of $100,000?

   (b) to make a profit of 15¢/lb?

   (c) to make a profit of 30% on sales?

9. A manufacturer's profit functions using two different production methods are given by

$$P_1 = .75x - 8400 \qquad \text{(method 1)}$$
$$P_2 = .90x - 10,800 \qquad \text{(method 2)}$$

where x is the number of units of output.

(a) How many units must be produced in order to break even under method 1? Under method 2?

(b) Draw a graph of the two profit functions. Which method is better for the manufacturer?

10. "Save the Lemurs," an ecology group, is planning a fund-raising dinner to be held in one of the two local hotels. The group must pay $1200 for the speaker and $200 for promotion. In addition, the Astor Hotel wants $400 plus $10 a plate to provide the meal, while the Belmont Hotel is asking $800 plus $7 a plate. Admission to the event is set at $15 a plate.

(a) How many people must attend for the event to break even, if it is held at the Astor? At the Belmont?

(b) Draw a graph of the two profit functions involved. Which hotel's offer is better?

11. An auto plant has fixed costs of $60,000 a day and a variable cost of $3000 for each car produced. The cars are sold to the distributor for $4250. A new computerized system could cut the variable cost by 20%, but installing the system would double the present fixed costs.

(a) How many cars must be produced daily to break even, under the present operation? Under the computerized system?

(b) Draw a graph of the two profit functions involved. Under what conditions would it pay to install the new system?

12. A corporation's annual operating costs and revenue (in dollars) are given by

$$C = 5x + 90,000$$
$$R = 20x$$

where x is the number of units of output per year. In addition to its operating costs the company must pay corporate tax. Under Tax Plan 1, the company pays a tax T equal to 15% of revenues received. Under Tax Plan 2, the tax T is equal to 40% of gross profit $(R - C)$. The company's *net* profit P, after taxes, is given by

$$P = R - C - T.$$

(a) Express the net profits, P_1 and P_2, under the two plans as linear functions of x.

(b) Compare these functions by graphing.

(c) What is the company's break-even point under Plan 1? Under Plan 2?

(d) For what values of x will Plan 1 be better for the company than Plan 2?

13. A private school is planning to produce a desk calendar to be mailed out to the community. Printer 1 charges $1000 for design and layout, plus $2.80 a copy. Printer 2 charges $4000 for design and layout, plus $1.80 a copy. Which offer is better, in terms of the number of copies the school needs? Illustrate with a graph.

14. Three different lawyers are willing to handle a certain lawsuit. Lawyer 1 charges $1000 plus 20% of the judgment to be awarded. Lawyer 2 charges $500 plus 40%

of the award. Lawyer 3 charges 60% of the award, with no fixed fee. Let x be the dollar amount awarded to the plaintiff, and let P_1, P_2, and P_3 be his net gain (award minus legal costs) if he accepts Lawyer 1, 2, or 3, respectively.

(a) Express P_1, P_2, and P_3 as linear functions of x.

(b) Compare these functions by graphing.

(c) What is the plaintiff's break-even point with each of the three lawyers?

(d) For what values of x will each lawyer's offer be the best?

15. Taxi rates are not regulated in Baker County, and they vary among the three competing companies. Red Cab charges a fixed fee of 65¢ plus a variable fee of 20¢ a mile. White Cab charges 85¢ plus 15¢ a mile, and Blue Cab charges $1.15 plus 12¢ a mile. A sales representative who uses cabs frequently would like a simple rule to determine which cab is least expensive, given the number x of miles she has to travel. Solve her problem, and illustrate with a graph.

16. A woman is comparing three hospitalization insurance plans. Plan 1 pays all hospital charges for an annual premium of $750. Plan 2 pays 90% of the charges for a premium of $500. Plan 3 pays 60% of the charges for a premium of $200. Let x be the hospital charges the woman might incur during the year. Express her net loss (premium plus uncovered charges) as a linear function of x, under each of the three plans. For what values of x will each plan be best? Illustrate with a graph.

17. Three recording companies are bidding on production rights for a demo featuring country singer Doily Barton. The first company has offered a flat payment of $1 million. The second has offered $100,000 plus 20% of gross receipts. The third has offered $250,000 plus 10% of gross receipts. Which offer is best, in terms of the expected gross receipts x? Illustrate with a graph.

18. Hans Kimmelman, the famous diplomat, is trying to decide between two publishers interested in his memoirs. Publisher 1 offers royalties of 90¢ per copy after the first 2000 copies sold in a given year, with no royalties paid on the first 2000 copies. Publisher 2 offers 50¢ per copy after the first 1000 copies sold. Which offer is better, in terms of the number x of copies he expects to be sold per year? Illustrate with a graph.

Chapter 1   Summary

Important Terms

1.2 system of linear equations
substitution method
elimination method
inconsistent system
redundant system

1.4 fixed cost
variable cost
break-even point
profit function

Important Formulas

1.1 General linear equation: $ax + by = c$

1.3 Slope: $m = \dfrac{y_2 - y_1}{x_2 - x_1}$ (undefined if L is vertical)

Slope-Intercept Form: $y = mx + k$

Point-Slope Form: $y - y_1 = m(x - x_1)$

Review Exercises

In Exercises 1 and 2, graph the given linear equation.

1. $3x - 2y = 36$

2. $3y = 45$

3. Find the point (if any) where the lines $2x + 5y = 50$ and $4x - 15y = 0$ intersect.

4. Find the point (if any) where the lines $x + \dfrac{1}{2}y = 1$ and $2x + y = 1$ intersect.

5. A breeding pond contains fish of types A and B. Each type A fish consumes 5 units of Nutrient 1 and 3 units of Nutrient 2 per day, while each type B fish consumes 9 units of Nutrient 1 and 7 units of Nutrient 2 per day. If 6000 units of Nutrient 1 and 4000 units of Nutrient 2 are added to the pond each day and all the nutrient is consumed, how many fish of each type does the pond contain?

6. A candy manufacturer has 130 lb of chocolate-covered cherries and 170 lb of mints in stock. He sells them in two standard one-pound assortments: Mixture A is half cherries and half mints, while Mixture B is one-third cherries and two-thirds mints. How many pounds of each mixture should he produce to use up all his present stock?

7. Find the slope and y-intercept of the line $5x - 3y = 12$.

8. Find the equation of the line through $(3, -1)$ and $(8, 5)$.

9. A piece of equipment worth $220,000 initially is depreciated linearly over a 25-year period, at the end of which it has a scrap value of $80,000. What will it be worth after ten years? After fifteen years?

10. A manufacturer finds that the demand for a certain product is a linear function of its price. When the price is $16 per unit there is a demand for 45,000 units, but when the price is $21 per unit the demand decreases to 37,000 units. At what price will the demand exactly match the company's present production capacity of 29,000 units?

11. A weekly magazine has fixed production costs of $275,000 per week and a variable cost of 18¢ per copy. If the magazine sells for 50¢, how many copies must be sold each week

    (a) to break even?

    (b) to make a profit of $150,000?

    (c) to make a profit of 10¢ per copy?

    (d) to make a profit of 14% on sales?

12. The owner of a hotel has daily fixed costs of $3600 and a variable cost of $19 for each room occupied. If the rooms rent for $35 a day, how many rooms must be occupied

    (a) to break even?

    (b) to make a profit of $1200 per day?

    (c) to make a profit of 10% on rental income?

13. A toy company produces a magic cube which sells for $10. The fixed cost of the operation is $11,000 per week, with a variable cost of $4 per cube. By automating the assembly of the cubes, the variable cost can be cut to $3.60, but the fixed cost will increase to $13,500 per week.

    (a) How many cubes must be produced per week to break even, under the present operation? Under the automated plan?

    (b) Under what conditions will it pay for the company to automate?

14. A sales representative has received three job offers. Company 1 will pay $2200 a month, with no commission on sales. Company 2 will pay $1900 a month, plus a 20% commission on sales. Company 3 will pay $1700 a month, plus a 30% commision on sales. Let x be the individual's expected monthly sales. For what values of x will each company's offer be best?

2

The Geometry of Linear Programming

Possibly the most widely applied mathematical theory in modern business and industry is one that did not even exist fifty years ago: the theory of *linear programming*. Today it is estimated that fully 25 percent of scientific computer use is devoted to linear programming, and the range of its applications seems almost unlimited.

In this chapter we study linear programming from a geometric viewpoint and present some basic applications of the theory. In Chapter 5 we shall reexamine the subject from an algebraic viewpoint and derive the simplex algorithm, a computational technique for solving such problems efficiently and methodically.

2.1 Formulation of Linear Programs

Many practical problems involve **optimization:** how can a certain quantity—profit or cost, for instance—be maximized or minimized, subject to given conditions? Questions of this type arise in every phase of industrial production. For example, materials must be obtained at minimum cost, subject to fixed production quotas; limited resources must be allocated in the most profitable way; finished goods must be transported as economically as possible, given fixed supplies and demands in different locations.

During World War II many optimization problems arose in connection with military logistics, and teams of research scientists were formed to investigate possible methods of solution. They noticed that many of their problems shared the same features, once they were formulated in mathematical terms. The name ''linear program'' was used to describe any abstract problem of this type, and a new mathematical theory was born. By the

1950s linear programming was being applied far beyond its original setting, to problems in business, industry, agriculture, and the natural and social sciences.

Mathematical Modeling   The discovery of linear programming illustrates the role that real-world problems play in the development of mathematics. The first step in this process—modeling a concrete problem mathematically—is the most important, since it forces us to pose the question in a clear and unambiguous way. In this section we consider several "word" problems, our goal being to *translate each problem into mathematical terms*. In later sections we shall develop methods that will enable us to solve not only these problems but many others as well.

Example 1   Albert's Smoke Shop has in stock 30 lb of Virginia and 45 lb of Latakia tobacco leaf. Albert sells two popular house blends in a standard 12-oz can: Blend A (Albert's Mixture) is two parts Virginia to one part Latakia, and sells at $16. Blend B (Balkan Intrigue) is one part Virginia to three parts Latakia, and sells at $12. How many cans of each blend should Albert make in order to maximize his total sales revenue?

Formulation   We assume that Albert can sell both blends at the stated prices, so that his problem is not limited demand but rather has limited supply of tobacco leaf. The unknowns in the problem are

$$x = \text{number of cans of Blend A produced}$$
$$y = \text{number of cans of Blend B produced.}$$

The total sales revenue R will depend on the values of x and y. Since he charges $16 per can for Blend A and $12 per can for Blend B, Albert's total revenue R, in dollars, is given by

$$R = 16x + 12y.$$

It is this quantity that Albert wants to maximize.

According to the information given, each 12-oz can of Blend A is two-thirds Virginia leaf and one-third Latakia leaf. Hence each can contains $(2/3)(12) = 8$ oz of Virginia and $(1/3)(12) = 4$ oz of Latakia. Similarly, each can of Blend B contains $(1/4)(12) = 3$ oz of Virginia and $(3/4)(12) = 9$ oz of Latakia. Albert has in stock a total of $30(16) = 480$ oz of Virginia and $45(16) = 720$ oz of Latakia (using the conversion 1 lb = 16 oz). This information is summarized in the table below, which shows the total supplies of Virginia and Latakia at the bottom of the corresponding columns.

Blend	Virginia per can	Latakia per can	Price per can	Number of cans
A	8 oz	4 oz	$16	x
B	3 oz	9 oz	$12	y
Available	480 oz	720 oz		

We see from the table that x cans of Blend A will use up $8x$ oz of Virginia, and y cans of Blend B will use up $3y$ oz of Virginia. In order not to exceed the supply of Virginia, it follows that x and y must satisfy the inequality

$$8x + 3y \le 480 \qquad (1)$$

which says that the total amount of Virginia leaf used is less than or equal to the available supply. Similarly, the inequality

$$4x + 9y \le 720 \qquad (2)$$

limits the amount of Latakia used to 720 oz, which is the available supply. We must also include the conditions

$$x \ge 0 \qquad (3)$$
$$y \ge 0 \qquad (4)$$

since we are only interested in nonnegative values of x and y. The original problem, therefore, takes the following mathematical form:

Maximize $R = 16x + 12y$
subject to:

$$8x + 3y \le 480 \qquad (1)$$
$$4x + 9y \le 720 \qquad (2)$$
$$x \ge 0 \qquad (3)$$
$$y \ge 0. \qquad (4) \quad \diamond$$

A problem of this type is called a **linear program.** The unknowns x and y are the **decision variables,** the quantity R to be optimized is the **objective function,** and conditions (1)–(4) are the **constraints.**

A pair of numbers (x, y) is called **feasible** if it satisfies all the constraints. The feasibility of a given pair (x, y) can be determined very quickly by direct substitution.

Example 1, Continued   Test the ordered pairs (30, 50) and (50, 30) for feasibility in the linear program of Example 1.

Solution   Set $x = 30$, $y = 50$ in constraints (1)–(4).

$$8x + 3y = 8(30) + 3(50) = 390 \le 480$$
$$4x + 9y = 4(30) + 9(50) = 570 \le 720$$
$$x = 30 \ge 0$$
$$y = 50 \ge 0$$

Since all the constraints are satisfied, the pair (30, 50) is feasible. This pair gives the value

$$R = 16x + 12y = 16(30) + 12(50) = 1080$$

for the objective function. Thus, Albert can make $1080 by producing 30 cans of Blend A and 50 cans of Blend B. The pair (50, 30), however, is *not* feasible, because constraint (1) is violated:

$$8x + 3y = 8(50) + 3(30) = 490 > 480.$$

In other words, the supply of Virginia leaf is not enough to produce 50 cans of Blend A and 30 cans of Blend B.   ◆

As it happens, (30, 50) is not the only feasible pair, nor is it the "best" one in terms of maximizing revenue. For example, the pairs (25, 60) and (45, 40) are also feasible, with respective revenues $1120 and $1200. The goal, therefore, can be stated as follows:

To find, among all feasible pairs (x, y), the one that gives the optimal value of the objective function. This pair is called the **optimal solution** of the linear program.

We shall return to this problem again in Section 2.4, when we have the necessary mathematical tools to solve it. For the moment, the important thing to observe is that the problem has been translated into precise mathematical terms.

Example 2 Nutritionists indicate that an adequate daily diet should provide at least 75 gm of carbohydrate, 60 gm of protein, and 60 gm of fat. An ounce of Food A contains 3 gm of carbohydrate, 2 gm of protein, and 4 gm of fat and costs 10¢, while an ounce of Food B contains 3 gm of carbohydrate, 4 gm of protein, and 2 gm of fat and costs 15¢. How many ounces of each food should be combined per day to meet the nutritional requirements at lowest cost?

Formulation The unknowns, or decision variables, are:

$$x = \text{number of ounces of Food A per day}$$
$$y = \text{number of ounces of Food B per day}.$$

The information is summarized in the table below.

Food	Carbohydrate per oz	Protein per oz	Fat per oz	Cost per oz	Number of ounces
A	3 gm	2 gm	4 gm	10¢	x
B	3 gm	4 gm	2 gm	15¢	y
Available	75 gm	60 gm	60 gm		

The objective function is the total cost

$$C = 10x + 15y \quad \text{(in cents)}$$

which we want to minimize. The total carbohydrate yield, in grams, is given by $3x + 3y$, which must be at least 75 to meet the daily requirement. In other words,

$$3x + 3y \geq 75.$$

Similarly, the protein and fat requirements lead to the following constraints:

$$2x + 4y \geq 60$$
$$4x + 2y \geq 60$$

Including the nonnegativity requirements $x \geq 0$, $y \geq 0$, we arrive at the following mathematical formulation of the problem:

$$\text{Minimize} \quad C = 10x + 15y$$
subject to:
$$3x + 3y \geq 75$$
$$2x + 4y \geq 60$$
$$4x + 2y \geq 60$$
$$x \geq 0$$
$$y \geq 0.$$

This is an example of a *minimum* program. The problem is to find, among all pairs (x, y) satisfying the above constraints, the one for which the objective function C takes its *smallest* value. This will be done in Section 2.4. ◆

Remark to Reader Be especially careful about the direction of the inequality sign in formulating constraints. The phrase "*a* is at least *b*" means $a \geq b$, while "*a* is at most *b*" means $a \leq b$. In some problems both types of inequality may occur, as the following example illustrates.

Example 3* The heating plant for a large government building must produce at least 300 million British thermal units (Btu) of heat per hour. The plant is equipped to burn any combination of fuel oil and coal. Each ton of oil costs \$290 and produces 40 million Btu, while each ton of coal costs \$65 and produces 20 million Btu. The plant manager wants to produce the required amount of heat at minimum cost, but there is another consideration. The government has set pollution limits of 300 lb and 125 lb on the respective amounts of sulfur dioxide (SO_2) and nitrogen dioxide (NO_2) the plant can emit per hour. When a ton of oil is burned, 20 lb of SO_2 and 5 lb of NO_2 are produced. For a ton of coal the figures are 30 lb and 20 lb, respectively. What should the plant manager do?

---

*The author wishes to thank Mr. William de Boisblanc of the Bay Area Air Quality Management District (San Francisco) for providing information on which this problem is based.

Formulation   First, the decision variables in the problem must be defined.

$$x = \text{number of tons of oil per hour}$$
$$y = \text{number of tons of coal per hour}$$

The data can be represented as in the table below.

Fuel	Heat (million Btu per ton)	SO_2 (lb per ton)	NO_2 (lb per ton)	Cost (dollars per ton)	Number of tons
Oil	40	20	5	290	x
Coal	20	30	20	65	y
Available	300	300	125		

Notice that oil gives 40 million Btu for \$290, or 40 million $\div$ 290, which is about 138,000 Btu per dollar, while coal gives 20 million $\div$ 65, or about 308,000 Btu per dollar. Hence coal is the cheaper energy source. But in order to get 300 million Btu we would need 300 million $\div$ 20 million or 15 tons of coal per hour. This would produce 450 lb of SO_2 and 300 lb of NO_2, both of which exceed the pollution limits. Using oil alone 7.5 tons would be required to get 300 million Btu, and this would produce only 150 lb of SO_2 and 37.5 lb of NO_2, which is acceptable. However, the cost of the oil would be \$2175. By using a suitable combination of oil and coal, we should be able to lower the cost while staying within the pollution limits.

The total heat (in millions of Btu) produced by x tons of oil and y tons of coal is given by $40x + 20y$. The heat requirement thus becomes

$$40x + 20y \geq 300.$$

The pollution limits give rise to the following inequalities in which the sign is reversed:

$$20x + 30y \leq 300$$
$$5x + 20y \leq 125$$

The objective function is the total hourly fuel cost (in dollars).

$$C = 290x + 65y$$

As usual, neither x nor y can be negative. The problem can thus be formulated as the following linear program with five constraints:

$$\text{Minimize} \quad C = 290x + 65y$$
$$\text{subject to:}$$
$$40x + 20y \geq 300$$
$$20x + 30y \leq 300$$
$$5x + 20y \leq 125$$
$$x \geq 0$$
$$y \geq 0. \quad \blacklozenge$$

In Section 2.4 we shall see how to find the optimal solution to the linear programs presented here. Our immediate objective, however, has been achieved: each of our "word problems" has been successfully translated into a problem in pure mathematics.

2.1 Exercises

1. In Example 2 in the text (the nutrition problem), represent each of the following combinations as an ordered pair (x, y). Determine which combinations are feasible, and compare their cost.

   (a) 10 oz of Food A and 20 oz of Food B

   (b) 25 oz of Food A and 2 oz of Food B

   (c) 15 oz of Food A and 15 oz of Food B

   (d) 35 oz of Food A and 0 oz of Food B

2. In Example 3 in the text (the heating plant problem), represent each of the following combinations as an ordered pair (x, y). Determine which combinations are feasible, and compare their cost.

   (a) 7 tons of oil and 3 tons of coal

   (b) 7.5 tons of oil and 0 tons of coal

   (c) 5 tons of oil and 6 tons of coal

   (d) 4 tons of oil and 7 tons of coal

3. In Example 1 in the text (the tobacco problem), how does it change the linear program if we add the following conditions? (a) Albert must produce at least 27 cans of Blend A, to satisfy his regular customers; and (b) he can produce at most 80 cans of blend altogether, because of limited shelf space.

4. In Example 3 in the text (the heating plant problem), how does it change the linear program if we add the following conditions? (a) Only 10 tons of oil per hour are available, due to fuel rationing; and (b) the plant is not equipped to burn more than 3 tons of coal per hour.

In Exercises 5–14, formulate the given problem in mathematical terms.

5. The House of Coffee sells two different mixtures of Mocha, Java, and Santos beans. Each pound of Mixture A contains 1/4 lb Mocha, 1/2 lb Java, and 1/4 lb Santos and sells for $3.50. Each pound of Mixture B contains 3/5 lb Mocha, 1/5 lb Java, and 1/5 lb Santos and sells for $4.00. If the store has a supply of 300 lb of Mocha, 250 lb of Java, and 150 lb of Santos beans, how many pounds of each mixture should be made in order to maximize total revenue?

6. A steel mill makes two grades of stainless steel, both of which are sold in 100-lb bars. The standard grade bar contains 90% steel and 10% chromium by weight, and sells for $110. The high-grade bar contains 80% steel and 20% chromium, and sells for $120. If the company has a supply of 36 tons (72,000 lb) of steel and 5 tons of chromium, how many bars of each grade should it produce to maximize total revenue?

7. A farmer has a 2000-acre plot of land on which he intends to plant some combination of alfalfa and beets. Alfalfa requires $20 capital and 25 inches of water

for each acre planted, while beets require $30 capital and 75 inches of water for each acre planted. The net profit per acre is $30 for alfalfa and $40 for beets. If the farmer has a total of $45,000 in capital and a total water allotment of 90,000 acre-inches, how many acres of each crop should he plant to maximize total profit? (Note: An acre-inch is the amount of water needed to irrigate an acre of land to a depth of one inch.)

8. A gem dealer buys rough-cut amethysts and beryls of uniform size, which she finishes for sale to retail jewelers. Each amethyst requires one hour on a grinder, one hour on a sander, and two hours on a polisher and yields a net profit of $75. Each beryl requires two hours on a grinder, five hours on a sander, and two hours on a polisher and yields a profit of $100. If the grinder, sander, and polisher are each available for 40 hours a week, how many amethysts and beryls should be finished each week to maximize total profit?

9. A cannery has decided to replace its old machines with newer equipment. The manager has been allocated $300,000 capital for the new machines. He has 36 available operators and space for at most 28 new units. Of the two models under consideration, Unit A costs $10,000, requires one operator, and produces 600 cans per hr, while Unit B costs $15,000, requires two operators, and produces 1000 cans per hr. How many units of each type should be purchased in order to maximize can production?

10. World Travel Service has agreed to transport a group of 1800 football fans from Los Angeles to Chicago for one of the season's big games. The agency can charter two different types of aircraft for the trip. Type A has 20 first-class seats and 80 economy seats and costs $3000 to operate for one trip. Type B has 40 first-class seats and 60 economy seats and costs $4000 to operate. If 600 of the passengers want to go first-class while the remaining 1200 have ordered economy seats, how many planes of each type should be assigned to the charter to minimize total operating cost?

11. A printing house orders newsprint in bulk rolls. The rolls supplied by Company A cost $100; each roll gives a two-day supply of paper and occupies 35 sq ft of storage space. The rolls supplied by Company B cost $300; each roll gives a five-day supply of paper and occupies 25 sq ft of storage space. The printing house has a total of 10,500 sq ft of storage space and needs enough paper for 1000 days. How can this be achieved at lowest cost?

12. The Exxess Oil Company operates two eastern refineries. The Albany facility produces 100 bbl of high-grade oil and 300 bbl of medium-grade oil per day. The Boston facility produces 200 bbl of high-grade and 100 bbl of medium-grade per day. The company has determined that the combined total output of high-grade oil should fall somewhere between 12,000 and 18,000 bbl (inclusive), and the combined output of medium-grade should fall between 24,000 and 30,000 bbl. If the Albany and Boston refineries cost $2000 and $4200 per day to operate, respectively, how many days should each be run to meet the requirements at lowest cost?

13. The Heine-Borel Corporation is trying to decide between two comprehensive insurance policies covering fire, theft, and liability. The Albatross Insurance Company offers a policy that provides $10,000, $5000, and $20,000 of fire, theft, and liability coverage, respectively, for each unit bought, at a premium of $500 per unit. The Baltimore Insurance Company offers a policy that provides $10,000

each of fire, theft, and liability coverage for each unit bought, at a premium of $450 per unit. The corporation needs coverage of at least $400,000 each for fire and theft and $1,000,000 for liability. How many units of each policy should be purchased to minimize cost?

14. A chemical firm has been trying to reduce sulfur oxide emissions from its two Chicago plants. Plant I produces 10 tons of sulfuric acid, 6 tons of ammonia, and 7.5 tons of ammonium sulfate per day, with 100 lb of sulfur oxide emitted. Plant II produces no sulfuric acid, 12 tons of ammonia, and 5 tons of ammonium sulfate per day, with 72 lb of sulfur oxide emitted. How many days should each plant operate in order to produce at least 600 tons of each of the three chemicals with a minimum total emission of sulfur oxide?

2.2   Linear Inequalities

The constraints in a linear program are **inequalities,** expressing conditions to the effect that one quantity is *at most* or *at least* equal to another. As a first step toward solving the problems formulated in the preceding section, we now examine the properties of inequalities.

Laws of Inequality   We assume that the reader is familiar with the four **inequality signs** below.

$$< \quad \text{less than} \qquad \leq \quad \text{less than or equal to}$$
$$> \quad \text{greater than} \qquad \geq \quad \text{greater than or equal to}$$

The signs on the left are called **strict,** those on the right **weak.** Thus, for example, the inequalities

$$2 < 5 \qquad 5 > -4 \qquad 3 \leq 4 \qquad 5 \geq -2$$
$$0 < 7 \qquad -1 > -2 \qquad 2 \leq 2 \qquad -1 \geq -4$$

are all true mathematical statements. We can use algebraic techniques to manipulate inequalities in much the same way that we manipulate equations. These techniques are based on the following laws:

Laws of Inequality

1. If $a \leq b,$ then $a + c \leq b + c$ and $a - c \leq b - c.$

2. If $a \leq b$ and $c > 0,$ then $ac \leq bc$ and $\dfrac{a}{c} \leq \dfrac{b}{c}.$

3. If $a \leq b$ and $c < 0,$ then $ac \geq bc$ and $\dfrac{a}{c} \geq \dfrac{b}{c}.$

The first law says that we can add any number to both sides of an inequality, or subtract it from both sides. It follows that we can *transpose* any term in an inequality; that is, transfer it from one side to the other and change its sign. The second law permits us to multiply or divide an inequality by a positive number, and the third says we can multiply or divide by a negative number *provided that we reverse the direction of the inequality*. These laws also hold when weak inequality is replaced by strict inequality throughout; that is, when $\leq$ and $\geq$ are replaced by $<$ and $>$, respectively.

Example 1   Simplify the inequality $2x + 1 \leq 16 - 3x$.

Solution   First, to collect the x-terms we add $3x$ to both sides of the inequality. This is justified by the first law.

$$2x + 1 \leq 16 - 3x$$
$$2x + 1 + 3x \leq 16 - 3x + 3x \qquad \text{(adding } 3x)$$
$$5x + 1 \leq 16$$

Next, we can subtract 1 from both sides to isolate the x-term on the left.

$$5x + 1 \leq 16$$
$$5x + 1 - 1 \leq 16 - 1 \qquad \text{(by the first law)}$$
$$5x \leq 15$$

Finally, we can apply the second law to obtain x alone on the left.

$$5x \leq 15$$
$$\frac{5x}{5} \leq \frac{15}{5} \qquad \text{(dividing by 5)}$$
$$x \leq 3$$

This is the simplified form of the inequality. From it we see that the original condition is satisfied by all numbers less than or equal to 3, and by no others.   ◆

Example 2   Simplify the inequality

$$\frac{7}{3}x - \frac{2}{5}(5y - 16) \leq 3\left[x + \frac{1}{9}y - \frac{1}{5}\right].$$

Solution   We first multiply out.

$$\frac{7}{3}x - 2y + \frac{32}{5} \leq 3x + \frac{1}{3}y - \frac{3}{5}$$

We can multiply through by 15 to eliminate fractions, using the second law.

$$35x - 30y + 96 \leq 45x + 5y - 9$$

Next, we transpose the terms containing variables to the left and the constant terms to the right as in the preceding example. That is, we subtract $45x$,

5y, and 96 from both sides and apply the first law. The result, after collecting terms, is the inequality

$$-10x - 35y \leq -105.$$

Finally, we can divide through by -5 and reverse the direction of the inequality (by the third law) to obtain

$$2x + 7y \geq 21.$$

This is the simplified form of the original inequality.   ◆

Linear Inequalities   A mathematical condition of one of the forms

$$ax + by < c, \qquad ax + by \leq c, \qquad ax + by > c, \qquad ax + by \geq c,$$

where a, b, and c are constants and at least one of the coefficients a, b is nonzero, is called a **linear inequality** in x and y. In the preceding examples, the simplified form of the inequality is linear in x and y. Specifically, in Example 1 the inequality has the form $ax + by \leq c$, with $a = 1$, $b = 0$, and $c = 3$, while the inequality in Example 2 has the form $ax + by \geq c$, with $a = 2$, $b = 7$, and $c = 21$.

Linear inequalities are similar in form to linear equations, the only difference being the presence of an inequality sign in place of an equals sign. To see how they are related, let us look at a specific example.

Example 3   The linear equation $2x + 3y = 6$ describes the straight line pictured in Figure 1. What inequalities are associated with this line?

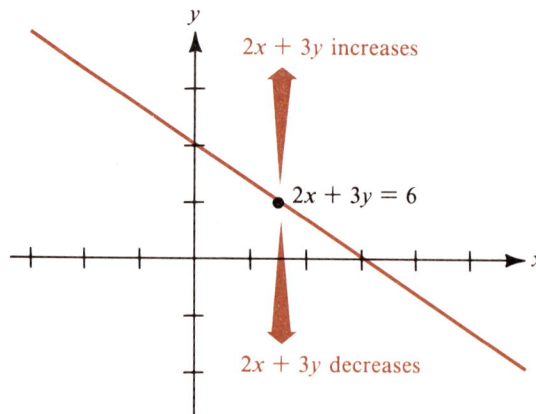

y

$2x + 3y$ increases

$2x + 3y = 6$

x

$2x + 3y$ decreases

Figure 1

Discussion   The quantity $2x + 3y$ takes a value at each point (x, y). For any point lying on the line this value is 6. As the point moves upward from the line its y-coordinate increases; therefore the value of $2x + 3y$ also increases. As the point moves downward y decreases and thus $2x + 3y$ decreases. It follows that $2x + 3y$ takes values *greater* than 6 above the line and *less* than 6 below it. (See Figure 2.)

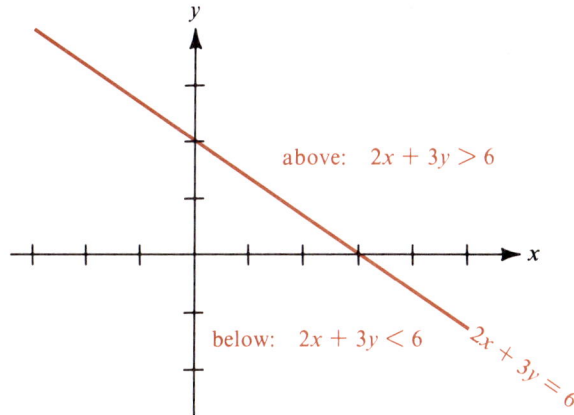

above: $2x + 3y > 6$

below: $2x + 3y < 6$

$2x + 3y = 6$

Figure 2

Similar reasoning leads to the following general observation:

Let $ax + by = c$ be any straight line. On one side of the line the inequality $ax + by < c$ holds, on the other side $ax + by > c$ holds.

The graph of a linear inequality is therefore a **half-plane,** the region on one side of a line. If the inequality is weak ($\leq$ or $\geq$) the half-plane is **closed;** that is, the line itself is part of the graph. If the inequality is strict ($<$ or $>$) the half-plane is **open,** and points on the line are excluded.

Our result yields the following simple method for graphing a linear inequality.

Graphing a Linear Inequality

Given a condition with one of the following forms:

$$ax + by \leq c \qquad ax + by < c \qquad ax + by \geq c \qquad ax + by > c$$

1. Graph the bounding line $ax + by = c$. Use a solid line if the inequality is weak, a broken line if the inequality is strict.
2. Test a point not on the bounding line to determine which half-plane the inequality determines.

Example 4 Graph the inequality $x - y \geq 1$.

Solution By the discussion above, we know that the graph of $x - y \geq 1$ is one of the half-planes determined by the line $x - y = 1$. If we graph this bounding line as in Figure 3, we can find out which side is which by testing a particular point not on the line. At the origin $(0, 0)$, for example, we find that

$$x - y = 0 - 0 = 0 < 1$$

so that the inequality is false. Since the origin lies above the line, the graph of $x - y \geq 1$ must consist of all points lying on or below the line. This is a closed half-plane.

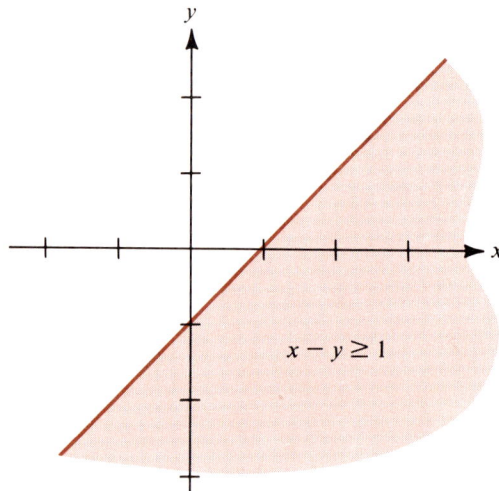

Figure 3

Example 5 Graph the inequality $5x + 4y > 0$.

Solution The bounding line $5x + 4y = 0$ goes through the origin, as shown in Figure 4. We therefore test the inequality at some other point, for instance, the point $(1, 1)$.

$$5x + 4y = 5(1) + 4(1) = 9 > 0$$

The inequality is satisfied. Since $(1, 1)$ lies above the line, the graph of $5x + 4y > 0$ must be the upper half-plane determined by the line in Figure 4. Since the inequality is strict, the half-plane is open; that is, the bounding line itself is not part of the graph. For this reason the bounding line is shown as a broken line.

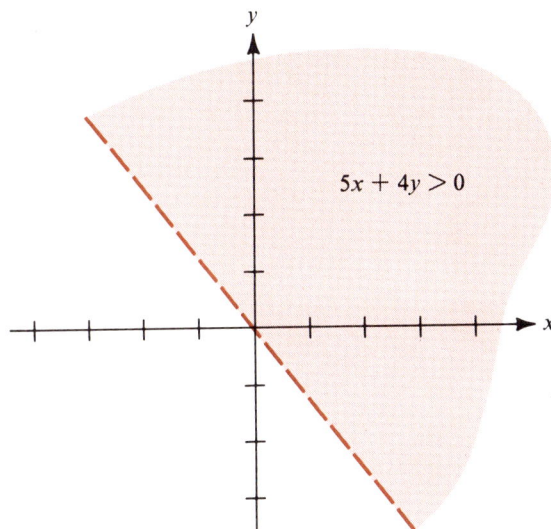

$5x + 4y > 0$

Figure 4

◆

Example 6   Graph the inequality $-3x \leq 5$.

Solution   The bounding line $-3x = 5$ is vertical, crossing the x-axis at $(-5/3, 0)$, as shown in Figure 5. Since the origin $(0, 0)$ satisfies the inequality, the graph is the closed half-plane to the right of the bounding line.

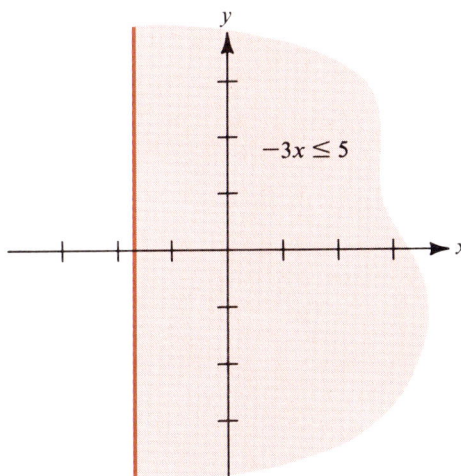

$-3x \leq 5$

Figure 5

◆

Of special interest are the inequalities $x \geq 0$ and $y \geq 0$, since they appear as constraints in almost all linear programs. These conditions have as their graphs the half-planes shown in Figures 6a and 6b. Note that in order to satisfy *both* inequalities, a point must lie in the first quadrant (Figure 6c).

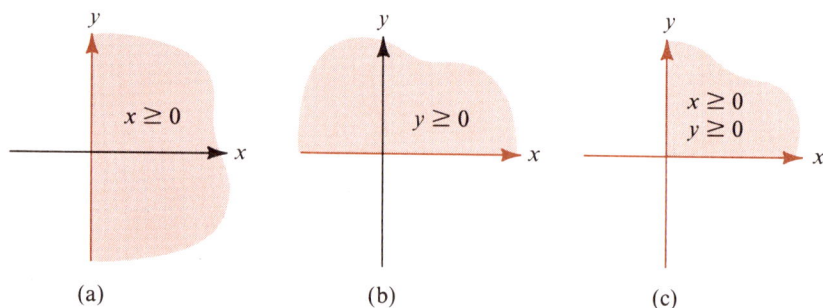

(a)   (b)   (c)

Figure 6

Applications   Many practical problems lead to constraints that seem mathematically complicated, but turn out to be linear inequalities after simplification. The next example is of a type found in many blending problems.

Example 7   A large grain silo contains 500 tons of grain, to which two other grains, A and B, are to be added to form a new mixture. The grain already in the silo contains 5% impurities by weight. Grains A and B contain 4% and 1% impurities, respectively. If x tons of Grain A and y tons of Grain B are added, what condition must x and y satisfy if the mixture is to contain at most 3% impurities?

Solution   We summarize the information in a table.

Grain	Impurities per ton	Number of tons
In silo	5%	500
A	4%	x
B	1%	y

The total amount of impurities in the mixture, in tons, is the sum of the amounts present in the three grains.

$$\text{total impurities} = .05(500) + .04(x) + .01(y)$$
$$= 25 + .04x + .01y \quad \text{(tons)}$$

The condition that the mixture contain at most 3% impurities can be expressed as

$$\text{total impurities} \leq .03 \text{ (total amount of mixture)}$$
$$\text{or} \quad 25 + .04x + .01y \leq .03(500 + x + y).$$

If we multiply out and collect terms in this inequality, we obtain

$$25 + .04x + .01y \leq 15 + .03x + .03y$$
$$.01x - .02y \leq -10.$$

This can be further simplified to

$$x - 2y \leq -1000$$
$$\text{or} \quad -x + 2y \geq 1000.$$

The graph of this last linear inequality is the half-plane pictured in Figure 7a. Of course, x and y must also be nonnegative to have physical significance in terms of the problem; that is, we would only be interested in the part of the half-plane which overlaps the first quadrant (Figure 7b).

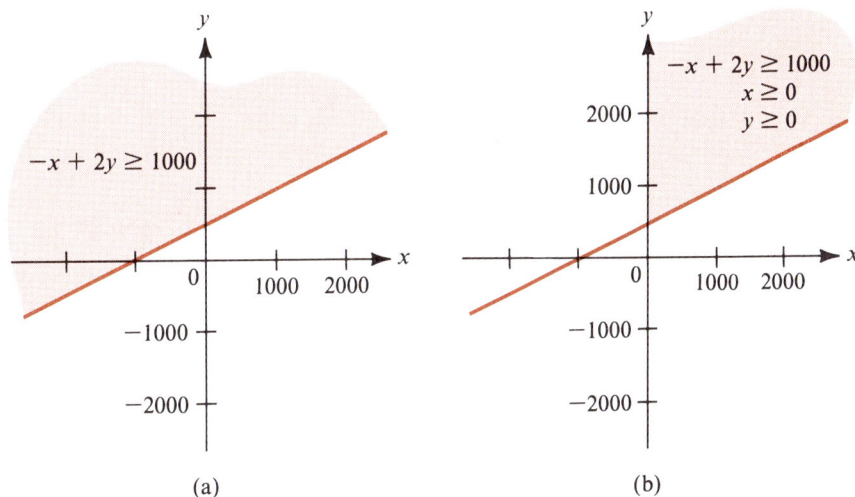

Figure 7

Note that the region in Figure 7b is defined by a *system* of linear inequalities. In the next section we shall study such systems in more detail.

2.2 Exercises *In Exercises 1–8, indicate which of the six points*

$A(2, 3) \quad B(-1, 4) \quad C(0, -2) \quad D(3/4, 5/4) \quad E(5, 0) \quad F(-4, -1)$

satisfy the given linear inequality and which do not.

1. $2x + 5y \geq 10$
2. $-x + y < 1$
3. $x + 4y \leq 0$
4. $2x \geq 3$
5. $5x - 2y > 4$
6. $y < 0$
7. $2x - 3y \geq 12$
8. $x + y < 10$

In Exercises 9–22, graph the linear inequality.

9. $5x + 3y \geq 60$

10. $4x - 3y < 10$

11. $x + y \leq 2000$

12. $16x - 9y > 150$

13. $10x + 9y \geq 0$

14. $y \leq 5$

15. $-x + 2y \leq 0$

16. $-2x \leq 0$

17. $2x \geq 3$

18. $10x + 100y > 1000$

19. $2x - 2y < 1$

20. $x + y \geq 0$

21. $y > -1$

22. $\frac{1}{2}x + \frac{1}{3}y \geq \frac{1}{4}$

In Exercises 23–28, express the given condition equivalently in the form of a linear inequality. Simplify the result as much as possible.

23. $2x + y \leq x + 2y + 1$

24. $x \geq -\frac{3}{7}y + 6$

25. $.20x + .30y < .05(x + y)$

26. $-\frac{1}{5}(x + 2y + 1) + \frac{2}{5}(3x - 1) \leq 2y + \frac{27}{5}$

27. $.4(.05x + .07y + 195) + .3(.1x + .04y + 140) > .2(x + y + 300)$

28. $\frac{2}{3}(x - 5y + 1) \geq \frac{1}{2}[(2x + y) - \frac{2}{3}(x - 2y - 11)]$

29. The owner of a nut shop has 50 lb of a mixture which is 40% cashews by weight. To this amount she adds x lb of Mixture A, which is 50% cashews, and y lb of Mixture B, which is 80% cashews. Express mathematically the condition that the entire mixture is at least 60% cashews. Simplify as much as possible.

30. A farmer's crop consists of x acres of alfalfa and y acres of beets. Alfalfa requires 25 inches of water for each acre planted, while beets require 75 inches of water per acre. Express mathematically the condition that the average water requirement of the crop is at most 35 inches per acre. Simplify as much as possible.

31. Bronze alloy A is 95% copper, 4% tin, and 1% zinc, while bronze alloy B is 90% copper, 2% tin, and 8% zinc by weight. Suppose that x tons of alloy A and y tons of alloy B are mixed together in a smelting kiln. Express each of the following conditions as a linear inequality in x and y, simplifying as much as possible.

    (a) The amount of copper in the mixture is at least 30 times the amount of tin.

    (b) The mixture is at most 5% zinc.

    (c) The mixture is at least 3% tin.

32. An oil storage tank contains 12,000 barrels (bbl) of low-grade crude with a sulfur content of 225 gm per bbl. Two higher grades of crude are added to the tank to lower the sulfur content: Grade A contains 125 gm of sulfur per bbl, while Grade B contains 100 gm per bbl. Suppose that x bbl of Grade A and y bbl of Grade B are added to the tank. Express each of the following conditions as a linear inequality in x and y, simplifying as much as possible.

    (a) The tank can hold a maximum of 50,000 bbl of crude.

    (b) The amount of sulfur in the tank is at most 6,000,000 gm.

    (c) The sulfur content of the mixture is at most 150 gm/bbl.

2.3 Systems of Linear Inequalities

A single linear inequality in x and y has as its graph a half-plane. We now investigate what happens when we impose not one but several such conditions at the same time, as we do in a linear program. For simplicity, we restrict our attention to systems of *weak* inequalities ($ax + by \leq c$ or $ax + by \geq c$).

Joint Constraints A system consisting of one or more weak linear inequalities will be called, for short, a **system of joint constraints.** A point (x, y) is a **feasible point** if it satisfies all the constraints. Together the feasible points form the **constraint set,** which is a certain region in the xy-plane.

Example 1 Graph the following system:
$$x + y \geq 3$$
$$x - y \leq 1.$$

Solution Each constraint describes the region on one side of a line (a half-plane). We begin by numbering the constraints and their corresponding bounding lines.

Constraint		Bounding Line	
$x + y \geq 3$	(1)	$x + y = 3$	L_1
$x - y \leq 1$	(2)	$x - y = 1$	L_2

The lines L_1 and L_2 are shown in Figure 1. Testing the origin in constraints (1) and (2) we find that (1) fails but (2) is satisfied. Hence, in order to be feasible a point (x, y) must lie on or above L_1 and *also* on or above L_2. The constraint set is thus the region where these two half-planes overlap. This region is shaded in Figure 1.

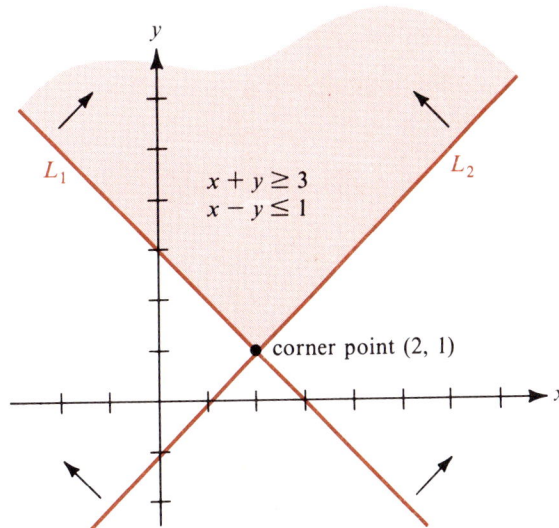

Figure 1

Note that the constraint set is an *unbounded* region, extending upward indefinitely. It is V-shaped, with a **corner point** at (2, 1) where the bounding lines meet. This point is itself a member of the constraint set, since it satisfies both inequalities.

In general, each additional constraint in a joint system has the geometric effect of "slicing off" part of the region determined by the earlier constraints. The final constraint set therefore always has a polygonal shape.

Example 2   Graph the system

$$5x + 2y \leq 30$$
$$-x + 4y \leq 16$$
$$x \geq 0$$
$$y \geq 0.$$

Solution   We list the constraints and bounding lines.

Constraint	*Bounding Line*	
$5x + 2y \leq 30$	$5x + 2y = 30$	L_1
$-x + 4y \leq 16$	$-x + 4y = 16$	L_2
$x \geq 0$	$x = 0$	L_3
$y \geq 0$	$y = 0$	L_4

The bounding lines are shown in Figure 2. To be feasible, a point (x, y) must lie on the same side of L_1 and L_2 as the origin, and in the first quadrant. The constraint set is thus the bounded, four-sided region shown in Figure 2.

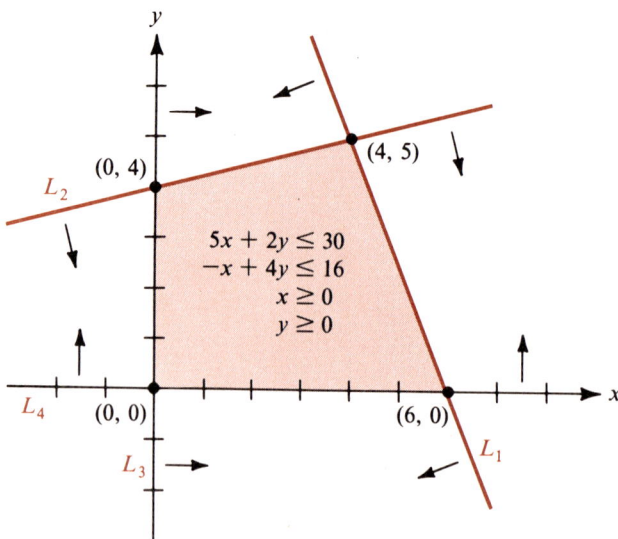

Figure 2

Each of the four corner points of the region is determined by a pair of bounding lines, and can therefore be found by solving two equations in two unknowns. For example, the corner point (4, 5) is the intersection of L_1 and L_2, and the corner point (0, 0) is the intersection of L_3 and L_4. ◆

Example 3   Graph the system
$$-x + 2y \leq 4$$
$$x - 2y \leq 2.$$

Solution   The bounding lines L_1 and L_2 for the constraints are pictured in Figure 3. These lines are parallel, with common slope 1/2. The constraint set consists of all points (x, y) which lie on or below L_1 and on or above L_2. This is a **parallel strip,** which is unbounded and has no corner points.

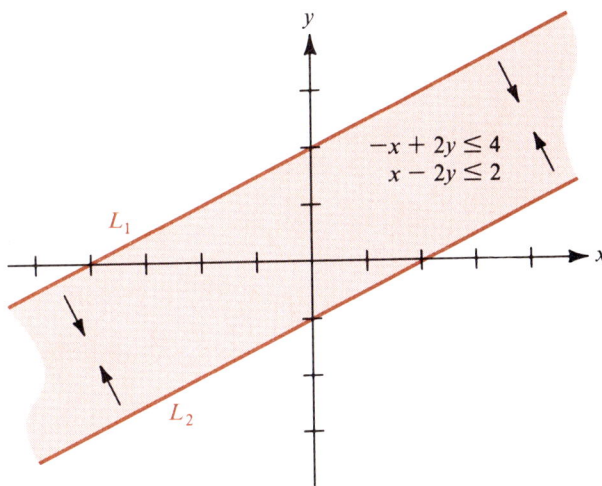

$$-x + 2y \leq 4$$
$$x - 2y \leq 2$$

Figure 3   ◆

Example 4   Graph the system
$$x \geq 3$$
$$y \geq 2$$
$$x + y \geq 3.$$

Solution   The constraint set is the unbounded region in Figure 4 on the next page, with a single corner point at (3, 2). The third constraint has no effect on the graph; the first and second constraints alone describe exactly the same region. This means that the third inequality is implied by the other two; in other words, the third constraint is **redundant.** We can omit redundant constraints if we wish, but there is no harm in retaining them.

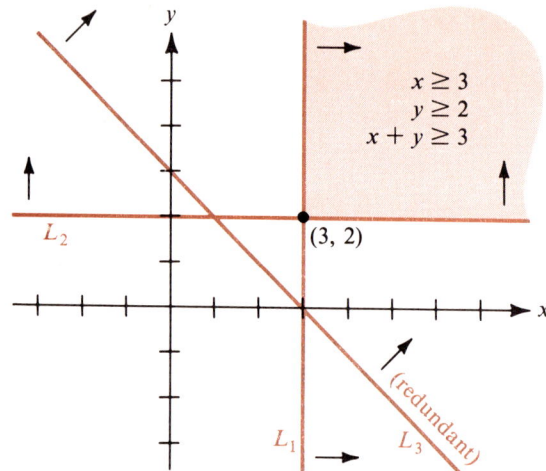

Figure 4

The graph of a system of joint constraints may occasionally degenerate into a line, a line segment, a single point, or even the empty set. For example, the joint constraints

$$2x + 3y \leq 6$$
$$2x + 3y \geq 6$$

are equivalent to the single equation $2x + 3y = 6$, which describes a straight line; and the joint constraints

$$2x + 3y \leq 2$$
$$2x + 3y \geq 6$$

have an empty graph, since they are inconsistent with each other (see Figures 5a and 5b). Of course, such exceptional cases rarely arise in practical problems.

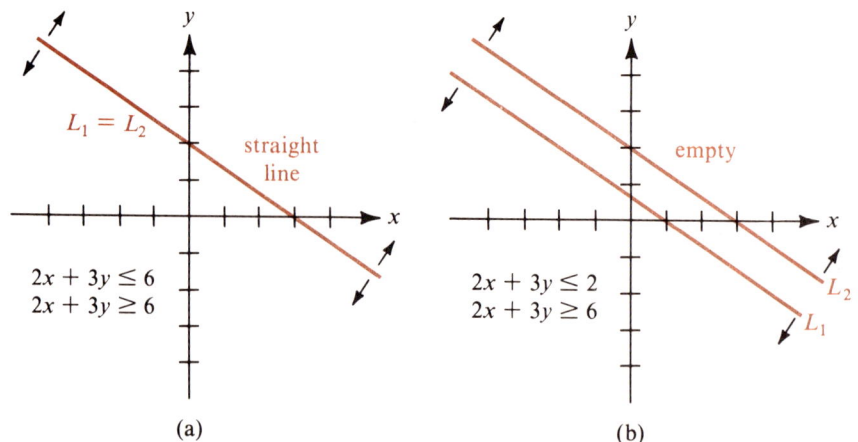

(a)   (b)

Figure 5

Convex Polygonal Sets   The constraint sets we have seen are all examples of **convex polygonal sets.** The term **convex** describes a set S that "bulges outward" and has no holes or indentations. More precisely: if P and Q are any points in a convex set S, the entire line segment PQ is contained in S. Figure 6 shows two sets in the plane, one convex and one nonconvex. (Why?)

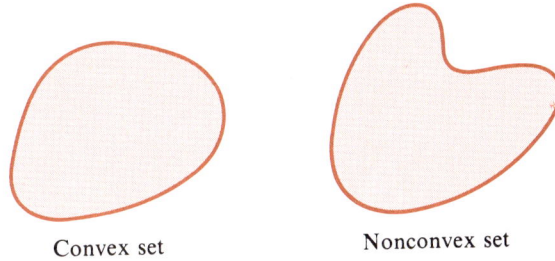

Convex set   Nonconvex set

Figure 6

A **polygonal set** S is a set whose boundary is made up of a finite number of lines or line segments. Figure 7 shows two polygonal sets, one convex and one nonconvex.

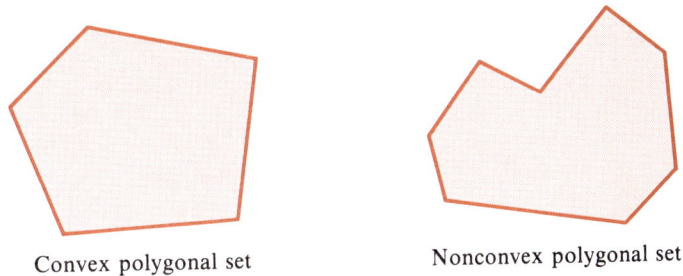

Convex polygonal set   Nonconvex polygonal set

Figure 7

These constraint sets are also **closed,** meaning that points on the boundary are included in the set. (This is because the constraints here are *weak* inequalities.) Taken together, these three geometric properties characterize constraint sets completely, as a result of the following fact:

Any system of joint constraints in two variables has as its graph a closed convex polygonal set. Conversely, any such set is the graph of a system of joint constraints.

As we have seen in our examples, the constraint region may be either bounded (of finite extent) or unbounded.

Corner Points and Basic Points   The corner points of the constraint set play a key role in the theory of linear programming, as we shall see in the next section. In the examples presented earlier we were able to find the corner points directly from the graph, using the fact that each corner point is the intersection of two bounding lines. In some cases, however—for example, when we have many bounding lines that intersect close to one another—it is difficult to determine the corner points in this way. We now describe an alternative procedure for finding the corner points that does not rely on pictures at all.

Definition

A **basic point** for a system of joint constraints is a point where two bounding lines intersect.

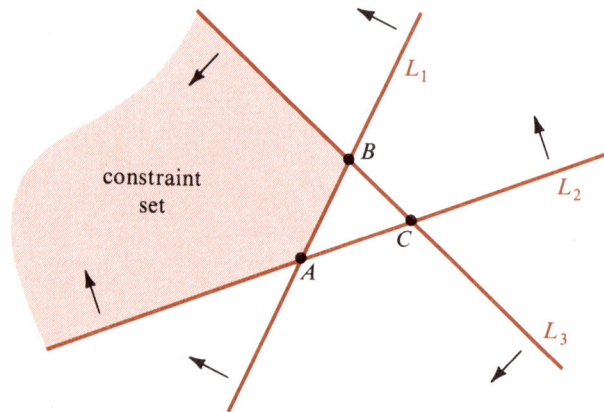

Figure 8

The meaning of the definition is illustrated in Figure 8. Points *A*, *B*, and *C* are all basic points, but only *A* and *B* are corner points. The reason *C* fails to be a corner point is that it falls outside the constraint set—in other words, it is not feasible. In general, *a corner point is just a feasible basic point*. We can therefore find the corner points of the constraint set as follows:

Method of Basic Points

To find the corner points of the constraint set:

1. List all possible pairs of bounding lines.
2. For each pair of lines, find the point of intersection by solving two equations in two unknowns.
3. Test the resulting point in the constraints to see whether it is feasible.

Since the feasible basic points are the corner points, we will eventually find them all in this way.

Example 5 Given the system

$$x - y \leq 2$$
$$x + y \leq 6$$
$$x \geq 1$$
$$y \leq 3$$

find all corner points of the constraint set, using the method of basic points.

Solution There are four bounding lines, which we number in the usual way. Taking them two at a time, there are six possible combinations; they are listed in the following table.

Bounding lines	Basic point	Feasible?
L_1, L_2	$(4, 2)$	Yes
L_1, L_3	$(1, -1)$	Yes
L_1, L_4	$(5, 3)$	No
L_2, L_3	$(1, 5)$	No
L_2, L_4	$(3, 3)$	Yes
L_3, L_4	$(1, 3)$	Yes

For example, the third row of the table indicates that bounding lines L_1 and L_4 intersect at point $(5, 3)$, which we discover by solving the system $x - y = 2$ and $y = 3$. The basic point $(5, 3)$ is not feasible, however, since it violates the second constraint. Thus it is not a corner point. (See Figure 9.)

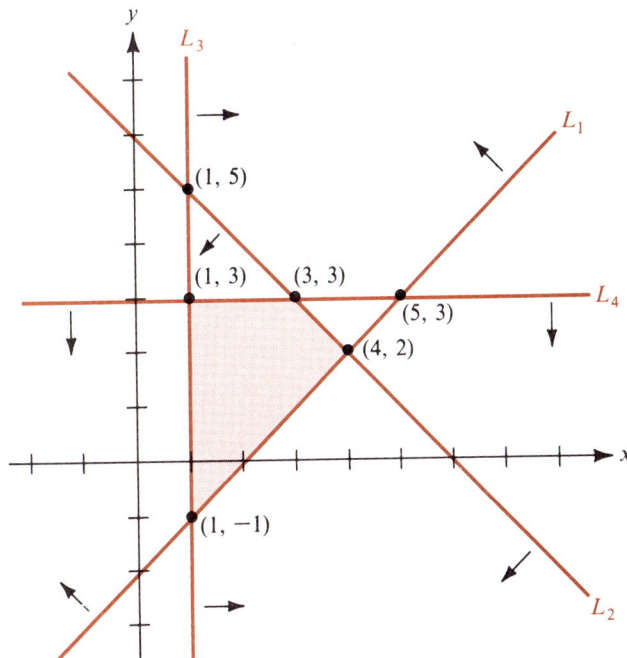

Figure 9

From the table we see that the constraint set has four corner points: (4, 2), (1, −1), (3, 3), and (1, 3). This is confirmed by the graph of the system, which appears in Figure 9. ◆

2.3 Exercises

In Exercises 1–18, graph the system of joint constraints. Identify all corner points of the constraint set.

1. $x - y \leq 1$
   $x + y \leq 3$

2. $-3x + 2y \leq 6$
   $x - 4y \leq 8$

3. $4x + 3y \geq 12$
   $x \leq 2$

4. $3x - 2y \geq 0$
   $5x + 2y \geq 0$

5. $7x + 3y \leq 210$
   $7x + 3y \geq 84$

6. $2x - 3y \leq 1$
   $-2x + 3y \geq 6$

7. $4x + y \leq 100$
   $x + 4y \leq 100$
   $x + y \geq 10$

8. $5x - 3y \leq 0$
   $x + y \leq 800$
   $x \geq 150$

9. $2x + 5y \geq 100$
   $x \geq 0$
   $y \geq 0$

10. $x + y \geq 4$
    $x - y \leq 4$
    $x - y \geq 0$

11. $2x - 3y \geq 0$
    $x + y \geq 6$
    $x \leq 3$

12. $2x + y \leq 5$
    $x + 2y \geq -2$
    $x + y \geq 1$

13. $2x + y \leq 4$
    $x + y \geq 1$
    $x \geq 0$
    $y \geq 0$

14. $x - y \leq 10$
    $x + y \geq 10$
    $y \geq 10$
    $x \leq 30$

15. $x + 2y \leq 8$
    $3x + y \leq 9$
    $x - 3y \leq 3$
    $x \geq 0$

16. $4x + 3y \leq 12$
    $4x + 3y \geq 12$
    $x \geq 0$
    $y \geq 0$

17. $3x + y \leq 5$
    $2x - y \leq 0$
    $x + y \leq 3$
    $x \leq 1$
    $y \geq 0$

18. $x - y \geq -2$
    $x + y \leq 4$
    $x \leq 3$
    $x \geq 0$
    $y \geq 0$

In Exercises 19–24, use the method of basic points to find all basic points and corner points of the constraint set. Confirm your results by graphing.

19. $x + y \geq 1$
    $x - y \leq 1$
    $y \leq 1$

20. $4x - y \geq 0$
    $x - 4y \leq 0$
    $x + y \geq 5$

21. $x + y \leq 6$
    $x - y \geq 0$
    $x \geq 1$
    $y \geq 0$

22. $5x - 3y \leq 0$
    $x + y \leq 800$
    $x + y \geq 200$
    $x \leq 150$

23. $x + 2y \leq 24$
     $2x + y \leq 24$
     $x + y \leq 15$
     $x \geq 0$
     $y \geq 0$

24. $x + y \geq 2$
     $x - y \geq 0$
     $y \geq 1$
     $y \leq 2$
     $x \geq 1$

In Exercises 25–28, find a system of joint constraints whose graph is the given convex polygonal set.

25. The triangle with corner points (1, 0), (0, 1), and (2, 2)

26. The triangle with corner points (0, 2), (3, 4), and (2, −2)

27. The square with corner points (1, 0), (0, 1), (1, 2), and (2, 1)

28. The quadrilateral with corner points (0, 0), (3, 0), (0, 5), and (4, 1)

2.4 Geometric Solution of Linear Programs

With the mathematical tools developed in the last two sections we can now return to the question of solving the problems we formulated in Section 2.1.

Linear Programs   A **linear program** is a problem in which we want to maximize or minimize a function

$$F = ax + by$$

where a and b are constants, subject to a system of joint constraints. A function F of this form is called a **linear function** of x and y.

Example 1   Albert's Smoke Shop has available 30 lb of Virginia and 45 lb of Latakia tobacco leaf. Albert sells two popular house blends in a standard 12-oz can. Blend A is two parts Virginia to one part Latakia, and sells at $16. Blend B is one part Virginia to three parts Latakia, and sells at $12. How many cans of each blend should Albert make to maximize his total revenue?

Discussion   In Section 2.1 we formulated this problem mathematically as follows:

$$\text{Maximize}\quad R = 16x + 12y$$
subject to:
$$8x + 3y \leq 480$$
$$4x + 9y \leq 720$$
$$x \geq 0$$
$$y \geq 0.$$

where x and y represent the number of cans of Blend A and B, respectively, and R represents the total revenue in dollars.

The graph of the four constraints is the convex polygonal set S shown in Figure 1, with corner points $O(0, 0)$, $A(60, 0)$, $B(36, 64)$, and $C(0, 80)$. The problem is to *maximize the objective function over this set;* in other words, to find the point (x, y) in S for which the revenue R is greatest.

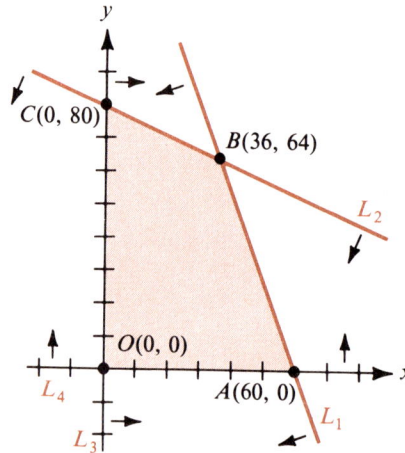

Figure 1

Isolines   If we fix the revenue R at a particular value c, we obtain a linear equation

$$16x + 12y = c$$

whose graph is called an **isoline** of the revenue function. All points (x, y) along this line yield the same revenue. Several of these isolines are shown in Figure 2.

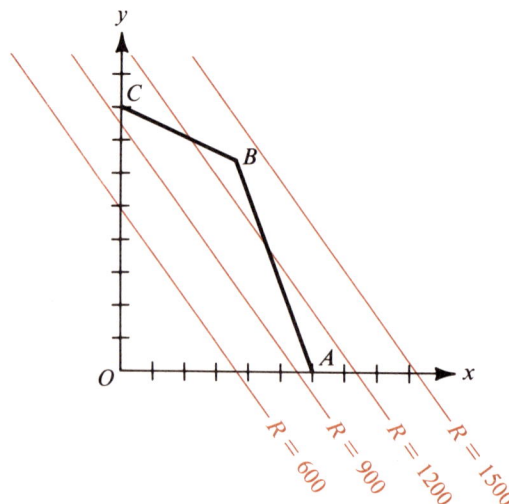

Figure 2

Note that all the isolines are parallel, with common slope $-4/3$. As c increases the lines move farther from the origin, until finally they miss the constraint set entirely. For example, we see that no feasible point (x, y) will yield $R = 1500$. From the picture it is intuitively clear that *the last isoline to touch the constraint set will pass through the corner point B(36, 64)*. Since $R = 16(36) + 12(64) = 1344$ at this point, we have found the optimal solution:

$$\text{maximum } R = 1344 \quad \text{when} \quad x = 36 \quad \text{and} \quad y = 64.$$

In other words, Albert should make 36 cans of Blend A and 64 cans of Blend B, which will give him the maximum possible profit, $1344. ◆

The Corner Point Theorem The method used in Example 1 can be generalized to any linear program. The constraint set will always be a closed convex polygonal set S, and the objective function will be of the form $F = ax + by$. The isolines $F = c$ will be parallel, and they will move steadily in one direction as c increases.

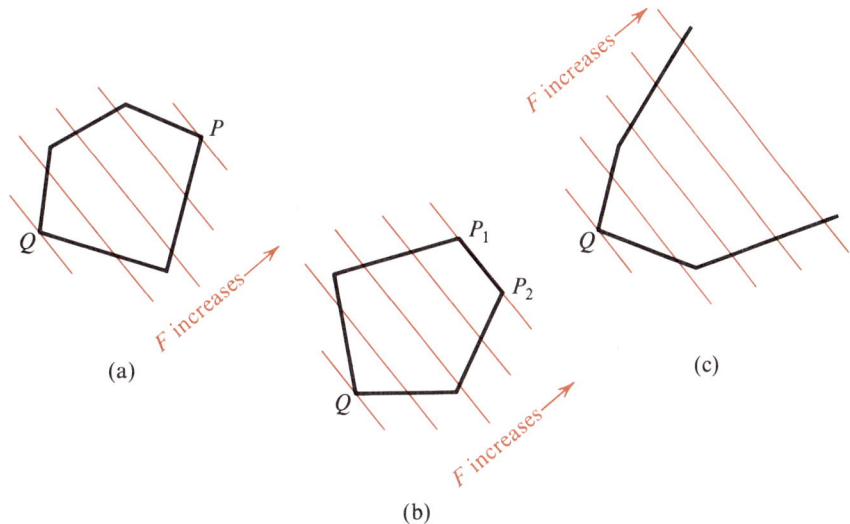

Figure 3

Figure 3 illustrates the possibilities. (In all three cases we assume the isolines $F = c$ move from left to right as c increases.) In Figure 3a the constraint set S is bounded, and the maximum and minimum values of F occur at P and Q respectively. In Figure 3b the last isoline to touch the constraint set happens to pass through *two* adjacent corner points, P_1 and P_2. In this case F is maximized along the entire edge P_1P_2. In Figure 3c the constraint set is unbounded. In this case, the minimum value of F occurs at Q, but no point in S will give a maximum value for F; that is, F increases without bound on the set S.

We can summarize these observations in the following statement, which is the fundamental theorem of linear programming.

The Corner Point Theorem

If a linear program has an optimal solution, it must occur at a corner point of the constraint set.

From our analysis we see that an optimal solution—that is, a maximum or minimum value of F on the set S—will *always* exist when S is bounded. When S is unbounded, as in Figure 3c, an optimal solution may or may not exist, depending on the program. Also, two adjacent corner points may *both* be optimal, in which case any point along the segment joining them is too.

In view of the Corner Point Theorem, we can find the optimal solution to a linear program, when it exists, as follows:

Geometric Solution of a Linear Program

1. Identify all corner points of the constraint set, either by graphing or by the method of basic points.
2. Test the value of the objective function at each corner point. The one for which the objective function assumes its greatest (or least) value is the optimal solution to the linear program.

Note that it is not necessary to graph any isolines. We merely compare the values of the objective function at the corner points, and use the Corner Point Theorem.

Example 2   Nutritionists indicate that an adequate daily diet should provide at least 75 gm of carbohydrate, 60 gm of protein, and 60 gm of fat. An ounce of Food A contains 3 gm of carbohydrate, 2 gm of protein, and 4 gm of fat and costs 10¢, while an ounce of Food B contains 3 gm of carbohydrate, 4 gm of protein, and 2 gm of fat and costs 15¢. How many ounces of each food should be combined per day to meet the nutritional requirements at lowest cost?

Solution   This problem was formulated in Section 2.1 as the minimum program:

$$\text{Minimize}\quad C = 10x + 15y$$
$$\text{subject to:}$$
$$3x + 3y \geq 75$$
$$2x + 4y \geq 60$$
$$4x + 2y \geq 60$$
$$x \geq 0$$
$$y \geq 0.$$

Here x and y represent the number of ounces of Food A and Food B, respectively, and C represents the total daily cost, in cents.

The constraint set S is the unbounded region shown in Figure 4, with corner points (0, 30), (5, 20), (20, 5), and (30, 0). The cost function $C = 10x + 15y$ must attain a minimum value on S, since it increases when we move away from the origin. By the Corner Point Theorem, we have only to compare the values of C at the four corner points.

Corner Point	Value of $C = 10x + 15y$
(0, 30)	450
(5, 20)	350
(20, 5)	275 ← minimum
(30, 0)	300

The minimum cost is therefore $C = 275¢ = \$2.75$, obtained by combining 20 oz of Food A and 5 oz of Food B per day.

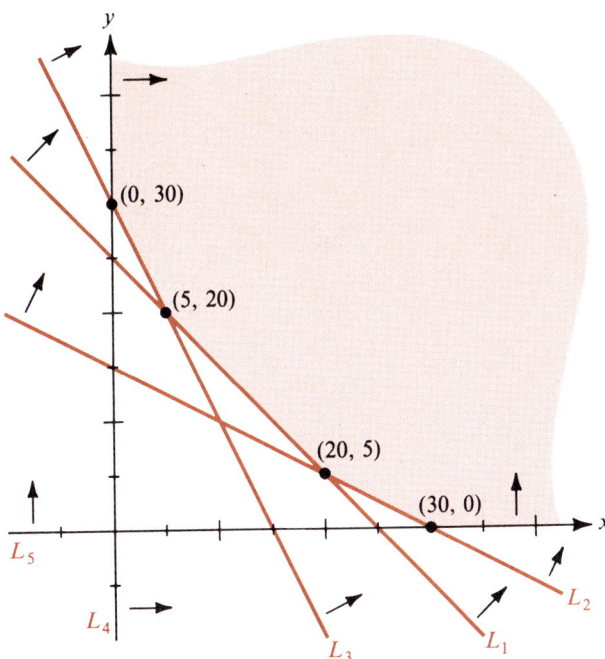

Figure 4

Example 3 The heating plant for a large government building must produce at least 300 million British thermal units (Btu) of heat per hour. The plant is equipped to burn any combination of fuel oil and coal. Each ton of oil costs $290 and produces 40 million Btu, while each ton of coal costs $65 and produces 20 million Btu. The plant manager wants to produce the required amount of heat at minimum cost, but there is another consideration: The government has set pollution limits of 300 lb and 125 lb on the respective amounts of sulfur dioxide (SO_2) and nitrogen dioxide (NO_2) the plant can emit per hour. When a ton of oil is burned, 20 lb of SO_2 and 5 lb of NO_2 are produced. For a ton of coal the figures are 30 lb and 20 lb, respectively. What should the plant manager do?

Solution As we saw in Section 2.1, this problem becomes the following linear program:

$$\text{Minimize}\quad C = 290x + 65y$$
$$\text{subject to:}$$
$$40x + 20y \geq 300$$
$$20x + 30y \leq 300$$
$$5x + 20y \leq 125$$
$$x \geq 0$$
$$y \geq 0.$$

In this case, x and y represent the number of tons of oil and coal burned per hour, respectively, and C represents the total hourly fuel cost in dollars.

If we solve the program by the method of basic points, we must consider ten possible combinations of the five bounding lines, as shown in the table below.

Bounding lines	Basic point	Feasible?	$C = 290x + 65y$
L_1, L_2	(3.75, 7.5)	No	
L_1, L_3	(5, 5)	Yes	1775 ← minimum
L_1, L_4	(0, 15)	No	
L_1, L_5	(7.5, 0)	Yes	2175
L_2, L_3	(9, 4)	Yes	2870
L_2, L_4	(0, 10)	No	
L_2, L_5	(15, 0)	Yes	4350
L_3, L_4	(0, 6.25)	No	
L_3, L_5	(25, 0)	No	
L_4, L_5	(0, 0)	No	

Of the ten basic points, only four turn out to be feasible and hence corner points. The objective function takes its smallest corner point value ($C = 1775$) at (5, 5). This point therefore represents the optimal solution to the linear program.

$$\text{minimum } C = \$1775 \quad \text{when} \quad x = 5 \text{ and } y = 5$$

The constraint set is illustrated in Figure 5.

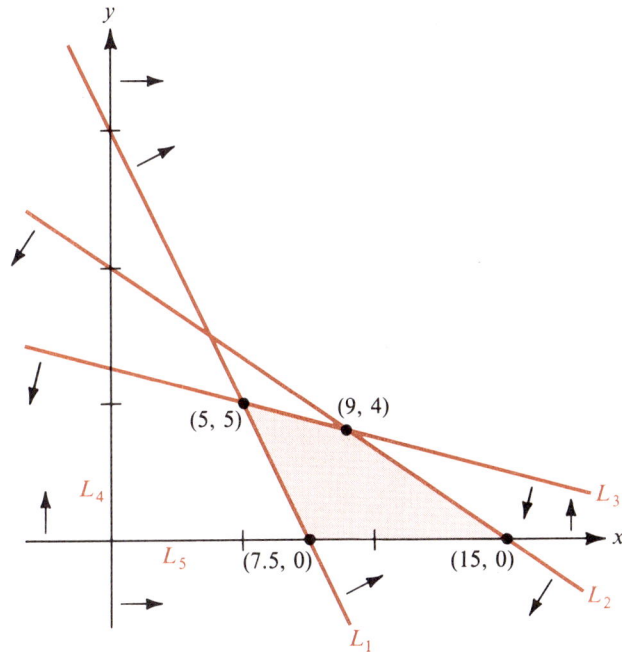

Figure 5

The method of basic points is somewhat tedious, and it gives us no assurance that the linear program actually *has* an optimal solution. On the other hand, it has the advantage that it does not require a picture. Since it is entirely algebraic, it could be carried out—in principle, at least—on a computer.

2.4 Exercises

1. In Example 2 in this section (the nutrition problem), find the optimal solution if Food A costs 14¢ per oz and Food B costs 12¢ per oz.

2. In Example 3 in this section (the heating plant problem), find the optimal solution if oil costs $130 per ton and coal costs $65 per ton.

In Exercises 3–14, use the method of graphing to solve the linear program given in the exercises for Section 2.1.

3. Exercise 3	4. Exercise 4	5. Exercise 5	6. Exercise 6
7. Exercise 7	8. Exercise 8	9. Exercise 9	10. Exercise 10
11. Exercise 11	12. Exercise 12	13. Exercise 13	14. Exercise 14

In Exercises 15–20, use the method of basic points to solve the linear program, without drawing a graph. Assume the existence of an optimal solution in each case.

15. Maximize $F = 3x + 2y$
  subject to:
$$4x + y \leq 100$$
$$x + 4y \leq 100$$
$$x + y \geq 10.$$

16. Minimize $G = 4x + 5y$
  subject to:
$$5x - 3y \leq 0$$
$$x + y \leq 800$$
$$x \geq 150.$$

17. Minimize $G = 2x + 2y$
  subject to:
$$2x + y \leq 4$$
$$x + y \geq 1$$
$$x \geq 0$$
$$y \geq 0.$$

18. Maximize $F = 10x - 5y$
  subject to:
$$x - y \leq 10$$
$$x + y \geq 10$$
$$y \geq 10$$
$$x \leq 30.$$

19. Maximize $F = 5x + 2y$
  subject to:
$$3x + y \leq 5$$
$$2x - y \leq 0$$
$$x + y \leq 3$$
$$x \leq 1$$
$$y \geq 0.$$

20. Minimize $G = 2x + 3y$
  subject to:
$$x - y \geq -2$$
$$x + y \leq 4$$
$$x \leq 3$$
$$x \geq 0$$
$$y \geq 0.$$

2.5 Geometric Solution in Higher Dimensions

When a linear program involves more than two variables, we cannot represent its constraint set as a region in the *xy*-plane. In this section we discuss the geometry of a simple three-variable program and show how to generalize the method of basic points to find its optimal solution.

Formation of Linear Programs The technique of formulating a three-variable problem as a linear program is essentially the same as in the two-variable case.

Example 1 A garment maker manufactures three different styles of men's suits. The profit on each, as well as the labor required for cutting and sewing, are given in the following table:

Style	Profit per suit	Labor per suit (work-hours)	
		Cutting	*Sewing*
A	$250	2 wh	2 wh
B	$400	3 wh	2 wh
C	$300	1 wh	3 wh

If the manufacturer has available a total of 60 wh (work-hours) of cutting labor and 96 wh of sewing labor per day, how many suits of each type should be produced to maximize total profit?

Formulation There are three unknowns, or decision variables.

$$x = \text{number of style A suits produced per day}$$
$$y = \text{number of style B suits produced per day}$$
$$z = \text{number of style C suits produced per day}$$

Formulating the problem mathematically, we obtain a three-variable linear program with five constraints:

$$\text{Maximize} \quad P = 250x + 400y + 300z$$
subject to:
$$2x + 3y + z \leq 60$$
$$2x + 2y + 3z \leq 96$$
$$x \geq 0$$
$$y \geq 0$$
$$z \geq 0. \quad \diamond$$

A triple (x, y, z) of numbers is **feasible** if it satisfies all the constraints. The problem in Example 1, to which we will return, is to find the feasible triple that gives the greatest value of P.

Three-Dimensional Space   In order to represent a constraint set graphically, we must use an xyz-coordinate system in three-dimensional space. Such a system is depicted in Figure 1.

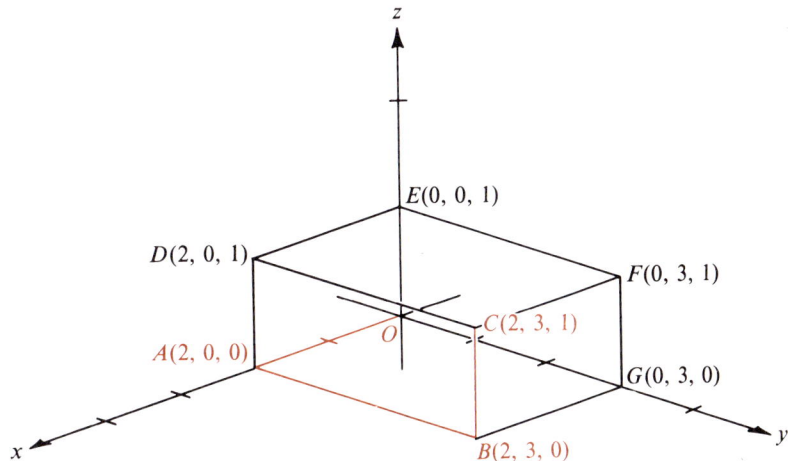

Figure 1

The *x*, *y*, and *z* axes are perpendicular to each other and meet at the origin *O*. The positive direction on each axis is indicated by the arrow. Once such a system is fixed, each ordered triple (*x*, *y*, *z*) determines a unique point *P* in space. For example, to plot the triple (2, 3, 1), we start at the origin and move 2 units in the *x*-direction, then 3 units in the *y*-direction, then 1 unit in the *z*-direction, following the path *OABC* in Figure 1.

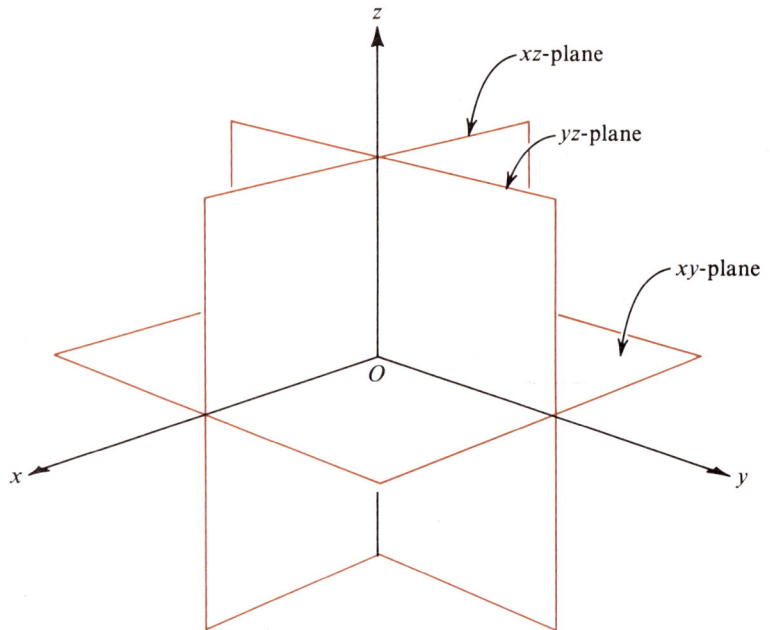

Figure 2

The three axes determine three **coordinate planes,** as shown in Figure 2. These planes divide space into eight regions, called **octants.** The **first octant** consists of all points—like *C* in Figure 1—that satisfy $x > 0$, $y > 0$, and $z > 0$. In Figure 2 we are viewing the origin from the perspective of a point somewhere in the first octant. There is no standard numbering of the other seven octants, but each one can be characterized by specifying the sign of *x*, *y*, and *z*.

Linear Equations   An equation of the form

$$ax + by + cz = d \tag{1}$$

where *a*, *b*, *c*, and *d* are constants and at least one of the coefficients *a*, *b*, *c* is nonzero, is called a **linear equation** in *x*, *y*, and *z*. The term *linear* is meant to emphasize the formal similarity with the two-variable case, but it is somewhat misleading, because a "linear" equation in *x*, *y*, and *z* describes not a line but a *plane*.

The graph of a linear equation $ax + by + cz = d$ is a plane in xyz-space, and conversely any plane is the graph of such an equation. Points on one side of the plane satisfy the inequality $ax + by + cz < d$, while points on the other side satisfy $ax + by + cz > d$.

Example 2 Graph the equation $4x + 3y + 6z = 12$.

Solution This equation is of the form (1), with $a = 4$, $b = 3$, $c = 6$, and $d = 12$. To find the **x-intercept** of the plane, we set $y = z = 0$ in the equation and solve for x.

$$4x + 3(0) + 6(0) = 12$$
$$4x = 12$$
$$x = 3$$

The plane therefore crosses the x-axis at the point $(3, 0, 0)$. Similarly, setting $x = z = 0$ gives $y = 4$ (the y-intercept), and setting $x = y = 0$ gives $z = 2$ (the z-intercept). If we plot these three points and join them by lines, as in Figure 3, we get a partial graph of the plane $4x + 3y + 6z = 12$. (Although the plane extends infinitely, the picture shows only the triangular part lying in the first octant.)

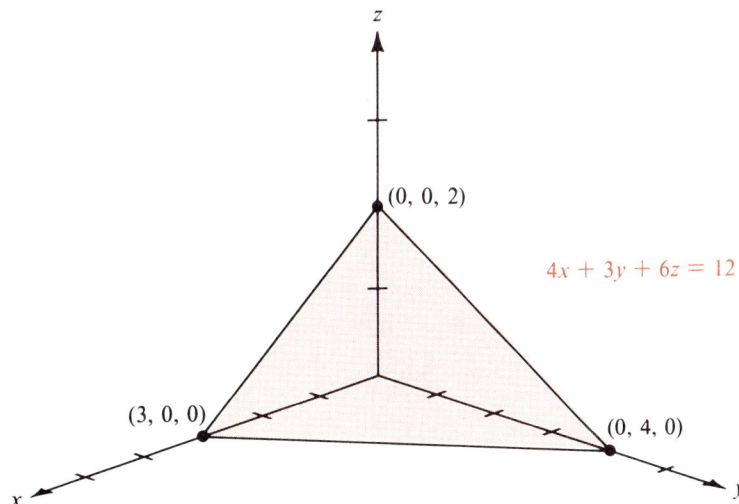

Figure 3

The inequality $4x + 3y + 6z \leq 12$ describes the origin side of the plane in Figure 3—in other words, all points (x, y, z) lying *on or behind* the plane from our point of perspective. ◆

Solution of Linear Programs The constraint set for a three-variable linear program is a solid region bounded by planes. The equations of these planes come from the constraining inequalities in the usual way.

Example 1,
Continued

Solve the linear program of Example 1.

Maximize $P = 250x + 400y + 300z$
subject to: *Bounding Planes*

$$2x + 3y + z \leq 60 \qquad\qquad 2x + 3y + z = 60 \qquad p_1$$
$$2x + 2y + 3z \leq 96 \qquad\qquad 2x + 2y + 3z = 96 \qquad p_2$$
$$x \geq 0 \qquad\qquad\qquad\qquad x = 0 \qquad p_3$$
$$y \geq 0 \qquad\qquad\qquad\qquad y = 0 \qquad p_4$$
$$z \geq 0. \qquad\qquad\qquad\qquad z = 0 \qquad p_5$$

Discussion The bounding planes p_1, p_2, p_3, p_4, and p_5 are pictured in Figure 4. Note that plane p_1 *(ABH)* crosses plane p_2 *(FGD)* along the line *EC*. Planes p_3, p_4, and p_5 are just the coordinate planes; for example, p_3 has the equation $x = 0$, which is satisfied by every point in the yz-plane.

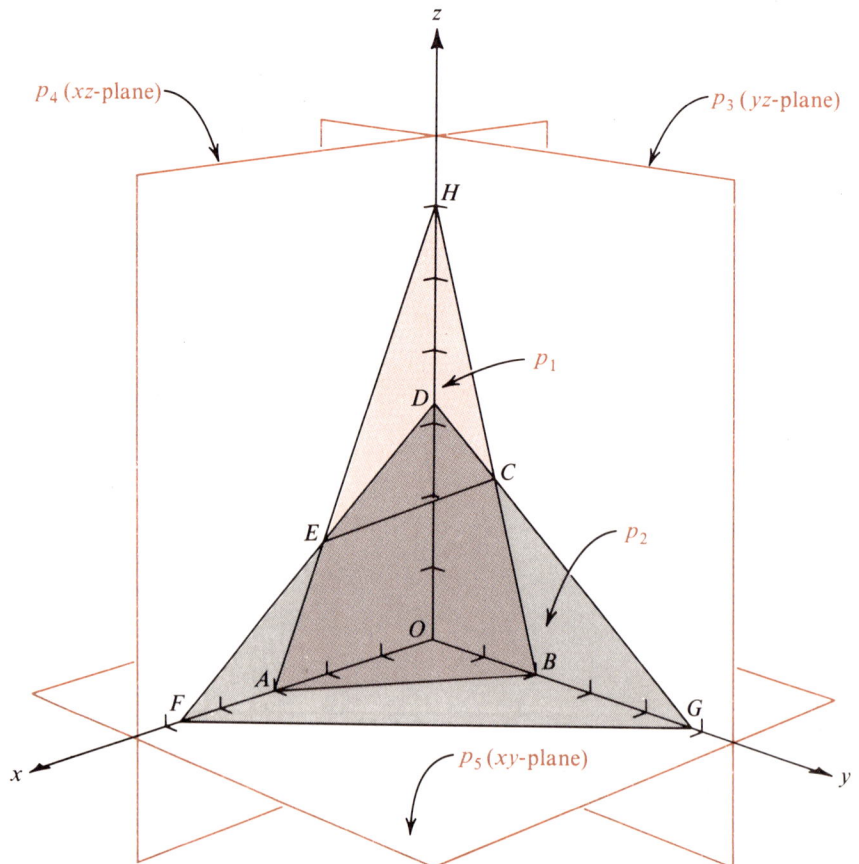

Figure 4

In order to satisfy all five constraints, a point (x, y, z) must lie on or behind p_1, on or behind p_2, and in the first octant. The constraint set is thus the solid, five-sided region shown in Figure 5.

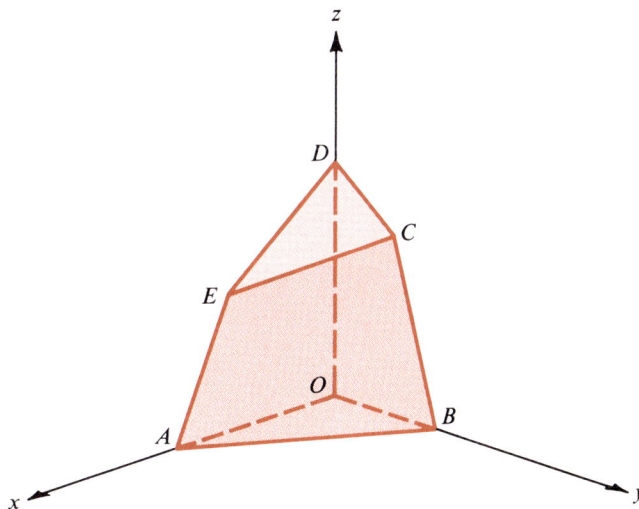

Figure 5

This set S is a **convex polyhedral set,** a three-dimensional convex set whose boundary is made up of a finite number of plane faces. Our experience with the two-dimensional case suggests that the optimal solution to our linear program should be one of the six corner points labeled in Figure 5. This conjecture turns out to be correct.

Corner Point Theorem

The constraint set S for a three-variable linear program is always a closed convex polyhedral set. If the program has an optimal solution, it must occur at a corner point of S.

We can find the coordinates of each corner point by solving three equations in three unknowns. For example, the point $C(x, y, z)$ is the intersection of bounding planes p_1, p_2, and p_3 (as we see from Figure 4) and its coordinates must therefore satisfy all three equations.

$$
\begin{aligned}
2x + 3y + z &= 60 & p_1 \\
2x + 2y + 3z &= 96 & p_2 \\
x &= 0 & p_3
\end{aligned}
$$

We can use the third equation to eliminate x from the first two.

$$3y + z = 60$$
$$2y + 3z = 96$$

Solving this pair of equations as usual, we obtain $y = 12$, $z = 24$. C is therefore the point $(0, 12, 24)$.

As in the two-variable case, we can find the corner points without drawing the graph at all. Since each corner point is determined by three bounding planes, we have only to consider all the possible combinations as in the table below.

Bounding planes	Basic point	Feasible?	Value of P
p_1, p_2, p_3	$(0, 12, 24)$	Yes	$12{,}000 \leftarrow$ maximum
p_1, p_2, p_4	$(21, 0, 18)$	Yes	$10{,}650$
p_1, p_2, p_5	$(84, -36, 0)$	No	
p_1, p_3, p_4	$(0, 0, 60)$	No	
p_1, p_3, p_5	$(0, 20, 0)$	Yes	8000
p_1, p_4, p_5	$(30, 0, 0)$	Yes	7500
p_2, p_3, p_4	$(0, 0, 32)$	Yes	9600
p_2, p_3, p_5	$(0, 48, 0)$	No	
p_2, p_4, p_5	$(48, 0, 0)$	No	
p_3, p_4, p_5	$(0, 0, 0)$	Yes	0

There are ten possible combinations of bounding planes, when taken three at a time. Each combination determines a **basic point,** which can be found by solving three equations in three unknowns. Of the ten basic points shown in the table, only six turn out to be feasible when tested in the constraints. These are the six corner points shown in Figure 5. From the table we see that point C is the optimal solution to the linear program:

maximum $P = \$12{,}000$ when $x = 0$, $y = 12$,   and $z = 24$.   ◆

The Geometry of the Invisible   When a linear program involves more than three variables, we can no longer use the method of graphing. For example, a program with four decision variables—for instance, x, y, z, and w—has a constraint set in four-dimensional space. Since our visual intuition is based on the physical world, which has only three spatial dimensions, we cannot hope to represent such a problem pictorially.

Fortunately, the method of basic points helps us through the darkness. In the two-variable case we needed two bounding equations to determine a basic point, and in the three-variable case we needed three bounding equations. *In the n-variable case, a basic point will be determined by n bounding equations.* Once we have a basic point, we can determine whether it is a corner point by testing it in the constraints as usual. Thus, we could solve an n-variable linear program as follows:

Geometric Solution of an *n*-Variable Program

1. List the equations corresponding to each of the constraints.
2. List all possible combinations of these equations, taken *n* at a time.
3. For each combination, find the basic point it determines by solving *n* equations in *n* unknowns.
4. Test the resulting point in the constraints to see whether it is feasible.
5. Compare the values of the objective function at all feasible basic points. The one for which the objective function assumes its greatest (or least) value is the optimal solution to the linear program.

This is the most general version of our method of basic points. Although it could easily be programmed on a computer, this method is extremely impractical, since the number of combinations of bounding equations is generally enormous. For example, a linear program with 15 variables and 30 constraints—and problems of this size often arise in applications—would produce over 155 million basic points. For each of these it would be necessary to solve 15 equations in 15 unknowns. Even if a computer could determine a single basic point in .01 second, it would take almost 18 days to find them all!

For this reason it is desirable to have a procedure for solving linear programs which does not require all the intermediate steps used in the geometric approach. In Chapter 5 we shall see such a procedure, called the *simplex algorithm*. When we study this method we shall see that the geometric concepts developed in this chapter, such as basic points and corner points, continue to play an important role.

2.5 Exercises

In Exercises 1–8, graph the linear equation as a plane in xyz-space.

1. $5x + 3y + 2z = 60$
2. $4x + 2y + 5z = 100$
3. $x + y + z = 1$
4. $4x - 3y + 6z = 12$
5. $2x + 3z = 6$ (no *y*-intercept)
6. $y + z = 20$ (no *x*-intercept)
7. $x = 2$ (no *y*- or *z*-intercept)
8. $y = -1$ (no *x*- or *z*-intercept)

In Exercises 9–12, graph the system of joint constraints as a convex polyhedral set.

9. $x + y + z \leq 1$
    $x \geq 0$
    $y \geq 0$
    $z \geq 0$

10. $2x + 6y + 3z \leq 12$
    $y \geq 1$
    $x \geq 0$
    $z \geq 0$

11. $2x + y + z \leq 12$
    $x + 2y + 2z \leq 12$
    $x \geq 0$
    $y \geq 0$
    $z \geq 0$

12. $x + y + z \geq 1$
    $x + y + z \leq 2$
    $x \geq 0$
    $y \geq 0$
    $z \geq 0$

In Exercises 13–16, formulate the given problem as a linear program and use the method of basic points to find the optimal solution.

13. Albert's Smoke Shop has in stock 30 lb of Virginia and 45 lb of Latakia tobacco. Albert sells three house blends in a standard 12-oz can. The composition and selling price of each blend are given in the following table.

Blend	Virginia per can	Latakia per can	Price per can
A	8 oz	4 oz	$16
B	3 oz	9 oz	$12
C	6 oz	6 oz	$15

How many cans of each blend should Albert make to maximize his total revenue?

14. A steel mill makes three grades of stainless steel, each of which is sold in 100-lb bars. The composition by weight and selling price of each grade are shown below.

Grade	Steel	Chromium	Price per bar
Standard	90%	10%	$110
High-grade	80%	20%	$120
Supreme	75%	25%	$150

If the company has a supply of 36 tons (72,000 lb) of steel and 5 tons of chromium, how many bars of each grade should it produce to maximize total revenue?

15. World Travel Service has agreed to transport a group of 1800 football fans from Los Angeles to Chicago for one of the season's big games. The agency can charter three different types of aircraft. Their respective operating costs for the trip, and their seating capacities, are as follows:

	Number of seats per plane		
Type	*First-class*	*Economy*	Cost
A	20	80	$3000
B	40	60	$4000
C	50	50	$5000

If 600 of the passengers want to go first-class while the remaining 1200 have ordered economy seats, how many planes of each type should be assigned to the charter to minimize the total operating cost?

16. The Allegheny Mining Company operates three mines which produce both high-grade and low-grade ore. The daily yield of each mine, and the daily operating cost, are given in the table below.

	Ore yield per day		
Mine	*High-grade*	*Low-grade*	Cost per day
1	2 tons	1 ton	$300
2	1 ton	2 tons	$200
3	0 tons	2 tons	$100

The company must fill an order for a total of 40 tons of high-grade ore and 50 tons of low-grade ore. How many days should each mine be operated in order to fill the order at lowest total cost?

Chapter 2   Summary

Important Terms

2.1 linear program
decision variable
objective function
optimal solution

2.2 linear inequality

2.3 system of joint constraints
feasible point
constraint set
method of basic points

2.4 Corner Point Theorem

Important Formulas

2.2 Laws of Inequality

1. If $a \leq b$, then $a + c \leq b + c$ and $a - c \leq b - c$.

2. If $a \leq b$ and $c > 0$, then $ac \leq bc$ and $\dfrac{a}{c} \leq \dfrac{b}{c}$.

3. If $a \leq b$ and $c < 0$, then $ac \geq bc$ and $\dfrac{a}{c} \geq \dfrac{b}{c}$.

Review Exercises

In Exercises 1 and 2, graph the linear inequality.

1. $6x + 7y > 42$

2. $5x - 3y \leq 0$

3. An investor puts x dollars into a "safe" investment A, which yields a 12% return, and y dollars into a "risk" investment B, which yields a 20% return. Express mathematically the condition that the return on the combined investment (A and B) is at least 15%. Simplify your answer as much as possible.

4. Assembly lines A, B, and C produce 2%, 4%, and 8% defectives, respectively. Suppose that x items from line A and y items from line B are added to a shipment that initially contains 3500 items from line C. Express mathematically the condition that the entire shipment contains at most 5% defectives. Simplify your answer as much as possible.

In Exercises 5 and 6, graph the system of joint constraints.

5. $5x + 4y \geq 200$
$2x - y \leq 15$
$y \leq 65$
$x \geq 0$

6. $-4x + 5y \leq 60$
$x + y \geq 12$
$2x - y \geq 0$
$y \geq 0$

In Exercises 7 and 8, solve the linear program by the method of graphing.

7. Maximize $F = 3x + 4y$
subject to:
$2x + 3y \leq 72$
$x + 5y \geq 50$
$y \leq 20$
$x \geq 0$.

8. Minimize $G = 4x + 5y$
subject to:
$3x + 2y \geq 60$
$x + 2y \geq 40$
$y \geq 10$
$x \geq 0$.

In Exercises 9–12, formulate the given problem as a linear program and solve it by the method of graphing.

9. A management consulting firm has 40 statisticians and 90 systems analysts available. The firm has found that for small-scale projects the most efficient problem-solving teams consist either of one statistician and two systems analysts (Type A) or one statistician and three systems analysts (Type B). A Type A team can complete five such projects per month, on the average, while a Type B team can complete eight projects. How many teams of each type should be formed in order to complete the maximum number of projects per month?

10. Jay needs 200 8-penny nails, 150 10-penny nails, and 80 16-penny nails for a weekend repair project. Since the hardware store is closed he must buy the nails at the local supermarket, which only carries two prepackaged assortments. Assortment A contains 10 8-penny nails, 15 10-penny nails, and 10 16-penny nails, while Assortment B contains 20 8-penny nails, 15 10-penny nails, and 5 16-penny nails. Both assortments sell for $2. How many packages of each type should Jay buy to meet his needs at lowest cost?

11. An electrical contractor does wiring installation in new single-family homes and condominium units. A single-family home requires 4 crew-days of work, on the average, and yields a profit of $14,000, while a condo unit requires 1 crew-day of work and yields a profit of $3000. The contractor has orders to do 30 single-family homes and 70 condo units during the coming month. Some or all of these orders can be accepted, the rest declined. If 150 crew-days of labor are available for the month, how many wiring projects of each type should be undertaken for maximum profit?

12. Each unit of Dogfood A provides a dog with 10% of the minimum daily requirement of vitamins and 5% of the minimum daily requirement of protein, and costs 30¢, while each unit of Dogfood B provides 4% of the required vitamins and 7% of the required protein, and costs 10¢. Also, a dog can eat at most 22 units of dog food per day. How much of each dog food should be used daily to meet these requirements at lowest cost?

3

Systems of Linear Equations

To find the point where two lines meet, or three planes meet, a system of linear equations must be solved. But the importance of linear systems goes far beyond these geometric applications: such systems arise not only in pure mathematics but also in the physical and social sciences, in business and engineering, and in statistics.

In this chapter we shall see how to solve arbitrary systems of linear equations. The ideas presented here will also play an essential role in later chapters, when matrix inversion and the simplex algorithm are discussed.

3.1 Gauss-Jordan Elimination

In Chapter 1 we reviewed two different methods for solving a system of two linear equations in two unknowns, substitution and elimination. While the substitution technique can be applied to larger systems as well, it becomes increasingly cumbersome. The elimination idea, however, can be generalized to give a systematic, efficient procedure known as the *Gauss-Jordan elimination* method.* In this section this method is used to solve a system of n linear equations in n unknowns.

---

*Carl Friedrich Gauss (1777–1855) was the greatest mathematician of all time. Camille Jordan (1838–1921), while not of Gauss's rank, made distinguished contributions to algebra, topology and analysis.

Linear Systems   An equation of the form

$$a_1x_1 + a_2x_2 + \cdots + a_nx_n = c$$

where $a_1, \ldots, a_n$, and c are constants and at least one of the coefficients a_i is nonzero, is called a **linear equation** in the unknowns $x_1, \ldots, x_n$. A system of m linear equations in n unknowns is called an **$m \times n$ linear system** (read "m by n"). For instance, the systems

$$
\begin{array}{ccc}
\begin{aligned}
2x + 3y &= 9 \\
3x - 5y &= 7
\end{aligned}
&
\begin{aligned}
x + 2y - z &= 4 \\
3x - y + 5z &= 6 \\
2x + y - 3z &= 1
\end{aligned}
&
\begin{aligned}
5x + y + z &= 8 \\
-x + 2y + 2z &= 5
\end{aligned}
\end{array}
$$

are 2×2, 3×3, and 2×3, respectively.

When a system has the same number of equations as unknowns (when $m = n$) it will normally have a unique solution $(x_1, x_2, \ldots, x_n)$. For the case $n = 2$ this is reflected in the fact that two lines usually intersect in a unique point. We now take up the general problem of finding the solution of an $n \times n$ linear system when it exists.

Elementary Operations   One approach to solving a system of linear equations is to transform it into an algebraically equivalent system, that is, a system having exactly the same solutions. The types of transformations we shall need, called **elementary operations,** are the following:

Type 1   An equation can be multiplied by a nonzero number.

Type 2   A multiple of one equation can be added to another.

Type 3   Two equations can be interchanged.

The Gauss-Jordan elimination method is based on the following strategy. Using a sequence of elementary operations, we attempt to *eliminate each unknown from all but one equation* of the system. Once the unknowns have been "untangled" in this way, the solution to the system will be obvious.

Example 1   Solve the following 3×3 linear system:

$$
\begin{aligned}
2x + 4y + 2z &= 6 \\
-2x + y + 3z &= 4 \\
3x + 2y + z &= 5
\end{aligned}
\tag{1}
$$

Discussion   The first task is to eliminate the unknown x from all but the first equation in the system. We multiply the first equation by 1/2 in order to change the coefficient of x to $+1$. This Type-1 operation transforms (1) into the following system:

$$
\begin{aligned}
x + 2y + z &= 3 \\
-2x + y + 3z &= 4 \\
3x + 2y + z &= 5
\end{aligned}
\tag{2}
$$

Note that the x-term in the first equation now has coefficient $+1$. Next, we add 2 times the (new) first equation to the second, in order to eliminate the x-term in the second equation. The scratch work for this Type-2 operation follows.

$$
\begin{array}{ll}
\textit{Scratch work} \quad \begin{array}{r} -2x + y + 3z = 4 \\ \underline{2x + 4y + 2z = 6} \\ 5y + 5z = 10 \end{array} & \begin{array}{l} \text{second equation} \\ 2 \times \text{first equation} \\ \text{sum} \end{array}
\end{array}
$$

This transforms (2) into the equivalent system (3).

$$
\begin{array}{rl}
x + 2y + z = 3 & \\
5y + 5z = 10 & \quad (3) \\
3x + 2y + z = 5 &
\end{array}
$$

Remark Notice that it is the second equation, not the first, which is changed by this operation. We merely *use* the first equation to eliminate x in the second equation.

Working now with the system (3), we add -3 times the first equation to the third to eliminate the x-term in the third equation.

$$
\begin{array}{ll}
\textit{Scratch work} \quad \begin{array}{r} 3x + 2y + z = 5 \\ \underline{-3x - 6y - 3z = -9} \\ -4y - 2z = -4 \end{array} & \begin{array}{l} \text{third equation} \\ (-3) \times \text{first equation} \\ \text{sum} \end{array}
\end{array}
$$

The sum gives us the new third equation, so that (3) becomes the following system:

$$
\begin{array}{rl}
x + 2y + z = 3 & \\
5y + 5z = 10 & \quad (4) \\
-4y - 2z = -4 &
\end{array}
$$

We have achieved our first objective: the unknown x has now been isolated so that it occurs in one and only one equation of the system. Our next goal is to isolate y in one of the remaining equations; we will use the second. First, to simplify the y-coefficient, we multiply the second equation by 1/5.

$$
\begin{array}{rl}
x + 2y + z = 3 & \\
y + z = 2 & \quad (5) \\
-4y - 2z = -4 &
\end{array}
$$

Next, we use the second equation to eliminate the y-terms in the other two. Specifically, we add -2 times the second equation to the first and add 4 times the second equation to the third. These Type-2 operations transform (5) into the following system in which both x and y have been isolated.

$$
\begin{array}{rl}
x - z = -1 & \\
y + z = 2 & \quad (6) \\
2z = 4 &
\end{array}
$$

To isolate z, multiply the third equation by 1/2 to simplify the z-coefficient.

$$\begin{aligned} x \quad - z &= -1 \\ y + z &= 2 \\ z &= 2 \end{aligned} \qquad (7)$$

We then add (1 times) the third equation to the first and add -1 times the third equation to the second in (7). The result is the system

$$\begin{aligned} x \quad &= 1 \\ y \quad &= 0 \\ z &= 2 \end{aligned} \qquad (8)$$

which announces its own solution, the ordered triple (1, 0, 2). Since all the systems (1) through (8) are algebraically equivalent, we conclude that (1, 0, 2) is the unique solution to the original system. The reader should check this result by direct substitution in (1).   ◇

Tableau Form   The elimination procedure can be streamlined by writing the given system in **tableau form,** as follows:

Linear system	Tableau form
$2x + 4y + 2z = 6$	$\begin{array}{ccc\|c} 2 & 4 & 2 & 6 \end{array}$
$-2x + y + 3z = 4$	$\begin{array}{ccc\|c} -2 & 1 & 3 & 4 \end{array}$
$3x + 2y + z = 5$	$\begin{array}{ccc\|c} 3 & 2 & 1 & 5 \end{array}$

This is merely a shorthand device which enables us to avoid rewriting the variables x, y, z and the equals sign after each transformation. The *detached coefficients* of the variables are listed to the left of the vertical line. The first column corresponds to x, the second to y, and the third to z. The three rows of the tableau correspond to the three equations in the system.

Our earlier operations on equations correspond to **elementary row operations,** defined as follows:

Elementary Row Operations

Type 1   A row can be multiplied by a nonzero number.

Type 2   A multiple of one row can be added to another.

Type 3   Two rows can be interchanged.

Example 1, Revisited   Solve the linear system in Example 1 using the tableau form.

Solution   The sequence of transformations (1)–(8) in Example 1 can be carried out in tableau form, as shown below. (The symbolic annotation to the right will be explained presently.) The first tableau represents the original system (1).

$$\begin{bmatrix} 2 & 4 & 2 & | & 6 \\ -2 & 1 & 3 & | & 4 \\ 3 & 2 & 1 & | & 5 \end{bmatrix}$$

Multiply the first row (equation) by 1/2.   *Annotation*

$$\begin{array}{ccc|c} 1 & 2 & 1 & 3 \\ -2 & 1 & 3 & 4 \\ 3 & 2 & 1 & 5 \end{array} \qquad R_1 = \tfrac{1}{2}R_1$$

Add 2 times the first row to the second row.

$$\begin{array}{ccc|c} 1 & 2 & 1 & 3 \\ 0 & 5 & 5 & 10 \\ 3 & 2 & 1 & 5 \end{array} \qquad R_2 = R_2 + 2R_1$$

Add -3 times the first row to the third row.

$$\begin{array}{ccc|c} 1 & 2 & 1 & 3 \\ 0 & 5 & 5 & 10 \\ 0 & -4 & -2 & -4 \end{array} \qquad R_3 = R_3 + (-3)R_1$$

Multiply the second row by 1/5.

$$\begin{array}{ccc|c} 1 & 2 & 1 & 3 \\ 0 & 1 & 1 & 2 \\ 0 & -4 & -2 & -4 \end{array} \qquad R_2 = \tfrac{1}{5}R_2$$

Add -2 times the second row to the first row.

Add 4 times the second row to the third row.

$$\begin{array}{ccc|c} 1 & 0 & -1 & -1 \\ 0 & 1 & 1 & 2 \\ 0 & 0 & 2 & 4 \end{array} \qquad \begin{array}{l} R_1 = R_1 + (-2)R_2 \\ \\ R_3 = R_3 + 4R_2 \end{array}$$

Multiply the third row by 1/2.

$$\begin{array}{ccc|c} 1 & 0 & -1 & -1 \\ 0 & 1 & 1 & 2 \\ 0 & 0 & 1 & 2 \end{array} \qquad R_3 = \tfrac{1}{2}R_3$$

Add the third row to the first row.

Add -1 times the third row to the second row.

$$\begin{array}{ccc|c} 1 & 0 & 0 & 1 \\ 0 & 1 & 0 & 0 \\ 0 & 0 & 1 & 2 \end{array} \qquad \begin{array}{l} R_1 = R_1 + R_3 \\ R_2 = R_2 + (-1)R_3 \end{array}$$

This final tableau corresponds to the linear system (8). The solution is therefore $x = 1$, $y = 0$, $z = 2$. ◆

The annotations written to the right of each tableau describe the row operations in a convenient symbolic form. The first annotation

$$R_1 = \frac{1}{2}R_1$$

is read: "New first row equals 1/2 times previous first row." This describes the operation of multiplying each entry in the first row by 1/2. The second annotation

$$R_2 = R_2 + 2R_1$$

is read: "New second row equals previous second row plus 2 times first row." This describes the operation of adding 2 times the first row to the second, term by term, as follows:

-2	1	3	4	previous second row	(R_2)
2	4	2	6	$2 \times$ first row	$(2R_1)$
0	5	5	10	sum	$(R_2 + 2R_1)$

The sum gives us the second row of the new tableau, next to which we write the annotation.*

The Pivot Transformation   When a linear system is written in tableau form, the process of isolating its unknowns can be carried out as follows: In each column, one by one, we choose a nonzero entry a. Using row operations, we change this entry to $+1$ and change the other entries in the column to 0. This combination of row operations is called a **pivot transformation** on the entry a. If we call the row and column of a the **pivot row** and **pivot column,** respectively, we can describe this transformation as follows:

Pivot Transformation on a Nonzero Entry a

Step 1   If $a \neq 1$, multiply the pivot row by $\frac{1}{a}$.

Step 2   For every other nonzero entry b in the pivot column, add $-b$ times the pivot row to the row in which it occurs.

The pivot transformation is the heart of the Gauss-Jordan elimination method. Before going further, we illustrate the procedure with an example.

---

*These annotations are not equations in the true sense, but rather symbolic instructions of the kind used in FORTRAN or BASIC programming.

Example 2 Perform a pivot transformation on the entry -2 (circled) in the following tableau:

$$\begin{array}{ccc|c} 3 & 1 & -1 & 3 \\ \boxed{-2} & 4 & 2 & 10 \\ 4 & 1 & 0 & 6 \end{array}$$

Solution Since the pivot entry is in the second row and first column, we designate row 2 as the pivot row and column 1 as the pivot column.

Step 1 The pivot entry is not equal to 1. Hence, we multiply the second row by $-1/2$.

$$\begin{array}{ccc|cl} 3 & 1 & -1 & 3 & \textit{Annotation} \\ \boxed{1} & -2 & -1 & -5 & R_2 = \left(-\frac{1}{2}\right)R_2 \\ 4 & 1 & 0 & 6 & \end{array}$$

Notice that this changes the pivot entry to $+1$.

Step 2 There are two other nonzero entries in the pivot column, namely 3 (first row) and 4 (third row). Hence, we add -3 times the second row to the first row, and add -4 times the second row to the third row.

$$\begin{array}{ccc|cl} 0 & 7 & 2 & 18 & R_1 = R_1 + (-3)R_2 \\ 1 & -2 & -1 & -5 & \\ 0 & 9 & 4 & 26 & R_3 = R_3 + (-4)R_2 \end{array}$$

This completes the pivot. Note that the pivot entry is now $+1$ and all other entries in the pivot column are zero. This means that the first unknown is now isolated (with coefficient 1) in the second equation of the system. ◆

Gauss-Jordan Elimination To solve an $n \times n$ linear system by Gauss-Jordan elimination we perform a sequence of pivots until finally all the unknowns have been isolated. The procedure is as follows:

Gauss-Jordan Elimination

To solve an $n \times n$ linear system:

1. Write the system in tableau form.
2. Perform a sequence of pivots on nonzero entries, using a different row and column for each pivot.
3. If the system has a unique solution this process will terminate after n pivots. The system is then solved.

Note that the pivot entries can be chosen arbitrarily, subject only to the condition that we use a new row and new column each time. Whenever possible it is desirable to use 1 as a pivot entry, since Step 1 of the pivot transformation then becomes unnecessary.

Example 3   Solve the linear system.

$$
\begin{aligned}
2y + 3z &= 8 \\
x + 2y + z &= 6 \\
3x + 2y &= 8
\end{aligned}
$$

Solution   This 3×3 system has the following tableau form:

$$
\left[\begin{array}{ccc|c}
0 & 2 & 3 & 8 \\
① & 2 & 1 & 6 \\
3 & 2 & 0 & 8
\end{array}\right]
$$

For the first pivot, we choose the entry 1 (circled) in the first column. Step 1, transforming the pivot entry to 1, is unnecessary. We proceed to Step 2, transforming the other nonzero entries in the pivot column to zero. Add -3 times the second row to the third row.

$$
\left[\begin{array}{ccc|c}
0 & ② & 3 & 8 \\
1 & 2 & 1 & 6 \\
0 & -4 & -3 & -10
\end{array}\right]
\quad
\begin{array}{l}
\textit{Annotation} \\
\\
R_3 = R_3 + (-3)R_2
\end{array}
$$

For the second pivot we must use an entry in a different row and column. For the entry 2 (circled), Step 1 of the pivot is to multiply the first row by 1/2.

$$
\left[\begin{array}{ccc|c}
0 & ① & \frac{3}{2} & 4 \\
1 & 2 & 1 & 6 \\
0 & -4 & -3 & -10
\end{array}\right]
\quad R_1 = \tfrac{1}{2}R_1
$$

Step 2 needs two operations. Add -2 times the first row to the second, and add 4 times the first row to the third.

$$
\left[\begin{array}{ccc|c}
0 & 1 & \frac{3}{2} & 4 \\
1 & 0 & -2 & -2 \\
0 & 0 & ③ & 6
\end{array}\right]
\quad
\begin{array}{l}
\\
R_2 = R_2 + (-2)R_1 \\
R_3 = R_3 + 4R_1
\end{array}
$$

For the last pivot we *must* use the circled entry 3. (Why?) For Step 1, multiply the third row by 1/3.

$$
\left[\begin{array}{ccc|c}
0 & 1 & \frac{3}{2} & 4 \\
1 & 0 & -2 & -2 \\
0 & 0 & ① & 2
\end{array}\right]
\quad R_3 = \tfrac{1}{3}R_3
$$

For Step 2, add $-3/2$ times the third row to the first, and add 2 times the third row to the second.

$$
\left[\begin{array}{ccc|c}
0 & 1 & 0 & 1 \\
1 & 0 & 0 & 2 \\
0 & 0 & 1 & 2
\end{array}\right]
\quad
\begin{array}{l}
R_1 = R_1 + \left(-\frac{3}{2}\right)R_3 \\
R_2 = R_2 + 2R_3
\end{array}
$$

The system is now solved. However, for the sake of appearance we *interchange the first and second rows*, so the pivot entries lie along the diagonal.

$$\begin{array}{ccc|c}
1 & 0 & 0 & 2 \\
0 & 1 & 0 & 1 \\
0 & 0 & 1 & 2
\end{array} \qquad \begin{array}{l} R_1 = R_2 \\ R_2 = R_1 \end{array}$$

This tableau corresponds to the linear system

$$\begin{aligned}
x \quad\quad &= 2 \\
y \quad &= 1 \\
z &= 2
\end{aligned}$$

which has the obvious solution $(2, 1, 2)$. The reader can check this result by direct substitution in the original equations. ◆

Remark to Reader The Gauss-Jordan elimination procedure—and the pivot transformation, on which it is based—will be needed for several later topics. For this reason *you should practice the technique until you have thoroughly mastered it*. Here are some suggestions.

1. Annotations are the key to success. Before performing a row operation, *write down its annotation*. Then carry out the actual calculation as scratch work and transfer the result to the new tableau.
2. Don't try to combine Steps 1 and 2 of the pivot transformation. As in the examples above, *first* do Step 1, *then* do all the Step 2 operations.
3. Be especially wary of mistakes involving signs ($+$ or $-$) and fractions. Bear in mind that one careless error throws the entire procedure off.
4. Always check your final answer by substitution in the original equations.

3.1 Exercises *Exercises 1–12 refer to the following three tableaux:*

A:

$$\begin{array}{cc|c}
2 & 3 & 12 \\
4 & -1 & 3
\end{array}$$

B:

$$\begin{array}{ccc|c}
1 & -2 & \frac{1}{2} & 4 \\
0 & \frac{1}{2} & -1 & -5 \\
2 & 0 & -\frac{2}{3} & 6
\end{array}$$

C:

$$\begin{array}{cccc|c}
2 & -1 & 1 & 0 & 6 \\
0 & \frac{1}{2} & \frac{5}{4} & -1 & 2 \\
\frac{3}{4} & -1 & 1 & 2 & 2 \\
-1 & 0 & 1 & 1 & -5
\end{array}$$

In Exercises 1–6, perform the indicated row operation on the given tableau.

1. A: $R_2 = \frac{1}{4}R_2$
2. A: $R_2 = R_2 + \left(-\frac{1}{2}\right)R_1$
3. B: $R_3 = R_3 + (-2)R_1$
4. B: $R_1 = R_1 + (-1)R_3$
5. C: $R_2 = 2R_2$
6. C: $R_4 = R_4 + 4R_2$

In Exercises 7–12, perform a pivot on the indicated entry in the given tableau.

7. A: pivot on -1 in row 2, column 2.
8. A: pivot on 3 in row 1, column 2.

9. B: pivot on 1/2 in row 1, column 3.

10. B: pivot on -2 in row 1, column 2.

11. C: pivot on 2 in row 3, column 4.

12. C: pivot on 3/4 in row 3, column 1.

Give a symbolic annotation for each of the row operations described in Exercises 13–18.

13. Multiply the third row by -1.

14. Add the first row to the fourth row.

15. Multiply the second row by 2/3.

16. Add 3 times the second row to the first row.

17. Add $-2/3$ times the third row to the second row.

18. Interchange the second and fourth rows.

In Exercises 19–28, solve the linear system by Gauss-Jordan elimination. Annotate each row operation and indicate pivot entries with a circle.

19. $x + 2y = 18$
    $5x + 4y = 60$

20. $3x + y = 7$
    $-2x + 4y = 7$

21. $.5x + y + .5z = 4$
    $x + 2y - .5z = 2$
    $-.5x \quad\quad + z = 3$

22. $x + y \quad\quad = 1$
    $x \quad\quad + z = 2$
    $y + z = 3$

23. $x + 2y - z = 2$
    $-x + y + 2z = 1$
    $x + y - z = 1$

24. $x + 2y + 3z = 3$
    $3x + y + 2z = 7$
    $2x + 3y + z = 2$

25. $x + y + z = 6$
    $x - y + z = 0$
    $x + y - z = 2$

26. $x + 2y + z = 1$
    $2x + 5y + z = 2$
    $y + 2z = 3$

27. $x + y + z \quad\quad = 0$
    $x + y \quad\quad + w = 1$
    $x \quad\quad + z + w = 2$
    $y + z + w = 3$

28. $-x + y + z + w = 20$
    $x - y + z + w = 40$
    $x + y - z + w = 60$
    $x + y + z - w = 80$

29. If a tableau T is transformed into T' by a Type-1 row operation $R_i = aR_i$ ($a \neq 0$), what row operation will transform T' back into T?

30. If a tableau T is transformed into T' by a Type-2 row operation $R_i = R_i + aR_j$ ($i \neq j$), what operation will transform T' back into T?

31. Show that the effect of a Type-3 operation ($R_i = R_j$, $R_j = R_i$) can be achieved by a sequence of four operations of Types 1 and 2.

Computer Projects

A. The PIVOT program performs a pivot transformation on a designated nonzero entry in a tableau. This program can be used in Exercises 19–28.

B. Use the PIVOT program to solve the linear system on the next page.

$$7x + 3y - 5z + u + 4v = 120$$
$$10x + 2y + 4z - 3u + v = 225$$
$$9x - 8y + z + 5u + 3v = 345$$
$$6x + y + 8z + 9u + 5v = 830$$
$$-2x + 7y + 3z + 4u - v = 325$$

3.2 Analysis of Linear Systems

It often happens that a system of linear equations fails to have a unique solution. For an $n \times n$ system this is somewhat exceptional, but when the system contains fewer (or more) equations than unknown it is almost always the case. In this section we consider the general $m \times n$ linear system, and show that in all cases—unique solution or not—the Gauss-Jordan elimination procedure tells us everything we might want to know.

Inconsistency and Redundancy Beneath a healthy looking system of linear equations there may lurk several types of pathology, as we see in the following example.

Example 1 Solve the following system:

$$x - y + z = 1$$
$$x + 2y - z = 4$$
$$3x + 3y - z = 6$$

Solution We write the system in tableau form and proceed by Gauss-Jordan elimination (pivot entries are circled).

$$\begin{array}{ccc|c} ① & -1 & 1 & 1 \\ 1 & 2 & -1 & 4 \\ 3 & 3 & -1 & 6 \end{array}$$

Add -1 times the first row to the second row.

Add -3 times the first row to the third row.

$$\begin{array}{ccc|c} 1 & -1 & 1 & 1 \\ 0 & ③ & -2 & 3 \\ 0 & 6 & -4 & 3 \end{array} \qquad \begin{array}{l} \textit{Annotation} \\ R_2 = R_2 + (-1)R_1 \\ R_3 = R_3 + (-3)R_1 \end{array}$$

Multiply the second row by 1/3.

$$\begin{array}{ccc|c} 1 & -1 & 1 & 1 \\ 0 & ① & -\frac{2}{3} & 1 \\ 0 & 6 & -4 & 3 \end{array} \qquad R_2 = \frac{1}{3}R_2$$

Add the second row to the first row.

Add -6 times the second row to the third row.

$$\begin{array}{ccc|c}
1 & 0 & \frac{1}{3} & 2 \\
0 & 1 & -\frac{2}{3} & 1 \\
0 & 0 & 0 & -3
\end{array} \qquad \begin{array}{l} R_1 = R_1 + R_2 \\ \\ R_3 = R_3 + (-6)R_2 \end{array}$$

For the final pivot, in the third column, we are now forced to use the entry in the third row, since the first and second rows have already been used as pivot rows. But *zero cannot be used as a pivot entry,* so there is no way to complete the Gauss-Jordan procedure.

If we translate the final tableau back into the language of linear equations, we can see where the problem lies.

$$1x + 0y + \frac{1}{3}z = 2$$

$$0x + 1y - \frac{2}{3}z = 1$$

$$0x + 0y + 0z = -3$$

The third equation in this system says $0 = -3$, which is false no matter what values x, y, and z represent. Since this system is equivalent to the one we started with, we conclude that the original system is an **inconsistent system;** that is, it has no solution (x, y, z).

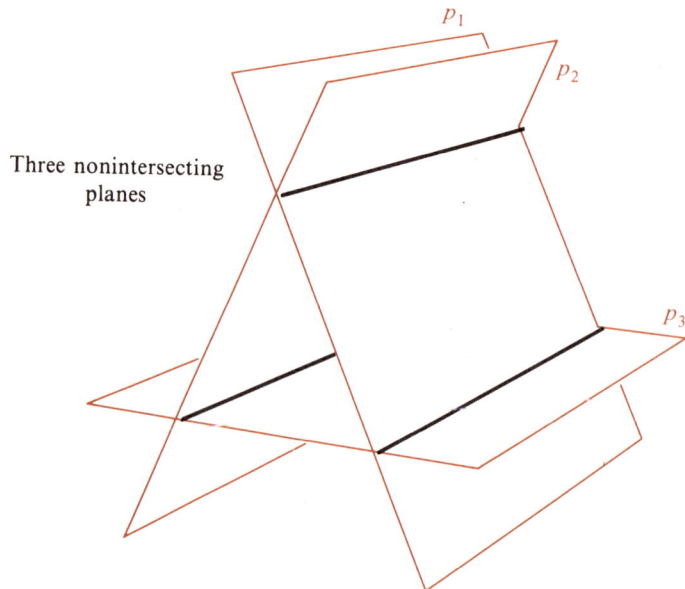

Three nonintersecting planes

Figure 1

Remark to Reader   Since each of the equations in the system of Example 1 defines a plane in *xyz*-space (see Section 2.5), the inconsistency of the system means that there is no point (x, y, z) that lies on all three planes. Figure 1 illustrates a situation of this sort: each pair of planes intersects along a line which runs parallel to the third plane. Because of this, no point can lie on all three planes.

The other type of exceptional behavior for an $n \times n$ system is *redundancy*, which we encounter in the following example.

Example 2   Solve the system below.

$$-x + 3y - 2z = 1$$
$$2x - 5y + 2z = 0$$
$$2x - y - 6z = 8$$

Solution   Again, we proceed by Gauss-Jordan elimination.

$$\begin{array}{ccc|c} \boxed{-1} & 3 & -2 & 1 \\ 2 & -5 & 2 & 0 \\ 2 & -1 & -6 & 8 \end{array}$$

Multiply the first row by -1.   *Annotation*

$$\begin{array}{ccc|c} \boxed{1} & -3 & 2 & -1 \\ 2 & -5 & 2 & 0 \\ 2 & -1 & -6 & 8 \end{array} \quad R_1 = -R_1$$

Add -2 times the first row to the second row.

Add -2 times the first row to the third row.

$$\begin{array}{ccc|c} 1 & -3 & 2 & -1 \\ 0 & \boxed{1} & -2 & 2 \\ 0 & 5 & -10 & 10 \end{array} \quad \begin{array}{l} \\ R_2 = R_2 + (-2)R_1 \\ R_3 = R_3 + (-2)R_1 \end{array}$$

Add 3 times the second row to the first row.

Add -5 times the second row to the third row.

$$\begin{array}{ccc|c} 1 & 0 & -4 & 5 \\ 0 & 1 & -2 & 2 \\ 0 & 0 & 0 & 0 \end{array} \quad \begin{array}{l} R_1 = R_1 + 3R_2 \\ \\ R_3 = R_3 + (-5)R_2 \end{array}$$

Once again the method breaks down. We cannot use zero as a pivot in the third column, and the other two entries are in rows that have already been used for pivots. This time, the third row represents the equation

$$0x + 0y + 0z = 0$$

which is true for *every* triple (x, y, z). Since it tells us nothing, we can delete this row from the final tableau. Its appearance indicates that the system is a **redundant** system; that is, one of the equations is implied by the others. (In our case, it happens that the third equation is equal to 8 times the first plus 5 times the second.)

Deleting the third row from the final tableau, we conclude that the original system is equivalent to the system

$$x \quad - 4z = 5$$
$$y - 2z = 2$$

of two equations in three unknowns. We can solve these equations for x and y in terms of z.

General solution
$$x = 5 + 4z$$
$$y = 2 + 2z$$

Note that z can be assigned any value we please in the general solution to yield a specific solution of the original system. For instance,

$$z = 0 \quad \text{gives} \quad x = 5, \quad y = 2 \quad \text{solution} \quad (5, 2, 0)$$
$$z = 2 \quad \text{gives} \quad x = 13, \quad y = 6 \quad \text{solution} \quad (13, 6, 2)$$
$$z = -1 \quad \text{gives} \quad x = 1, \quad y = 0 \quad \text{solution} \quad (1, 0, -1)$$

The original system therefore has *infinitely many solutions* (x, y, z), one for each value of z.   ◆

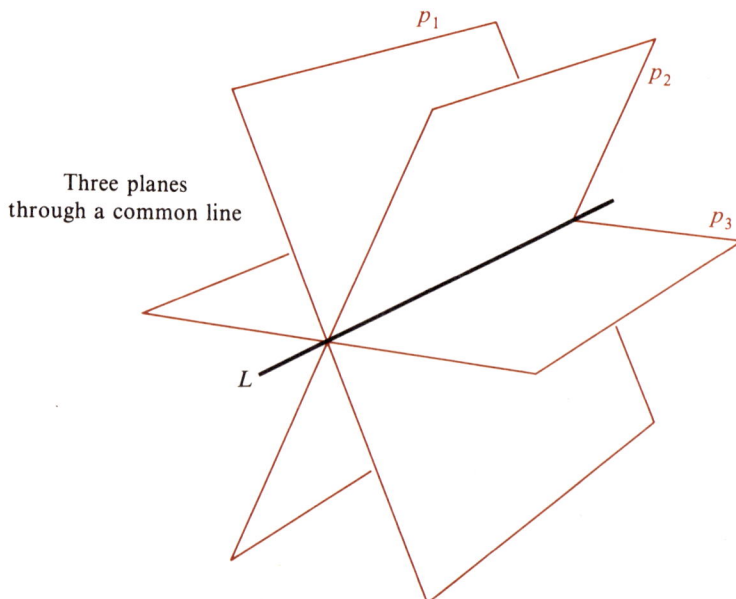

Three planes through a common line

Figure 2

Remark to Reader The geometric meaning of this is illustrated in Figure 2, which shows three planes intersecting along a common line L. Any point on this line lies on all three planes and therefore satisfies all three equations.

The preceding examples, together with those of the previous section, illustrate the three possibilities that can arise in solving a system of n equations in n unknowns.

Every $n \times n$ linear system leads to exactly one of the following three cases:

Typical case: The system has a unique solution, and can be solved for all the unknowns.

Inconsistent case: The system has no solution.

Redundant case: The system has infinitely many solutions, and can be solved for some of the unknowns in terms of the others.

To analyze a given $n \times n$ system we apply the Gauss-Jordan elimination procedure as far as possible, using a different row and column for each successive pivot. Since zero cannot be used as a pivot entry we may be unable to solve for all n unknowns. When that is the case, we first inspect the final tableau for rows of the form

$$0 \quad 0 \quad 0 \quad \cdots \quad 0 \mid a$$

where $a \neq 0$. If any such row appears, the system is inconsistent. If no such row appears, there will be one or more rows of the form

$$0 \quad 0 \quad 0 \quad \cdots \quad 0 \mid 0$$

in which case the system is redundant. We delete all such rows from the final tableau and use the remaining equations to solve for some of the unknowns in terms of the others. (Note that we check for inconsistency *first*. If an inconsistent row appears, it does not matter whether there are also redundant rows.)

Basic and Free Variables When an $n \times n$ linear system is redundant, there is more than one way to represent its general solution. Thus, in Example 2 we saw that the redundant system

$$
\begin{aligned}
-x + 3y - 2z &= 1 \\
2x - 5y + 2z &= 0 \\
2x - y - 6z &= 8
\end{aligned}
$$

has general solution

$$
\begin{aligned}
x &= 5 + 4z \\
y &= 2 + 2z.
\end{aligned}
$$

In this form of the solution, x and y are called **basic variables** and z is called a **free variable** (or **parameter**). We can also solve the given system for x and z in terms of y, or for y and z in terms of x. For instance, to make x and z the basic variables, we first write the equations of the general solution as a linear system

$$x - 4z = 5$$
$$y - 2z = 2$$

or in tableau form.

$$\begin{array}{cc c|c} 1 & 0 & -4 & 5 \\ 0 & 1 & \boxed{-2} & 2 \end{array}$$

A single pivot, on the entry -2, will now reverse the roles of y and z, making z basic and y free:

$$\begin{array}{cc c|c} 1 & 0 & -4 & 5 \\ 0 & 1 & -2 & 2 \end{array}$$

Multiply the second row by $-1/2$.

$$\begin{array}{cc c|c} 1 & 0 & -4 & 5 \\ 0 & -\frac{1}{2} & \boxed{1} & -1 \end{array} \qquad \begin{array}{l} \textit{Annotation} \\ R_2 = \left(-\frac{1}{2}\right)R_2 \end{array}$$

Add 4 times the second row to the first row.

$$\begin{array}{cc c|c} 1 & -2 & 0 & 1 \\ 0 & -\frac{1}{2} & 1 & -1 \end{array} \qquad R_1 = R_1 + 4R_2$$

If we write this tableau in equational form and transpose the y-terms to the right-hand side, we obtain

General solution
$$x = 1 + 2y$$
$$z = -1 + \frac{1}{2}y.$$

This is the general solution of the original system with x and z basic, y free. If we let y take every possible value above, we generate all solutions (x, y, z) of the system. Hence, the two general solutions are equivalent.

When we solve a linear system for certain variables (basic) in terms of the others (free), the system will always have infinitely many particular solutions. One of these will play a special role when we discuss the simplex algorithm in Chapter 5.

Definition

The **basic solution** is obtained by setting all free variables equal to zero in the general solution.

Thus, the basic solution corresponding to the first general solution above is $(5, 2, 0)$, obtained by setting $z = 0$; while the basic solution corresponding to the second is $(1, 0, -1)$, obtained by setting $y = 0$.

Solution of $m \times n$ Systems   To this point we have been concerned only with $n \times n$ systems. The same method, however, can be used to analyze *any* linear system, regardless of the number of equations or unknowns. We summarize the procedure as follows:

Gauss-Jordan Elimination

To solve an $m \times n$ linear system:

1. Write the system in tableau form.
2. Perform a sequence of pivots on nonzero entries, using a different row and column for each pivot. After k pivots $(k \le n)$, this process will terminate.
3. Look for rows of the form $0\,0 \cdots 0 \mid a$, where $a \neq 0$. If any such row appears, the system has no solution.
4. Look for rows of the form $0\,0 \cdots 0 \mid 0$. Delete all such rows from the final tableau.
5. If $k = n$ the system has a unique solution, which can be read from the final tableau.
6. If $k < n$ the system has infinitely many solutions. Use the final tableau to solve for the basic variables in terms of the free variables.

The basic variables will be those unknowns corresponding to pivot columns in Step 2. In the case $k < n$ it is usually possible to solve for any k unknowns in terms of the others.*

Example 3   Solve the following 3×5 system:

$$\begin{aligned} x - 2y + z &= 1 \\ y - 2z + u &= 2 \\ z - 2u + v &= 3 \end{aligned}$$

Discussion   Since we have five unknowns and only three equations, the system is **underdetermined,** that is, it does not provide enough information to enable us to solve uniquely for all five values. The best we can hope for, barring inconsistency or redundancy, is to solve for three of the unknowns in terms of the other two. In tableau form, the system becomes

---

*There are exceptions, however; see Exercise 20.

$$
\begin{array}{ccccc}
x & y & z & u & v \\
1 & -2 & 1 & 0 & 0 \,\big|\, 1 \\
0 & 1 & -2 & 1 & 0 \,\big|\, 2 \\
0 & 0 & 1 & -2 & 1 \,\big|\, 3.
\end{array}
$$

We must first decide which three unknowns to make the basic variables. If we designate x, y, and z as basic, we should choose our pivot entries from these three columns. The x-pivot is unnecessary (x is already isolated), so we proceed to the y-column.

$$
\begin{array}{ccccc}
1 & -2 & 1 & 0 & 0 \,\big|\, 1 \\
0 & ① & -2 & 1 & 0 \,\big|\, 2 \\
0 & 0 & 1 & -2 & 1 \,\big|\, 3
\end{array}
$$

Add 2 times the second row to the first row.

$$
\begin{array}{ccccc}
1 & 0 & -3 & 2 & 0 \,\big|\, 5 \\
0 & 1 & -2 & 1 & 0 \,\big|\, 2 \\
0 & 0 & ① & -2 & 1 \,\big|\, 3
\end{array}
$$

Annotation
$R_1 = R_1 + 2R_2$

Add 3 times the third row to the first row.

Add 2 times the third row to the second row.

$$
\begin{array}{ccccc}
1 & 0 & 0 & -4 & 3 \,\big|\, 14 \\
0 & 1 & 0 & -3 & 2 \,\big|\, 8 \\
0 & 0 & 1 & -2 & 1 \,\big|\, 3
\end{array}
$$

$R_1 = R_1 + 3R_3$
$R_2 = R_2 + 2R_3$

From this tableau we obtain the general solution, with the basic variables x, y, and z expressed in terms of the free variables u and v.

General solution
$$
\begin{aligned}
x &= 14 + 4u - 3v \\
y &= 8 + 3u - 2v \\
z &= 3 + 2u - v
\end{aligned}
$$

The corresponding basic solution is $(14, 8, 3, 0, 0)$; that is, $x = 14$, $y = 8$, $z = 3$, $u = 0$, and $v = 0$. This is obtained by setting the free variables u and v equal to zero in the general solution. Of course, the basic solution is just one among infinitely many solutions generated by the general solution.

If we now wish to make x, z, and v the basic variables, with y and u free, we have only to perform one more pivot, on the entry 2 in our last tableau.

$$
\begin{array}{ccccc}
1 & 0 & 0 & -4 & 3 \,\big|\, 14 \\
0 & 1 & 0 & -3 & ② \,\big|\, 8 \\
0 & 0 & 1 & -2 & 1 \,\big|\, 3
\end{array}
$$
— why her

Multiply the second row by 1/2.

Annotation

$$\begin{array}{ccccc|c} 1 & 0 & 0 & -4 & 3 & 14 \\ 0 & \frac{1}{2} & 0 & -\frac{3}{2} & ① & 4 \\ 0 & 0 & 1 & -2 & 1 & 3 \end{array} \qquad R_2 = \frac{1}{2}R_2$$

Add -3 times the second row to the first row.

Add -1 times the second row to the third row.

$$\begin{array}{ccccc|c} 1 & -\frac{3}{2} & 0 & \frac{1}{2} & 0 & 2 \\ 0 & \frac{1}{2} & 0 & -\frac{3}{2} & 1 & 4 \\ 0 & -\frac{1}{2} & 1 & -\frac{1}{2} & 0 & -1 \end{array} \qquad \begin{array}{l} R_1 = R_1 + (-3)R_2 \\ \\ R_3 = R_3 + (-1)R_2 \end{array}$$

This gives us the general solution in the following form:

General solution

$$x = 2 + \tfrac{3}{2}y - \tfrac{1}{2}u$$
$$z = -1 + \tfrac{1}{2}y + \tfrac{1}{2}u$$
$$v = 4 - \tfrac{1}{2}y + \tfrac{3}{2}u$$

The corresponding basic solution is $(2, 0, -1, 0, 4)$. Of course, we could designate any other three unknowns as the basic variables, and obtain solutions similar to those above. ◆

Example 4 Solve the following 4×3 system:

$$x - y + z = 5$$
$$2x + y - z = 1$$
$$-x + 3y + 2z = 4$$
$$x - 5y + z = 3$$

Solution Here the system is **overdetermined:** there are more equations than unknowns. The Gauss-Jordan method produces the following sequence of tableaus:

$$\begin{array}{ccc|c} x & y & z & \\ ① & -1 & 1 & 5 \\ 2 & 1 & -1 & 1 \\ -1 & 3 & 2 & 4 \\ 1 & -5 & 1 & 3 \end{array}$$

Add -2 times the first row to the second row.

Add the first row to the third row.

Add -1 times the first row to the fourth row.

$$\begin{array}{rrr|r}
1 & -1 & 1 & 5 \\
0 & ③ & -3 & -9 \\
0 & 2 & 3 & 9 \\
0 & -4 & 0 & -2
\end{array}$$

Annotation

$R_2 = R_2 + (-2)R_1$
$R_3 = R_3 + R_1$
$R_4 = R_4 + (-1)R_1$

Multiply the second row by 1/3.

$$\begin{array}{rrr|r}
1 & -1 & 1 & 5 \\
0 & ① & -1 & -3 \\
0 & 2 & 3 & 9 \\
0 & -4 & 0 & -2
\end{array}$$

$R_2 = \frac{1}{3}R_2$

Add the second row to the first row.

Add -2 times the second row to the third row.

Add 4 times the second row to the fourth row.

$$\begin{array}{rrr|r}
1 & 0 & 0 & 2 \\
0 & 1 & -1 & -3 \\
0 & 0 & ⑤ & 15 \\
0 & 0 & -4 & -14
\end{array}$$

$R_1 = R_1 + R_2$

$R_3 = R_3 + (-2)R_2$
$R_4 = R_4 + 4R_2$

Multiply the third row by 1/5.

$$\begin{array}{rrr|r}
1 & 0 & 0 & 2 \\
0 & 1 & -1 & -3 \\
0 & 0 & ① & 3 \\
0 & 0 & -4 & -14
\end{array}$$

$R_3 = \frac{1}{5}R_3$

Add the third row to the second row.

Add 4 times the third row to the fourth row.

$$\begin{array}{rrr|r}
1 & 0 & 0 & 2 \\
0 & 1 & 0 & 0 \\
0 & 0 & 1 & 3 \\
0 & 0 & 0 & -2
\end{array}$$

$R_2 = R_2 + R_3$

$R_4 = R_4 + 4R_3$

Each column has been used as a pivot column, so the process terminates. From the last row we see that the given system is inconsistent, that is, it has no solution. ◆

3.2 Exercises   *Each of the linear systems in Exercises 1–16 fails to have a unique solution. Use Gauss-Jordan elimination to find if there are no solutions or infinitely many.*

1. $3x - 14y = 12$
    $-9x + 42y = 30$

2. $.2x + y = 1$
    $x + 5y = 2$

3. $1.5x - .4y = .5$
   $-7.5x + 2y = -2.5$

4. $6x + 9y = 15$
   $4x + 6y = 10$

5. $x + 2y = 1$
   $2x - 3y = 9$
   $3x + 2y = 3$

6. $x - 3y = 1$
   $-2x + 6y = 2$
   $5x - 9y = 5$

7. $x - y \quad\quad = 1$
   $x \quad\quad - z = 2$
   $\quad\quad y - z = 3$

8. $x + y + z = 1$
   $2x + 3y + 4z = 0$
   $3x + 4y + 5z = 0$

9. $2x - 3y + z = 6$
   $x + 2y - 3z = 10$
   $3x - y - 2z = 16$

10. $.2x + .3y + .4z = 2$
    $x + 1.5y + 2z = 10$
    $.5x + .75y + z = 5$

11. $x + y + z = 1$
    $x + 2y + 5z = 8$
    $3x + 4y + 7z = 10$
    $x + 3y + 9z = 15$

12. $x - 2y + z = 1$
    $x + y + z = 7$
    $-x + 5y - z = 5$
    $2x - y + 2z = 8$

13. $x + 2y + 3z + 4w = 1$
    $x + 3y + 5z + 7w = 2$
    $x + 4y + 7z + 10w = 3$
    $x + 5y + 9z + 13w = 4$

14. $x \quad + z \quad\quad = 30$
    $\quad y \quad\quad + w = 30$
    $x \quad\quad\quad + w = 20$
    $\quad y + z \quad\quad = 40$

15. $x + y + z + w = 2$
    $x - y - z + w = 2$
    $x + 2y + 2z + w = 3$
    $2x + y + z + 2w = 3$

16. $x - y + z - w = 1$
    $-x + y - z + w = -1$
    $2x - 2y + 2z - 2w = 2$
    $-3x + 3y - 3z + 3w = -2$

17. Use Gauss-Jordan elimination to find the general and basic solutions of the 2×3 system

$$x + 2y + 3z = 2$$
$$3x + 2y + z = 10$$

(a) with x and y basic, z free

(b) with x and z basic, y free

(c) with y and z basic, x free.

18. Use Gauss-Jordan elimination to find the general and basic solutions of the 2×4 system

$$x + 2y + 3z + 4w = 5$$
$$x + 3y + 5z + 7w = 9$$

(a) with x and y basic, z and w free

(b) with x and z basic, y and w free

(c) with z and w basic, x and y free.

19. Use Gauss-Jordan elimination to find the general and basic solutions of the 3×5 system

$$x + y - z + u + v = 10$$
$$x - y + z + u \quad\quad = 20$$
$$-x + y + z \quad\quad + v = 30$$

(a) with x, y, and z basic, u and v free

(b) with x, y, and u basic, z and v free

(c) with y, u, and v basic, x and z free.

20. Show that the 3×5 system

$$
\begin{aligned}
x - y + z \qquad\quad &= 1 \\
y - z + u \quad &= 2 \\
z - u + v &= 3
\end{aligned}
$$

can be solved for x, y, and z in terms of u and v; or for y, z, and u in terms of x and v; but *not* for x, z, and u in terms of y and v. Where does the Gauss-Jordan method break down when we try to make x, z, and u basic?

Computer Projects

A. The PIVOT program performs a pivot transformation on a designated nonzero entry in a tableau. This program can be used in Exercises 1–20.

B. Use the PIVOT program to find the general and basic solution of the 4×6 system

$$
\begin{aligned}
4x + 6y + 3z - u + 21v + 20w &= 197 \\
3x + y + 6z - 4u - 5v + 32w &= 109 \\
7x + 9y + 7z - 2u + 30v + 38w &= 343 \\
x + 8y - 6z + 7u + 46v - 31w &= 138
\end{aligned}
$$

(a) with x, y, z, and u basic, v and w free

(b) with z, u, v, and w basic, x and y free.

3.3 Applications of Linear Systems

We conclude this chapter with some examples of problems that are solved by using systems of linear equations. Linear systems have many applications besides those here, some of which we shall see later. For the most part, more advanced applications rely on the matrix form for linear systems, which will be developed in the next chapter.

Product-Mix Problems   There is a large class of applications referred to as **product-mix** problems, since they involve finding a suitable combination of products or available resources in order to satisfy certain given conditions. These problems generalize some of the 2×2 problems we encountered in Section 1.2.

Example 1   A flooring company makes linoleum in a standard size roll. These large rolls are cut into small, medium, and large tiles according to one of three available cutting patterns.

Number of tiles per roll

Pattern	Small	Medium	Large
A	20	25	50
B	80	90	0
C	40	100	20

How many rolls should be cut according to each pattern, if the company wants to fill an order for 2000 small tiles, 3400 medium tiles, and 1400 large tiles, without wasting any linoleum?

Solution The three unknowns in the problem are:

$$x = \text{number of rolls cut according to Pattern A}$$
$$y = \text{number of rolls cut according to Pattern B}$$
$$z = \text{number of rolls cut according to Pattern C.}$$

The condition that we fill the order without any waste—that is, without any tiles left over—translates into a 3×3 system of linear equations.

$$20x + 80y + 40z = 2000 \quad \text{(small tiles)}$$
$$25x + 90y + 100z = 3400 \quad \text{(medium tiles)}$$
$$50x \qquad\quad + 20z = 1400 \quad \text{(large tiles)}$$

We write the system in tableau form and apply Gauss-Jordan elimination.

$$\begin{array}{ccc|c} ⓐ20 & 80 & 40 & 2000 \\ 25 & 90 & 100 & 3400 \\ 50 & 0 & 20 & 1400 \end{array}$$

Multiply the first row by 1/20.

Annotation

$$\begin{array}{ccc|c} ①1 & 4 & 2 & 100 \\ 25 & 90 & 100 & 2400 \\ 50 & 0 & 20 & 1400 \end{array} \qquad R_1 = \frac{1}{20}R_1$$

Add -25 times the first row to the second row.

Add -50 times the first row to the third row.

$$\begin{array}{ccc|c} 1 & 4 & 2 & 100 \\ 0 & \boxed{-10} & 50 & 900 \\ 0 & -200 & -80 & -3600 \end{array} \qquad \begin{array}{l} R_2 = R_2 + (-25)R_1 \\ R_3 = R_3 + (-50)R_1 \end{array}$$

Multiply the second row by $-1/10$.

$$\begin{array}{ccc|c} 1 & 4 & 2 & 100 \\ 0 & ①1 & -5 & -90 \\ 0 & -200 & -80 & -3600 \end{array} \qquad R_2 = \left(-\frac{1}{10}\right)R_2$$

Add -4 times the second row to the first row.

Add 200 times the second row to the third row.

$$\left[\begin{array}{ccc|c} 1 & 0 & 22 & 460 \\ 0 & 1 & -5 & -90 \\ 0 & 0 & \boxed{-1080} & -21{,}600 \end{array}\right] \quad \begin{array}{l} R_1 = R_1 + (-4)R_2 \\ \\ R_3 = R_3 + 200R_2 \end{array}$$

Multiply the third row by $-1/1080$.

$$\left[\begin{array}{ccc|c} 1 & 0 & 22 & 460 \\ 0 & 1 & -5 & -90 \\ 0 & 0 & \boxed{1} & 20 \end{array}\right] \quad R_3 = \left(-\dfrac{1}{1080}\right)R_3$$

Add -22 times the third row to the first row.

Add 5 times the third row to the second row.

$$\left[\begin{array}{ccc|c} 1 & 0 & 0 & 20 \\ 0 & 1 & 0 & 10 \\ 0 & 0 & 1 & 20 \end{array}\right] \quad \begin{array}{l} R_1 = R_1 + (-22)R_3 \\ R_2 = R_2 + 5R_3 \end{array}$$

The solution is therefore $(20, 10, 20)$; that is, the company should cut 20 rolls according to Pattern A, 10 according to Pattern B, and 20 according to Pattern C. ◆

In the next example we are unable to solve for all the unknowns, because the linear system is underdetermined—it contains fewer equations than unknowns.

Example 2   A cereal manufacturer has four different grains available. The fiber content (by weight) and cost of each grain are shown in the table below.

Grain	Fiber per ton	Cost per ton
1	5%	$150
2	10%	$250
3	20%	$500
4	25%	$800

A new high-fiber cereal is to be made by combining the four grains in certain proportions. The new cereal is to contain 15% fiber and cost $400 per ton. Can this be done? Is there more than one way to do it? Analyze the problem mathematically.

Solution   Let x_1, x_2, x_3, and x_4 represent the number of tons of grains 1, 2, 3, and 4, respectively, that are combined to form the new cereal. The condition that the mixture contains 15% fiber can be expressed as follows.

total amount fiber = .15 (total amount of mixture)
$$.05x_1 + .10x_2 + .20x_3 + .25x_4 = .15(x_1 + x_2 + x_3 + x_4)$$

Multiplying out and collecting terms, we obtain a linear equation in the four unknowns.

$$-.10x_1 - .05x_2 + .05x_3 + .10x_4 = 0$$

Similarly, the condition that the mixture costs \$400 per ton becomes

total cost of mixture = 400 (total amount of mixture)
$$150x_1 + 250x_2 + 500x_3 + 800x_4 = 400(x_1 + x_2 + x_3 + x_4).$$

This simplifies to the linear equation

$$-250x_1 - 150x_2 + 100x_3 + 400x_4 = 0.$$

The two conditions therefore translate into a 2×4 system of linear equations.

$$-.10x_1 - .05x_2 + .05x_3 + .10x_4 = 0$$
$$-250x_1 - 150x_2 + 100x_3 + 400x_4 = 0$$

Using Gauss-Jordan elimination, we can solve for two of the unknowns in terms of the others.

$$\left[\begin{array}{cccc|c} \boxed{-.10} & -.05 & .05 & .10 & 0 \\ -250 & -150 & 100 & 400 & 0 \end{array}\right]$$

Multiply the first row by -10.

$$\left[\begin{array}{cccc|c} \boxed{1} & .5 & -.5 & -1 & 0 \\ -250 & -150 & 100 & 400 & 0 \end{array}\right] \quad \begin{array}{l} Annotation \\ R_1 = -10R_1 \end{array}$$

Add 250 times the first row to the second row.

$$\left[\begin{array}{cccc|c} 1 & .5 & -.5 & -1 & 0 \\ 0 & \boxed{-25} & -25 & 150 & 0 \end{array}\right] \quad R_2 = R_2 + 250R_1$$

Multiply the second row by $-1/25$.

$$\left[\begin{array}{cccc|c} 1 & .5 & -.5 & -1 & 0 \\ 0 & \boxed{1} & 1 & -6 & 0 \end{array}\right] \quad R_2 = \left(-\frac{1}{25}\right)R_2$$

Add $-.5$ times the second row to the first row.

$$\left[\begin{array}{cccc|c} 1 & 0 & -1 & 2 & 0 \\ 0 & 1 & 1 & -6 & 0 \end{array}\right] \quad R_1 = R_1 + (-.5)R_2$$

From this tableau we obtain the general solution, with x_1 and x_2 basic, x_3 and x_4 free.

General Solution
$$x_1 = x_3 - 2x_4$$
$$x_2 = -x_3 + 6x_4$$

For any values of x_3 and x_4, these equations yield a solution of the initial system. For a solution to be feasible, however, we must have

$$x_3 - 2x_4 \geq 0$$
$$-x_3 + 6x_4 \geq 0$$

which together are equivalent to the single inequality

$$2x_4 \leq x_3 \leq 6x_4.$$

For example, with $x_3 = 500$ and $x_4 = 100$ we obtain the feasible solution (300, 100, 500, 100), but with $x_3 = 400$ and $x_4 = 300$ we obtain the infeasible solution $(-200, 1400, 400, 300)$. ◆

Curve-Fitting We are often faced with the problem of finding a mathematical relationship between two variables x and y. In most cases we begin with some empirical data, in the form of ordered pairs (x, y), and we try to find an equation satisfied by the given values. In geometric terms, this means finding a curve of a certain type that passes through the given points.

Of special interest in such curve-fitting problems are **polynomial functions,** such as

$$y = ax + b \qquad \text{(first degree)}$$
$$y = ax^2 + bx + c \qquad \text{(second degree)}$$
$$y = ax^3 + bx^2 + cx + d \qquad \text{(third degree)}$$

and so on. The **degree** of the polynomial is the highest exponent to which x appears.* First-degree polynomials are just the **linear functions** that we studied in Section 1.3. Polynomials of second and third degree are called **quadratic** and **cubic** functions, respectively.

It was shown in Section 1.3 that two data points uniquely determine a linear function. In the same way, three points determine a quadratic function, four points determine a cubic function, and so on.

Let $n + 1$ points be given, no two of which have the same x-coordinate. Then there exists a unique polynomial of degree at most n, whose graph passes through them all.

Example 3 Find the quadratic polynomial $y = ax^2 + bx + c$ whose graph passes through the points (1, 2), (2, 3), and (3, 6).

---

*We assume here that the leading coefficient a is not zero.

Solution Since the curve passes through the point $(1, 2)$, its equation must be satisfied when $x = 1$ and $y = 2$; thus,

$$2 = a(1)^2 + b(1) + c$$

or $\quad a + b + c = 2.$

Similarly, the points $(2, 3)$ and $(3, 6)$ must lie on the curve, and we are led to a system of three equations in the unknown coefficients a, b, and c.

$$a + b + c = 2$$
$$4a + 2b + c = 3$$
$$9a + 3b + c = 6$$

We can solve the above system by Gauss-Jordan elimination.

$$\begin{array}{ccc|c} ① & 1 & 1 & 2 \\ 4 & 2 & 1 & 3 \\ 9 & 3 & 1 & 6 \end{array}$$

Add -4 times the first row to the second row.

Add -9 times the first row to the third row.

Annotation

$$\begin{array}{ccc|c} 1 & 1 & 1 & 2 \\ 0 & ⊝2 & -3 & -5 \\ 0 & -6 & -8 & -12 \end{array} \quad \begin{array}{l} R_2 = R_2 + (-4)R_1 \\ R_3 = R_3 + (-9)R_1 \end{array}$$

Multiply the second row by $-1/2$.

$$\begin{array}{ccc|c} 1 & 1 & 1 & 2 \\ 0 & ① & \frac{3}{2} & \frac{5}{2} \\ 0 & -6 & -8 & -12 \end{array} \quad R_2 = \left(-\frac{1}{2}\right)R_2$$

Add -1 times the second row to the first row.

Add 6 times the second row to the third row.

$$\begin{array}{ccc|c} 1 & 0 & -\frac{1}{2} & -\frac{1}{2} \\ 0 & 1 & \frac{3}{2} & \frac{5}{2} \\ 0 & 0 & ① & 3 \end{array} \quad \begin{array}{l} R_1 = R_1 + (-1)R_2 \\ \\ R_3 = R_3 + 6R_2 \end{array}$$

Add $1/2$ times the third row to the first row.

Add $-3/2$ times the third row to the second row.

$$\begin{array}{ccc|c} 1 & 0 & 0 & 1 \\ 0 & 1 & 0 & -2 \\ 0 & 0 & 1 & 3 \end{array} \quad \begin{array}{l} R_1 = R_1 + \frac{1}{2}R_3 \\ R_2 = R_2 + \left(-\frac{3}{2}\right)R_3 \end{array}$$

The solution is $a = 1$, $b = -2$, $c = 3$. These values give us the curve $y = x^2 - 2x + 3$, which is shown in Figure 1 on the following page. (A curve of this type is called a **parabola**.)

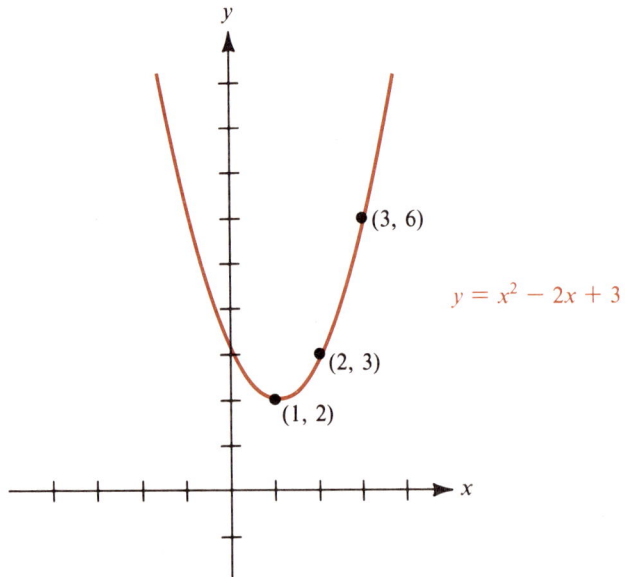

$$y = x^2 - 2x + 3$$

(3, 6)

(2, 3)

(1, 2)

Figure 1

A related problem arises when we want to fit $n + 1$ data points with a polynomial of degree *less* than n. In such a case we cannot expect to obtain a perfect fit, as in the example above, but we may still be interested in finding the "best" fit possible. For example, Figure 2 shows five data points, together with a linear function $y = ax + b$ which gives a fairly good fit to the observed values.

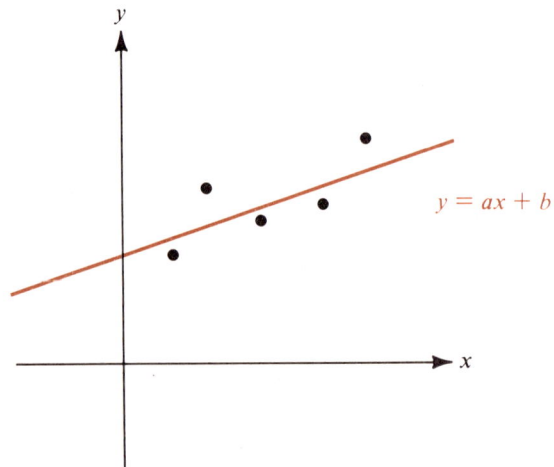

$$y = ax + b$$

Figure 2

There are various ways to give a precise meaning to the term "best fit." The most commonly used is the **least-squares** criterion, which guarantees that the approximating curve will come fairly close to most of the data points. It can be shown, using calculus, that for any set of data points there exists, for each k, a best-fitting polynomial of degree $\leq k$. The coefficients of this polynomial satisfy a system of $k + 1$ linear equations in $k + 1$ unknowns—known as the *normal equations*—which can be solved by Gauss-Jordan elimination. (See Exercises 13–14.)

3.3 Exercises

1. The Allegheny Mining Company operates three mines which produce high-, medium-, and low-grade ore. The daily yield of each mine is given in the table below.

Ore yield per day

Mine	High-grade	Medium-grade	Low-grade
1	2 tons	2 tons	1 ton
2	1 ton	3 tons	2 tons
3	0 tons	1 ton	2 tons
	40	75	50

The company must fill an order for a total of 40 tons of high-grade, 75 tons of medium-grade, and 50 tons of low-grade ore. How many days should each mine be operated to fill this order exactly?

2. The Consolidated Steel Corporation has been ordered to reduce the three main pollutants—particulates, sulfur oxide, and hydrocarbons—emitted at its Pittsburgh plant, by 5400, 6000, and 9600 units per month, respectively. Three filtering methods are available. The monthly reduction in emissions, per dollar spent, for each of the filtering methods is given below.

Monthly reduction per dollar spent

Method	Particulates	Sulfur oxide	Hydrocarbons
A	9 units	15 units	15 units
B	10 units	10 units	20 units
C	10 units	5 units	15 units

(a) How much must the company allocate per month to each method, in order to meet the reduction order exactly? (b) Is this the right question for the company to be asking?

3. An agricultural supply house makes three types of fertilizer. Type A contains 10% nitrogen and 2% phosphates by weight. Type B contains 5% nitrogen and 1% phosphates, and Type C contains 8% nitrogen and 4% phosphates. The company has received an order for 600 lb of fertilizer which is to contain 9% nitrogen and 2% phosphates. Can the order be filled by mixing together the three available types?

4. Each ounce of Food A contains 3 gm of carbohydrate, 2 gm of protein, and 4 gm of fat, while an ounce of Food B contains 3 gm of carbohydrate, 4 gm of protein,

and 2 gm of fat. By combining these foods, is it possible to obtain exactly 75 gm of carbohydrate, 60 gm of protein, and 60 gm of fat?

5. Four grades of crude oil are to be blended in an oil storage tank. The cost and sulfur content of each grade are given in the table below.

Grade	Cost per bbl	Sulfur per bbl
A	$20	125 gm
B	$25	100 gm
C	$40	75 gm
D	$50	25 gm

The tank initially contains 12,000 barrels of crude which costs $15 per bbl and has a sulfur content of 225 gm per bbl. Suppose that x, y, z, and w barrels of the four respective grades are added to the tank, to produce a blend which costs $25 per bbl and contains 150 gm per bbl of sulfur.

(a) What equations must x, y, z, and w satisfy?

(b) Solve these equations for x and y in terms of z and w.

(c) What values of z and w will yield a feasible solution?

6. A metallurgist has melted down three bronze alloys, whose composition by weight is given in the table below.

Alloy	Copper	Tin	Zinc
A	95%	4%	1%
B	90%	2%	8%
C	80%	10%	10%

A new alloy is to be formed by adding x, y, and z tons of the three respective alloys to a smelting kiln that already contains 100 tons of Alloy A.

(a) Is it possible to obtain a mixture whose composition is 90% copper, 5% tin, and 5% zinc?

(b) Is the solution unique?

7. Find the quadratic polynomial $y = ax^2 + bx + c$ whose graph passes through the points $(1, 4)$, $(2, 2)$, and $(3, 2)$.

8. Find the quadratic polynomial $y = ax^2 + bx + c$ whose graph passes through the points $(-1, 1)$, $(0, 2)$, and $(1, 1)$.

9. Find the cubic polynomial $y = ax^3 + bx^2 + cx + d$ whose graph passes through the points $(-1, 1)$, $(0, 1)$, $(1, 3)$, and $(2, 19)$.

10. Find the cubic polynomial $y = ax^3 + bx^2 + cx + d$ whose graph passes through the points $(0, 9)$, $(1, 15)$, $(2, 9)$, and $(3, 15)$.

11. The general equation of a circle in the plane is

$$x^2 + y^2 + ax + by + c = 0.$$

Find the equation of the circle through the points $(-1, 1)$, $(3, 1)$, and $(2, -2)$.

12. Find the equation of the circle through the points (1, 0), (1, 2), and (3, 2), as in Exercise 11. .

Least-Squares Approximation It can be shown that for n given data points (x, y) there exists a best-fitting linear function y = ax + b. The coefficients a and b satisfy the linear equations

$$(\Sigma x^2)a + (\Sigma x)b = \Sigma xy$$
$$(\Sigma x)a + nb = \Sigma y$$

where the symbol Σ denotes summation.

13. The table below shows four data points (x, y), together with the corresponding values of x^2 and xy. Sums are shown at the bottom of each column.

x	y	x^2	xy
-1	1	1	-1
0	2	0	0
1	1	1	1
2	3	4	6
Sum Σ: 2	7	6	6

Find the best-fitting linear function for this data.

14. As in Exercise 13, find the best-fitting linear function for the five data points (x, y) show below.

x	y	x^2	xy
-2	1	4	-2
-1	2	1	-2
0	2	0	0
1	3	1	3
2	4	4	8
Sum Σ: 0	12	10	7

Computer Projects   The PIVOT program performs a pivot transformation on a designated nonzero entry in a tableau. This program can be used in Exercises 1–14.

Chapter 3 Summary

Important Terms

3.1 $m \times n$ linear system
  tableau form
  elementary row operations
  pivot transformation
  Gauss-Jordan elimination method

3.2 inconsistent system
  redundant system
  basic variable
  free variable
  general solution
  basic solution

Review Exercises

In Exercises 1–8, use Gauss-Jordan elimination to find all solutions of the linear system.

1. $2x + y = 8$
    $3x + 2y = 9$

2. $x - 2y = 5$
    $-3x + 6y = 2$

3. $x - y + 2z = 1$
    $2x - 3y + z = 3$
    $3x - 5y - z = 6$

4. $2x + 3y = 5$
    $x - 2y + 4z = 10$
    $\phantom{x +{}} y - z = -2$

5. $x - 2y + z - w = 1$
    $2x - y + 5z + w = 5$
    $4x - 5y + 7z - w = 7$
    $-x + 5y + 2z + 4w = 2$

6. $x + y + z + w = 1$
    $2x + 3y + 4z + 8w = 4$
    $7x + 5y + 4z - 2w = 3$
    $-2x + 2y + 5z + 20w = 7$

7. $x + 2y + 3z = 5$
    $x + y + 3z = 4$
    $4x + 3y + 6z = 3$
    $3x + 2y + 4z = 1$

8. $x + 2y + z = 3$
    $2x + 9y - 3z = 1$
    $x + 3y - 2z = -2$
    $-2x - 7y + 3z = 2$

9. Use Gauss-Jordan elimination to find the general and basic solution of the linear system

$$2x + 5y + z + u + v = 0$$
$$3x + 2y - z + v = -2$$
$$6x + 3y - 3z - u + 2v = -1$$

where z, u, and v are basic, x and y free.

10. Use Gauss-Jordan elimination to find the general and basic solution of the linear system

$$x - 2y + 3z + 2u - v = 6$$
$$x - 5y + 4z + 4u - 2v = 8$$
$$-x + 4y - z - 4u + v = -4$$

where x, z, and v are basic, y and u free.

11. Three major cities have submitted requests for government urban renewal grants. The cities intend to allocate their grants as follows:

City	Low-cost housing	Job programs	Other
A	20%	50%	30%
B	40%	40%	20%
C	50%	10%	40%

The government has decided that a total of $120 million should be granted to the three cities, and that 40% of this amount should go toward low-cost housing and 30% toward job programs. How much should each city receive?

12. The general equation of a sphere in *xyz*-space is

$$x^2 + y^2 + z^2 + ax + by + cz + d = 0.$$

Find the equation of the sphere through the four points $(1, 1, 1)$, $(0, -2, 1)$, $(1, 0, 2)$, and $(3, 1, -1)$.

4

Matrix Algebra

The discussion of linear systems can be greatly simplified by using *matrices,* or rectangular arrays of numbers. We have already made limited use of matrices—for example, when we represented linear systems in tableau form—but heretofore they have served only as a convenient notational device. We now propose to elevate matrices to a new status as mathematical objects in their own right. By treating matrices as though they were numbers—adding and multiplying them, using them in equations, and so on—we enter the domain of *matrix algebra,* a discipline which has practical applications in many areas.

4.1   Matrices and Vectors

We begin by introducing some standard terminology and notation. The rest of this section will deal with the simpler aspects of matrix algebra and its applications.

Definitions

A **matrix** is a rectangular array of numbers,* which are called its **entries.** An array with m rows and n columns is called an $m \times n$ **matrix,** or a matrix of **dimension** $m \times n$ (read ''m by n''). If $m = n$ the matrix is called **square.**

---

*The term *matrix* is due to J. J. Sylvester (1850), but matrix algebra was developed chiefly by W. R. Hamilton (1853) and A. Cayley (1858).

The general form of an $m \times n$ matrix is

$$A = \begin{bmatrix} a_{11} & a_{12} & a_{13} & \cdots & a_{1n} \\ a_{21} & a_{22} & a_{23} & \cdots & a_{2n} \\ \vdots & \vdots & \vdots & & \vdots \\ a_{m1} & a_{m2} & a_{m3} & \cdots & a_{mn} \end{bmatrix}$$

which is abbreviated by writing

$$A = [a_{ij}]. \qquad (A \text{ is } m \times n)$$

The rows are numbered from top to bottom and the columns from left to right, so that for any subscripts i and j, the entry in row i and column j is denoted by a_{ij}. For example, the entry in the second row, third column of A is a_{23}. Note that the order of the subscripts is important: the *first* subscript represents the row, the *second* the column, in which the entry occurs.*

Example 1   Consider the matrices A, B, and C.

$$A = \begin{bmatrix} 3 & 2 & 6 & 0 \\ 0 & \frac{1}{2} & 1 & -5 \\ 2 & 1 & 3 & 2 \end{bmatrix}, \qquad B = \begin{bmatrix} 3 & -5 \\ 2 & 1 \end{bmatrix}, \qquad C = \begin{bmatrix} 5 & 1 \\ 4 & 7 \\ 0 & 2 \\ 4 & -2 \end{bmatrix}$$

Discussion   Here A is 3×4, B is 2×2 (square), and C is 4×2. If we write $A = [a_{ij}]$, $B = [b_{ij}]$, and $C = [c_{ij}]$, then $a_{13} = 6$, $a_{31} = 2$, $b_{11} = 3$, $b_{22} = 1$, $c_{31} = 0$, $c_{42} = -2$.   ◆

With matrices, the term **scalar** means an ordinary real number (as opposed to a matrix). As in the example above, we reserve uppercase letters (A, B, etc.) for matrices and lowercase letters (a, b, etc.) for scalars, and generally try to use the same letter (for example, A and a) for a matrix and its scalar entries.

Definition

A **vector** is a matrix consisting of either a single row or a single column. Specifically, a *k*-termed **row vector** is just a $1 \times k$ matrix, and a *k*-termed **column vector** is just a $k \times 1$ matrix.

---

*This double-subscript notation is very widely used. When i or j has more than one digit we write $a_{i,j}$ instead of a_{ij} to avoid confusion.

Example 2 What are the dimensions of the matrices A and B?

$$A = [2 \quad -1 \quad 3 \quad 5], \qquad B = \begin{bmatrix} 4 \\ \frac{2}{5} \\ 0 \\ -1 \\ 3 \end{bmatrix}$$

Discussion The dimension of A is 1×4, since it has one row and four columns (one for each entry). A is thus a 4-termed row vector. Similarly, the 5×1 matrix B is a 5-termed column vector. ◆

Row and column vectors play a prominent role in matrix algebra, which is why they are given a special name.* Notice that a 1×1 matrix is just a scalar, enclosed in brackets.

Equality of Matrices Matrices A and B are considered **equal,** and we write $A = B$, only when they look exactly the same; that is, when they have the same dimensions and exactly the same corresponding entries. Thus, a 3×2 matrix can only be equal to another 3×2 matrix, and we have

$$A = \begin{bmatrix} 3 & 5 \\ 1 & -4 \\ 0 & 2 \end{bmatrix} \neq \begin{bmatrix} 3 & 5 \\ 1 & -4 \\ 0 & 4 \end{bmatrix} = B$$

because not all corresponding entries are equal (in particular, $a_{32} = 2$ but $b_{32} = 4$). For equality to hold, two matrices must have exactly the same entries in exactly the same positions.

Example 3 Find scalars x, y, z, w which satisfy the matrix equation

$$\begin{bmatrix} y & 2z \\ 4 & w \end{bmatrix} = \begin{bmatrix} 1 & 6 \\ x & 2 \end{bmatrix}.$$

Solution Since corresponding entries must be equal, we are led to four conditions.

$$y = 1 \qquad 2z = 6 \qquad 4 = x \qquad w = 2$$

These equations have the obvious solution $x = 4, y = 1, z = 3, w = 2$. ◆

This example illustrates an important fact we shall use later, namely that a *system* of scalar equations can be summarized in a *single* matrix equation.

---

*In physics, a vector quantity is one having both magnitude and direction, such as force or velocity. There is a connection (which we shall not pursue) between this notion of a vector and the present definition.

Operations on Matrices   We now define the first of several operations that form the basis of matrix algebra.

Definition

The **transpose** of a matrix A, written A^t, is the matrix obtained by changing the rows of A to columns, in order.

Example 4   Find the transpose of the matrix A.

$$A = \begin{bmatrix} 4 & 2 & 6 & 1 \\ 5 & 0 & 1 & 3 \end{bmatrix}$$

Solution   Writing the rows of A as columns, we obtain the transpose matrix A^t.

$$A^t = \begin{bmatrix} 4 & 5 \\ 2 & 0 \\ 6 & 1 \\ 1 & 3 \end{bmatrix}$$

Note that the first and second rows of A become, respectively, the first and second columns of A^t, and at the same time the columns of A become the rows of A^t. The dimensions of A and A^t are 2×4 and 4×2, respectively. More generally, if a matrix has dimension $m \times n$ then its transpose has dimension $n \times m$; hence, in particular, the transpose of a row vector is a column vector, and vice-versa.   ◆

There are four fundamental algebraic operations on matrices: (1) addition of two matrices, (2) taking the negative of a matrix, (3) multiplication of a matrix by a scalar, and (4) multiplication of two matrices. The last of these—matrix multiplication—is somewhat complicated, so we defer its discussion for the next section. The other three operations are quite straightforward.

Definition

The **sum** of two $m \times n$ matrices A and B, written $A + B$, is the matrix obtained by adding corresponding entries.

The **negative** of a matrix A, written $-A$, is the matrix obtained by taking the negative (changing the sign) of each entry of A.

The **product of a scalar** c **and a matrix** A, written cA, is the matrix obtained by multiplying each entry of A by c.

In the double-subscript notation, these definitions can be stated very succinctly as follows:

sum: $\quad [a_{ij}] + [b_{ij}] = [a_{ij} + b_{ij}]$

negative: $\quad -[a_{ij}] = [-a_{ij}]$

scalar product: $\quad c[a_{ij}] = [ca_{ij}]$

Example 5   Given the matrices

$$A = \begin{bmatrix} 3 & 0 \\ -2 & 1 \\ 6 & -5 \end{bmatrix} \quad \text{and} \quad B = \begin{bmatrix} 1 & 5 \\ 3 & -1 \\ -4 & 2 \end{bmatrix},$$

calculate $A + B$, $-B$, and $3A$.

Solution

$$A + B = \begin{bmatrix} 3 & 0 \\ -2 & 1 \\ 6 & -5 \end{bmatrix} + \begin{bmatrix} 1 & 5 \\ 3 & -1 \\ -4 & 2 \end{bmatrix}$$

$$= \begin{bmatrix} 3 + 1 & 0 + 5 \\ -2 + 3 & 1 + (-1) \\ 6 + (-4) & -5 + 2 \end{bmatrix} = \begin{bmatrix} 4 & 5 \\ 1 & 0 \\ 2 & -3 \end{bmatrix}$$

$$-B = -\begin{bmatrix} 1 & 5 \\ 3 & -1 \\ -4 & 2 \end{bmatrix} = \begin{bmatrix} -1 & -5 \\ -3 & 1 \\ 4 & -2 \end{bmatrix}$$

$$3A = 3\begin{bmatrix} 3 & 0 \\ -2 & 1 \\ 6 & -5 \end{bmatrix} = \begin{bmatrix} 9 & 0 \\ -6 & 3 \\ 18 & -15 \end{bmatrix} \quad \blacklozenge$$

Remark to Reader   Note that *only matrices of the same dimension can be added*. When A and B are of different dimensions, their sum is undefined.

Definition

The **difference** of two $m \times n$ matrices A and B is defined by the formula

$$A - B = A + (-B).$$

For the matrices A and B of Example 5, we have

$$A - B = \begin{bmatrix} 3 & 0 \\ -2 & 1 \\ 6 & -5 \end{bmatrix} - \begin{bmatrix} 1 & 5 \\ 3 & -1 \\ -4 & 2 \end{bmatrix}$$

$$= \begin{bmatrix} 3 & 0 \\ -2 & 1 \\ 6 & -5 \end{bmatrix} + \begin{bmatrix} -1 & -5 \\ -3 & 1 \\ 4 & -2 \end{bmatrix} = \begin{bmatrix} 2 & -5 \\ -5 & 2 \\ 10 & -7 \end{bmatrix}.$$

Notice that this just amounts to subtracting each entry in B from the corresponding entry in A.

Definition

A matrix all of whose entries are zero is called a **zero matrix,** and denoted by the letter O.

There is a different zero matrix for each dimension $m \times n$. For example,

$$\begin{bmatrix} 0 & 0 \\ 0 & 0 \end{bmatrix}, \quad \begin{bmatrix} 0 & 0 \\ 0 & 0 \\ 0 & 0 \end{bmatrix}, \quad [0 \quad 0 \quad 0], \quad \begin{bmatrix} 0 \\ 0 \\ 0 \end{bmatrix}$$

are zero matrices of dimension 2×2, 3×2, 1×3 and 3×1, respectively. Although we use the same symbol O to denote any zero matrix, the intended dimension is generally clear from context. For example, the matrix identity $A + O = A$ only makes sense when A and O have the same dimension.

Laws of Matrix Algebra   From our definitions of the matrix operations it is easy to prove the following general laws:

Laws of Matrix Algebra

M1. $A + B = B + A$

M2. $A + (B + C) = (A + B) + C$

M3. $A + O = A$

M4. $A + (-A) = O$

M5. $c(A + B) = cA + cB$

M6. $(c + d)A = cA + dA$

M7. $c(dA) = (cd)A$

M8. $1A = A$

These laws hold whenever both sides are defined; for example, in M1 we assume that A and B are matrices of the same dimension. From M1–M8 we can derive other obvious principles of matrix algebra, such as the laws $(-1)A = -A$ and $0A = O$. Since these laws are the same as the usual algebraic rules for real numbers, we can—for the time being, at least—manipulate matrices algebraically just as though they were numbers.

Example 6 Find a 2×2 matrix X satisfying the equation

$$2(B + 3X) = 3B - 4(A - X)$$

where

$$A = \begin{bmatrix} 1 & -5 \\ 3 & 0 \end{bmatrix} \quad \text{and} \quad B = \begin{bmatrix} 8 & -6 \\ 4 & 2 \end{bmatrix}.$$

Solution First, solve the equation for the unknown X, just as usual.

$$2(B + 3X) = 3B - 4(A - X)$$
$$2B + 6X = 3B - 4A + 4X$$
$$2X = B - 4A$$
$$X = \frac{1}{2}(B - 4A) = \frac{1}{2}B - 2A$$

Substitute the matrices A and B to compute the solution.

$$X = \frac{1}{2}B - 2A$$
$$= \frac{1}{2}\begin{bmatrix} 8 & -6 \\ 4 & 2 \end{bmatrix} - 2\begin{bmatrix} 1 & -5 \\ 3 & 0 \end{bmatrix}$$
$$= \begin{bmatrix} 4 & -3 \\ 2 & 1 \end{bmatrix} - \begin{bmatrix} 2 & -10 \\ 6 & 0 \end{bmatrix} = \begin{bmatrix} 2 & 7 \\ -4 & 1 \end{bmatrix} \quad \blacklozenge$$

We remark (without proof) that each of the steps used in solving the equation—such as multiplying out and transposing—can be justified on the basis of the laws M1–M8.

Applications Matrices provide a useful means for classifying data, since they enable us to manipulate several items of information simultaneously.

Example 7 A furniture manufacturer has plants in countries X and Y that produce three styles of chairs for export to the U.S. The costs of labor, materials, and transportation for each chair vary between the two plants. The figures, in dollars per unit, are given in the table below.

	Labor		Materials		Transport	
Style	X	Y	X	Y	X	Y
Antique	$39	$44	$10	$5	$5	$10
Baroque	$61	$70	$15	$5	$7	$12
Classic	$22	$26	$15	$10	$4	$12

Discussion   We can represent the labor cost per unit as a 3×2 matrix.

$$L = \begin{array}{c} \\ A \\ B \\ C \end{array} \begin{array}{cc} X & Y \\ \left[\begin{array}{cc} 39 & 44 \\ 61 & 70 \\ 22 & 26 \end{array}\right] \end{array} \quad \text{(dollars/chair)}$$

In the same way, the materials cost and transportation cost can be represented by 3×2 matrices.

$$M = \left[\begin{array}{cc} 10 & 5 \\ 15 & 5 \\ 15 & 10 \end{array}\right], \quad T = \left[\begin{array}{cc} 5 & 10 \\ 7 & 12 \\ 4 & 12 \end{array}\right]$$

The total cost per unit, including labor, materials, and transportation, is given by the sum of the three matrices.

$$C = L + M + T$$

$$= \left[\begin{array}{cc} 39 & 44 \\ 61 & 70 \\ 22 & 26 \end{array}\right] + \left[\begin{array}{cc} 10 & 5 \\ 15 & 5 \\ 15 & 10 \end{array}\right] + \left[\begin{array}{cc} 5 & 10 \\ 7 & 12 \\ 4 & 12 \end{array}\right] = \begin{array}{c} \\ A \\ B \\ C \end{array} \begin{array}{cc} X & Y \\ \left[\begin{array}{cc} 54 & 59 \\ 83 & 87 \\ 41 & 48 \end{array}\right] \end{array}$$

For instance, the total cost of producing and shipping a Baroque chair from Country Y is $70 + 5 + 12 = \$87$. If the three chairs sell in the U.S. for \$95, \$140, and \$85, respectively, then the matrix

$$R = \begin{array}{c} \\ A \\ B \\ C \end{array} \begin{array}{cc} X & Y \\ \left[\begin{array}{cc} 95 & 95 \\ 140 & 140 \\ 85 & 85 \end{array}\right] \end{array}$$

represents the company's revenue per unit, and the matrix difference

$$P = R - C = \left[\begin{array}{cc} 95 & 95 \\ 140 & 140 \\ 85 & 85 \end{array}\right] - \left[\begin{array}{cc} 54 & 59 \\ 83 & 87 \\ 41 & 48 \end{array}\right] = \begin{array}{c} \\ A \\ B \\ C \end{array} \begin{array}{cc} X & Y \\ \left[\begin{array}{cc} 41 & 36 \\ 57 & 53 \\ 44 & 37 \end{array}\right] \end{array}$$

represents the company's profit per unit. Suppose now that all materials costs increase uniformly by 20%. Since $m + .2m = 1.2m$, the effect on the matrix M will be to multiply each entry by a scalar factor 1.2. Hence, the new materials cost matrix will be

$$1.2 M = 1.2 \left[\begin{array}{cc} 10 & 5 \\ 15 & 5 \\ 15 & 10 \end{array}\right] = \begin{array}{c} \\ A \\ B \\ C \end{array} \begin{array}{cc} X & Y \\ \left[\begin{array}{cc} 12 & 6 \\ 18 & 6 \\ 18 & 12 \end{array}\right] \end{array}. \quad \blacklozenge$$

We could apply the same technique, of course, if there were plants in ten countries, each producing fifteen styles of chairs. In such large-scale applications the advantage of using matrices as an organizing device can be readily appreciated.

4.1 Exercises

In Exercises 1–6, determine the dimension of the matrix. Is it a row vector? A column vector?

1. $\begin{bmatrix} 3 & 2 \\ -1 & 0 \end{bmatrix}$
2. $[5 \quad 1 \quad 2]$
3. $\begin{bmatrix} 8 \\ 6 \end{bmatrix}$

4. $\begin{bmatrix} 0 & 2 \\ 1 & -1 \\ 2 & 0 \end{bmatrix}$
5. $\begin{bmatrix} 5 & 2 & -1 \\ 4 & 7 & 3 \end{bmatrix}$
6. $\begin{bmatrix} 3 \\ 4 \\ 2 \end{bmatrix}$

7. Given the 3×3 matrix

$$A = \begin{bmatrix} 2 & 5 & 9 \\ 7 & -4 & 1 \\ 3 & 0 & 6 \end{bmatrix}$$

   (a) identify the entries a_{21}, a_{22}, a_{23}
   (b) identify the entries a_{13}, a_{23}, a_{33}.

8. Given the 4×4 matrix

$$A = \begin{bmatrix} 5 & -1 & 2 & 3 \\ 0 & 1 & 4 & 0 \\ 2 & -2 & -1 & 5 \\ 4 & 3 & 8 & 2 \end{bmatrix}$$

   (a) identify the entries $a_{11}, a_{22}, a_{33}, a_{44}$
   (b) identify the entries $a_{14}, a_{23}, a_{32}, a_{41}$.

9. Find scalars x, y, z for which

$$\begin{bmatrix} 2x & 5 \\ 4 & 3z \end{bmatrix} = \begin{bmatrix} 6 & 5 \\ y & x \end{bmatrix}.$$

10. Find scalars x, y for which

$$\begin{bmatrix} 4x + 2y \\ 5x + 3y \end{bmatrix} = \begin{bmatrix} 18 \\ 25 \end{bmatrix}.$$

11. Given the matrices

$$A = \begin{bmatrix} 2 & 3 & 1 \\ 4 & 5 & 2 \end{bmatrix}, \quad B = \begin{bmatrix} 2 & 1 & -1 \\ 5 & 3 & 0 \end{bmatrix}, \quad C = \begin{bmatrix} 3 & 4 \\ 5 & 1 \\ -2 & 2 \end{bmatrix},$$

   calculate the following (if possible):

   (a) $A + B$
   (b) $A - B$
   (c) A^t
   (d) $\frac{1}{2}C$
   (e) $3A - 2B$
   (f) $2B^t + C$

12. Given the vectors

$$A = \begin{bmatrix} 2 \\ 5 \\ 4 \end{bmatrix}, \qquad B = \begin{bmatrix} 1 \\ -1 \\ 3 \end{bmatrix}, \qquad C = [4 \quad 2 \quad 5],$$

calculate the following (if possible):

(a) $10A$ (c) $A^t + B^t + C$ (e) $\frac{1}{2}A + \frac{1}{2}B$

(b) $2B + 3C^t$ (d) $(A - B)^t + C$ (f) $A + C$

13. Generic Motors divides its national sales territory into four geographic regions. The projected first-quarter demand for the company's new sedan, the Albatross, is 25,000 for the East; 20,000, Midwest; 30,000, West; and 15,000, South. The projected demand for the company's new compact, the Bracer, over the same period is 40,000 for the East; 15,000, Midwest; 35,000, West; and 10,000, South.

(a) Represent the projected first-quarter demand for the two cars, by region, as a 2×4 matrix D_1.

(b) If second-quarter demand is expected to be 15% higher in each region, calculate the second-quarter demand matrix D_2 as a scalar multiple of D_1.

(c) Calculate the matrix D, representing projected demand over the first six months, by matrix addition.

14. An interacting economic system has four producing sectors: agriculture (A), manufacturing (M), energy (E), and services (R). The annual output of each sector is measured in standard units, where one unit represents $1 million worth of goods. For the year just ended, total (gross) output in the four sectors was 220, 310, 240, and 170 units, respectively. However, of these amounts 60, 75, 130, and 80 units, respectively, were used up internally in the production process itself.

(a) Represent the total (gross) production output of the economy as a column vector P.

(b) Represent the amount of output used up internally as a column vector U.

(c) Calculate the column vector S, representing the surplus (net) production output, by means of vector subtraction.

(d) If P and U both increase uniformly by 40%, what is the effect on S?

In Exercises 15–18, verify the given law for the 2 × 2 case by letting

$$A = \begin{bmatrix} 2 & 4 \\ -3 & 5 \end{bmatrix}, \qquad B = \begin{bmatrix} 1 & -2 \\ 2 & 3 \end{bmatrix}, \qquad C = \begin{bmatrix} 0 & 4 \\ -1 & 0 \end{bmatrix},$$

and showing that both sides of the equation represent the same matrix.

15. $A + B = B + A$ 16. $A + (B + C) = (A + B) + C$

17. $A + O = A$ 18. $A + (-A) = O$

In Exercises 19–22, prove the given law for the 2 × 2 case by letting

$$A = \begin{bmatrix} a_{11} & a_{12} \\ a_{21} & a_{22} \end{bmatrix}, \qquad B = \begin{bmatrix} b_{11} & b_{12} \\ b_{21} & b_{22} \end{bmatrix},$$

and showing that both sides of the equation represent the same matrix.

19. $c(A + B) = cA + cB$

20. $(c + d)A = cA + dA$

21. $c(dA) = (cd)A$

22. $1A = A$

23. Solve the matrix equation

$$3(A + 2X) = 2(2X + B - 2A)$$

for the unknown matrix X, where

$$A = \begin{bmatrix} 2 & 4 \\ 0 & -2 \\ -2 & 6 \end{bmatrix}, \quad B = \begin{bmatrix} 8 & 10 \\ 3 & -2 \\ 1 & 12 \end{bmatrix}.$$

24. Solve the vector equation

$$3(B - (A + X)) = 2(A - X)$$

for the unknown vector X, where

$$A = \begin{bmatrix} 1 & -3 & 5 & -7 \end{bmatrix},$$
$$B = \begin{bmatrix} 2 & -4 & 6 & -8 \end{bmatrix}.$$

25. Letting A and B be arbitrary 2×2 matrices, as in Exercises 19–22, prove the following laws involving the transpose operator:

(a) $(A + B)^t = A^t + B^t$

(c) $(-A)^t = -(A^t)$

(b) $(cA)^t = c(A^t)$

(d) $(A^t)^t = A$

26. Making use of the laws for transpose (Exercise 25), solve the equation

$$2(B^t + 2X^t - A)^t = (-A^t + 2B)^t + 2X$$

for the unknown matrix X, where

$$A = \begin{bmatrix} 4 & 6 \\ 0 & 2 \end{bmatrix}, \quad B = \begin{bmatrix} 1 & -1 \\ 3 & 4 \end{bmatrix}.$$

4.2 Matrix Multiplication

We have seen that the algebra of matrices has many points in common with the algebra of real numbers. The two systems come to a sharp parting of the ways, however, when we introduce the operation of matrix multiplication. Although the definition of multiplication seems "unnatural" at first, it is this operation, more than any other, which gives matrix algebra its unique power as a mathematical tool.

Dot Product In order to define the product AB of two matrices, we first consider the case where A is a row vector and B is a column vector.

Definition

Let A be a row vector and B a column vector, each with the same number of entries. The **dot product** $A \cdot B$ is the sum of the products of their corresponding entries. That is, if

$$A = [a_1 \, a_2 \cdots a_n] \quad \text{and} \quad B = \begin{bmatrix} b_1 \\ b_2 \\ \cdot \\ \cdot \\ \cdot \\ b_n \end{bmatrix}$$

then

$$A \cdot B = a_1 b_1 + a_2 b_2 + \cdots + a_n b_n.$$

Notice that we take the dot product of a *row* vector with a *column* vector, in that order, and that the two vectors must have the same number of entries. Note also that the dot product of two vectors is a *scalar,* not a vector.

Example 1   (a) $[2 \quad 4] \cdot \begin{bmatrix} 8 \\ 3 \end{bmatrix} = 2(8) + 4(3) = 28$

(b) $\left[\dfrac{1}{2} \quad 4 \quad -5\right] \cdot \begin{bmatrix} 6 \\ \frac{3}{2} \\ 2 \end{bmatrix} = \left(\dfrac{1}{2}\right)(6) + 4\left(\dfrac{3}{2}\right) + (-5)(2) = -1$ ◆

Dot products arise very commonly in applications, even in the simplest settings.

Example 2   A furniture manufacturer has established daily production quotas of 15, 25, and 20 chairs of the Antique, Baroque, and Classic styles, respectively. The three chairs require 2, 3, and 1 work-hours of construction each, respectively. How many work-hours of construction will be needed to fill the quota?

Solution   The given figures can be summarized in a *production vector* (row) and a *construction-labor vector* (column) as follows:

$$\begin{array}{ccc} & A \quad B \quad C & \\ \text{Production} = [15 & 25 & 20] \end{array} \quad \text{(chairs)}$$

$$\text{Construction-labor} = \begin{array}{c} A \\ B \\ C \end{array}\begin{bmatrix} 2 \\ 3 \\ 1 \end{bmatrix} \quad \text{(work-hours/chair)}$$

The dot product of these vectors gives the total amount of construction labor needed.

$$[15 \quad 25 \quad 20] \cdot \begin{bmatrix} 2 \\ 3 \\ 1 \end{bmatrix} = 15(2) + 25(3) + 20(1) = 125 \qquad \text{(work-hours)}$$

Matrix multiplication is essentially just a way of recording many dot products, instead of merely one as in the example above. To illustrate, suppose now that the manufacturer in Example 2 has two plants, in countries X and Y, with daily production quotas as shown in the following *production matrix:*

$$\text{Production} \quad P = \begin{array}{c} \\ X \\ Y \end{array} \begin{array}{ccc} A & B & C \\ \begin{bmatrix} 15 & 25 & 20 \\ 30 & 20 & 35 \end{bmatrix} \end{array} \qquad \text{(chairs)}$$

Suppose also that each chair requires two stages in its manufacture—construction and finishing—and that the labor required per chair (at either plant) is given by the following *labor matrix:*

$$\text{Labor} \quad L = \begin{array}{c} \\ A \\ B \\ C \end{array} \begin{bmatrix} 2 & 3 \\ 3 & 5 \\ 1 & 2 \end{bmatrix} \qquad \text{(work-hours/chair)}$$

with upper labels: ⌐Construct, ⌐Finish

From the matrices P and L we can calculate the total amount of labor needed per day for each state of manufacture, at each of the two plants. The result is the *total labor matrix* shown below.

$$\text{Total labor} \quad T = \begin{array}{c} \\ X \\ Y \end{array} \begin{bmatrix} 125 & 210 \\ 155 & 260 \end{bmatrix} \qquad \text{(work-hours)}$$

with upper labels: ⌐Construct, ⌐Finish

Each entry in T is computed by taking the dot product of a certain *row* of P with a certain *column* of L.

$$\begin{array}{l} \text{Total construction labor} \\ \text{needed at plant X} \end{array} = [15 \quad 25 \quad 20] \cdot \begin{bmatrix} 2 \\ 3 \\ 1 \end{bmatrix} = 125$$

Production at plant X   Construction labor

$$\begin{array}{l} \text{Total finishing labor} \\ \text{needed at plant X} \end{array} = [15 \quad 25 \quad 20] \cdot \begin{bmatrix} 3 \\ 5 \\ 2 \end{bmatrix} = 210$$

Production at plant X   Finishing labor

$$\begin{array}{l} \text{Total construction labor} \\ \text{needed at plant Y} \end{array} = \begin{bmatrix} 30 & 20 & 35 \end{bmatrix} \cdot \begin{bmatrix} 2 \\ 3 \\ 1 \end{bmatrix} = 155$$

Production at plant Y   Construction labor

$$\begin{array}{l} \text{Total finishing labor} \\ \text{needed at plant Y} \end{array} = \begin{bmatrix} 30 & 20 & 35 \end{bmatrix} \cdot \begin{bmatrix} 3 \\ 5 \\ 2 \end{bmatrix} = 260$$

Production at plant Y   Finishing labor

This matrix T is called the *product* of the matrices P and L.

$$PL = \begin{bmatrix} 15 & 25 & 20 \\ 30 & 20 & 35 \end{bmatrix} \begin{bmatrix} 2 & 3 \\ 3 & 5 \\ 1 & 2 \end{bmatrix} = \begin{bmatrix} 125 & 210 \\ 155 & 260 \end{bmatrix} = T \quad \blacklozenge$$

With this example in mind as motivation, we now present the general definition.

Matrix Multiplication   To multiply matrices A and B, we multiply each *row* of A by each *column* of B, as follows:

Definition

Let A and B be matrices of dimension $m \times n$ and $n \times k$, respectively. Then the **matrix product** AB exists and has dimension $m \times k$. The entry in row i, column j of AB is the dot product of the ith row of A with the jth column of B.

According to the definition, we can only form the product AB when the number of columns of A equals the number of rows of B; that is, when the *inner dimensions are equal* in the following diagram.

$$A \quad \times \quad B \qquad = \qquad AB$$
$$m \times n \quad n \times k \qquad \qquad m \times k$$

equal

The outer dimensions then determine the size of AB, which has the same number of rows, m, as the left factor A, and the same number of columns, k, as the right factor B.

   If we let A_i represent the ith row of A and B_j the jth column of B, the definition can be summarized in the following diagram:

$$\begin{bmatrix} A_1 \\ \hline A_2 \\ \vdots \\ \hline A_m \end{bmatrix} \times \begin{bmatrix} B_1 & B_2 & \cdots & B_k \end{bmatrix} = \begin{bmatrix} A_1 \cdot B_1 & A_1 \cdot B_2 & \cdots & A_1 \cdot B_k \\ A_2 \cdot B_1 & A_2 \cdot B_2 & \cdots & A_2 \cdot B_k \\ \vdots & \vdots & & \vdots \\ A_m \cdot B_1 & A_m \cdot B_2 & \cdots & A_m \cdot B_k \end{bmatrix}$$

$$A \qquad\qquad B \qquad\qquad\qquad AB$$

Note that we divide the *left* factor into rows and the *right* factor into columns. The order of the factors A, B is crucial in matrix multiplication.

Example 3 Calculate AB, where

$$A = \begin{bmatrix} 3 & 2 & 6 & 5 \\ 1 & 4 & 2 & 3 \\ 5 & 2 & 1 & 2 \end{bmatrix}, \qquad B = \begin{bmatrix} 2 & 3 \\ 4 & 2 \\ 1 & 5 \\ 3 & 4 \end{bmatrix}.$$

Solution We first check the dimensions: since A is 3×4 and B is 4×2, the product AB is defined and has dimension 3×2. Dividing the left factor A into rows and the right factor B into columns, we write

$$\begin{bmatrix} 3 & 2 & 6 & 5 \\ \hline 1 & 4 & 2 & 3 \\ \hline 5 & 2 & 1 & 2 \end{bmatrix} \times \begin{bmatrix} 2 & 3 \\ 4 & 2 \\ 1 & 5 \\ 3 & 4 \end{bmatrix} = \begin{bmatrix} ? & ? \\ ? & ? \\ ? & ? \end{bmatrix}$$

$$A \qquad\quad B \qquad = \qquad AB$$

$$3 \times 4 \quad 4 \times 2 \qquad\qquad 3 \times 2$$

$$\underset{\text{equal}}{\underbrace{\qquad\qquad}}$$

We then compute the first row of AB by taking the dot product of the first row of A with each column of B.

$$A_1 \cdot B_1 = \begin{bmatrix} 3 & 2 & 6 & 5 \end{bmatrix} \cdot \begin{bmatrix} 2 \\ 4 \\ 1 \\ 3 \end{bmatrix} = 3(2) + 2(4) + 6(1) + 5(3) = 35$$

$$A_1 \cdot B_2 = \begin{bmatrix} 3 & 2 & 6 & 5 \end{bmatrix} \cdot \begin{bmatrix} 3 \\ 2 \\ 5 \\ 4 \end{bmatrix} = 3(3) + 2(2) + 6(5) + 5(4) = 63$$

The second and third rows of AB are computed similarly, but using A_2 and A_3 instead of A_1. The result is shown on the next page.

$$AB = \begin{bmatrix} 3 & 2 & 6 & 5 \\ 1 & 4 & 2 & 3 \\ 5 & 2 & 1 & 2 \end{bmatrix} \begin{bmatrix} 2 & 3 \\ 4 & 2 \\ 1 & 5 \\ 3 & 4 \end{bmatrix} = \begin{bmatrix} 35 & 63 \\ 29 & 33 \\ 25 & 32 \end{bmatrix} \quad \blacklozenge$$

Notice that if we reverse the order of the factors A and B, the inner dimensions no longer match.

$$\underset{\substack{B \\ 4 \times 2}}{\begin{bmatrix} 2 & 3 \\ 4 & 2 \\ 1 & 5 \\ 3 & 4 \end{bmatrix}} \times \underset{\substack{A \\ 3 \times 4}}{\begin{bmatrix} 3 & 2 & 6 & 5 \\ 1 & 4 & 2 & 3 \\ 5 & 2 & 1 & 2 \end{bmatrix}} \quad \text{undefined}$$

unequal

The product BA is undefined. In general, the matrix products AB and BA need not be equal, even when they both exist and have the same dimension.

Example 4 Calculate AB and BA, where

$$A = \begin{bmatrix} 0 & 1 \\ 1 & 2 \end{bmatrix}, \qquad B = \begin{bmatrix} 2 & 1 \\ 0 & 1 \end{bmatrix}.$$

Solution Since A and B are both 2×2, the products AB and BA are both defined and have dimension 2×2.

$$AB = \begin{bmatrix} 0 & 1 \\ 1 & 2 \end{bmatrix} \begin{bmatrix} 2 & 1 \\ 0 & 1 \end{bmatrix} = \begin{bmatrix} 0 & 1 \\ 2 & 3 \end{bmatrix}$$

$$BA = \begin{bmatrix} 2 & 1 \\ 0 & 1 \end{bmatrix} \begin{bmatrix} 0 & 1 \\ 1 & 2 \end{bmatrix} = \begin{bmatrix} 1 & 4 \\ 1 & 2 \end{bmatrix} \quad \blacklozenge$$

This example shows that in matrix algebra the *commutative law* $AB = BA$ is *not* generally valid.

Laws of Matrix Algebra Despite the failure of the commutative law, there are many ways in which matrix multiplication does resemble ordinary multiplication of numbers. Specifically, we can add the following laws to our earlier list M1–M8 in Section 4.1:

M9. $A(BC) = (AB)C$

M10. $A(B + C) = AB + AC$

M11. $(A + B)C = AC + BC$

M12. $(cA)B = c(AB) = A(cB)$

As before, these equations hold whenever both sides are defined. The general proof of these laws is complex, so we shall take them for granted.

Example 5 Verify the *associative law* $A(BC) = (AB)C$, using the following matrices:

$$A = \begin{bmatrix} -1 & 1 \\ 2 & -1 \end{bmatrix}, \qquad B = \begin{bmatrix} 0 & 1 & 2 \\ 1 & 2 & 0 \end{bmatrix}, \qquad C = \begin{bmatrix} 3 \\ 2 \\ 1 \end{bmatrix}$$

Solution We can calculate both sides of the equation.

$$BC = \begin{bmatrix} 0 & 1 & 2 \\ 1 & 2 & 0 \end{bmatrix} \begin{bmatrix} 3 \\ 2 \\ 1 \end{bmatrix} = \begin{bmatrix} 4 \\ 7 \end{bmatrix}$$

$$A(BC) = \begin{bmatrix} -1 & 1 \\ 2 & -1 \end{bmatrix} \begin{bmatrix} 4 \\ 7 \end{bmatrix} = \begin{bmatrix} 3 \\ 1 \end{bmatrix}$$

$$AB = \begin{bmatrix} -1 & 1 \\ 2 & -1 \end{bmatrix} \begin{bmatrix} 0 & 1 & 2 \\ 1 & 2 & 0 \end{bmatrix} = \begin{bmatrix} 1 & 1 & -2 \\ -1 & 0 & 4 \end{bmatrix}$$

$$(AB)C = \begin{bmatrix} 1 & 1 & -2 \\ -1 & 0 & 4 \end{bmatrix} \begin{bmatrix} 3 \\ 2 \\ 1 \end{bmatrix} = \begin{bmatrix} 3 \\ 1 \end{bmatrix}$$

The law is satisfied for these matrices, since

$$A(BC) = \begin{bmatrix} 3 \\ 1 \end{bmatrix} = (AB)C. \quad \blacklozenge$$

The real number system has the important principle that $a \cdot 1 = a$; that is, multiplication by 1 leaves any real number unchanged. It turns out that some matrices act like the number 1 with respect to matrix multiplication.

Definition

An **identity matrix** I is a square matrix with entries 1 along the main diagonal (upper-left to lower-right) and 0 elsewhere.

There is a different identity matrix I_n for each square dimension $n \times n$. For example,

$$I_2 = \begin{bmatrix} 1 & 0 \\ 0 & 1 \end{bmatrix}, \qquad I_3 = \begin{bmatrix} 1 & 0 & 0 \\ 0 & 1 & 0 \\ 0 & 0 & 1 \end{bmatrix}, \qquad I_4 = \begin{bmatrix} 1 & 0 & 0 & 0 \\ 0 & 1 & 0 & 0 \\ 0 & 0 & 1 & 0 \\ 0 & 0 & 0 & 1 \end{bmatrix}.$$

If a matrix A is multiplied on either side by an identity matrix of the same dimension, it remains unchanged. That is,

---

M13. If A has dimension $m \times n$, then

$$I_m A = A = A I_n$$

Since the dimension of I is always clear from context, we can express M13 in the simpler form

$$IA = A = AI.$$

Example 6   Verify that $IA = A$ and $AI = A$, where $A = \begin{bmatrix} 1 & 2 & 3 \\ 4 & 5 & 6 \end{bmatrix}$.

Solution   Since A has dimension 2×3 we must use a 2×2 identity matrix I in forming IA, and a 3×3 identity matrix I in forming AI.

$$IA = \begin{bmatrix} 1 & 0 \\ 0 & 1 \end{bmatrix} \begin{bmatrix} 1 & 2 & 3 \\ 4 & 5 & 6 \end{bmatrix} = \begin{bmatrix} 1 & 2 & 3 \\ 4 & 5 & 6 \end{bmatrix} = A$$

$$AI = \begin{bmatrix} 1 & 2 & 3 \\ 4 & 5 & 6 \end{bmatrix} \begin{bmatrix} 1 & 0 & 0 \\ 0 & 1 & 0 \\ 0 & 0 & 1 \end{bmatrix} = \begin{bmatrix} 1 & 2 & 3 \\ 4 & 5 & 6 \end{bmatrix} = A$$

As predicted, A remains unchanged when multiplied by I on either side.   ◆

In the next section we shall see how the laws of matrix algebra can be used to solve systems of linear equations.

4.2 Exercises   *In Exercises 1–6, compute the indicated dot product.*

1. $[2 \quad 3] \cdot \begin{bmatrix} 4 \\ 5 \end{bmatrix}$

2. $[3 \quad 4] \cdot \begin{bmatrix} 3 \\ 4 \end{bmatrix}$

3. $[3 \quad -4 \quad 2] \cdot \begin{bmatrix} 5 \\ 3 \\ -1 \end{bmatrix}$

4. $[1 \quad 3 \quad -2] \cdot \begin{bmatrix} 0 \\ 2 \\ 3 \end{bmatrix}$

5. $[.2 \quad .3 \quad .1] \cdot \begin{bmatrix} 10 \\ 40 \\ 20 \end{bmatrix}$

6. $\begin{bmatrix} \frac{1}{4} & \frac{1}{2} & \frac{1}{4} \end{bmatrix} \cdot \begin{bmatrix} 4 \\ 5 \\ 2 \end{bmatrix}$

7. Let A, B, and C be matrices of dimension 2×3, 3×2, and 2×4, respectively. Which of the following matrix products are defined? For those that are, what is the dimension of the product?

(a) AB (c) AC (e) A^tC

(b) ABC (d) BB (f) $ABAB$

8. Let A, B, and C be matrices of dimension 3×5, 5×1, and 1×3, respectively. Which of the following matrix products are defined? For those that are, what is the dimension of the product?

(a) CA (c) BC (e) AA^t

(b) BA (d) ABC (f) CAB

In Exercises 9–12, calculate the products AB *and* BA *for the given matrices.*

9. $A = \begin{bmatrix} 1 & 2 \\ 3 & 4 \end{bmatrix}, \quad B = \begin{bmatrix} 4 & 3 \\ 2 & 1 \end{bmatrix}$

10. $A = \begin{bmatrix} 1 & 2 & 3 & 2 \\ 4 & 3 & 2 & 1 \end{bmatrix}, \quad B = \begin{bmatrix} -1 & 2 \\ 0 & -1 \\ 1 & 0 \\ 2 & 1 \end{bmatrix}$

11. $A = [2 \quad 5], \quad B = \begin{bmatrix} 3 \\ 2 \end{bmatrix}$

12. $A = \begin{bmatrix} 1 & 0 & 1 \\ 0 & 2 & 0 \\ 1 & 0 & 1 \end{bmatrix}, \quad B = \begin{bmatrix} 1 & 0 & -1 \\ 0 & 2 & 0 \\ -1 & 0 & 1 \end{bmatrix}$

13. In Example 2 in the text we introduced the labor matrix

$$L = \begin{array}{c} \\ A \\ B \\ C \end{array} \begin{array}{c} \text{Construct} \quad \text{Finish} \\ \begin{bmatrix} 2 & 3 \\ 3 & 5 \\ 1 & 2 \end{bmatrix} \end{array} \quad \text{(work-hours/chair)}$$

Suppose the hourly pay-rates for construction and finishing in each of the two plants (X and Y) are as shown below.

$$R = \begin{array}{r} \\ \text{Construct} \\ \text{Finish} \end{array} \begin{array}{c} \text{X} \quad \text{Y} \\ \begin{bmatrix} 12 & 10 \\ 5 & 8 \end{bmatrix} \end{array} \quad \text{(dollars/hr)}$$

For example, finishers are paid \$8 per hr at Plant Y. Calculate the product matrix LR, and interpret its meaning.

14. The House of Coffee sells two different mixtures of Mocha, Java, and Santos beans. The composition of each mixture is shown in the matrix below.

$$C = \begin{array}{r} \\ \text{Mix A} \\ \text{Mix B} \end{array} \begin{array}{c} \text{Mocha} \quad \text{Java} \quad \text{Santos} \\ \begin{bmatrix} \frac{1}{4} & \frac{1}{2} & \frac{1}{4} \\ \frac{3}{5} & \frac{1}{5} & \frac{1}{5} \end{bmatrix} \end{array}$$

For example, each pound of Mixture A contains 1/4 lb Mocha, 1/2 lb Java, and 1/4 lb Santos. The production vector

$$\begin{array}{cc} \text{Mix A} & \text{Mix B} \end{array}$$
$$P = [\ 80 \qquad 75\,] \qquad \text{(pounds)}$$

indicates how many pounds of each mixture are to be produced. Calculate the product matrix PC, and interpret its meaning.

In Exercises 15–18, A^2 denotes AA and A^3 denotes AAA, where A is a square matrix. I denotes the identity matrix.

15. Show that the matrix

$$X = \begin{bmatrix} 4 & -5 \\ 2 & -3 \end{bmatrix}$$

satisfies the polynomial equation $X^2 - X - 2I = O$.

16. Show that the matrix

$$X = \begin{bmatrix} 1 & 0 & 1 \\ 0 & 1 & 0 \\ 1 & 0 & 1 \end{bmatrix}$$

satisfies the polynomial equation $X^3 - 3X^2 + 2X = O$.

17. Find an example to show that the identity

$$(A + B)^2 = A^2 + 2AB + B^2$$

does not hold for all square matrices A, B.

18. Find an example to show that the identity

$$A^2 - B^2 = (A + B)(A - B)$$

does not hold for all square matrices A, B.

In Exercises 19–22, verify the given law for the 2 × 2 case, by letting

$$A = \begin{bmatrix} 2 & 4 \\ -3 & 5 \end{bmatrix}, \quad B = \begin{bmatrix} 1 & -2 \\ 2 & 3 \end{bmatrix}, \quad C = \begin{bmatrix} 0 & 4 \\ -1 & 0 \end{bmatrix},$$

and showing that both sides of the equation represent the same matrix.

19. $A(BC) = (AB)C$ 20. $A(B + C) = AB + AC$

21. $(A + B)C = AC + BC$ 22. $IA = A = AI$

23. Letting A and B be arbitrary 2 × 2 matrices, as in Exercises 19–22 of Section 4.1, prove the following law for the transpose of a matrix product:

$$(AB)^t = B^t A^t$$

24. Verify the transpose law of Exercise 23, using the matrices A and B.

$$A = \begin{bmatrix} 2 & 3 \\ 4 & 0 \\ -1 & 2 \end{bmatrix}, \quad B = \begin{bmatrix} 1 & 2 \\ -1 & 1 \end{bmatrix}$$

25. In matrix algebra it is possible to have $AB = O$ even though $A \neq O$ and $B \neq O$. Verify this using the matrices A and B below.

$$A = \begin{bmatrix} 2 & 1 \\ 6 & 3 \end{bmatrix}, \qquad B = \begin{bmatrix} -1 & 2 \\ 2 & -4 \end{bmatrix}$$

26. In matrix algebra it is possible to have $AB = AC$ and $A \neq O$, and yet $B \neq C$. Verify this using the matrices A, B, and C.

$$A = \begin{bmatrix} 1 & 1 \\ -1 & -1 \end{bmatrix}, \qquad B = \begin{bmatrix} 0 & 1 \\ 2 & 3 \end{bmatrix}, \qquad C = \begin{bmatrix} 2 & 0 \\ 0 & 4 \end{bmatrix}$$

Computer Projects

A. Most versions of BASIC include matrix operations. Write a BASIC program to confirm your answer in Exercises 9–26.

B. Write a BASIC program to verify the following laws for several different choices of the matrices A, B, and C.

(a) $A(BC) = (AB)C$

(b) $A(B + C) = AB + AC$

(c) $(A + B)C = AC + BC$

(d) $(AB)^t = B^t A^t$

EXAM #1 3.1 - 4.2.

4.3 Matrix Solution of Linear Systems

It was shown earlier that an entire system of scalar equations can be summarized in a single matrix equation. This fact has important applications to systems of linear equations, of the type we studied in Section 3.1. By writing an $n \times n$ linear system in matrix form, we can apply the laws of matrix algebra—together with a new technique known as *matrix inversion*—to obtain the solution in a new way.

Linear Systems in Matrix Form The matrix-theoretic approach is illustrated in the following example involving a system of three equations in three unknowns.

Example 1 Consider the 3×3 system

$$\begin{aligned} x - 2y + 3z &= 1 \\ 2x - y + 4z &= 25 \\ -x + 3y - 4z &= 5. \end{aligned} \tag{1}$$

Discussion If we let

$$A = \begin{bmatrix} 1 & -2 & 3 \\ 2 & -1 & 4 \\ -1 & 3 & -4 \end{bmatrix}, \qquad X = \begin{bmatrix} x \\ y \\ z \end{bmatrix}, \qquad B = \begin{bmatrix} 1 \\ 25 \\ 5 \end{bmatrix}$$

then the system (1) is equivalent to the single matrix equation

$$AX = B \tag{2}$$

because (2) says that

$$AX = \begin{bmatrix} 1 & -2 & 3 \\ 2 & -1 & 4 \\ -1 & 3 & -4 \end{bmatrix} \begin{bmatrix} x \\ y \\ z \end{bmatrix} = \begin{bmatrix} x - 2y + 3z \\ 2x - y + 4z \\ -x + 3y - 4z \end{bmatrix} = \begin{bmatrix} 1 \\ 25 \\ 5 \end{bmatrix} = B$$

and these two 3×1 vectors can be equal only if all three corresponding entries are equal; that is, only if equations (1) hold.   ◇

Equation (2) is called the **matrix form of the linear system.** A is the **matrix of coefficients,** X is the **vector of unknowns,** and B is the **vector of right-hand constants.** Note that X and B are both written as *column* vectors.

Any $m \times n$ linear system can be expressed in matrix form as

$$AX = B,$$

where

A = matrix of coefficients $(m \times n)$
X = vector of unknowns $(n \times 1)$
B = vector of right-hand constants $(m \times 1)$.

The Equation $AX = B$ We saw in Section 3.1 that an $n \times n$ linear system normally has a unique solution. In other words, when A is square we expect the matrix equation $AX = B$ to have one and only one solution vector X. Before we discuss solving this equation by matrix methods, let us review briefly the Gauss-Jordan elimination procedure as applied to the linear system (1).

The first step is to write the system in tableau form.

$$\begin{array}{ccc|c} 1 & -2 & 3 & 1 \\ 2 & -1 & 4 & 25 \\ -1 & 3 & -4 & 5 \end{array}$$

Except for the omission of brackets, this is just the matrix $[A \mid B]$, which is called the **augmented matrix** of the system. It consists of the matrix A of coefficients, "augmented" by the vector B of right-hand constants.

The Gauss-Jordan procedure amounts to applying elementary row operations to $[A \mid B]$, until finally the identity matrix I appears on the left side of the line.

Example 1,
Continued

Use the Gauss-Jordan procedure on the following tableau.

$$\begin{array}{ccc|c} ① & -2 & 3 & 1 \\ 2 & -1 & 4 & 25 \\ -1 & 3 & -4 & 5 \end{array}$$

Solution   Add -2 times the first row to the second row.

Add the first row to the third row.

Annotation

$$
\begin{array}{ccc|c}
1 & -2 & 3 & 1 \\
0 & 3 & -2 & 23 \\
0 & \textcircled{1} & -1 & 6
\end{array}
\qquad
\begin{array}{l}
\\
R_2 = R_2 + (-2)R_1 \\
R_3 = R_3 + R_1
\end{array}
$$

Add 2 times the third row to the first row.

Add -3 times the third row to the second row.

$$
\begin{array}{ccc|c}
1 & 0 & 1 & 13 \\
0 & 0 & \textcircled{1} & 5 \\
0 & 1 & -1 & 6
\end{array}
\qquad
\begin{array}{l}
R_1 = R_1 + 2R_3 \\
R_2 = R_2 + (-3)R_3 \\
\\
\end{array}
$$

Add -1 times the second row to the first row.

Add the second row to the third row.

$$
\begin{array}{ccc|c}
1 & 0 & 0 & 8 \\
0 & 0 & 1 & 5 \\
0 & 1 & 0 & 11
\end{array}
\qquad
\begin{array}{l}
R_1 = R_1 + (-1)R_2 \\
\\
R_3 = R_3 + R_2
\end{array}
$$

Interchange the second and third rows.

$$
\begin{array}{ccc|c}
1 & 0 & 0 & 8 \\
0 & 1 & 0 & 11 \\
0 & 0 & 1 & 5
\end{array}
\qquad
\begin{array}{l}
\\
R_2 = R_3 \\
R_3 = R_2
\end{array}
$$

The solution is therefore $x = 8$, $y = 11$, $z = 5$, or in vector form,

$$
X = \begin{bmatrix} x \\ y \\ z \end{bmatrix} = \begin{bmatrix} 8 \\ 11 \\ 5 \end{bmatrix}. \qquad \diamondsuit
$$

Note that the final tableau has the form $[I \,|\, C]$, where I is the identity matrix and $X = C$ is the unique solution. More generally, we have the result:

Theorem 1

Consider the equation

$$AX = B \tag{3}$$

where A is a square matrix. If $[A \,|\, B]$ can be transformed into $[I \,|\, C]$ by elementary row operations, then (3) has the unique solution $X = C$.

In fact, it is not difficult to show that Theorem 1 holds not only when X and B are column vectors, but more generally whenever A is $n \times n$ and X and B are both $n \times k$ matrices. The next example illustrates this for $n = k = 2$.

Example 2   Solve the matrix equation

$$\begin{bmatrix} 1 & 2 \\ 3 & 2 \end{bmatrix} \begin{bmatrix} x & y \\ z & w \end{bmatrix} = \begin{bmatrix} 7 & 4 \\ 9 & 8 \end{bmatrix}.$$

Solution   This has the form $AX = B$, where X is an unknown 2×2 matrix. Accordingly, we form the augmented matrix $[A \mid B]$ and try to reach $[I \mid C]$ by elementary row operations:*

$$A \quad \begin{array}{cc|cc} ① & 2 & 7 & 4 \\ 3 & 2 & 9 & 8 \end{array} \quad B$$

Add -3 times the first row to the second row.

$$\begin{array}{cc|cc} 1 & 2 & 7 & 4 \\ 0 & -4 & -12 & -4 \end{array}$$

Multiply the second row by $-1/4$.

$$\begin{array}{cc|cc} 1 & 2 & 7 & 4 \\ 0 & ① & 3 & 1 \end{array}$$

Add -2 times the second row to the first row.

$$I \quad \begin{array}{cc|cc} 1 & 0 & 1 & 2 \\ 0 & 1 & 3 & 1 \end{array} \quad C$$

Since the identity appears on the left, the solution appears on the right.

$$X = \begin{bmatrix} x & y \\ z & w \end{bmatrix} = \begin{bmatrix} 1 & 2 \\ 3 & 1 \end{bmatrix}$$

Check $$AX = \begin{bmatrix} 1 & 2 \\ 3 & 2 \end{bmatrix} \begin{bmatrix} 1 & 2 \\ 3 & 1 \end{bmatrix} = \begin{bmatrix} 7 & 4 \\ 9 & 8 \end{bmatrix} = B \quad \blacklozenge$$

Matrix Inversion   Let us now consider the matrix equation $AX = B$ from the point of view of matrix algebra. Since matrices behave somewhat like numbers, we should be able to "divide" both sides of the equation by the matrix A, to obtain the solution as $X = B/A$. We must proceed carefully, though, because matrix division is not defined in general, and even matrix multiplication does not obey all the rules of ordinary algebra, as we have seen.

The first step is to define the "reciprocal" of a matrix A. Since the identity matrix I plays the role of the number 1 in matrix algebra, the definition is as follows:

---

*Through the remainder of this section formal annotations are omitted.

Definition

Let A be a square matrix. The **inverse of A**, written A^{-1}, is a square matrix satisfying the conditions

$$AA^{-1} = I \quad \text{and} \quad A^{-1}A = I.$$

If such a matrix exists we say that A is **invertible.**

As we shall see (Example 4), there are noninvertible square matrices. However, it is an important fact that *the inverse of a square matrix is unique when it exists.* That is, no matrix A can have two different inverses (see Exercise 23).

Since the inverse of A is a solution of the matrix equation $AX = I$, we can find it by the method of Theorem 1. This procedure is called **matrix inversion.**

Matrix Inversion

To find the inverse of an $n \times n$ matrix A:

1. Form the augmented matrix $[A|I]$, where I is the $n \times n$ identity matrix.
2. Apply elementary row operations until I appears to the left of the line. The inverse A^{-1} then appears to the right of the line.

To illustrate the method, let us apply it to the coefficient matrix A in Example 1.

Example 1, Revisited   Invert the matrix

$$A = \begin{bmatrix} 1 & -2 & 3 \\ 2 & -1 & 4 \\ -1 & 3 & -4 \end{bmatrix}.$$

Solution   We form the augmented matrix $[A \mid I]$ and try to reach $[I \mid A^{-1}]$ by elementary row operations.

$$A \quad \begin{bmatrix} ① & -2 & 3 & | & 1 & 0 & 0 \\ 2 & -1 & 4 & | & 0 & 1 & 0 \\ -1 & 3 & -4 & | & 0 & 0 & 1 \end{bmatrix} \quad I$$

Add -2 times the first row to the second row.

Add the first row to the third row.

$$\begin{bmatrix} 1 & -2 & 3 & | & 1 & 0 & 0 \\ 0 & 3 & -2 & | & -2 & 1 & 0 \\ 0 & ① & -1 & | & 1 & 0 & 1 \end{bmatrix}$$

Add 2 times the third row to the first row.

Add -3 times the third row to the second row.

$$\begin{array}{ccc|ccc} 1 & 0 & 1 & 3 & 0 & 2 \\ 0 & 0 & ① & -5 & 1 & -3 \\ 0 & 1 & -1 & 1 & 0 & 1 \end{array}$$

Add -1 times the second row to the first row.

Add the second row to the third row.

$$\begin{array}{ccc|ccc} 1 & 0 & 0 & 8 & -1 & 5 \\ 0 & 0 & 1 & -5 & 1 & -3 \\ 0 & 1 & 0 & -4 & 1 & -2 \end{array}$$

Interchange the second and third rows.

$$I \quad \begin{array}{ccc|ccc} 1 & 0 & 0 & 8 & -1 & 5 \\ 0 & 1 & 0 & -4 & 1 & -2 \\ 0 & 0 & 1 & -5 & 1 & -3 \end{array} \quad A^{-1}$$

Since the identity now appears on the left side of the line, the inverse appears on the right.

$$A^{-1} = \begin{bmatrix} 8 & -1 & 5 \\ -4 & 1 & -2 \\ -5 & 1 & -3 \end{bmatrix}$$

Check:

$$AA^{-1} = \begin{bmatrix} 1 & -2 & 3 \\ 2 & -1 & 4 \\ -1 & 3 & -4 \end{bmatrix} \begin{bmatrix} 8 & -1 & 5 \\ -4 & 1 & -2 \\ -5 & 1 & -3 \end{bmatrix} = \begin{bmatrix} 1 & 0 & 0 \\ 0 & 1 & 0 \\ 0 & 0 & 1 \end{bmatrix} = I \quad \blacklozenge$$

It can be shown that for a square matrix A, the equation $AX = I$ implies $XA = I$, so that only the "one-sided" check of the inverse is necessary.

Matrix Solution of $AX = B$ If A is invertible, the equation $AX = B$ can be solved very simply. All we have to do is multiply both sides on the left by A^{-1} and use some laws of matrix algebra.

$$AX = B$$
$$A^{-1}(AX) = A^{-1}B$$
$$(A^{-1}A)X = A^{-1}B \qquad \text{(by M9)}$$
$$IX = A^{-1}B \qquad \text{(definition of inverse)}$$
$$X = A^{-1}B \qquad \text{(by M13)}$$

For the system $AX = B$ of Example 1, this gives

$$X = A^{-1}B = \begin{bmatrix} 8 & -1 & 5 \\ -4 & 1 & -2 \\ -5 & 1 & -3 \end{bmatrix} \begin{bmatrix} 1 \\ 25 \\ 5 \end{bmatrix} = \begin{bmatrix} 8 \\ 11 \\ 5 \end{bmatrix}$$

which agrees with the result we obtained by Gauss-Jordan elimination.

Note that *only a single matrix multiplication is required to solve AX = B*, once the inverse of A has been calculated.

Matrix Solution of $AX = B$

If A is an invertible matrix, then the equation $AX = B$ has the unique solution $X = A^{-1}B$.

Question: Why would anyone use matrix inversion to solve $AX = B$, when Gauss-Jordan elimination requires fewer steps?

Answer: It is true that the Gauss-Jordan method is more efficient, if we are only interested in solving a *single* system $AX = B$. But many situations call for the solution of *several* systems $AX = B$, each having the same coefficient matrix A but different constants B. In such cases, matrix inversion saves a good deal of computation.

Example 3 The Allegheny Mining Company produces two grades of ore. Grade A contains 25% copper and 50% iron by weight, while Grade B contains 30% copper and 20% iron. How many tons of each grade should be sent to the refinery to fill the following orders exactly, without wasting any copper or iron?
(a) 1000 tons of copper, 1000 tons of iron
(b) 1000 tons of copper, 1500 tons of iron
(c) 2000 tons of copper, 1000 tons of iron

Solution If the company sends x tons of Grade A ore and y tons of Grade B ore, the amount of copper and iron in the shipment is given by

$$\text{copper} = .25x + .30y \qquad \text{(tons)}$$
$$\text{iron} = .50x + .20y. \qquad \text{(tons)}$$

To fill an order for b_1 tons of copper and b_2 tons of iron, x and y must satisfy the equations

$$.25x + .30y = b_1$$
$$.50x + .20y = b_2$$

or, in matrix form,

$$AX = \begin{bmatrix} .25 & .30 \\ .50 & .20 \end{bmatrix} \begin{bmatrix} x \\ y \end{bmatrix} = \begin{bmatrix} b_1 \\ b_2 \end{bmatrix} = B.$$

Note that *the matrix A stays the same for all three orders; only the vector B changes*. While we could solve all three systems by Gauss-Jordan elimination, it makes more sense to invert the matrix A at the outset. The computation runs as follows:

$$A \quad \begin{array}{cc|cc} \fbox{.25} & .30 & 1 & 0 \\ .50 & .20 & 0 & 1 \end{array} \quad I$$

Multiply the first row by 4.

$$\begin{array}{cc|cc} \fbox{1} & 1.2 & 4 & 0 \\ .5 & .2 & 0 & 1 \end{array}$$

Add $-.5$ times the first row to the second row.

$$\begin{array}{cc|cc} 1 & 1.2 & 4 & 0 \\ 0 & \fbox{$-.4$} & -2 & 1 \end{array}$$

Multiply the second row by -2.5.

$$\begin{array}{cc|cc} 1 & 1.2 & 4 & 0 \\ 0 & \fbox{1} & 5 & -2.5 \end{array}$$

Add -1.2 times the second row to the first row.

$$I \quad \begin{array}{cc|cc} 1 & 0 & -2 & 3 \\ 0 & 1 & 5 & -2.5 \end{array} \quad A^{-1}$$

The inverse is therefore

$$A^{-1} = \begin{bmatrix} -2 & 3 \\ 5 & -2.5 \end{bmatrix}.$$

Substituting the appropriate vector B in the equation $X = A^{-1}B$, we now obtain each of the three solutions.

(a) $X = A^{-1}B = \begin{bmatrix} -2 & 3 \\ 5 & -2.5 \end{bmatrix} \begin{bmatrix} 1000 \\ 1000 \end{bmatrix} = \begin{bmatrix} 1000 \\ 2500 \end{bmatrix} \begin{array}{l} = x \\ = y \end{array}$

(b) $X = A^{-1}B = \begin{bmatrix} -2 & 3 \\ 5 & -2.5 \end{bmatrix} \begin{bmatrix} 1000 \\ 1500 \end{bmatrix} = \begin{bmatrix} 2500 \\ 1250 \end{bmatrix} \begin{array}{l} = x \\ = y \end{array}$

(c) $X = A^{-1}B = \begin{bmatrix} -2 & 3 \\ 5 & -2.5 \end{bmatrix} \begin{bmatrix} 2000 \\ 1000 \end{bmatrix} = \begin{bmatrix} -1000 \\ 7500 \end{bmatrix} \begin{array}{l} = x \\ = y \end{array}$

Note that the solution in (c) calls for a negative amount of Grade A ore, which means that this order cannot be filled exactly. That is, there is no way to provide exactly 2000 tons of copper and 1000 tons of iron. ◆

Noninvertible Matrices   The method of matrix inversion can only be used to solve $n \times n$ linear systems, because only a square matrix can have an inverse. But we may even encounter problems in the $n \times n$ case if the coefficient matrix A is noninvertible. When we try to invert such a matrix, our procedure will eventually produce a row in which *all the entries to the left of the line are zero*. This tells us that the matrix has no inverse.

Example 4 Invert the matrix

$$A = \begin{bmatrix} 1 & -1 & 0 \\ 1 & 0 & -1 \\ 0 & 1 & -1 \end{bmatrix}.$$

Solution We form $[A \mid I]$, as usual, and try to obtain an identity matrix I on the left side.

$$A \quad \begin{array}{ccc|ccc} ① & -1 & 0 & 1 & 0 & 0 \\ 1 & 0 & -1 & 0 & 1 & 0 \\ 0 & 1 & -1 & 0 & 0 & 1 \end{array} \quad I$$

Add -1 times the first row to the second row.

$$\begin{array}{ccc|ccc} 1 & -1 & 0 & 1 & 0 & 0 \\ 0 & ① & -1 & -1 & 1 & 0 \\ 0 & 1 & -1 & 0 & 0 & 1 \end{array}$$

Add the second row to the first row.

Add -1 times the second row to the third row.

$$\begin{array}{ccc|ccc} 1 & 0 & -1 & 0 & 1 & 0 \\ 0 & 1 & -1 & -1 & 1 & 0 \\ 0 & 0 & 0 & 1 & -1 & 1 \end{array}$$

We cannot continue, because the third row of the left-hand matrix consists entirely of zeros. The matrix A is therefore noninvertible.* ◆

The above procedure for matrix inversion can be carried out on a computer, and for the large (100×100, for example) matrices that arise in practice the computer is essential. But even with the aid of a computer, the task of matrix inversion can present some annoying problems. For example, the matrices

$$A_1 = \begin{bmatrix} 1 & 3 \\ 2 & 6.01 \end{bmatrix}, \quad A_2 = \begin{bmatrix} 1 & 3 \\ 2 & 6.02 \end{bmatrix}$$

are almost identical, differing only by .01 in a single entry. But their inverses,

$$A_1^{-1} = \begin{bmatrix} 601 & -300 \\ -200 & 100 \end{bmatrix}, \quad A_2^{-1} = \begin{bmatrix} 301 & -150 \\ -100 & 50 \end{bmatrix},$$

---

*In such cases (when A is square but noninvertible), it can be shown that the equation $AX = B$ will have either no solution or infinitely many, depending on the vector B.

are strikingly different! Matrices like A_1 and A_2 are called **ill-conditioned.**
Normally we expect a small change in the entries of A to produce a corresponding small change in the entries of A^{-1}, but for ill-conditioned matrices
this is not the case. Since computer calculations always involve a roundoff
error, such matrices present a serious problem in numerical linear algebra.

4.3 Exercises

In Exercises 1–4, express the linear system in matrix form (AX = B).

1. $x + y = 4$
   $x - y = 2$

2. $x + 2y + 3z = 45$
   $3x + y + 2z = 90$
   $2x + 3y + z = 45$

3. $2x + 2y + z + w = -1$
   $x + 2y + 2z + w = 0$
   $x + y + 2z + 2w = 1$

4. $x + y \quad\quad = 0$
   $\quad\ y + z = 0$
   $x \quad\quad + z = 2$
   $x + y + z = 1$

*In Exercises 5–8 a matrix equation AX = B is given, where A is square. Solve the
equation by transforming [A | B] into [I | C] and using Theorem 1. Check your result
by matrix multiplication.*

5. $\begin{bmatrix} 1 & 2 \\ 3 & 4 \end{bmatrix} \begin{bmatrix} x \\ y \end{bmatrix} = \begin{bmatrix} 6 \\ 8 \end{bmatrix}$

6. $\begin{bmatrix} 1 & 2 \\ 2 & 1 \end{bmatrix} \begin{bmatrix} x & y \\ z & w \end{bmatrix} = \begin{bmatrix} 1 & -1 \\ -1 & 1 \end{bmatrix}$

7. $\begin{bmatrix} 0 & 1 & 1 \\ 1 & 0 & 1 \\ 1 & 1 & 0 \end{bmatrix} \begin{bmatrix} x \\ y \\ z \end{bmatrix} = \begin{bmatrix} 50 \\ 40 \\ 30 \end{bmatrix}$

8. $\begin{bmatrix} 3 & 0 & 0 \\ 2 & 3 & 0 \\ 1 & 2 & 3 \end{bmatrix} \begin{bmatrix} x & y \\ z & w \\ u & v \end{bmatrix} = \begin{bmatrix} 3 & 6 \\ 2 & 7 \\ 7 & 4 \end{bmatrix}$

*In Exercises 9–16, find the inverse of the given matrix, if it exists. Check your result
by matrix multiplication.*

9. $\begin{bmatrix} 3 & 5 \\ 1 & 2 \end{bmatrix}$

10. $\begin{bmatrix} 2 & 3 \\ 4 & 5 \end{bmatrix}$

11. $\begin{bmatrix} 1 & 3 & 2 \\ 2 & 7 & 2 \\ -1 & -4 & 1 \end{bmatrix}$

12. $\begin{bmatrix} -1 & 1 & 0 \\ 1 & -1 & 1 \\ 0 & 1 & -1 \end{bmatrix}$

13. $\begin{bmatrix} 1 & 3 & 2 \\ 2 & 6 & 9 \\ 3 & 9 & 8 \end{bmatrix}$

14. $\begin{bmatrix} 0 & 2 & -1 \\ 1 & 0 & 3 \\ 2 & 1 & 5 \end{bmatrix}$

15. $\begin{bmatrix} 0 & 1 & 0 & 0 \\ 1 & 0 & 1 & 0 \\ 0 & 1 & 0 & 1 \\ 0 & 0 & 1 & 0 \end{bmatrix}$

16. $\begin{bmatrix} 1 & 1 & 0 & 0 \\ 1 & 0 & 1 & 0 \\ 0 & 1 & 0 & 1 \\ 0 & 0 & 1 & 1 \end{bmatrix}$

Use matrix inversion to solve the following linear systems:

17. $.9x - .2y = 50$
$-.3x + .9y = 75$

18. $2x \quad\quad - z = 1$
$2x - y \quad\quad = 3$
$2y - z = 5$

19. Use matrix inversion to solve the linear system

$$2x + 5y = b_1$$
$$3x + 7y = b_2$$

where b_1 and b_2 take the following values:

(a) $b_1 = 1, b_2 = 1$ \qquad (b) $b_1 = 1, b_2 = 0$ \qquad (c) $b_1 = 0, b_2 = 1$

20. Use matrix inversion to solve the linear system

$$-x + y + z = b_1$$
$$x - y + z = b_2$$
$$x + y - z = b_3$$

where b_1, b_2 and b_3 take the following values:

(a) $b_1 = 1, b_2 = 0,$ \qquad (b) $b_1 = 1, b_2 = 1,$ \qquad (c) $b_1 = 1, b_2 = 1,$
$\quad b_3 = 0$ \qquad\qquad\qquad $b_3 = 0$ \qquad\qquad\qquad $b_3 = 1$

21. Each ounce of Food A contains 2 gm of carbohydrate and 2 gm of protein, while each ounce of Food B contains 4 gm of carbohydrate and 2 gm of protein. By combining x ounces of Food A and y ounces of Food B it is desired to obtain a total of b_1 gm of carbohydrate and b_2 gm of protein.

(a) Express the conditions as a matrix equation $AX = B$.

(b) Using matrix inversion, solve for x and y in terms of b_1 and b_2.

22. An aluminum manufacturer makes rolls of aluminum foil in a standard 10-ft width. These 10-ft rolls are then cut into smaller widths of 12 and 18 inches for sale to retail distributors. Each 10-ft roll is cut into either four 12-in and four 18-in rolls (Pattern 1), or else seven 12-in and two 18-in rolls (Pattern 2). By cutting x_1 rolls according to Pattern 1 and x_2 rolls according to Pattern 2, it is desired to obtain a total of exactly b_1 12-in rolls and b_2 18-in rolls.

(a) Express the conditions as a matrix equation $AX = B$.

(b) Using matrix inversion, solve for x_1 and x_2 in terms of b_1 and b_2.

23. Prove that no matrix A can have *two* inverses by showing that if X and Y are matrices such that $AX = I$, $XA = I$, $AY = I$, and $YA = I$, then $X = Y$. (*Hint:* Consider the product $(XA)Y = X(AY)$.)

24. Prove: If A is invertible then the matrix equation $XA = B$ has the unique solution $X = BA^{-1}$.

25. Prove: If A is invertible and $AB = O$, then $B = O$. (See Exercise 25 in Section 4.2.)

26. Prove: If A is invertible and $AB = AC$, then $B = C$. (See Exercise 26 in Section 4.2.)

In Exercises 27–30, verify the given law for the 2 × 2 case, by letting

$$A = \begin{bmatrix} 2 & 1 \\ -1 & -1 \end{bmatrix}, \qquad B = \begin{bmatrix} 0 & 1 \\ -1 & 2 \end{bmatrix}, \qquad c = 2$$

and showing that both sides of the equation represent the same matrix.

27. $(A^{-1})^{-1} = A$ 28. $(AB)^{-1} = B^{-1}A^{-1}$

29. $(A^t)^{-1} = (A^{-1})^t$ 30. $(cA)^{-1} = (1/c)A^{-1}$

Inverse of a 2 × 2 Matrix   *Let* $A = \begin{bmatrix} a & b \\ c & d \end{bmatrix}$ *and let* $\Delta = ad - bc$.

If $\Delta \neq 0$ *then* $A^{-1} = \begin{bmatrix} d/\Delta & -b/\Delta \\ -c/\Delta & a/\Delta \end{bmatrix}$.

In Exercises 31 and 32, use this formula to find the inverse of the given 2 × 2 matrix.

31. $\begin{bmatrix} 3 & 5 \\ 1 & 2 \end{bmatrix}$ 32. $\begin{bmatrix} 1 & 2 \\ 3 & 4 \end{bmatrix}$

Computer Projects

A. The PIVOT program performs a pivot transformation on a designated nonzero entry in an augmented matrix. This program can be used in Exercises 5–16.

B. Most versions of BASIC include matrix operations. Write a BASIC program to confirm your answer in Exercises 17–22 and 27–30.

C. Write a BASIC program to verify the following laws for several different choices of the matrices A and B and the scalar c.

   (a) $(A^{-1})^{-1} = A$ (b) $(AB)^{-1} = B^{-1}A^{-1}$

   (c) $(A^t)^{-1} = (A^{-1})^t$ (d) $(cA)^{-1} = (1/c)A^{-1}$

D. Write a BASIC program to solve an arbitrary 5 × 5 linear system by matrix inversion. Use your program to solve the following system:

$$\begin{aligned} 7x + 3y - 5z + u + 4v &= 120 \\ 10x + 2y + 4z - 3u + v &= 225 \\ 9x - 8y + z + 5u + 3v &= 345 \\ 6x + y + 8z + 9u + 5v &= 830 \\ -2x + 7y + 3z + 4u - v &= 325 \end{aligned}$$

4.4 The Leontief Input-Output Model

The technique of matrix inversion gives us a powerful new tool for the solution of $n \times n$ linear systems. This method is particularly useful when we have a number of large systems to solve, all of which share the same coefficient matrix. In this section we encounter an important application of this type.

Introduction   The economies of modern industrial nations are extremely complex. When we break such systems down into their components we find hundreds or even thousands of different industries, each supplying the others with goods and services needed in the production process. The coal industry, for example, supplies the steel industry with one of its main raw materials, but at the same time steel—in the form of machinery, vehicles, and so on—is needed to mine and transport coal. It is this constant flow of goods between industries that characterizes an interacting economy, and at the same time constitutes one of the main obstacles to effective economic planning. If we set arbitrary production goals for each industry we may wind up with shortages, as some eastern European countries have painfully learned; and a price increase in one key industry may send a ripple through the entire economy.

Is it possible to describe mathematically the interaction of an economic system? This question occurred to Wassily Leontief, a Russian-born economist, while he was a student in Berlin in the 1920s. Using matrix algebra, Leontief developed a very elegant model for the workings of an economy, known as *input-output theory*.* After emigrating to the U.S. in 1931 he concentrated on the task of putting his theoretical ideas into practice. By the late 1940s, with the help of the newly developed computer, Leontief was able to distill the parameters he needed from a mountain of government statistics. His model proved to be remarkably accurate in predicting the needs of the postwar American economy, and today input-output theory is used to forecast the production needs of large corporations, as well as advanced and developing nations. The theory is also used to set prices for various commodities, to predict the effects of price changes, and to analyze the impact of production on the environment. In recognition of the widespread application of his ideas, Leontief was awarded the Nobel Prize in Economics in 1973.

Input-Output Theory   We begin by dividing an economy into a certain number of **producing sectors** $P_1, P_2, \ldots, P_n$, each of which is thought of as producing a single good, commodity, or service. In his famous study of the 1947 U.S. economy, Leontief identified 500 industries, which he combined into 42 producing sectors such as "textile mill products," "petroleum and coal products," "electrical machinery," and so on.†

Once the producing sectors are determined, we must decide on an appropriate unit of measurement for the output of each sector. These units may be either *physical* or *monetary;* for example, one unit of steel might represent 10,000 tons (physical) or $1 million worth (monetary). The advantage of monetary units is that the output of different sectors can be directly compared, but physical units are sometimes essential, as in discussions of pric-

---

*The theory is set forth in Leontief's work, *The Structure of American Economy, 1919–1939,* 2nd ed. (1951; reprint ed., Armonk, N.Y.: M. E. Sharpe, 1976).

†See Wassily W. Leontief, "Input-Output Economics," *Scientific American* 185, no. 4 (October 1951), pp. 15–21.

ing. For our present discussion it does not matter which type of unit we adopt, so we shall use the term *unit* without specifying whether it is physical or monetary.

The central element of Leontief's theory is the input-output ratio.

Definition

If the sectors $P_1, \ldots, P_n$ produce respective goods $G_1, \ldots, G_n$ then the **input-output ratio** a_{ij} specifies the amount of G_i needed to produce one unit of G_j.

The $n \times n$ matrix $A = [a_{ij}]$ whose entries are the input-output ratios is called the **input-output matrix** (or **technology matrix**) for the economy.

The meaning of the input-output matrix is illustrated in the next example.

Example 1   Consider a highly simplified economy with three producing sectors: agriculture (P_1), manufacturing (P_2), and energy (P_3). Suppose the input-output matrix is as follows:

$$
\begin{array}{c}
\text{To produce one unit of} \\
\begin{array}{ccc}
\text{Agr} & \text{Mfg} & \text{Engy}
\end{array}
\end{array}
$$

$$
A = \begin{bmatrix} .2 & .1 & .3 \\ .4 & .2 & .1 \\ .3 & .5 & .5 \end{bmatrix} \begin{array}{l} \text{Agr} \\ \text{Mfg} \\ \text{Engy} \end{array} \quad \begin{array}{c} \text{Input} \\ \text{units} \\ \text{needed} \end{array}
$$

Discussion   The entries in each *column* indicate how many input units are needed from each of the three sectors in order to produce one unit of output. For example, the second column (manufacturing) shows that we need .1, .2, and .5 units of input from the agriculture, manufacturing, and energy sectors, respectively, to produce one unit of manufactured goods. If the units are monetary, this means that for each $1 worth of manufactured goods produced, the manufacturing sector consumes 10¢ worth of agricultural products, 20¢ worth of (other) manufactured goods, and 50¢ worth of energy.   ◇

In practice, these input-output ratios could be determined by looking at how much each sector buys from the others. For instance, if the manufacturing sector bought $45 billion worth of energy in a given year and produced $90 billion worth of goods, then each $1 worth of manufactured goods required $45/90 = .5 = 50¢$ worth of energy. This is essentially how Leontief calculated his input-output ratios.

It is a fundamental assumption of the input-output model that *these ratios remain the same regardless of the levels of production*. Thus, if one unit of manufactured output requires .5 units of energy, then 1000 units of output should require 500 units of energy, and more generally x units of output should require $.5x$ units of energy. This assumption must be carefully investigated in real applications. For example, for large values of x there may be a more energy-efficient manufacturing process available. In practice, however, the assumption has been found to hold fairly true, provided that production levels stay within certain limits. Barring any sudden technological changes, therefore—such as the discovery of a new, cheap energy source—the input-output matrix will be a fairly stable feature of the economy.

Intensity and Surplus Returning to Example 1, let us represent the total (gross) output of the three sectors by x_1, x_2, x_3, respectively.

x_1 = total units of G_1 produced (agriculture)
x_2 = total units of G_2 produced (manufactured goods)
x_3 = total units of G_3 produced (energy)

Then the flow of goods in the economy can be represented graphically by means of the diagram in Figure 1.

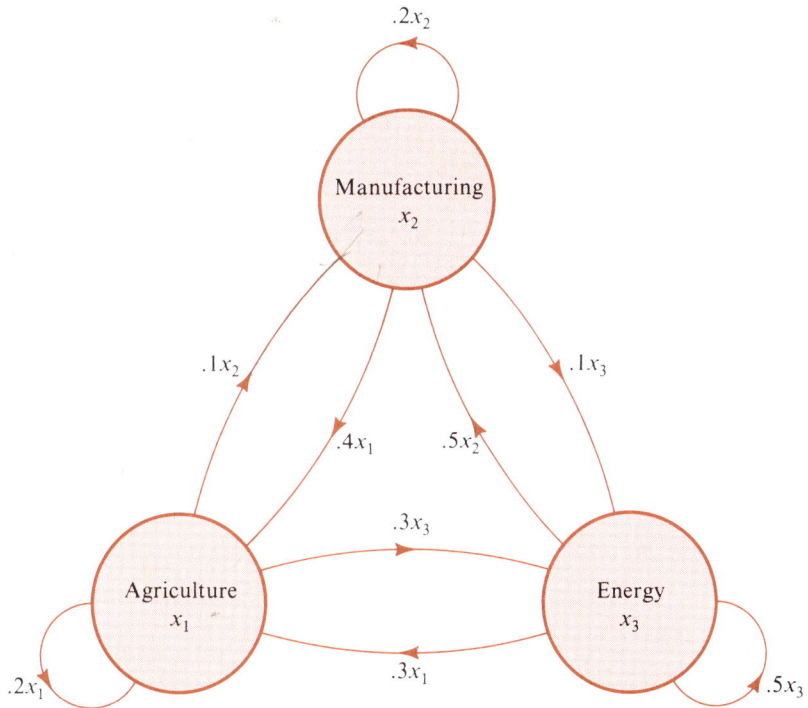

Figure 1

In addition to the three sectors and their gross outputs, the diagram shows the number of units of output sent from each sector to the others. For example, since we are producing x_1 units of agriculture, the input-output matrix shows that the agriculture sector will need as input $.2x_1$ units from itself, $.4x_1$ units from manufacturing, and $.3x_1$ units from energy, as indicated in Figure 1. The other paths are labeled similarly, corresponding to the input-output matrix A.

Looking at the three outgoing paths from the agriculture sector, we see that the amount of agricultural output *used up internally* (the number of units needed for the production process itself) is given by the following equation:

$$u_1 = .2x_1 + .1x_2 + .3x_3$$

There are similar equations for the other two sectors.

$$u_2 = .4x_1 + .2x_2 + .1x_3$$
$$u_3 = .3x_1 + .5x_2 + .5x_3$$

In matrix terms these three equations say that

$$\begin{bmatrix} u_1 \\ u_2 \\ u_3 \end{bmatrix} = \begin{bmatrix} .2 & .1 & .3 \\ .4 & .2 & .1 \\ .3 & .5 & .5 \end{bmatrix} \begin{bmatrix} x_1 \\ x_2 \\ x_3 \end{bmatrix}$$

or $$U = AX \qquad (1)$$

where

$$U = \begin{bmatrix} u_1 \\ u_2 \\ u_3 \end{bmatrix} = \textbf{internal consumption} \text{ vector}$$

$$X = \begin{bmatrix} x_1 \\ x_2 \\ x_3 \end{bmatrix} = \textbf{intensity} \text{ or } \textbf{gross production} \text{ vector}$$

$$A = \begin{bmatrix} .2 & .1 & .3 \\ .4 & .2 & .1 \\ .3 & .5 & .5 \end{bmatrix} = \text{input-output matrix.}$$

If we are given the gross production X, we can calculate $U = AX$ and then find the surplus, or net production, by subtracting.

$$S = X - U = \textbf{surplus} \text{ or } \textbf{net production} \text{ vector} \qquad (2)$$

The entries of S represent the amount of output available for consumer demand or export.

Example 1, Continued

Suppose the gross production of the economy is 800 units of agriculture, 900 units of manufactured goods and 1500 units of energy. What will the net production be?

Solution

The intensity vector is given by

$$X = \begin{bmatrix} x_1 \\ x_2 \\ x_3 \end{bmatrix} = \begin{bmatrix} 800 \\ 900 \\ 1500 \end{bmatrix}.$$

Substitute this into formula (1).

$$U = AX = \begin{bmatrix} .2 & .1 & .3 \\ .4 & .2 & .1 \\ .3 & .5 & .5 \end{bmatrix} \begin{bmatrix} 800 \\ 900 \\ 1500 \end{bmatrix} = \begin{bmatrix} 700 \\ 650 \\ 1440 \end{bmatrix}$$

Hence, by (2),

$$\text{net production} \quad S = X - U = \begin{bmatrix} 800 \\ 900 \\ 1500 \end{bmatrix} - \begin{bmatrix} 700 \\ 650 \\ 1440 \end{bmatrix} = \begin{bmatrix} 100 \\ 250 \\ 60 \end{bmatrix} \begin{matrix} = s_1 \\ = s_2 \\ = s_3. \end{matrix}$$

In other words, at the given intensity the economy will be able to keep itself running and have 100 units of agriculture, 250 units of manufactured goods and 60 units of energy left over.* ◇

Meeting a Given Demand We can also turn the problem around. Suppose we want the economy to produce just enough (net, or surplus) to meet a given **demand vector**.

$$D = \begin{bmatrix} d_1 \\ d_2 \\ d_3 \end{bmatrix} \quad \begin{matrix} \text{units of } G_1 & \text{(agriculture)} \\ \text{units of } G_2 & \text{(manufactured goods)} \\ \text{units of } G_3 & \text{(energy)} \end{matrix}$$

At what intensity X must the economy operate?
To answer this, we first observe that by (1) and (2),

$$\text{surplus} \quad S = X - U$$
$$= X - AX.$$

Setting the surplus (net production) exactly equal to the demand, we obtain the matrix equation

$$X - AX = D$$

in which A and D are known and X is to be determined. Using some tricks from matrix algebra we can combine the X-terms.

$$X - AX = D$$
$$IX - AX = D \quad \text{(by M13)}$$
$$(I - A)X = D \quad \text{(by M11)}$$

If we multiply both sides on the left by $(I - A)^{-1}$, we arrive at the solution

$$X = (I - A)^{-1}D \tag{3}$$

provided, of course, that the matrix $I - A$ is invertible.

_____

*Fortunately, these values are all positive. If some "surplus" came out negative it would mean that the economy was consuming more of that commodity than it produced. The missing input (called *primary* input) would then have to be imported from a source outside the economy.

Example 1,
Continued

At what intensity must the economy operate to meet a demand for 100 units of agriculture, 200 units of manufactured goods and 150 units of energy?

Solution

The demand vector is given by

$$D = \begin{bmatrix} 100 \\ 200 \\ 150 \end{bmatrix}.$$

To find the intensity vector X using formula (3) we first compute the matrix $I - A$.

$$I - A = \begin{bmatrix} 1 & 0 & 0 \\ 0 & 1 & 0 \\ 0 & 0 & 1 \end{bmatrix} - \begin{bmatrix} .2 & .1 & .3 \\ .4 & .2 & .1 \\ .3 & .5 & .5 \end{bmatrix} = \begin{bmatrix} .8 & -.1 & -.3 \\ -.4 & .8 & -.1 \\ -.3 & -.5 & .5 \end{bmatrix}$$

Inverting this matrix by the method of Section 4.3, we obtain

$$(I - A)^{-1} = \begin{bmatrix} 2.8 & 1.6 & 2 \\ 1.84 & 2.48 & 1.6 \\ 3.52 & 3.44 & 4.8 \end{bmatrix}.$$

Formula (3) now gives

$$X = (I - A)^{-1}D = \begin{bmatrix} 2.8 & 1.6 & 2 \\ 1.84 & 2.48 & 1.6 \\ 3.52 & 3.44 & 4.8 \end{bmatrix} \begin{bmatrix} 100 \\ 200 \\ 150 \end{bmatrix} = \begin{bmatrix} 900 \\ 920 \\ 1760 \end{bmatrix} \begin{matrix} = x_1 \\ = x_2 \\ = x_3. \end{matrix}$$

That is, the three sectors must produce 900, 920, and 1760 units of output, respectively, in order to meet the specified demands. If we put these values for x_1, x_2, and x_3 in the diagram in Figure 1, we can see exactly how much output is flowing from sector to sector in Figure 2. For instance, the agriculture sector is producing 900 units of output (gross), of which 180 units will be used up in the agriculture sector, 92 units in manufacturing, and 528 units in energy production. This leaves a surplus of $900 - 180 - 92 - 528 = 100$ units, which is exactly the given demand. ◆

The derivation of equation (3) can be generalized to any n-sector economy as follows:

Theorem 1

Suppose an economy has input-output matrix A. Then in order to meet a given demand vector D, the economy must operate at intensity X, where

$$X = (I - A)^{-1}D.$$

Matrix inversion makes good practical sense here. While we could in principle solve the system $(I - A)X = D$ directly by Gauss-Jordan elimination, we would have to go through the whole procedure again for every new demand vector D. Since the number of producing sectors, and hence the ma-

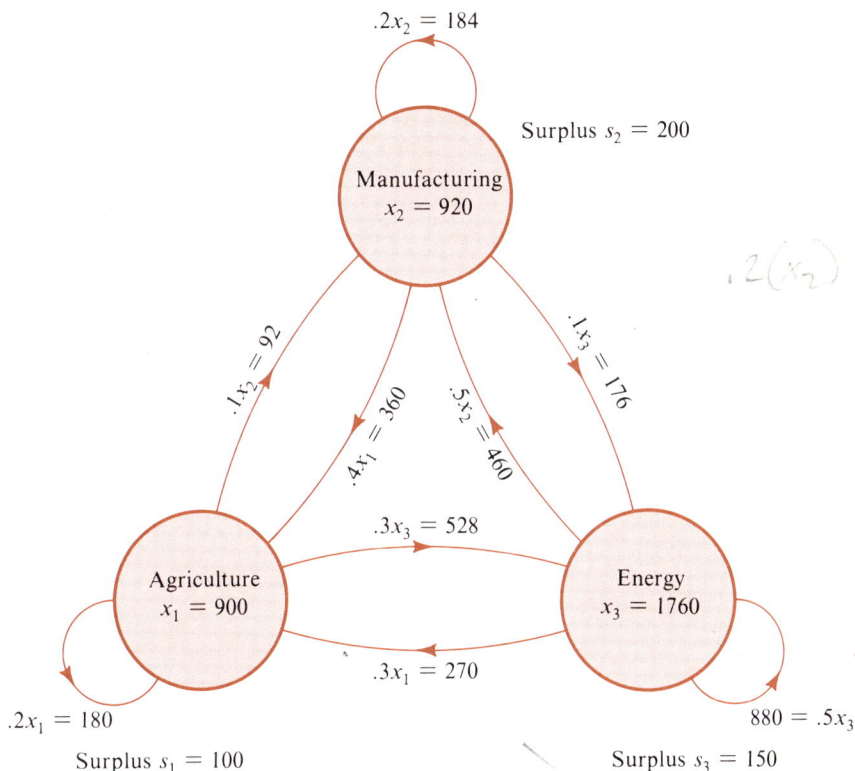

$.2x_2 = 184$

Surplus $s_2 = 200$

Manufacturing
$x_2 = 920$

$.2(x_2)$

$.1x_2 = 92$

$.4x_1 = 360$

$.5x_2 = 460$

$.1x_3 = 176$

$.3x_3 = 528$

Agriculture
$x_1 = 900$

Energy
$x_3 = 1760$

$.3x_1 = 270$

$.2x_1 = 180$

$880 = .5x_3$

Surplus $s_1 = 100$

Surplus $s_3 = 150$

Figure 2

trix $(I - A)$, may be very large in actual applications, it is far more efficient to invert $(I - A)$ once and for all and then recalculate $(I - A)^{-1}D$ by a single matrix multiplication whenever the demand changes.

Remark to Reader   Note that it is the matrix $I - A$ which is inverted in this formula, *not* the input-output matrix A itself.

The next example illustrates Theorem 1, this time involving the "interacting economy" of a large corporation.

Example 2   Amalgamated Business Products, Inc., has a computer division (P_1) and a machinery division (P_2). To produce one unit of output the computer division needs .5 units of machinery, while the machinery division needs .4 units of computers for each unit of output produced. Neither division consumes any of its own output. How much must each division produce (gross) to meet a demand for:
(a) 120 units of computers and 100 units of machinery?
(b) 100 units of computers and 130 units of machinery?

Solution   The input-output matrix is shown below.

To produce one unit of   Computers ⌐   ⌐ Machinery

$$A = \begin{bmatrix} 0 & .4 \\ .5 & 0 \end{bmatrix} \begin{matrix} \text{Computers} \\ \text{Machinery} \end{matrix} \quad \begin{matrix} \text{Input} \\ \text{needed} \end{matrix}$$

By Theorem 1 we have $X = (I - A)^{-1}D$, where

$$X = \begin{bmatrix} x_1 \\ x_2 \end{bmatrix}, \quad D = \begin{bmatrix} d_1 \\ d_2 \end{bmatrix}$$

are the intensity and demand vectors, respectively. The reader can check that the inverse of

$$I - A = \begin{bmatrix} 1 & 0 \\ 0 & 1 \end{bmatrix} - \begin{bmatrix} 0 & .4 \\ .5 & 0 \end{bmatrix} = \begin{bmatrix} 1 & -.4 \\ -.5 & 1 \end{bmatrix}$$

is the matrix

$$(I - A)^{-1} = \begin{bmatrix} 1.25 & .5 \\ .625 & 1.25 \end{bmatrix}.$$

We can thus solve for X in (a) and (b) as follows:

(a) $X = (I - A)^{-1}D = \begin{bmatrix} 1.25 & .5 \\ .625 & 1.25 \end{bmatrix} \begin{bmatrix} 120 \\ 100 \end{bmatrix} = \begin{bmatrix} 200 \\ 200 \end{bmatrix} \begin{matrix} = x_1 \\ = x_2 \end{matrix}$

(b) $X = (I - A)^{-1}D = \begin{bmatrix} 1.25 & .5 \\ .625 & 1.25 \end{bmatrix} \begin{bmatrix} 100 \\ 130 \end{bmatrix} = \begin{bmatrix} 190 \\ 225 \end{bmatrix} \begin{matrix} = x_1 \\ = x_2 \end{matrix}$

The two solutions are illustrated in Figure 3.

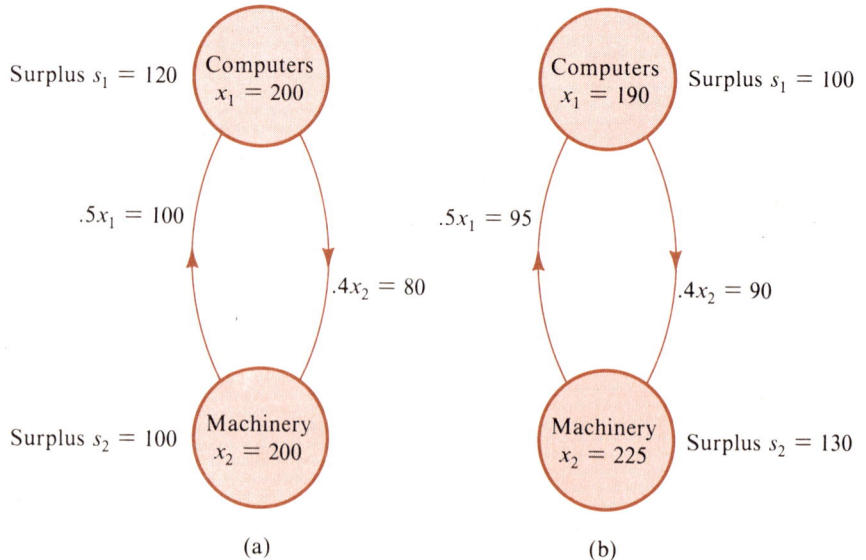

Surplus $s_1 = 120$ Computers $x_1 = 200$ Computers $x_1 = 190$ Surplus $s_1 = 100$

$.5x_1 = 100$ $.5x_1 = 95$

$.4x_2 = 80$ $.4x_2 = 90$

Surplus $s_2 = 100$ Machinery $x_2 = 200$ Machinery $x_2 = 225$ Surplus $s_2 = 130$

(a)   (b)

Figure 3

These examples give only a small indication of the scope of Leontief's methods in modern economics. While matrix algebra has been widely used by mathematicians and physicists for many years, it has become increasingly applicable in the social sciences also, as input-output theory illustrates.

4.4 Exercises

Exercises 1–4 refer to an interacting economy with four producing sectors: metals (P_1), nonmetals (P_2), energy (P_3), and services (P_4). Measuring output in monetary units, the input-output matrix is as follows:

To produce one unit of

	Metals	Nonmetals	Energy	Services	
Metals	.32	.04	.09	.02	
Nonmetals	.05	.22	.41	.06	Input
Energy	.14	.03	.12	.02	needed
Services	.07	.11	.08	.05	

1. On what sector is the energy sector most dependent? Least dependent?

2. Which sector is most dependent on energy? Least dependent on energy?

3. If the nonmetals sector produces $7.5 million worth of output, how much input must it consume from each of the four sectors?

4. If the services sector produces $5.8 million worth of output, how much input must it consume from each of the four sectors?

Exercises 5–8 refer to an interacting economy with five producing sectors: agriculture (P_1), textiles (P_2), chemicals (P_3), machinery (P_4), and transportation (P_5). Measuring output in monetary units, the input-output matrix is as follows:

To produce one unit of

	Agr	Tex	Chem	Mach	Trns	
Agr	.25	.21	.12	0	0	
Tex	0	.13	0	0	0	
Chem	.02	.09	.35	.03	0	Input
Mach	.05	.03	.01	.08	.14	needed
Trns	.07	.10	.02	.04	.10	

5. On what sector is the textiles sector most dependent? Least dependent?

6. Which sector is most dependent on machinery? Least dependent on machinery?

7. If the agriculture sector produces $20 billion worth of output, how much input must it consume from each of the five sectors?

8. If the machinery sector produces $12 billion worth of output, how much input must it consume from each of the five sectors?

9. An economy has two interacting sectors, manufacturing (P_1) and energy (P_2). It takes .3 units of manufacturing and .4 units of energy to produce one unit of manufactured goods, and it takes .2 units of manufacturing and .1 units of energy to produce one unit of energy. What is the input-output matrix for this economy?

10. An economy has two interacting sectors, agriculture (P_1) and chemicals (P_2). In a typical year, the agriculture sector spends a total of \$30 billion on its own goods and \$6 billion on chemicals and produces \$60 billion worth of output, while the chemicals sector spends \$20 billion on its own products and \$10 billion on agricultural goods and produces \$50 billion worth of output. Using monetary units, what is the input-output matrix for this economy?

11. A two-sector economy has the following input-output matrix:

$$A = \begin{bmatrix} .02 & .08 \\ .12 & .03 \end{bmatrix}$$

The gross production of the economy is $x_1 = 1250$, $x_2 = 1800$.

(a) Compute the intensity vector X.

(b) Compute the internal consumption vector U.

(c) Compute the surplus, or net production, S.

(d) Make a diagram like those in Figure 3, showing the flow of output in this economy.

12. A three-sector economy has the following input-output matrix:

$$A = \begin{bmatrix} 0 & .11 & .05 \\ .15 & .02 & .08 \\ .09 & .14 & .01 \end{bmatrix}$$

The gross production of the economy is $x_1 = 1500$, $x_2 = 1400$, $x_3 = 1700$.

(a) Compute the intensity vector X.

(b) Compute the internal consumption vector U.

(c) Compute the surplus, or net production, S.

(d) Make a diagram like those in Figure 3, showing the flow of output in this economy.

13. A two-sector economy has the following input-output matrix:

$$A = \begin{bmatrix} .2 & .3 \\ .2 & .3 \end{bmatrix}$$

At what intensity must the economy operate in order to meet a demand for

(a) 50 units of G_1 and 60 units of G_2?

(b) 200 units of G_1 and 100 units of G_2?

14. A two-sector economy has the following input-output matrix:

$$A = \begin{bmatrix} 0 & .6 \\ .5 & .3 \end{bmatrix}$$

At what intensity must the economy operate in order to meet a demand for

(a) 2000 units of G_1 and 2000 units of G_2?

(b) 0 units of G_1 and 4000 units of G_2?

15. A three-sector economy has the following input-output matrix:

$$A = \begin{bmatrix} .1 & .1 & .1 \\ .1 & .2 & .2 \\ .3 & .2 & .2 \end{bmatrix}$$

(a) Use matrix multiplication to verify the following:

$$(I - A)^{-1} = \begin{bmatrix} 1.2 & .2 & .2 \\ .28 & 1.38 & .38 \\ .52 & .42 & 1.42 \end{bmatrix}$$

(b) At what intensity must the economy operate in order to meet a demand for 2500 units of G_1, 3500 units of G_2, and 2000 units of G_3?

(c) Make a diagram like that of Figure 2, showing the flow of output in this economy.

16. A three-sector economy has the following input-output matrix:

$$A = \begin{bmatrix} 0 & .1 & .2 \\ .4 & .4 & 0 \\ .2 & 0 & .6 \end{bmatrix}$$

(a) Use matrix multiplication to verify the following:

$$(I - A)^{-1} = \begin{bmatrix} 1.2 & .2 & .6 \\ .8 & 1.8 & .4 \\ .6 & .1 & 2.8 \end{bmatrix}$$

(b) At what intensity must the economy operate in order to meet a demand for 500 units of each of the three commodities?

(c) Make a diagram like that of Figure 2, showing the flow of output in this economy.

*For Exercises 17–20 we define a **Leontief matrix** to be a square matrix A satisfying the following conditions:*

(1) $A \geq O$ (all entries of A are nonnegative)
(2) $(I - A)$ is invertible
(3) $(I - A)^{-1} \geq O$

17. If the input-output matrix for an economy is a Leontief matrix, it is possible to find an intensity $X \geq O$ to meet *any* given demand vector $D \geq O$. Explain why this is so.

18. It can be shown that *a square matrix A $\geq$ O is always a Leontief matrix, provided that the entries in each column add up to less than one.** Explain why the input-output matrix for an economy can be expected to satisfy this condition if output is measured in *monetary* units.

_____

*The converse, however, is not true: there are Leontief matrices with column-sums greater than one.

19. Which of the following are Leontief matrices?

(a) $\begin{bmatrix} .8 & .4 \\ .4 & .2 \end{bmatrix}$ 　　(b) $\begin{bmatrix} .3 & .3 \\ .3 & .3 \end{bmatrix}$ 　　(c) $\begin{bmatrix} .4 & .01 \\ 2 & .3 \end{bmatrix}$ 　　(d) $\begin{bmatrix} .3 & .9 \\ .6 & .3 \end{bmatrix}$

20. Prove: If A is a Leontief matrix then so is A^t.

Computer Projects

A. Most versions of BASIC include matrix operations. Write a BASIC program to confirm your answer in Exercises 11–16.

B. Write a BASIC program to calculate the intensity X needed to meet a given demand vector D, for any five-sector economy. Apply your program to the economy with input-output matrix (a) and demand vectors (b) and (c).

(a) $\begin{bmatrix} .29 & 0 & .12 & .09 & 0 \\ .08 & .21 & .05 & .14 & .02 \\ .03 & .02 & .04 & .07 & .01 \\ .01 & .03 & .15 & .12 & 0 \\ .12 & .07 & .08 & .19 & .10 \end{bmatrix}$ 　　(b) $\begin{bmatrix} 1376 \\ 699 \\ 1245 \\ 1708 \\ 906 \end{bmatrix}$ 　　(c) $\begin{bmatrix} 2363 \\ 4133 \\ 5450 \\ 4119 \\ 2489 \end{bmatrix}$

Chapter 4 Summary

Important Terms

4.1 $m \times n$ matrix
row vector
column vector
transpose A^t
sum $A + B$
negative $-A$
scalar product cA
difference $A - B$
zero matrix O

4.2 dot product $A \cdot B$
matrix product AB

identity matrix I

4.3 matrix form of a linear system
inverse A^{-1}
matrix solution of $AX = B$

4.4 input-output ratio a_{ij}
input-output matrix A
internal consumption U
intensity X
surplus, or net production S
demand D

Important Formulas

Laws of Matrix Algebra

4.1 M1. $A + B = B + A$
M2. $A + (B + C) = (A + B) + C$
M3. $A + O = A$
M4. $A + (-A) = O$
M5. $c(A + B) = cA + cB$
M6. $(c + d)A = cA + dA$
M7. $c(dA) = (cd)A$

M8. $1A = A$
4.2 M9. $A(BC) = (AB)C$
M10. $A(B + C) = AB + AC$
M11. $(A + B)C = AC + BC$
M12. $(cA)B = c(AB) = A(cB)$
M13. $IA = A = AI$

Review Exercises

In Exercises 1–8, perform the indicated matrix operation (if possible) using the following matrices:

$$A = \begin{bmatrix} 4 & -1 \\ 2 & 3 \end{bmatrix}, \quad B = \begin{bmatrix} 5 & -2 & 4 \end{bmatrix}, \quad C = \begin{bmatrix} 2 \\ -2 \end{bmatrix}$$

$$D = \begin{bmatrix} 3 & 0 \\ -1 & 2 \\ 1 & 4 \end{bmatrix}, \quad E = \begin{bmatrix} 2 & -1 & 0 \\ 3 & 5 & -4 \end{bmatrix}, \quad F = \begin{bmatrix} 1 \\ -2 \\ 1 \end{bmatrix}$$

1. $A + C$

2. $D + E^t$

3. $2D^t + 3E$

4. $3F^t - 2B$

5. AE

6. ED

7. $AC + EF$

8. $(BDA)^t$

In Exercises 9–12, solve the given equation for the unknown 2×2 matrix X, where A, B, and I are the following matrices:

$$A = \begin{bmatrix} 2 & 1 \\ 5 & 3 \end{bmatrix}, \quad B = \begin{bmatrix} 2 & 3 \\ 3 & 4 \end{bmatrix}, \quad I = \begin{bmatrix} 1 & 0 \\ 0 & 1 \end{bmatrix}$$

9. $3(X - A) = 2(X - B)$

10. $2(A + 2X) = B + 3(I + X)$

11. $AX = I$

12. $BX = A$

13. Use matrix inversion to solve the linear system

$$\begin{aligned} x + 3y &= b_1 \\ 2x + 7y &= b_2 \end{aligned}$$

for x and y in terms of b_1 and b_2.

14. Use matrix inversion to solve the linear system

$$\begin{aligned} x - 2y + 3z &= b_1 \\ -2x + 3y - 4z &= b_2 \\ x - 3y + 4z &= b_3 \end{aligned}$$

for x, y, and z in terms of b_1, b_2, and b_3.

15. An economy has two interacting sectors, goods (P_1) and services (P_2). It takes .1 units of goods and .5 units of services to produce one unit of goods, and it takes .19 units of goods and .2 units of services to produce one unit of services. How many units should be produced by each sector in order to meet a demand for 7000 units of goods and 2000 units of services?

16. An economy has three interacting sectors, energy (P_1), food (P_2), and goods (P_3). It takes .2 units of energy (but no food or goods) to produce one unit of energy; it takes .1 units of energy and .1 units of goods to produce one unit of food; and it takes .3 units of energy and .5 units of goods to produce one unit of goods. How many units should be produced by each sector in order to meet a demand for 400 units of energy, 500 units of food, and 200 units of goods?

5

The Algebra of Linear Programming

In Chapter 2 we saw how to solve linear programs geometrically by finding all corner points of the constraint set. In this chapter an algebraic method called the *simplex algorithm* is presented, which was developed by the American mathematician George B. Dantzig in the 1940s. The new method relies heavily on the theory of linear equations developed in Chapter 3.

5.1 The Simplex Algorithm

We begin by considering maximum programs that have certain standard features. In the next section the method is extended to cover minimum programs, as well as the vast majority of "nonstandard" maximum programs.

Standard Maximum Programs   Recall from Section 2.4 that a **linear program** is a problem in which we want to optimize a linear function subject to a system of joint constraints.

Example 1

$$\text{Maximize} \quad F = 4x + 5y$$
subject to:
$$2x + 3y \le 90$$
$$-x + y \le 20$$
$$x \le 30 \tag{1}$$
$$x \ge 0$$
$$y \ge 0.$$

Discussion   This is an example of a *maximum* program, with **decision variables** x, y and **objective function** $F = 4x + 5y$. The *constraints* can be divided into two types:

Functional constraints:
$$2x + 3y \leq 90$$
$$-x + y \leq 20$$
$$x \leq 30$$

Nonnegativity constraints:
$$x \geq 0$$
$$y \geq 0$$

Note that each of the functional constraints is of $\leq$ **type;** that is, each has the form $ax + by \leq c$, as opposed to $ax + by \geq c$. Also, the **right-hand constraints**—90, 20, and 30 in our example—are all nonnegative. For the remainder of this section, we restrict our attention to programs of this type. ◇

Definition

A **standard maximum program** is a linear program with these characteristics:

1. The objective function is to be maximized.
2. All functional constraints are of $\leq$ type.
3. All right-hand constants are nonnegative.
4. All decision variables are constrained to be ≥ 0.

The more general case will be discussed in the next section.

The Initial Tableau   The first step in the simplex algorithm is to express the functional constraints in equational form.

Example 1, Continued   Express the functional constraints in Example 1 as equations.

Formulation   We can convert the inequality

$$2x + 3y \leq 90$$

into an equation by *adding a nonnegative quantity* to the left-hand side.

$$2x + 3y + u = 90$$

We refer to u as a **slack variable,** since it "takes up the slack" between the left- and right-hand sides of the inequality. If slack variables $v \geq 0$ and $w \geq 0$ are introduced for the other functional constraints, the linear program can be written in **equational form.**

$$\text{Maximize} \quad F = 4x + 5y$$

subject to:

$$2x + 3y + u = 90$$
$$-x + y + v = 20 \tag{1'}$$
$$x + w = 30.$$
$$(x,\ y,\ u,\ v,\ w \geq 0)$$

A solution $(x,\ y,\ u,\ v,\ w)$ of this system of equations is **feasible** if all five values are nonnegative. Our goal is to find the **optimal** solution, that is, the feasible solution that gives the greatest value of F.

The equation defining the objective function can be written as

$$-4x - 5y + F = 0. \tag{2}$$

Together, the system $(1')$ and the equation (2) comprise a system of four linear equations in six unknowns, which we can write in tableau form as follows:

	Decision variables		Slack variables				
	x	y	u	v	w	F	
	2	3	1	0	0	0	90
	−1	1	0	1	0	0	20
	1	0	0	0	1	0	30
Indicators:	−4	−5	0	0	0	1	0

Right-hand constants

This is called the **initial simplex tableau.** The last row, corresponding to the objective function, is separated by a horizontal line since it plays a special role in the simplex algorithm. The numbers in this row (to the left of the vertical line) are called **indicators.** ◇

Our procedure to this point can be summarized as follows:

Forming the Initial Simplex Tableau

1. Convert each functional constraint into an equation by adding a slack variable to the left-hand side.
2. Transpose all variables to the left-hand side in the equation defining the objective function.
3. Write these equations in tableau form, separating the last row (objective function) by a horizontal line.

The Simplex Algorithm   Having formed the initial tableau, we now perform a sequence of pivot transformations, each of which replaces the given linear system by another one equivalent to it. The goal of this procedure is to *eliminate all negative indicators.* We shall see later how this leads to the optimal solution, and also what the simplex pivots mean intuitively.

Simplex Algorithm

1. Locate the *most negative* indicator. This determines the pivot column.
2. Divide each *positive* entry in this column into the right-hand constant in its row. Record the quotients.
3. Locate the positive entry that produced the *smallest* quotient. Perform a pivot transformation on this entry.
4. Repeat Steps 1–3 until all indicators are ≥ 0.

It should be emphasized that negative indicators may appear in columns headed by either decision variables or slack variables. Also, in applying the "smallest quotient" criterion (Step 3) we may be forced to use the same pivot row or column more than once.

Example 1, Continued

Apply the simplex algorithm to the linear program of Example 1 and read the optimal solution.

Solution

We begin by looking at the indicators in the initial tableau.

Tableau I

	x	y	u	v	w	F		Quotients
	2	3	1	0	0	0	90	$90/3 = 30$
	-1	①	0	1	0	0	20	$20/1 = 20 \leftarrow$
	1	0	0	0	1	0	30	
Indicators:	-4	-5	0	0	0	1	0	Smallest quotient

Most negative

The most negative indicator is -5, in the y-column. The positive entries in this column are 3 (in row 1) and 1 (in row 2). When we divide these entries into their respective right-hand constants, we obtain the quotients $90/3 = 30$ and $20/1 = 20$. Since the smallest quotient is 20, the entry in row 2 (circled) should be chosen as pivot. The result of the pivot transformation* is the tableau shown below.

Tableau II

	x	y	u	v	w	F		Quotients
	⑤	0	1	-3	0	0	30	$30/5 = 6 \leftarrow$
	-1	1	0	1	0	0	20	
	1	0	0	0	1	0	30	$30/1 = 30$
	-9	0	0	5	0	1	100	

---

*Annotation: $R_1 = R_1 + (-3)R_2$, $R_4 = R_4 + 5R_2$.

Since Tableau II still has a negative indicator (-9, in the x-column), we repeat Steps 1–3. The positive entries in the x-column are 5 (in row 1) and 1 (in row 3), which determine the quotients $30/5 = 6$ and $30/1 = 30$, respectively. Since 5 determines the smallest quotient, we pivot on this entry (circled). The result is shown below.

Tableau	x	y	u	v	w	F		Quotients
III	1	0	$\frac{1}{5}$	$-\frac{3}{5}$	0	0	6	
	0	1	$\frac{1}{5}$	$\frac{2}{5}$	0	0	26	$26/\left(\frac{2}{5}\right) = 65$
	0	0	$-\frac{1}{5}$	$\left(\frac{3}{5}\right)$	1	0	24	$24/\left(\frac{3}{5}\right) = 40 \leftarrow$
	0	0	$\frac{9}{5}$	$-\frac{2}{5}$	0	1	154	

The new tableau again has a negative indicator, in the v-column. The positive entries $2/5$ and $3/5$ determine the respective quotients 65 and 40, as shown. We therefore pivot on $3/5$ (circled), with the result shown below.

Tableau	x	y	u	v	w	F	
IV	1	0	0	0	1	0	30
	0	1	$\frac{1}{3}$	0	$-\frac{2}{3}$	0	10
	0	0	$-\frac{1}{3}$	1	$\frac{5}{3}$	0	40
	0	0	$\frac{5}{3}$	0	$\frac{2}{3}$	1	170

Since the indicators in Tableau IV are all nonnegative, we have reached a **terminal tableau.** The simplex algorithm is completed.

Reading the Optimal Solution   Each of the tableaus I–IV represents a system of four equations in the six unknowns x, y, u, v, w, F. In particular, the terminal Tableau IV represents the system

$$
\begin{aligned}
x &&&&+ w &&= 30 \\
y + \tfrac{1}{3}u &&&&- \tfrac{2}{3}w &&= 10 \\
-\tfrac{1}{3}u &+ v &&+ \tfrac{5}{3}w &&= 40 \\
\tfrac{5}{3}u &&&&\tfrac{2}{3}w + F &&= 170
\end{aligned}
$$

which is equivalent (by elementary row operations) to our original system. Notice that *this system is solved for the basic variables x, y, v, F in terms of the free variables, u, w.* All we have to do is transpose the u and w terms to the right-hand side.

$$
\begin{aligned}
x &= 30 &&- w \\
y &= 10 - \tfrac{1}{3}u &&+ \tfrac{2}{3}w \\
v &= 40 + \tfrac{1}{3}u &&- \tfrac{5}{3}w \\
F &= 170 - \tfrac{5}{3}u &&- \tfrac{2}{3}w
\end{aligned}
$$

The corresponding basic solution is obtained by setting the free variables u, w equal to zero.

$$x = 30, \quad y = 10, \quad u = 0, \quad v = 40, \quad w = 0, \quad F = 170$$

We now claim that *this gives the optimal solution to the linear program.* That is,

$$\text{maximum } F = 170 \quad \text{when} \quad x = 30, \quad y = 10.$$

To see this, we look at the equation

$$F = 170 - \frac{5}{3}u - \frac{2}{3}w.$$

This equation represents the value of F as 170 *minus* two nonnegative terms.* If we want to maximize F we must therefore let $u = w = 0$, which is exactly how we obtained the basic solution above. $\Diamond$

The optimal solution can be read from the terminal tableau, as follows:

Reading the Tableau

1. A variable is basic if one entry in its column is 1 and all others are 0. Otherwise it is free.
2. For each basic variable, locate the 1 in its column. Assign to the variable the right-hand constant in this row.
3. To each free variable, assign the value 0.

This technique is illustrated below for the terminal Tableau IV.

	basic	basic		basic		basic				Assigned values	
Tableau IV	x	y	u	v	w	F				**Basic**	**Free**
	①	0	0	0	1	0	(30)			$x = 30$	$u = 0$
	0	①	$\frac{1}{3}$	0	$-\frac{2}{3}$	0	(10)			$y = 10$	$w = 0$
	0	0	$-\frac{1}{3}$	①	$\frac{5}{3}$	0	(40)			$v = 40$	
	0	0	$\frac{5}{3}$	0	$\frac{2}{3}$	①	(170)			$F = 170$	

Meaning of the Simplex Pivots   We can get a better understanding of the simplex algorithm by reading not only the terminal tableau, but the others as well. Each of the tableaus I–IV represents a "solved" linear system, in which four variables are basic and two are free.

---

*Recall that $u, w \geq 0$ for any feasible solution.

Example 1,
Revisited

Read the basic solution associated with each of the tableaus I-IV.

Solution

Using the method described above, we obtain the results shown in Table 1.

			Basic solution				Corner point
Tableau	x	y	u	v	w	F	(x, y)
I	0	0	90	20	30	0	(0, 0)
II	0	20	30	0	30	100	(0, 20)
III	6	26	0	0	24	154	(6, 26)
IV	30	10	0	40	0	170	(30, 10)

Table 1

In the last column are listed just the values (x, y) of the decision variables, ignoring the slack variables and the objective function.

Now let us look at the constraint set S for the program of Example 1, which is graphed in Figure 1. We see that each of the basic solutions in Table 1 represents a corner point of S. Moreover, *each simplex pivot takes us from one corner point to an adjacent one, where the value of F is greater.* We thus follow the path $(0, 0) \rightarrow (0, 20) \rightarrow (6, 26) \rightarrow (30, 10)$ along the edges of the constraint set, increasing the value of F with each pivot. Since $(30, 10)$ is optimal, the algorithm terminates.

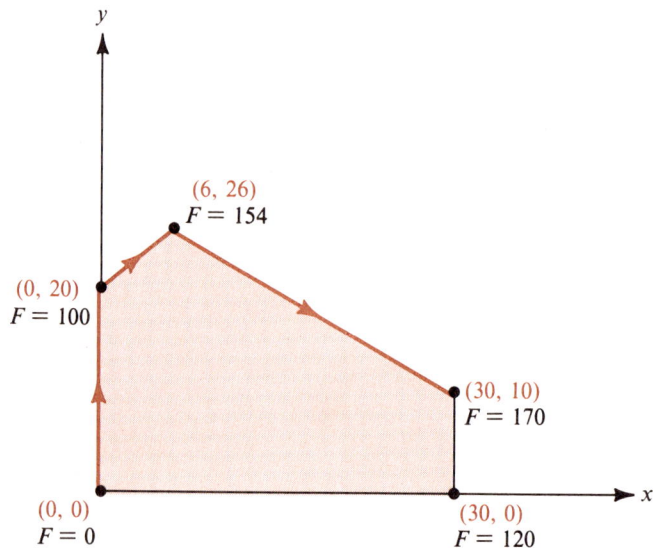

Figure 1

The simplex method can be applied to any standard maximum program, regardless of the number of decision variables or constraints.

Example 2

Maximize $F = 4x + 5y + 2z$
subject to

$$x + y + z \le 30$$
$$x + 2y + z \le 50.$$
$$(x, y, z \ge 0)$$

Solution   This is a standard maximum program with three decision variables and two functional constraints. Adding slack variables $u, v \ge 0$ we can write the program in equational form as follows:

Maximize $F = 4x + 5y + 2z$
subject to:

$$x + y + z + u \quad\quad = 30$$
$$x + 2y + z \quad\quad + v = 50.$$
$$(x, y, z, u, v \ge 0)$$

The initial simplex tableau therefore takes the following form:

x	y	z	u	v	F	
1	1	1	1	0	0	30
1	2	1	0	1	0	50
Indicators: -4	-5	-2	0	0	1	0

When we apply the simplex algorithm we obtain the sequence of tableaus shown below.

x	y	z	u	v	F		*Quotients*
1	1	1	1	0	0	30	$30/1 = 30$
1	②	1	0	1	0	50	$50/2 = 25 \leftarrow$
-4	-5	-2	0	0	1	0	

Basic solution: $x = 0, y = 0, z = 0, u = 30, v = 50, F = 0$.

x	y	z	u	v	F		*Quotients*
⓵⁄₂	0	$\frac{1}{2}$	1	$-\frac{1}{2}$	0	5	$5/(\frac{1}{2}) = 10 \leftarrow$
$\frac{1}{2}$	1	$\frac{1}{2}$	0	$\frac{1}{2}$	0	25	$25/(\frac{1}{2}) = 50$
$-\frac{3}{2}$	0	$\frac{1}{2}$	0	$\frac{5}{2}$	1	125	

Basic solution: $x = 0, y = 25, z = 0, u = 5, v = 0, F = 125$.

x	y	z	u	v	F	
1	0	1	2	-1	0	10
0	1	0	-1	1	0	20
0	0	2	3	1	1	140

Basic solution: $x = 10, y = 20, z = 0, u = 0, v = 0, F = 140$.

In two pivots we have reached a terminal tableau. The last basic solution is therefore optimal.

maximum $F = 140$ when $x = 10$, $y = 20$, $z = 0$.

The constraint set for this program is a convex polyhedral set in xyz-space, of the type described in Section 2.5. Our first pivot takes us from the corner point $(0, 0, 0)$, where $F = 0$, to the corner point $(0, 25, 0)$, where $F = 125$. The second pivot takes us from $(0, 25, 0)$ to the optimal corner point $(10, 20, 0)$, where $F = 140$. In carrying out the simplex algorithm we therefore follow a path along the edges of the constraint set, starting at the origin and ending at the optimal corner point.   ◆

Summary

Given a standard maximum program, we find its optimal solution as follows:

1. Form the initial simplex tableau.
2. Apply the simplex algorithm until all indicators are ≥ 0.
3. Read the optimal solution from the terminal tableau.

Question 1:   In carrying out Step 1 of the simplex algorithm, what if a tie occurs for the most negative indicator?

Answer: Choose either one. In fact, *any* negative indicator can be used to determine the pivot column (See Exercises 21 and 22).

Question 2:   How do we carry out Step 2 of the simplex algorithm if the pivot column contains no positive entries?

Answer: If at any stage all entries in the pivot column are zero or negative, it means that the program has no optimal solution.

Question 3:   In Step 3 of the simplex algorithm, what if a tie occurs for the positive entry producing the smallest quotient?

Answer: Choose either one as pivot. When such a tie occurs the program is called *degenerate*. It is theoretically possible for the simplex procedure to *cycle* in such a case, never reaching a terminal tableau, but this is relatively rare in practice.

Question 4:   Is there any way to tell whether or not the optimal solution is unique?

Answer: Yes. This is discussed in Exercises 23 and 24.

5.1 Exercises

In Exercises 1 and 2, write the initial simplex tableau for the given standard maximum program, but go no further.

1. Maximize $F = 3x + 7y$
   subject to:
   $$x + y \leq 10$$
   $$2x + 3y \leq 24$$
   $$y \leq 6.$$
   $$(x, y \geq 0)$$

2. Maximize $F = 15x + 10y + 20z$
   subject to:
   $$2x + y + z \leq 20$$
   $$x + 5y + 2z \leq 50$$
   $$3x - y + z \leq 30.$$
   $$(x, y, z \geq 0)$$

In Exercises 3 and 4 a terminal tableau is shown for a standard maximum program. What is the optimal solution?

3. Decision variables

x	y	u	v	w	F	
0	0	1	$-\frac{1}{2}$	$\frac{1}{2}$	0	1
1	0	0	$\frac{1}{2}$	$-\frac{3}{2}$	0	3
0	1	0	0	1	0	6
0	0	0	$\frac{3}{2}$	$\frac{5}{2}$	1	51

4. Decision variables

x	y	z	u	v	w	F	
2	1	1	1	0	0	0	20
-3	3	0	-2	1	0	0	10
1	-2	0	-1	0	1	0	10
25	10	0	20	0	0	1	400

In Exercises 5–10, solve the given standard maximum program by the simplex method and also by the method of graphing. Indicate on your graph the edge-path described by the simplex pivots.

5. Maximize $F = 5x + 2y$
   subject to:
   $$2x + y \leq 30$$
   $$x + 2y \leq 30.$$
   $$(x, y \geq 0)$$

6. Maximize $F = 10x + 15y$
   subject to:
   $$x + 2y \leq 4$$
   $$x + y \leq 3.$$
   $$(x, y \geq 0)$$

7. Maximize $F = 10x + 5y$
   subject to:
   $$x + y \leq 5$$
   $$x \leq 3$$
   $$y \leq 4.$$
   $$(x, y \geq 0)$$

8. Maximize $F = 2x - y$
   subject to:
   $$x + y \leq 10$$
   $$x - y \leq 4$$
   $$x \leq 6.$$
   $$(x, y \geq 0)$$

9. Maximize $F = 3x - 2y$
   subject to:
   $$x \leq 20$$
   $$y \leq 20$$
   $$x - y \leq 10$$
   $$-x + y \leq 10.$$
   $$(x, y \geq 0)$$

10. Maximize $F = 3x + 4y$
    subject to:
    $$2x + 3y \leq 180$$
    $$x - y \leq 15$$
    $$y \leq 50$$
    $$x \leq 30.$$
    $$(x, y \geq 0)$$

In Exercises 11–16, solve the standard maximum program by the simplex method.

11. Maximize $F = 2x + y + z$
    subject to:
    $$x + 2y + 3z \leq 80$$
    $$x + y + z \leq 60.$$
    $$(x, y, z \geq 0)$$

12. Maximize $F = 2x + 3y + 4z$
    subject to:
    $$2x + y + z \leq 40$$
    $$x + y + 2z \leq 50.$$
    $$(x, y, z \geq 0)$$

13. Maximize $F = 2x + y - 3z$
    subject to:
    $$x + y + z \le 4$$
    $$x - y + z \le 2$$
    $$-x + y + z \le 1.$$
    $$(x, y, z \ge 0)$$

14. Maximize $F = 5x - 4y + z$
    subject to:
    $$x - y + z \le 20$$
    $$x + y \le 10$$
    $$y + z \le 15.$$
    $$(x, y, z \ge 0)$$

15. Maximize $F = x + 2y - z + w$
    subject to:
    $$2x + y + 3z + w \le 10$$
    $$x + z + 2w \le 2$$
    $$y + 2z + w \le 4.$$
    $$(x, y, z, w \ge 0)$$

16. Maximize $F = x + 2y + 4z + w$
    subject to:
    $$x - y + z - w \le 1$$
    $$x - z \le 1$$
    $$y - w \le 1.$$
    $$(x, y, z, w \ge 0)$$

In Exercises 17–20, formulate the given problem as a linear program and solve by the simplex method.

17. The House of Coffee sells four different mixtures of Mocha, Java, and Santos beans. The composition and selling price of each mixture are shown below.

Mixture	Mocha per lb	Java per lb	Santos per lb	Price per lb
A	$\frac{1}{4}$ lb	$\frac{1}{2}$ lb	$\frac{1}{4}$ lb	$3.50
B	$\frac{3}{5}$ lb	$\frac{1}{5}$ lb	$\frac{1}{5}$ lb	$4.00
C	$\frac{1}{4}$ lb	$\frac{1}{4}$ lb	$\frac{1}{2}$ lb	$2.00
D	$\frac{1}{2}$ lb	$\frac{1}{2}$ lb	0 lb	$5.00

If the store has available 300 lb of Mocha, 250 lb of Java, and 150 lb of Santos, how many pounds of each mixture should be produced to maximize total revenue?

18. A steel mill makes three grades of stainless steel, each of which is sold in 100-lb bars. The composition by weight and selling price of each grade are shown below.

Grade	Steel	Chromium	Price per bar
Standard	90%	10%	$110
High-grade	80%	20%	$120
Supreme	75%	25%	$150

If the company has a supply of 36 tons (72,000 lb) of steel and 5 tons of chromium, how many bars of each grade should it produce to maximize total revenue?

19. A farmer has 1000 acres of land on which he intends to plant asparagus, brussels sprouts, and cauliflower. These three crops require, respectively, 15, 25, and 20 inches of water per acre to grow to maturity, and the farmer has been allotted a total of 20,000 acre-inches of water for the season. (An acre-inch is the amount of water needed to irrigate an acre of land to a depth of one inch.) The respective profits on the three crops are $200, $250, and $220 an acre. How much of each crop should be planted to maximize total profit?

20. The Zeno Corporation makes three models of stereo receivers. There are three stages in the manufacture of each unit, as shown in the following table:

Model	Assembly	Wiring	Testing
A	.2 wh	.2 wh	.1 wh
B	.3 wh	.2 wh	.2 wh
C	.1 wh	.1 wh	0 wh

Work-hours per unit

The respective profits on the three models are $50, $75, and $30 per unit. The company has available a total of 40 wh of labor per day for assembly, 28 for wiring, and 16 for testing. How many receivers of each type should be made to maximize total profit?

Negative Indicators *The claim is made in the text that **any** negative indicator can be used to determine the pivot column in Step 1 of the simplex algorithm. Exercises 21 and 22 illustrate this fact.*

21. Rework Example 1 in this section, this time using the x-column (indicator -4) for the first pivot. What edge-path does this describe in Figure 1?

22. Rework Example 2 in this section, this time using the x-column (indicator -4) for the first pivot. What edge-path does this describe in xyz-space?

More than One Optimal Solution *By looking at the terminal tableau one can determine whether or not there is more than one optimal solution, as follows: If every column headed by a free variable has a **positive** indicator then the optimal solution is unique. But if some such column has indicator **zero,** there is more than one optimal solution. Exercises 23 and 24 illustrate this fact.*

For the linear programs in Exercises 23 and 24:
(a) Use the simplex method to obtain a terminal tableau. What optimal solution does this tableau represent?

(b) Locate a free variable column having a zero indicator. Apply Steps 2–3 of the simplex algorithm to this column. What optimal solution does the new tableau represent?

23. Maximize $F = 15x + 25y$
subject to:
$$3x + 5y \le 120$$
$$x + y \le 30.$$
$$(x, y \ge 0)$$

24. Maximize $F = 10x + 9y + 8z$
subject to:
$$x + y + z \le 20$$
$$5x + 3y + z \le 40.$$
$$(x, y, z \ge 0)$$

Computer Projects

A. The PIVOT program performs a pivot transformation on a designated nonzero entry in a tableau. This program can be used in Exercises 5–24.

B. The SIMPL program computes the optimal solution to a linear program by the simplex method. Use this program to confirm your answer in Exercises 5–20.

C. Use the SIMPL program to solve the following standard maximum program:

$$\text{Maximize} \quad F = 4x + 5y + 2z + 2u + 3v$$

subject to:

$$2z + y + z \leq 85$$
$$y + 2z + u \leq 90$$
$$z + 2u + v \leq 40$$
$$x + 2y \leq 65$$
$$y + 2u \leq 35$$
$$u + 2v \leq 20.$$
$$(x, y, z, u, v \geq 0)$$

5.2 The Simplex Algorithm: Nonstandard Programs

In Section 5.1 the simplex algorithm for standard maximum programs was presented. Such programs satisfy the following conditions:

1. The objective function is to be maximized.
2. All functional constraints are of $\leq$ type.
3. All right-hand constants are nonnegative.
4. All decision variables are constrained to be ≥ 0.

In this section we extend the method to cover programs that fail to satisfy one or more of conditions (1)–(3). We shall continue to require condition (4), in order to simplify the discussion. As we have seen, the majority of linear programs arising from practical problems include nonnegativity constraints.

Minimum Programs   The first question that arises is how to apply the simplex algorithm to a *minimum* program. Fortunately, we have a simple way of turning any minimum program into a maximum program. If the objective, for example, is to

$$\text{minimize} \quad G = ax + by + cz$$

then instead we maximize the negative of $G;$ that is, we

$$\text{maximize} \quad F = -G = -ax - by - cz.$$

The optimal solution (x, y, z) remains the same, and the optimal values of F and G are just negatives of each other, as illustrated in Figure 1.

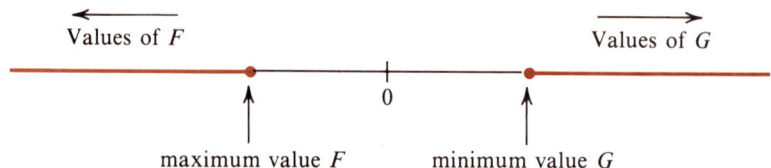

Figure 1

Constraints of $\geq$ Type If a linear program has constraints of $\geq$ type, we can convert them to constraints of $\leq$ type by taking negatives on both sides and reversing the inequality. Thus, the constraint

$$ax + by + cz \geq d$$

is algebraically equivalent to

$$-ax - by - cz \leq -d$$

which is of $\leq$ type. By rewriting some or all of the constraints as necessary, we can always arrange for condition (2) to be satisfied before starting the simplex algorithm.

Negative Constants The presence of negative right-hand constants is therefore the only real difficulty presented by nonstandard programs. The following example shows why the failure of condition (3) causes problems when we attempt to apply the simplex algorithm.

Example 1

$$\text{Minimize} \quad G = 5x + 4y$$
subject to:
$$6x + 5y \leq 300$$
$$2x + 3y \geq 60$$
$$2x + y \geq 40.$$
$$(x, y \geq 0)$$

Discussion We begin by converting the program to a maximum program, and writing all functional constraints in $\leq$ form.

$$\text{Maximize} \quad F = -G = -5x - 4y$$
subject to:
$$6x + 5y \leq 300$$
$$-2x - 3y \leq -60$$
$$-2x - y \leq -40.$$
$$(x, y \geq 0)$$

Notice that *two of the right-hand constants are negative*. Introducing slack variables $u, v, w \geq 0$, we can write the program in equational form as follows:

$$\text{Maximize} \quad F = -5x - 4y$$
subject to:
$$6x + 5y + u \qquad\qquad = 300$$
$$-2x - 3y \qquad + v \qquad = -60$$
$$-2x - y \qquad\qquad + w = -40.$$
$$(x, y, u, v, w \geq 0)$$

The initial simplex tableau is therefore

Tableau
I

x	y	u	v	w	F	
6	5	1	0	0	0	300
−2	−3	0	1	0	0	−60
−2	−1	0	0	1	0	−40
5	4	0	0	0	1	0

This initial tableau is very deceptive. It *looks* like a terminal tableau, since all indicators are nonnegative. Reading the tableau gives the basic solution

$$x = 0, \quad y = 0, \quad u = 300, \quad v = -60, \quad w = -40, \quad F = 0.$$

This corresponds to the origin $(0, 0)$ in the xy-plane. As we see from Figure 2, this point is not even feasible for the program in Example 1. The problem is caused by the negative right-hand constants, which give illegal negative values ($v = -60$, $w = -40$) to the slack variables v and w in the basic solution. The origin is a *basic* point for the given constraints, but it is not a corner point because it is not feasible.   ◇

Phase 1: Finding a Corner Point   Before we can begin the simplex algorithm we must eliminate all negative right-hand constants by means of a sequence of pivot transformations. These initial pivots are referred to as **Phase 1** of the simplex method, the goal of which is to move us from the origin to some corner point of the constraint set.

Example 1, Continued   Pivot on the entry −1 in the third row, y-column of the initial tableau.

Tableau
I

x	y	u	v	w	F	
6	5	1	0	0	0	300
−2	−3	0	1	0	0	−60
−2	(−1)	0	0	1	0	−40
5	4	0	0	0	1	0

Solution   The designated pivot entry is circled. After the pivot transformation the tableau becomes

Tableau
II

x	y	u	v	w	F	
−4	0	1	0	5	0	100
4	0	0	1	−3	0	60
2	1	0	0	−1	0	40
−3	0	0	0	4	1	−160

Notice that all three right-hand constants are now nonnegative, which means that Phase 1 is completed. The basic solution corresponding to II is

$$x = 0, \quad y = 40, \quad u = 100, \quad v = 60, \quad w = 0, \quad F = -160$$

which represents the corner point $(0, 40)$ in Figure 2.

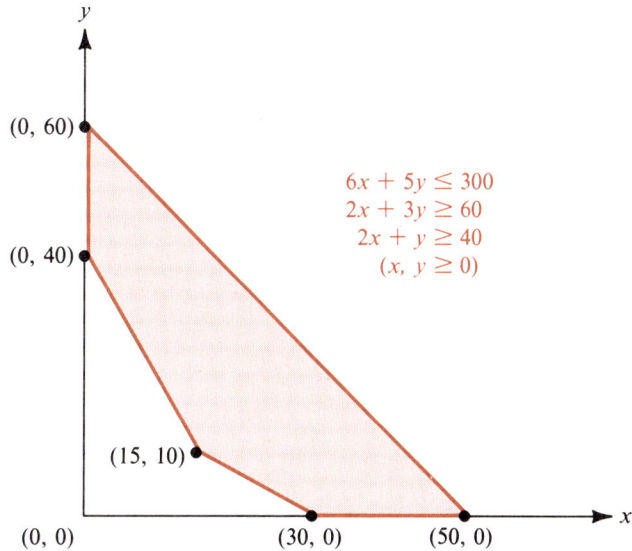

Figure 2

We can now proceed with **Phase 2** of the simplex method—reaching the optimal corner point—by applying the simplex algorithm of Section 5.1.

Tableau II	x	y	u	v	w	F		Quotients
	-4	0	1	0	5	0	100	
	④	0	0	1	-3	0	60	$60/4 = 15$ ←
	2	1	0	0	-1	0	40	$40/2 = 20$
	-3	0	0	0	4	1	-160	Smallest quotient

↑ Most negative indicator

Following the smallest-quotient rule, we pivot on the entry 4 in the second row, x-column.

Tableau III	x	y	u	v	w	F	
	0	0	1	1	2	0	160
	1	0	0	$\frac{1}{4}$	$-\frac{3}{4}$	0	15
	0	1	0	$-\frac{1}{2}$	$\frac{1}{2}$	0	10
	0	0	0	$\frac{3}{4}$	$\frac{7}{4}$	1	-115

Since III has no negative indicators it is terminal, and the corresponding basic solution is optimal.

$$\left. \begin{array}{l} \text{maximum } F = -115 \\ \text{minimum } G = 115 \end{array} \right\} \quad \text{when} \quad x = 15, y = 10 \quad (u = 160, v = w = 0)$$

This is the corner point $(15, 10)$ in Figure 2. ◆

Question 1:   How do we know which entries to use as pivots in Phase 1?

Answer: A complete version of the simplex algorithm* would include a step-by-step method for accomplishing Phase 1, using a device known as *artificial variables*. Since the description of this procedure is rather lengthy, and the linear programs discussed here involve only a few variables, we rely instead on trial-and-error. That is, we try to find *some* sequence of pivots that will eliminate the negative constants.

Question 2:   Can any nonzero entry be tried as a pivot in Phase 1?

Answer: Except for entries in the indicator row, *any* nonzero entry—positive or negative—can be tried as a pivot in Phase 1. Remember that *the goal of Phase 1 is to eliminate negative right-hand constants*. Only *after* this is accomplished do we look for negative indicators.

Question 3:   Why is it necessary to use pivot transformations to eliminate negative right-hand constants? Can't we just multiply a row by -1 to make the right-hand constant positive?

Answer: No, because when this is done a basic variable is lost and the tableau no longer represents a "solved" system of equations.

In some cases we need more than one pivot to accomplish Phase 1, as the next example illustrates.

Example 2

Minimize $G = 3x + y + 2z$
subject to:
$$x + y \leq 10$$
$$y + z \geq 15$$
$$x + z \leq 10.$$
$$(x, y, z \geq 0)$$

Solution

We first convert the program to a maximum program, and write the functional constraints in $\leq$ form.

Maximize $F = -G = -3x - y - 2z$
subject to:
$$x + y \leq 10$$
$$-y - z \leq -15$$
$$x + z \leq 10.$$
$$(x, y, z \geq 0)$$

Adding slack variables $u, v, w \geq 0$, we obtain the initial tableau.

---

*See for example David G. Luenberger, *Introduction to Linear and Nonlinear Programming* (Reading, Mass.: Addison-Wesley, 1973).

Tableau I

	x	y	z	u	v	w	F	
	1	1	0	1	0	0	0	10
	0	(−1)	−1	0	1	0	0	−15
	1	0	1	0	0	1	0	10
	3	1	2	0	0	0	1	0

Phase 1 Since a negative right-hand constant occurs in the initial Tableau I, we first try an arbitrary pivot—for instance, on the entry −1 in the y-column. The result is the following tableau:

Tableau II

	x	y	z	u	v	w	F	
	1	0	−1	1	1	0	0	−5
	0	1	1	0	−1	0	0	15
	1	0	(1)	0	0	1	0	10
	3	0	1	0	1	0	1	−15

Since II again contains a negative right-hand constant, we try another pivot —for instance, on the entry 1 in the third row, z-column.

Tableau III

	x	y	z	u	v	w	F	
	2	0	0	1	1	1	0	5
	−1	1	0	0	−1	−1	0	5
	1	0	1	0	0	1	0	10
	2	0	0	0	1	−1	1	−25

We are in luck: All three right-hand constants in III are nonnegative, so we have reached a corner point (namely, the point $(0, 5, 10)$ in xyz-space).

Phase 2 We can now apply the simplex algorithm as in Section 5.1, in order to reach the optimal corner point.

Tableau III

	x	y	z	u	v	w	F		Quotients
	2	0	0	1	1	(1)	0	5	5/1 = 5 ←
	−1	1	0	0	−1	−1	0	5	
	1	0	1	0	0	1	0	10	10/1 = 10
	2	0	0	0	1	−1	1	−25	

By the smallest-quotient rule, pivot on the entry 1 in the first row, w-column.

Tableau IV

	x	y	z	u	v	w	F	
	2	0	0	1	1	1	0	5
	1	1	0	1	0	0	0	10
	−1	0	1	−1	−1	0	0	5
	4	0	0	1	2	0	1	−20

Tableau IV is terminal, so the corresponding basic solution is optimal.

$$\left.\begin{array}{l} \text{maximum } F = -20 \\ \text{minimum } G = 20 \end{array}\right\} \quad \text{when} \quad x = 0, \quad y = 10, \quad z = 5 \quad \blacklozenge$$

Remark to Reader Although our examples above were both minimum programs, the problem of negative constants can arise with maximum programs as well. Geometrically, the presence of a negative constant means that the origin lies outside the constraint set. In such a case, our first task is to reach an initial corner point (Phase 1). Once there, the algorithm in Section 5.1 will always take us around the edges of the constraint set to the optimal corner point (Phase 2).

Summary

The two-phase simplex algorithm for nonstandard programs can be summarized as follows:

1. Convert the given program to a maximum program, if necessary, and write all functional constraints in $\leq$ form.

2. Add slack variables and form the initial tableau.

3. *Phase 1:* If negative right-hand constants occur, perform a sequence of pivots until all such constants have been eliminated.

4. *Phase 2:* If negative indicators occur, apply the simplex algorithm of Section 5.1 until all such indicators have been eliminated.

5. Read the optimal solution from the terminal tableau.

5.2 Exercises

For the following exercises, both odd- and even-numbered, hints on the selection of Phase 1 pivot entries can be found in the answer section at the back of the book.

In Exercises 1 and 2, write the initial simplex tableau for the given linear program, but go no further.

1. Minimize $G = 5x + 4y$
   subject to:
   $$-x + y \leq 5$$
   $$x + 2y \geq 20$$
   $$x + y \geq 15.$$
   $$(x, y \geq 0)$$

2. Minimize $G = x - y + 2z$
   subject to:
   $$2x + y + 4z \geq 20$$
   $$x - 2y + z \geq 40$$
   $$x + y + z \geq 30.$$
   $$(x, y, z \geq 0)$$

In Exercises 3 and 4 a terminal tableau is shown for a minimum program. What is the optimal solution?

3.

Decision variables

x	y	u	v	w	$-G$	
0	0	$\frac{1}{2}$	1	$-\frac{3}{2}$	0	5
1	0	$-\frac{1}{2}$	0	$-\frac{1}{2}$	0	5
0	1	$\frac{1}{2}$	0	$-\frac{1}{2}$	0	10
0	0	$\frac{1}{2}$	0	$\frac{9}{2}$	1	-65

4.

Decision variables

x	y	z	u	v	w	$-G$	
0	-5	-2	1	-2	0	0	60
1	-2	1	0	-1	0	0	40
0	-3	0	0	-1	1	0	10
0	1	1	0	1	0	1	-40

In Exercises 5–10, solve the given nonstandard program by the simplex method and also by the method of graphing. Indicate on your graph the path described by the simplex pivots.

5. Maximize $F = 4x + 7y$
   subject to:
   $$3x + y \geq 30$$
   $$x + y \leq 20.$$
   $$(x, y \geq 0)$$

6. Minimize $G = x + y$
   subject to:
   $$2x + y \geq 24$$
   $$x + 3y \geq 27.$$
   $$(x, y \geq 0)$$

7. Minimize $G = 6x + 2y$
   subject to:
   $$x + 4y \geq 40$$
   $$4x + y \geq 40$$
   $$x + y \geq 25.$$
   $$(x, y \geq 0)$$

8. Minimize $G = 10x + 20y$
   subject to:
   $$5x + 2y \geq 20$$
   $$y \geq 5$$
   $$x - y \leq 0.$$
   $$(x, y \geq 0)$$

9. Minimize $G = 5x + 4y$
   subject to:
   $$x + y \geq 20$$
   $$x - y \leq 10$$
   $$-x + y \leq 10.$$
   $$(x, y \geq 0)$$

10. Maximize $F = 3x + 2y$
    subject to:
    $$x + y \leq 4$$
    $$x \geq 1$$
    $$y \geq 1.$$
    $$(x, y \geq 0)$$

In Exercises 11–16, solve the given nonstandard program by the simplex method.

11. Minimize $G = 2x + 4y + z$
    subject to:
    $$x + 2y + 3z \geq 80$$
    $$x + y + z \geq 60.$$
    $$(x, y, z \geq 0)$$

12. Minimize $G = 5x + 4y + 3z$
    subject to:
    $$2x + y + z \geq 40$$
    $$x + y + 2z \geq 50.$$
    $$(x, y, z \geq 0)$$

13. Maximize $F = 4x - y + 2z$
    subject to:
    $$x + y + z \leq 8$$
    $$x - y + z \geq 4$$
    $$-x + y + z \geq 2.$$
    $$(x, y, z \geq 0)$$

14. Minimize $G = x + 4y + 3z$
    subject to:
    $$x + y + z \geq 20$$
    $$y + z \geq 15$$
    $$z \geq 10.$$
    $$(x, y, z \geq 0)$$

15. Minimize $G = 2x + y + z + 2w$
    subject to:
    $$z + w \leq 50$$
    $$x + z \geq 60$$
    $$y + w \geq 40.$$
    $$(x, y, z, w \geq 0)$$

16. Maximize $F = x + y + z + w$
    subject to:
    $$x - y \geq 1$$
    $$z - w \geq 2$$
    $$x + z + w \leq 5.$$
    $$(x, y, z, w \geq 0)$$

In Exercises 17–20, formulate the given problem as a linear program and solve by the simplex method.

17. The heating plant for a large government building must produce at least 300 million Btu's of heat per hour. The plant is equipped to burn any combination of fuel oil and coal, but it must observe government pollution limits of at most 300 lb of sulfur dioxide and 125 lb of nitrogen dioxide emitted per hour. The cost, heat

yield and pollution content per ton for each of the fuels are summarized in the following table:

Fuel	Heat (million Btu per ton)	SO$_2$ (lb per ton)	NO$_2$ (lb per ton)	Cost (dollars per ton)
Oil	40	20	5	290
Coal	20	30	20	65

How many tons of each fuel should the plant burn per hour in order to meet all the requirements at minimum cost?

18. Nutritionists indicate that an adequate daily diet should provide at least 75 gm of carbohydrate, 60 gm of protein, and 60 gm of fat. Two foods are available; their nutritional yields and costs, per ounce, are as follows:

Food	Carb per oz	Protein per oz	Fat per oz	Cost per oz
A	3 gm	2 gm	4 gm	10¢
B	3 gm	4 gm	2 gm	15¢

How many ounces of each food should be combined per day to meet the nutritional requirements at lowest cost?

19. World Travel Service has agreed to transport a group of 1800 football fans from Los Angeles to Chicago for one of the season's big games. The agency can charter three different types of aircraft. Their operating costs for the trip, and their seating capacities, are as follows:

Type	Number of seats per plane First-class	Economy	Cost
A	20	80	$3000
B	40	60	$4000
C	50	50	$5000

If 600 of the passengers want to go first-class while the remaining 1200 have ordered economy seats, how many planes of each type should be assigned to the charter to minimize the total operating cost?

20. The Allegheny Mining Company operates three mines which produce both high-grade and low-grade ore. The daily yield of each mine, and the daily operating cost, are given in the table below.

Mine	Ore yield per day High-grade	Low-grade	Cost per day
1	2 tons	1 ton	$300
2	1 ton	2 tons	$200
3	0 tons	2 tons	$100

The company must fill an order for a total of 40 tons of high-grade ore and 50 tons of low-grade ore. How many days should each mine be operated in order to fill the order at lowest total cost?

Computer Projects	A. The PIVOT program performs a pivot transformation on a designated nonzero entry in a tableau. This program can be used in Exercises 5–20.
	B. The SIMPL program computes the optimal solution to a linear program by the simplex method. Use this program to confirm your answer in Exercises 5–20.
	C. Use the SIMPL program to solve the following nonstandard program:

$$\text{Minimize} \quad G = 2x + 7y + 9z + 7u + 8v$$
$$\text{subject to:}$$
$$x + 2y + z + u + v \le 148$$
$$x + y + 2z + u + v \ge 177$$
$$x + y + z + 2u + v \ge 152$$
$$2x - y + 2z \le 127$$
$$2y - z + 2u \le 182$$
$$2z - u + 2v \ge 137.$$
$$(x, y, z, u, v \ge 0)$$

5.3 Linear Programming Models

With the help of the simplex algorithm we can now solve any problem that has the mathematical form of a linear program. For large-scale programs the algorithm can be carried out by computer, and this is what is done in practice. Given the initial tableau as input, the computer can perform the necessary pivot transformations with blinding speed, finally printing out the terminal tableau and the optimal solution.

Unfortunately, the computer is not equipped to deal with the ambiguities of real-world problems as they actually occur. We are still left with the most important part of the task, namely translating the given problem into mathematical terms. In this section we examine some problems of a more sophisticated type than those presented earlier. From these examples, and the problems provided in the exercises, the reader can get a somewhat better idea of the true scope of linear programming.

Blending Problems In many industrial applications we are interested in finding an optimal blend of certain known ingredients, subject to constraints on the composition of the mixture.

Example 1 A cereal manufacturer has a grain silo with a 200-ton capacity. Initially the silo contains 75 tons of grain which is 30% barley and 10% oats by weight and costs $80 per ton. A mixture is to be formed by adding other grains to the silo. Three different grains are available:

Grain	Barley per ton	Oats per ton	Cost per ton
1	40%	25%	$100
2	40%	20%	$300
3	45%	20%	$350

The mixture must contain at least 35% barley and at most 15% oats. How can this be achieved at lowest cost?

Formulation   We begin by identifying the decision variables, or unknowns. Let x_1, x_2, and x_3 represent the number of tons of grains 1, 2, and 3, respectively, added to the silo to form the new mixture. Our objective is to minimize the total cost, in dollars.

$$C = 100x_1 + 300x_2 + 350x_3 + 6000$$

where $6000 = 75(80)$ is the cost of the grain initially in the silo. This is an example of a **linear function with constant term**. Since C will be minimized when the "linear part" ($C = 6000$) is minimized, the simplex algorithm can be extended to cover functions of this type.

Since the capacity of the silo is 200 tons, the total amount of mixture cannot exceed this limit. Thus:

$$75 + x_1 + x_2 + x_3 \leq 200$$
$$\text{or} \quad x_1 + x_2 + x_3 \leq 125.$$

The condition that the mixture must contain at least 35% barley can be expressed in the form

total amount of barley $\geq .35$(total amount of mixture)

(See Example 6 in Section 2.2). Measuring these amounts in tons, we have

$$\text{total amount of barley} = .30(75) + .40x_1 + .40x_2 + .45x_3$$
$$\text{total amount of mixture} = 75 + x_1 + x_2 + x_3.$$

Hence, the barley constraint becomes

$$22.5 + .40x_1 + .40x_2 + .45x_3 \geq .35(75 + x_1 + x_2 + x_3).$$

Multiplying out on the right side and collecting terms, we obtain

$$.05x_1 + .05x_2 + .10x_3 \geq 3.75.$$

This inequality can be multiplied through by 20 to eliminate fractions.

$$x_1 + x_2 + 2x_3 \geq 75$$

Similarly, the condition that the mixture must contain at most 15% oats can be expressed as

total amount of oats $\leq .15$(total amount of mixture)

which takes the mathematical form

$$.10(75) + .25x_1 + .20x_2 + .20x_3 \leq .15(75 + x_1 + x_2 + x_3).$$

After simplification we arrive at the linear inequality

$$2x_1 + x_2 + x_3 \leq 75.$$

Our blending problem therefore translates into the following linear program:

$$\text{Minimize} \quad C = 100x_1 + 300x_2 + 350x_3 + 6000$$

subject to:

$$x_1 + x_2 + x_3 \le 125$$
$$x_1 + x_2 + 2x_3 \ge 75$$
$$2x_1 + x_2 + x_3 \le 75.$$
$$(x_1, x_2, x_3 \ge 0)$$

The initial simplex tableau for this program is the following:

x_1	x_2	x_3	s_1	s_2	s_3	$-C$	
1	1	1	1	0	0	0	125
-1	-1	-2	0	1	0	0	-75
2	1	1	0	0	1	0	75
100	300	350	0	0	0	1	-6000

A single pivot, on -1 in the x_2-column, will eliminate negative right-hand constants. Using the simplex algorithm, the optimal solution is found to be

$$\text{minimum } C = \$17,250 \quad \text{when} \quad x_1 = 25, \quad x_2 = 0, \quad x_3 = 25. \quad \blacklozenge$$

Transportation Problems   Often the key to problem formulation lies in making the right choice of decision variables. An example is the typical *transportation problem,* in which goods are to be shipped from fixed sources to fixed destinations at lowest cost.

Example 2   The Consolidated Steel Corporation has two plants, P_1 and P_2, and three distributors, D_1, D_2, and D_3. The two plants have supplies of 55,000 tons and 30,000 tons of steel, respectively. The three distributors need at least 30,000, 20,000, and 20,000 tons, respectively. The transportation cost varies depending on the plant and the distributor, as shown below. (Figures represent dollars per ton shipped.)

	To distributor		
	D_1	D_2	D_3
From plant $\quad P_1$	\$2	\$4	\$3
P_2	\$5	\$2	\$2

How can the company minimize its total transportation cost?

Formulation   We must determine how many tons of steel to send from each plant to each distributor. Thus, there are six decision variables.

$$x_{ij} = \text{number of tons sent from } P_i \text{ to } D_j \quad (i = 1, 2; j = 1, 2, 3)$$

These unknowns are shown explicitly in the following table:

		To distributor			
		D_1	D_2	D_3	Supply
Number of tons	P_1	x_{11}	x_{12}	x_{13}	55,000
sent from plant	P_2	x_{21}	x_{22}	x_{23}	30,000
Demand		30,000	20,000	20,000	

The total transportation cost, in dollars, is given by the linear function

$$C = 2x_{11} + 4x_{12} + 3x_{13} + 5x_{21} + 2x_{22} + 2x_{23}.$$

The total amount of steel shipped from plant P_1 cannot exceed the supply at P_1, which is 55,000 tons. Similarly, the total amount shipped from P_2 cannot exceed 30,000 tons. We thus have the following supply constraints:

$$x_{11} + x_{12} + x_{13} \le 55,000$$
$$x_{21} + x_{22} + x_{23} \le 30,000$$

In addition, we have demand constraints. The total amount of steel shipped to distributor D_1 must be at least 30,000 tons, with similar conditions for the other two distributors. Mathematically, these conditions say that

$$x_{11} + x_{21} \ge 30,000$$
$$x_{12} + x_{22} \ge 20,000$$
$$x_{13} + x_{23} \ge 20,000.$$

We thus arrive at the following six-variable linear program:

Minimize $C = 2x_{11} + 4x_{12} + 3x_{13} + 5x_{21} + 2x_{22} + 2x_{23}$
subject to:

$$x_{11} + x_{12} + x_{13} \le 55,000$$
$$x_{21} + x_{22} + x_{23} \le 30,000$$
$$x_{11} + x_{21} \ge 30,000$$
$$x_{12} + x_{22} \ge 20,000$$
$$x_{13} + x_{23} \ge 20,000.$$
$$(x_{ij} \ge 0)$$

The initial simplex tableau for this minimum program is shown below.

x_{11}	x_{12}	x_{13}	x_{21}	x_{22}	x_{23}	s_1	s_2	s_3	s_4	s_5	$-C$	
1	1	1	0	0	0	1	0	0	0	0	0	55,000
0	0	0	1	1	1	0	1	0	0	0	0	30,000
−1	0	0	−1	0	0	0	0	1	0	0	0	−30,000
0	−1	0	0	−1	0	0	0	0	1	0	0	−20,000
0	0	−1	0	0	−1	0	0	0	0	1	0	−20,000
2	4	3	5	2	2	0	0	0	0	0	1	0

Here several Phase 1 pivots are needed to eliminate the negative right-hand constants. The optimal solution turns out to be

$$\text{minimum } C = \$150,000 \quad \text{when}$$

$$x_{11} = 30,000 \qquad x_{21} = 0$$
$$x_{12} = 0 \qquad x_{22} = 20,000$$
$$x_{13} = 10,000 \qquad x_{23} = 10,000.$$

This optimal shipping plan is illustrated in Figure 1.

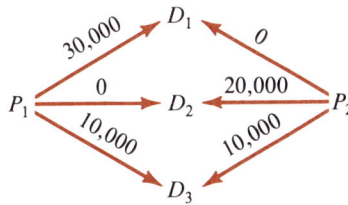

Figure 1

Since the tableaus arising in transportation problems have a very special form, methods of solution have been devised which are considerably more efficient for problems of this type than the simplex method. Of course, in realistic applications the number of sources and destinations may both be quite large, so the computer plays an essential role in determining optimal shipping plans.

Time-Period Problems   There are many practical situations in which we must choose between options covering different time periods. Here again, the key to formulation is defining the decision variables correctly. A simple leasing problem will illustrate.

Example 3   During the ski season, World Travel Service offers a one-month vacation in Aspen, including round-trip fare and accommodations at the Aspen Lodge. The number of rooms needed at the Lodge for each of the months December through March are as follows:

Month	Number of rooms needed
December	18
January	25
February	12
March	20

The Lodge leases its rooms to World Travel at a discount rate of $800 for two consecutive months or $1000 for three consecutive months. What leasing plan will be least expensive?

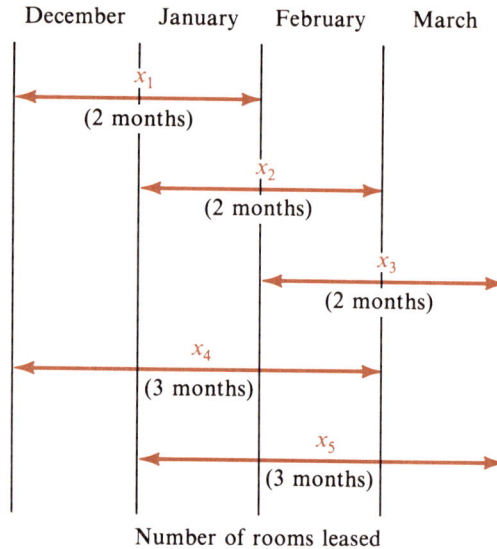

Figure 2

Formulation   The Travel Service must decide how many rooms to lease for two months and how many for three months, and also which periods the various leases should cover. The five unknowns are illustrated in Figure 2. For example, x_1 represents the number of rooms leased for two months beginning December 1, x_2 the number leased for two months beginning January 1, and so on. The total cost is given by

$$C = 800x_1 + 800x_2 + 800x_3 + 1000x_4 + 1000x_5.$$

From Figure 2 we see that the total number of rooms available for December will be $x_1 + x_4$. Thus, the condition that 18 rooms are needed for this month translates into the constraint

$$x_1 + x_4 \geq 18.$$

The requirements for the remaining three months lead in the same way to the conditions

$$x_1 + x_2 + x_4 + x_5 \geq 25$$
$$x_2 + x_3 + x_4 + x_5 \geq 12$$
$$x_3 + x_5 \geq 20.$$

The problem can thus be formulated as a five-variable minimum program.

Minimize $C = 800x_1 + 800x_2 + 800x_3 + 1000x_4 + 1000x_5$
subject to:
$$x_1 + x_4 \geq 18$$
$$x_1 + x_2 + x_4 + x_5 \geq 25$$
$$x_2 + x_3 + x_4 + x_5 \geq 12$$
$$x_3 + x_5 \geq 20.$$
$$(x_i \geq 0)$$

Using the simplex algorithm, the optimal solution is found to be

$$\text{minimum } C = \$31,800 \quad \text{when}$$
$$x_1 = 18, \quad x_2 = 0, \quad x_3 = 13, \quad x_4 = 0, \quad x_5 = 7.$$

Notice that under this plan, the Travel Service will have eight unused rooms during the month of February.   ◆

Equality Constraints   In formulating word problems as linear programs we sometimes encounter **equality constraints,** that is, conditions that correspond to equations rather than inequalities. For instance, in the transportation problem discussed earlier (Example 2), we might wish to impose the condition that distributor D_1 receives *exactly* 30,000 tons of steel. This would be expressed by the equality constraint

$$x_{11} + x_{21} = 30,000. \tag{1}$$

In practice such constraints are generally handled using the artificial-variable technique, which was mentioned in Section 5.2 in connection with Phase 1 of the simplex algorithm. For our purposes it is sufficient to use the fact that an equation is equivalent to a pair of inequalities. Thus, (1) can be replaced by the two conditions

$$x_{11} + x_{21} \leq 30,000$$
$$x_{11} + x_{21} \geq 30,000$$

which together say the same thing as (1).

Remark to Reader   In the exercises below you are asked to formulate various word problems as linear programs. The preceding examples illustrate some types of applications, but in general *each problem calls for its own mathematical analysis.* There is no "mechanical" method to follow. We can, however, offer a few suggestions.

1. Begin by introducing variables for the unknown quantities, specifying clearly what each one stands for.
2. Express the quantity to be optimized in terms of the decision variables.
3. Use a data table or figure (for example, Figure 2 in the text) where helpful.
4. Formulate each constraint as an inequality, simplifying the result as much as possible. Be careful to note the units (pounds, tons, dollars, or whatever) in which each quantity is measured.

Once formulated, these linear programs can be solved either by hand or with the aid of a computer simplex routine. Optimal solutions to the odd-numbered exercises are given in the answer section at the back of the book. For the reader interested in additional applications of linear programming we recommend Saul I. Gass, *An Illustrated Guide to Linear Programming* (New York: McGraw-Hill, 1970).

5.3 Exercises

Formulate each of the following problems as a linear program.

1. The owner of a nut shop has 50 lb of a mixture which is 40% cashews and 50% peanuts by weight. To this amount she adds certain quantities of the mixtures described below.

Mixture	Cashews	Peanuts	Cost per lb
A	50%	5%	$4
B	80%	10%	$6

The final blend is to contain at least 60% cashews and at most 20% peanuts. How can this be achieved at lowest cost?

2. A metallurgist wants to make a new bronze alloy by combining certain amounts of the three alloys described below (percentages by weight).

Alloy	Copper	Zinc	Cost per ton
A	95%	1%	$500
B	90%	8%	$450
C	75%	5%	$200

There must be at least 10 tons of the new alloy, and its composition must be at least 80% copper and at most 5% zinc. How can this be done at lowest cost?

3. The Exxess Oil Company has three grades of gasoline stored at its refinery.

Grade	Lead per gal	Cost per gal
A	1 mg	$.80
B	2 mg	$.50
C	5 mg	$.40

The company wants to blend together certain quantities of the three grades to obtain exactly 2000 gallons of mixture containing at most 1.5 mg of lead per gallon. How can this be achieved at lowest cost?

4. In Exercise 3, suppose the company wants to obtain exactly 2000 gallons of mixture costing at most 44¢ per gallon. What blend will have the lowest lead content?

5. The Hurts Car Rental Company must reallocate its rental cars between its agencies in various cities. At the moment, city A_1 has a surplus of 50 cars and city A_2 has a surplus of 30 cars, while cities B_1, B_2, and B_3 need at least 15, 20, and 25 cars, respectively. The mileage figures between the A cities and the B cities are given in the table below.

From city		To city	
	B_1	B_2	B_3
A_1	100	200	150
A_2	50	100	200

(number of miles)

How many cars should be driven from each of the A cities to each of the B cities in order to meet the requirements with minimum total mileage driven?

6. Gulp Oil has three North Sea pumping stations, S_1, S_2, and S_3, and two mainland refineries, R_1 and R_2. The cost of shipping a ton of crude oil from each station to each refinery is given below in dollars per ton.

		To refinery	
		R_1	R_2
From station	S_1	$50	$30
	S_2	$40	$75
	S_3	$20	$15

The three stations can produce a maximum of 40, 80, and 20 tons per day, respectively. The two refineries need at least 60 and 70 tons per day, respectively. How can the company meet these needs at lowest cost?

7. A transport company has two types of trucks available for overland shipping.

Type	Supply	Freight capacity
1	24	6 tons
2	60	4 tons

Cities A and B have placed urgent orders for 120 and 180 tons of freight, respectively. Because of the distances involved, no truck can be used more than once. The cost of sending a truck on one complete journey from the central terminal to each of the cities is given below.

		City A	City B
Truck type	1	$400	$500
	2	$300	$400

How can the company meet the cities' requirements at minimum total cost?

8. A shoe company buys a certain type of ski boot made in countries 1 and 2 and exports them to countries A, B, and C. The unit cost and available supplies in countries 1 and 2 are as follows:

Country	Cost per pair	Number available
1	$20	30,000
2	$15	50,000

The boots sell for $50, $60, and $75 in countries A, B, and C, respectively. Due to import quotas, the company cannot send more than 20,000 pairs to country A, 25,000 to country B, or 20,000 to country C. The unit cost of transportation is given in the following table:

		To country		
		A	B	C
From country	1	$10	$5	$15
	2	$5	$20	$25

How can the company maximize its total profit (revenue minus direct costs and transportation costs)?

9. A medical society is planning a convention at the Essex Hotel which will last for three days. The organizing committee has determined that 5 overhead projectors will be needed the first day, 20 the second day, and 10 the third day. The machines can be rented for any one day at a cost of $30, or for any two consecutive days at a cost of $50. What plan will be most economical?

10. Shipping tycoon Stavros Pythagoras has agreed to transport liquified natural gas (LNG) from Algeria to the U.S. during three winter months. The shipments will require 8 tankers in January, 14 in February, and 10 in March. Unfortunately, the firm has no LNG tankers of its own, so they must be leased on a short-term basis. The lease rates per tanker are $50,000 for one month, $90,000 for two consecutive months, and $120,000 for three consecutive months. Assume the round trip takes exactly one month, and lease periods begin on the first of each month. What leasing plan will minimize the total transportation cost?

11. In Exercise 10, suppose that any ship whose lease period begins on February 1 costs 50% more than usual, due to increased demand during that month. What leasing plan will be optimal?

12. An investor has $10,000 available at the beginning of Year 1, and four investment opportunities, as follows:

Investment	Available	Duration	Return	Maximum amount
A	Now	2 years	30%	$4000
B	1 year	2 years	30%	$4000
C	1 year	1 year	20%	$5000
D	2 years	1 year	10%	$5000

For example, Investment B will be available at the beginning of Year 2 for a maximum amount of $4000, and at the end of Year 3 it will yield a return of 30% on the amount invested. Find an optimal plan for this investor, that is, one that will maximize his or her capital at the end of Year 3.

13. A builder's supply company makes mineral-surface roofing paper in a standard roll which is 30 feet wide. These standard rolls are then cut into widths of 16, 12, and 8 feet for sale to retail distributors. There are five cutting patterns for the standard roll, as described below.

	Number of rolls			
Cutting plan	16 ft	12 ft	8 ft	Trim waste
1	1	1	0	2 ft
2	1	0	1	6 ft
3	0	2	0	6 ft
4	0	1	2	2 ft
5	0	0	3	6 ft

The company has received an order for 200 16-ft rolls, 400 12-ft rolls, and 300 8-ft rolls. What method should be used to fill the order, if the goal is to minimize the total amount of trim waste?

14. In Exercise 13, suppose the company must fill the order as soon as possible. In such a case the objective would be to minimize the total number of standard rolls that have to be cut, without regard to trim waste. What method should now be used to fill the order?

Computer Projects

The SIMPL program computes the optimal solution to a linear program by the simplex method. This program can be used in Exercises 1–14.

Chapter 5   Summary

Important Terms

5.1 standard maximum program
slack variable
initial simplex tableau

indicator
simplex algorithm

Review Exercises

In Exercises 1–6, solve the linear program by the simplex method.

1. Maximize $F = 5x + 2y$
   subject to:
   $$2x + y \leq 20$$
   $$3x + y \leq 24$$
   $$x \leq 6.$$
   $$(x, y \geq 0)$$

2. Maximize $F = 2x + 5y$
   subject to:
   $$3x + y \leq 90$$
   $$3x + 5y \leq 150$$
   $$-x + y \leq 22.$$
   $$(x, y \geq 0)$$

3. Minimize $G = 3x + 2y$
   subject to:
   $$x + y \geq 50$$
   $$x + 2y \leq 80$$
   $$x + 3y \leq 90.$$
   $$(x, y \geq 0)$$

4. Minimize $G = 4x + y$
   subject to:
   $$10x + 3y \geq 120$$
   $$12x + 5y \geq 120$$
   $$5x + y \geq 50.$$
   $$(x, y \geq 0)$$

5. Maximize $F = x + 2y + 3z$
   subject to:
   $$2x + y + z \leq 12$$
   $$x + 2y + z \leq 20$$
   $$x + y + 2z \leq 20.$$
   $$(x, y, z \geq 0)$$

6. Minimize $G = 5x + 3y + 2z$
   subject to:
   $$2x + y \geq 20$$
   $$x + y + 2z \geq 30$$
   $$x + 2z \geq 40.$$
   $$(x, y, z \geq 0)$$

In Exercises 7–10, formulate the problem as a linear program and solve it by the simplex method.

7. A mining company operates three mines which produce titanium ore. The ore from Mine A is 5% titanium by weight and requires 4 work-hours per ton to mine. The ore from Mine B is 3% titanium and requires 3 work-hours per ton, while the ore from Mine C is 2% titanium and requires 1 work-hour per ton. The company has 800 work-hours of mining labor available per day, and its refinery can

handle up to 500 tons of ore per day. How much ore should be taken daily from each mine in order to maximize the total amount of titanium produced?

8. An amusement park has "A" rides, which cost $1, and "B" rides, which cost $2. An economy book containing 10 A-ride tickets and 5 B-ride tickets can be purchased for $18 (a 10% discount), and a super economy book containing 25 A-ride tickets and 10 B-ride tickets can be purchased for $36 (a 20% discount). A teacher is bringing a class of 20 students to the park and each has been promised 5 A-rides and 3 B-rides. Thus, she must purchase at least 100 A-ride tickets and 60 B-ride tickets. How can this be done at lowest cost?

9. A large food chain buys onions from two farms, F_1 and F_2, and ships them to its two regional distributors, D_1 and D_2. The two farms can supply a maximum of 8 carloads and 15 carloads per day, respectively, and the distributors need at least 12 carloads and 7 carloads per day, respectively. The transportation cost, in dollars per carload, from each farm to each distributor is shown below:

		To distributor	
		D_1	D_2
From farm	F_1	$1200	$2500
	F_2	$1800	$900

How can the food chain minimize its total transportation cost?

10. A hotel must rent folding chairs for a five-day political convention. The hotel manager estimates that 450 chairs will be needed on Monday, 675 on Tuesday, 750 on Wednesday, 600 on Thursday, and 375 on Friday. The rental rate is $8 per chair for the entire week, or $5 per chair for any consecutive three-day period. There is no charge for delivery or pickup. What rental plan will be least expensive?

6

Sets and Counting

Counting the elements in a set is perhaps the most primitive of all mathematical skills, one which most of us develop even before we enter grade school. Some counting problems, though, continue to pose a challenge: finding the number of possible poker hands, for example, or the number of different genetic codes that can be carried by a DNA molecule. In this chapter are developed some sophisticated techniques for counting, which will play an important role in the study of probability theory (Chapter 7).

6.1 Elements of Set Theory

The notion of a set is basic not only in probability theory but throughout mathematics. In this section we introduce some elementary concepts and notation which will be useful in our later work.

A **set** is an arbitrary collection of objects, called its **elements** or **members.** If the object x is an element of the set A we write $x \in A$; if not, we write $x \notin A$. We can specify a set by listing its elements between braces. For example, the set

$$A = \{1, 3, 5, 7, 9\}$$

contains as elements just the five numbers indicated; hence $3 \in A$ and $5 \in A$ but $2 \notin A$. This set A can also be described using the *set-builder* notation, as

$$\{x \mid x \text{ is a positive odd integer less than 10}\}$$

which is read: "the set of all objects x such that x is a positive odd integer less than 10." Similarly, the set

$$\{x \mid x \text{ is a vowel in the English alphabet}\}$$

contains as elements the letters a, e, i, o, and u, and the set

$$\{x \mid x \text{ is a positive prime integer}\}$$

contains the so-called **prime numbers** 2, 3, 5, 7, 11, . . .; that is, those numbers > 1 which cannot be written as a product of two smaller numbers. Notice that in order to describe an infinite set such as the set of all prime numbers we *must* use this type of notation, since it is impossible to list all the elements of the set.

As indicated in the foregoing examples, the **set-builder** notation

$$\{x \mid P(x)\}$$

denotes the set of all objects x such that the condition $P(x)$ is true. (Here "$P(x)$" simply stands for a statement containing the symbol "x.") For the purposes of a given discussion, however, we limit ourselves to those objects x *of a certain type:* real numbers, for instance, or points in the plane.

Definitions

The set which contains *all* the objects under consideration is called the **universal set,** and is denoted by U.

The set which contains *none* of the objects under consideration is called the **empty set** or **null set,** and is denoted by $\varnothing$.

In general, the notation $\{x \mid P(x)\}$ denotes the set of all objects x in U such that the condition $P(x)$ is true.

Example 1 Find the elements in the universe

$$U = \text{the set of real numbers}$$

which belong to the following sets:

(a) $\{x \mid x^2 + x - 2 = 0\}$ (b) $\{x \mid x^2 - 2x + 1 \geq 0\}$ (c) $\{x \mid x^2 + 1 = 0\}$

Solution (a) The set in question consists of all real numbers x such that $x^2 + x - 2 = 0$. Since $x^2 + x - 2 = (x - 1)(x + 2)$, we see that this equation has solutions $x = 1$ and $x = -2$. Hence

$$\{x \mid x^2 + x - 2 = 0\} = \{1, -2\}.$$

(b) Since $x^2 - 2x + 1 = (x - 1)^2$ and the square of a real number is always nonnegative, we see that *every* real number satisfies the condition $x^2 - 2x + 1 \geq 0$. That is,

$$\{x \mid x^2 - 2x + 1 \geq 0\} = U.$$

(c) The equation $x^2 + 1 = 0$ has no real solution x. Hence

$$\{x \mid x^2 + 1 = 0\} = \varnothing. \quad \blacklozenge$$

Equality and Inclusion   The most important relation between sets is that of equality.

Definition

Sets A and B are considered **equal,** and we write $A = B$, when they have exactly the same elements.

Thus, for example, we have

$$\{2, 3, 5\} = \{5, 2, 3\} = \{2, 3, 3, 5\}$$

because all three sets contain exactly the same elements, namely 2, 3, and 5. Notice that the elements of a set can be listed in any order we please, with or without repetitions. In practice we usually delete repetitions in such a listing, since they have no effect.

The second fundamental set-theoretic relation is **inclusion.**

Definition

A is **included** in B, or A is a **subset** of B, when every element of A is also an element of B. In this case we write $A \subset B$.

For example, if we let

$$A = \{2, 3, 5\}, \quad B = \{1, 2, 3, 4, 5\}$$

then A is a subset of B, since all three elements of A belong to B as well; but B is not a subset of A, because there exist elements of B which are not elements of A (namely, 1 and 4).

Example 2   List all subsets of the set $A = \{a, b, c\}$.

Solution   There are eight subsets, containing 0, 1, 2, or 3 elements.

$$\varnothing, \{a\}, \{b\}, \{c\}, \{a, b\}, \{a, c\}, \{b, c\}, \{a, b, c\}$$

Notice that the empty set $\varnothing$ and the set A itself are listed among the subsets of A; so we have

$$\varnothing \subset A \quad \text{and} \quad A \subset A.$$

The first of these inclusions is justified by an indirect argument. If $\varnothing$ were *not* a subset of A then $\varnothing$ would contain some element not in A. But $\varnothing$ contains no elements at all, so this cannot be the case. The second inclusion just says that every element of A is an element of A.   ◆

Operations on Sets   We can combine sets in various ways to form new sets, in much the same way as we combine numbers in ordinary arithmetic. The operations we shall need are defined as follows:

Definition

The **union** of sets A and B, written $A \cup B$, is the set of all objects that belong *either* to A, *or* to B, or to both.

The **intersection** of sets A and B, written $A \cap B$, is the set of all objects that belong to *both A and B*.

The **complement** of a set A, written A^c, is the set of all objects (in the universe U) that do *not* belong to A.

In set-builder notation, these definitions become

$$
\begin{aligned}
\textbf{union:} && A \cup B &= \{x \mid x \in A \text{ or } x \in B\} \\
\textbf{intersection:} && A \cap B &= \{x \mid x \in A \text{ and } x \in B\} \\
\textbf{complement:} && A^c &= \{x \mid x \notin A\}.
\end{aligned}
$$

For three or more sets, the union and intersection are defined in the obvious way. For example,

$$A \cup B \cup C = \{x \mid x \in A \text{ or } x \in B \text{ or } x \in C\}.$$

To illustrate these operations, consider the sets

$$A = \{1, 2, 3, 4\}, \quad B = \{2, 4, 6\}, \quad C = \{1, 3, 5, 7\}$$

in the universe $\quad U = \{1, 2, 3, 4, 5, 6, 7, 8, 9, 10\}.$

Applying the definitions, we have, for instance,

$$
\begin{aligned}
A \cup B &= \{1, 2, 3, 4, 6\} \\
A \cap B &= \{2, 4\} \\
A^c &= \{5, 6, 7, 8, 9, 10\} \\
A \cap C &= \{1, 3\} \\
(A \cap C)^c &= \{2, 4, 5, 6, 7, 8, 9, 10\} \\
A \cup B \cup C &= \{1, 2, 3, 4, 5, 6, 7\}.
\end{aligned}
$$

Two sets are said to be **disjoint** if their intersection is empty; that is, they have no elements in common. The sets B and C above are disjoint, for instance, since $B \cap C = \varnothing$.

Venn Diagrams   We can illustrate the set operations by means of a pictorial device known as a **Venn diagram.*** Here sets are represented by simple regions within a rectangle U, which represents the universal set. Figure 1a shows two overlapping circles, representing "typical" sets A and B. The union, intersection and complement operations are illustrated by shading the corresponding regions as in Figures 1b–1d. Venn diagrams are helpful in visualizing the different logical combinations of several given properties, as we see in the next example.

---

*After John Venn (1834–1923), English clergyman and logician.

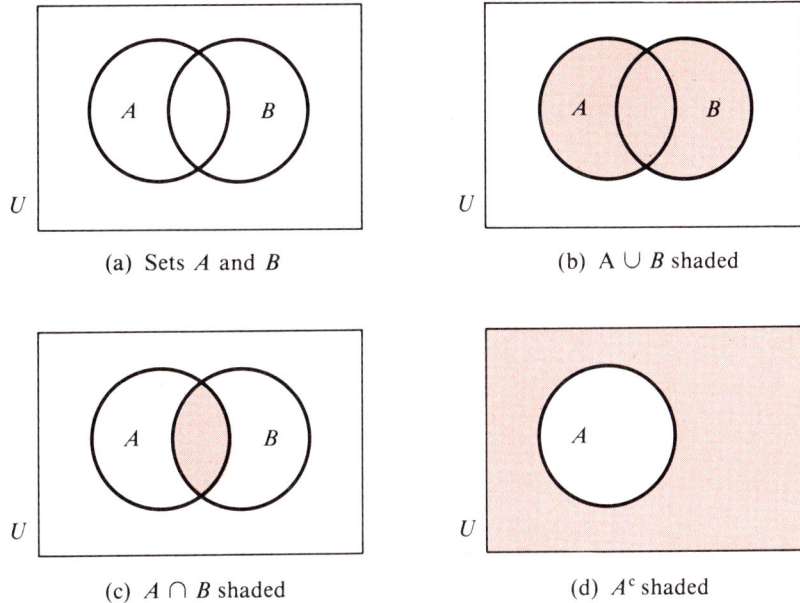

(a) Sets A and B

(b) $A \cup B$ shaded

(c) $A \cap B$ shaded

(d) A^c shaded

Figure 1

Example 3 Consider the sets

$$M = \text{set of all males}$$
$$A = \text{set of all adults}$$

in the universe

$$U = \text{set of all people.}$$

Discussion The sets M and A divide U into four nonoverlapping regions, numbered I–IV in the Venn diagram in Figure 2. Each of these regions can be described using the intersection and complement operations, as follows:

Region I $= M \cap A =$ set of all men (male adults)
Region II $= M \cap A^c =$ set of all boys (male nonadults)
Region III $= M^c \cap A =$ set of all women (nonmale adults)
Region IV $= M^c \cap A^c =$ set of all girls (nonmale nonadults)

Note that Region IV, consisting of all individuals who are neither male nor adults, can also be described as $(M \cup A)^c$. (See Exercise 15.)

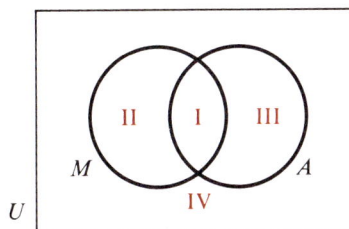

Figure 2

If we start with three sets rather than two, the corresponding Venn diagram will contain eight component regions instead of four.

Example 4   A nutritionist is conducting a survey of professional athletes to determine which vitamins, if any, they take regularly. Let U be the set of athletes surveyed, and let A, B, and C be the sets of those who take vitamins A, B, and C, respectively.

Discussion   The three sets are pictured as circular regions in the Venn diagram in Figure 3. Note that eight nonoverlapping regions are formed, numbered I–VIII in the diagram. For example,

$$\text{Region I} = A \cap B \cap C$$

consists of those athletes who regularly take all three vitamins, while

$$\text{Region III} = A \cap B^c \cap C$$

consists of those who take vitamins A and C but not B.

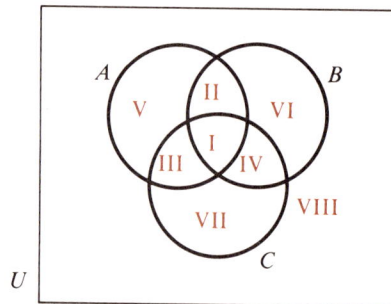

Figure 3

Example 4, Continued   Using set operations, describe the group consisting of those athletes who take
(a) at least one of the three vitamins
(b) exactly one of the three vitamins.

Solution   (a) The set in question contains those individuals who belong to set A or set B or set C (or more than one of the three). This is just the union $A \cup B \cup C$, which is pictured in Figure 4a.
(b) Those who take exactly one of the vitamins are represented in regions V, VI and VII of Figure 3. The set in question is thus the union of these regions, or

$$(A \cap B^c \cap C^c) \cup (A^c \cap B \cap C^c) \cup (A^c \cap B^c \cap C).$$

This set is shown in Figure 4b.

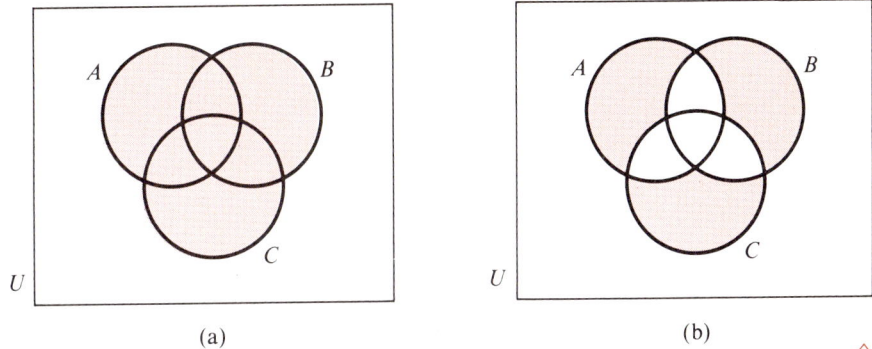

(a)   (b)

Figure 4   ◇

Applications to Counting   The method of Venn diagrams can be used to solve certain types of counting problems, in which various sets overlap.

Example 4,
Continued

Suppose that out of a total group of 150 athletes surveyed, it is found that

> 90 take vitamin A,
> 88 take vitamin B,
> 97 take vitamin C,
> 53 take both vitamin A and B,
> 55 take both vitamin A and C,
> 57 take both vitamin B and C,

and   32 take all three.

How many take none of the three vitamins?

Solution   We work backward from the given information. Since 32 athletes take all three vitamins, we place this number in the region $A \cap B \cap C$, as in Figure 5. Since 53 athletes take both A and B, and of these 32 also take C, there must be $53 - 32 = 21$ who take A and B but not C. Similarly, $55 - 32 = 23$ take A and C but not B, and $57 - 32 = 25$ take B and C but not A. We place these numbers in the corresponding regions, as in Figure 6.

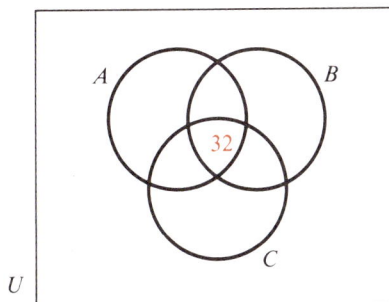

Figure 5

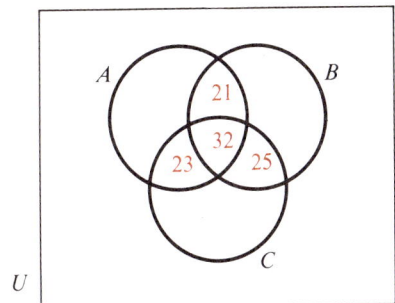

Figure 6

We have now accounted for $32 + 21 + 23 = 76$ of the 90 athletes who take vitamin A. This leaves $90 - 76 = 14$ who take A but not B or C. Similarly, $88 - 78 = 10$ take vitamin B alone, and $97 - 80 = 17$ take C alone. From Figure 7 we see that the union $A \cup B \cup C$ contains a total of

$$14 + 21 + 10 + 23 + 32 + 25 + 17 = 142$$

elements, leaving

$$150 - 142 = 8$$

athletes who take none of the three vitamins.

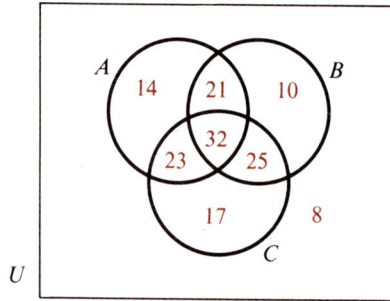

Figure 7

6.1 Exercises

In Exercises 1–8, let the universe U be the set of all positive integers,

$$U = \{1, 2, 3, \ldots\}.$$

List the elements in each of the following subsets of U:

1. $\{x \mid 2 < x \text{ and } x \le 8\}$
2. $\{x \mid x^2 - 5x + 6 = 0\}$
3. $\{x \mid 2x + 8 > 5x + 1 \text{ or } 3x + 5 = 20\}$
4. $\{x \mid \text{both 24 and 60 are divisible by } x\}$
5. $\{x \mid x + 10 \text{ is divisible by } x\}$
6. $\{x \mid x = 5n + 3 \text{ for some integer } n \le 4\}$
7. $\{x \mid x \text{ is odd and } x^2 \text{ is even}\}$
8. $\{x \mid x \le 5 \text{ or } 2x + 3 = 7\}$
9. Let $U = \{a, b, c, d\}$. List all subsets of U containing
   (a) exactly one element
   (b) exactly two elements
   (c) exactly three elements.

∧ intersection

10. Let $U = \{1, 2, 3, 4, 5\}$. List all subsets of U containing

   (a) no elements

   (b) exactly three elements

   (c) exactly five elements.

11. Consider the subsets $A = \{x, y, u, w\}$, $B = \{y, u, v\}$, and $C = \{x, z, v\}$ of the universe $U = \{x, y, z, u, v, w\}$. List the elements in each of the following sets:

   (a) $A \cap B$

   (b) $A \cup C$

   (c) $A \cap B \cap C$

   (d) B^c

   (e) $(A \cup B)^c$

   (f) $(A \cap B^c) \cup (A^c \cap B)$

   (g) $A \cup (B \cap C)$

   (h) $A \cap B^c \cap C$

12. Given the subsets $A = \{1, 2, 3, 4\}$, $B = \{5, 6, 7, 8\}$, and $C = \{1, 2, 7, 8\}$ of the universe $U = \{1, 2, 3, \ldots, 9, 10\}$, determine whether each of the following assertions is true or false:

   (a) $2 \in A \cap (B \cup C)$

   (b) $7 \in (A \cup B) \cap C^c$

   (c) $A \cap C \subset B^c$

   (d) $A \subset B \cup C$

   (e) $A \cup B \cup C = U$

   (f) $A \cap B \cap C = \varnothing$

   (g) A and B are disjoint

   (h) A^c and B^c are disjoint

13. Three midsemester exams are given in Math 125. Let U be the set of students enrolled in the course, and let P_1, P_2, and P_3 be the sets of students who pass exams 1, 2, and 3, respectively. Describe each of the following groups of students using set operations, and illustrate with a Venn diagram:

   (a) Those who pass all three exams

   (b) Those who fail all three exams

   (c) Those who pass at least one exam

   (d) Those who pass exactly two of the exams

14. A government agency classifies loan applicants according to sex, marital status, and employment status. Let U be the set of all applicants, and let F, M, and E be the sets of those who are female, married, and employed, respectively. Describe each of the following categories using set operations, and illustrate with a Venn diagram:

   (a) Unemployed bachelors

   (b) Working wives

   (c) Those who are either single or unemployed

   (d) Women who are either married or employed

In Exercises 15–18, verify the given law of set theory by showing that both sides of the equation represent the same region in a Venn diagram.

15. $(A \cup B)^c = A^c \cap B^c$

16. $(A \cap B)^c = A^c \cup B^c$

17. $A \cap (B \cup C) = (A \cap B) \cup (A \cap C)$

18. $A \cup (B \cap C) = (A \cup B) \cap (A \cup C)$

19. Two welfare-reform measures failed in the senate last year by close margins. In analyzing the votes it was found that 45 senators voted for Measure A and 47 for Measure B. Moreover, 39 senators voted for both measures. Senator Foghorn is proposing a compromise bill this year, which he is sure will be supported by all those who voted for either of the two earlier measures. If his bill needs 51 votes to pass, will it succeed?

20. A medical research team is comparing two different diagnostic tests for diabetes. When both tests were given to a group of known diabetics, 94% reacted to Test A and 91% reacted to Test B. Moreover, 87% reacted to both tests. What percentage of the group did not react to either test?

21. A criminologist interviewed a group of 250 prison inmates. When she asked how many were convicted of arson, 140 raised their hands. For burglary only 80 hands went up, while for car theft the count was 100. When she asked how many were convicted of both arson *and* burglary, 30 raised their hands. Similarly, the count for arson and car theft was 50, while for burglary and car theft it was 40. Only 12 had been convicted of all three crimes. How many in the group were in prison for some other reason?

22. A publisher is trying to persuade the author of a new chemistry text to add optional chapters on topics A, B, and C. In a market survey, instructors were asked to check those topics which they felt were essential in any book they would adopt. Of those surveyed, 24% checked Topic A, 19% Topic B, and 22% Topic C. Only 3% checked both A *and* B, 2% checked both A and C, and 1% checked both B and C. None of those surveyed checked all three topics. What fraction of the market does the book stand to lose if none of the three topics is included?

23. The standard classification of human blood is based on the presence or absence of three antigens, denoted by A, B, and Rh. Blood is said to be of *type* A, B, or AB according as it contains the A antigen, the B antigen, or both. Blood lacking both the A and B antigens is called *type* O. When the Rh antigen is present the blood type is *Rh-positive;* otherwise it is *Rh-negative.* Thus, blood of type A^+ contains the A and Rh antigens only, while blood of type B^- contains the B antigen only.

    In a hospital survey, blood samples were drawn from each of 750 patients. It was found that 360 samples contained the A antigen, 105 had the B antigen, and 645 had the Rh antigen. Both the A and B antigens were present in 30 of the samples; likewise, 315 samples contained both the A and Rh antigens, 90 contained both the B and Rh antigens, and 30 contained all three.

    (a) How many patients are there of each blood type?

    (b) What percentage of the total group is either type A^+ or O^+?

    (c) Of those with type B blood, what percentage are Rh-positive?

24. Last month Congressman Flipside hired an opinion research firm to survey voter attitudes in his district on several key issues. The survey results included the following figures: 80% of the voters interviewed favor gun control, 53% favor nuclear power, and 69% favor tighter pollution limits. Both gun control *and* nuclear power were favored by 21% of those surveyed; likewise, 46% favored both gun control and pollution limits, 34% favored both nuclear power and pollution limits, and 9% favored all three.

    After studying the report overnight, the congressman accused the research firm of incompetence, claiming that these survey results are impossible. How did he reach this conclusion?

25. In a sociological study, 500 married couples were asked whether they were "more or less satisfied" with their marriage. In 210 cases both the husband and wife were satisfied, while in 65 cases both were dissatisfied. Twice as many husbands as wives reported that they were satisfied. How many husbands were satisfied? How many wives? (Hint: Put unknowns in various regions of a Venn diagram.)

26. A group of 4000 taxpayers is randomly selected by the Internal Revenue Service. The returns of each member of this group for the years 1981, 1982, and 1983 are subjected to a computer audit, which either accepts or rejects each return. Suppose that 1150 of the 1981 returns, 1350 of the 1982 returns, and 1150 of the 1983 returns are rejected. Also, of the total group of taxpayers audited suppose that 1650 have no returns rejected and 350 have all three returns rejected. How many taxpayers in the group have exactly one return rejected? Exactly two? (Hint: Put unknowns in various regions of a Venn diagram.)

6.2 Techniques of Counting

Many counting problems involve a sequence of choices, each of which can be made in a number of ways. There is a simple mathematical rule known as the Counting Principle which can be applied in such cases, and which provides one of the most powerful general tools of elementary mathematics. In this section we examine this new technique.

Tree Diagrams At several points in our later study of probability theory we make use of a pictorial device called a **tree diagram.** One application of tree diagrams is to help keep track of the different ways in which a sequence of choices can be made.

Example 1 Generic Motors' new compact, the Bracer, is available in red, yellow, or blue, with either automatic or shift transmission. The buyer also has a choice of bench seats or bucket seats. How many different possibilities are represented by these three options?

Solution There are three choices of color. For each of these, there are two choices of transmission, followed by two choices of seat-type. Thus, in making these decisions the buyer follows one of twelve possible paths through a **decision tree** (Figure 1, next page). For instance, the buyer who selected a yellow car with automatic transmission and bucket seats follows the brown path in the figure. We can outline the different possible choice-sequences in a **counting scheme,** as follows:

$$
\begin{array}{lll}
C_1: & \text{Choose color} & 3 \text{ ways} \\
C_2: & \text{Choose transmission} & 2 \text{ ways} \\
C_3: & \text{Choose seat-type} & \underline{2 \text{ ways}} \\
& \text{Total} \quad 3 \cdot 2 \cdot 2 = & 12 \text{ ways}
\end{array}
$$

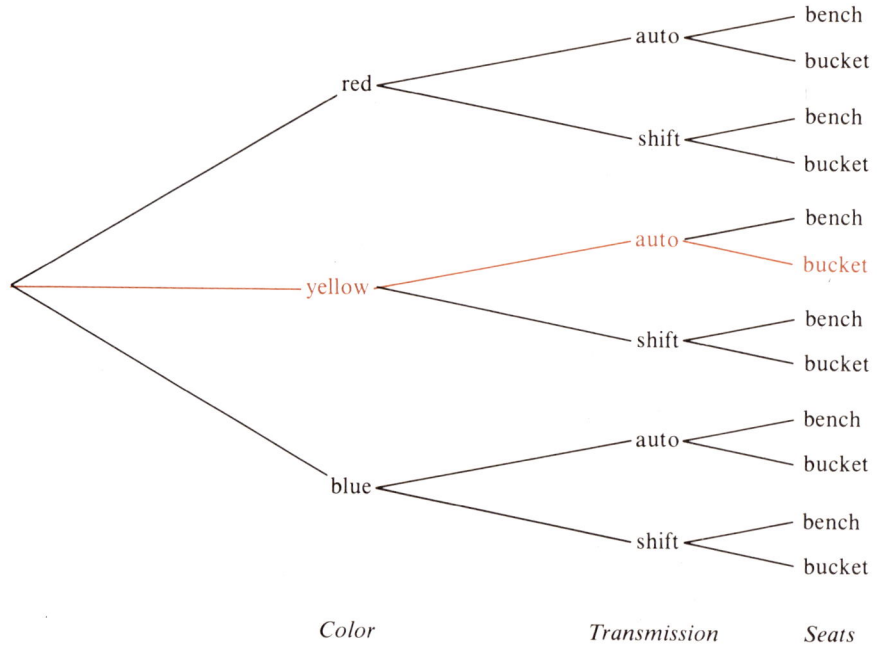

Figure 1   ◆

Note that we *multiply* the number of choices available at each stage to obtain the number of possible choice-sequences.

The Counting Principle   The reasoning used in Example 1 applies to many other situations in which we are faced with a sequence of choices. If we generalize this reasoning we arrive at the following very important principle:

The Counting Principle

Suppose there are

$$n_1 \text{ ways to make a choice } C_1,$$
$$\text{then, } n_2 \text{ ways to make a choice } C_2,$$
$$\text{then, } n_3 \text{ ways to make a choice } C_3, \text{ and so on.}$$

The product

$$n_1 n_2 n_3 \cdot \cdot \cdot$$

then represents the number of different ways in which the entire sequence of choices can be made.

We assume here that the choices $C_1, C_2, \ldots$ are made *in order*, and that the kth choice C_k can be made in n_k ways, *no matter how the earlier choices have been made*. In applying the Counting Principle we rarely display the full decision tree as in Figure 1, since diagrams of this kind are unwieldy. Instead, we rely on the counting scheme itself, as in the following example.

Example 2 A man has five different shirts, four ties, two jackets, and three pairs of slacks. How many different outfits are possible?

Solution An outfit is determined by a choice of shirt, tie, jacket, and slacks. The counting scheme is therefore

C_1: Choose a shirt   5 ways
C_2: Choose a tie   4 ways
C_3: Choose a jacket   2 ways
C_4: Choose slacks   3 ways
Total $5 \cdot 4 \cdot 2 \cdot 3 = 120$ ways

By the Counting Principle, the man can put together 120 different outfits. ◆

Example 3 Five rock music groups have volunteered their services for a benefit performance. Agents for the five groups are competing to determine which three of the five will be given billing in the ad promoting the show, and in what order their names will be listed. How many different billings are possible?

Solution We can describe the various possible billings as the different possible ways of choosing three of the five groups, in order.

C_1: Choose group for 1st billing   5 ways
C_2: Choose group for 2nd billing   4 ways
C_3: Choose group for 3rd billing   3 ways
Total $5 \cdot 4 \cdot 3 = 60$ ways

Explanation: There are five candidates for first billing, hence $n_1 = 5$ ways to make choice C_1. Once this choice is made—and *no matter how* it is made—there remain four candidates for second billing, hence $n_2 = 4$ ways to make choice C_2. By the time we choose third billing we have only $n_3 = 3$ candidates left. By the Counting Principle, the number of complete decision paths—and therefore the number of possible billings—is $5 \cdot 4 \cdot 3 = 60$. ◆

Example 4 A coin is tossed four times, and the outcome of each toss (heads or tails) is recorded. How many different patterns of heads and tails are possible?

Solution The various patterns (HTHH, TTTH, and so on) can be obtained by filling in each of the four blanks

    ‾‾‾‾   ‾‾‾‾   ‾‾‾‾   ‾‾‾‾
      1   2   3   4

with one of the symbols H or T, in every possible way. The counting scheme for this is as follows:

C_1: Fill in 1st blank   2 ways   (either H or T)
C_2: Fill in 2nd blank   2 ways
C_3: Fill in 3rd blank   2 ways
C_4: Fill in 4th blank   2 ways
        Total $2^4 = 16$ ways

Hence, by the Counting Principle, there are 16 possible patterns of heads and tails.   ◆

Example 5   A California license plate consists of three letters followed by three digits (for example, TBA 043). How many different license plates are possible?

Solution   A license plate is determined by choosing any three letters, in order, followed by any three digits (0 through 9). We assume that repetitions are allowed; that is, a letter or digit may occur more than once. The counting scheme is

C_1: Choose 1st letter   26 ways
C_2: Choose 2nd letter   26 ways
C_3: Choose 3rd letter   26 ways
C_4: Choose 1st digit   10 ways
C_5: Choose 2nd digit   10 ways
C_6: Choose 3rd digit   10 ways
        Total $26^3 \cdot 10^3 = 17{,}576{,}000$ ways

By the Counting Principle, the number of possible license plates is therefore 17,576,000.   ◆

Example 6   A Scrabble player has seven letter-tiles in her tray, forming the word NARPLES. In order to reach a triple word score she must use exactly five of these tiles.
(a) How many five-letter words (meaningful or not) can be formed using these seven tiles?
(b) How many of them begin with the letter S?
(c) How many of them contain both vowels?

Solution   (a) We form a five-letter word by filling in each of the blanks

——— ——— ——— ——— ———
 1   2   3   4   5

with a different letter-tile from among the seven given.

C_1:   Choose 1st letter   7 ways
C_2:   Choose 2nd letter   6 ways
C_3:   Choose 3rd letter   5 ways
C_4:   Choose 4th letter   4 ways
C_5:   Choose 5th letter   3 ways
Total $7 \cdot 6 \cdot 5 \cdot 4 \cdot 3 = 2520$ ways

(b) Using the same counting scheme as in part (a), we now have:

C_1:   Choose 1st letter   1 way   (must be S)
C_2:   Choose 2nd letter   6 ways
C_3:   Choose 3rd letter   5 ways
C_4:   Choose 4th letter   4 ways
C_5:   Choose 5th letter   3 ways
Total $1 \cdot 6 \cdot 5 \cdot 4 \cdot 3 = 360$ ways

(c) To count the number of words containing both vowels (A and E) we cannot use the counting scheme above. The reason is that the number of letters available to fill the fourth and fifth positions will depend on the sequence of choices made earlier. For instance, the letter following PEA
may be chosen in four ways, but the letter following SPL must be one of
the two vowels.

We can get around this difficulty by *first* deciding where to put the
vowels, then filling the three remaining positions.

C_1:   Choose position for A   5 ways   (any blank)
C_2:   Choose position for E   4 ways
C_3:   Choose letter to fill first vacancy   5 ways   (any consonant)
C_4:   Choose letter to fill next vacancy   4 ways
C_5:   Choose letter to fill last vacancy   3 ways
Total $5 \cdot 4 \cdot 5 \cdot 4 \cdot 3 = 1200$ ways

Thus, of the 2520 five-letter words that can be formed, exactly 1200 contain
both vowels.   ◆

Remark to Reader   As the last example illustrates, use of the Counting
Principle requires careful attention, and occasionally a bit of ingenuity.
While the principle itself is easy to understand, its application in particular
cases can be rather subtle, and errors—often of very large magnitude—are
easily made. Here are a few suggestions.

1. Be clear in your own mind exactly what kind of things you are trying to
   count. Make up a few typical examples.
2. Ask yourself how you could generate them all in a systematic way, by
   making a sequence of decisions. Be sure that each thing you are trying to
   count can be obtained by making these decisions in one and only one way.
3. Formulate the sequence of decisions as a counting scheme. Be careful to

specify precisely what is being chosen at each step. Do not start multiplying any numbers together until you are convinced that your counting scheme is a sound one. Also, bear in mind that there is often more than one "right" approach to a counting problem.

6.2 Exercises

1. Wendy's breakfast always consists of (1) either orange juice or grapefruit juice; (2) either cereal or else scrambled, fried, or soft-boiled eggs; and (3) either wheat toast or a bran muffin. Draw a decision tree showing all possible breakfasts. How many are there?

2. The five finalists in a contest are A, B, C, D, and E. From these five, the judges must select a winner and a runner-up. Draw a tree diagram showing all possible outcomes. How many are there?

3. A restaurant menu lists six appetizers, eight entrees, and four desserts. How many different complete meals are possible (one of each)?

4. A tourist can take one of three different routes from her hotel to the museum, and one of five routes from the museum to the park. In how many different ways can she go from her hotel to the park, stopping at the museum?

5. An organization with 25 members elects a president, a secretary, and a treasurer (all different). How many outcomes can the election have?

6. A consumer rating service tests twelve new cars and rates the three best ones in order of preference. How many different results can the test produce?

7. A telephone number is a sequence of seven digits (0 through 9), the first of which cannot be zero. How many telephone numbers are there?

8. A license plate in a certain state consists of any two letters followed by any four digits. How many different license plates are possible?

9. In how many different ways can a baseball manager arrange his nine players in a batting lineup? What if Garfield, the slugger, must bat fourth and Gondolf, the pitcher, must bat last?

10. A basketball coach has ten players available. How many different teams can she form, consisting of five players in assigned positions?

11. In how many different ways can each of ten numbered billiard balls be assigned to one of six pockets?

12. In how many different ways can each of six court cases be assigned to one of four judges, if a judge can hear more than one case?

13. How many five-letter words can be formed using any letters of the alphabet, if (a) repetitions are allowed; (b) repetitions are not allowed?

14. In how many ways can five rooms be painted one of ten colors, if (a) different rooms can be the same color; (b) each room must be a different color?

15. In a chess tournament with 20 players, each one must face every other player twice, once as white and once as black. How many games must be played?

16. In a league with twelve teams, every team must be scheduled to play every other team twice (once in each team's home court). How many games must be scheduled altogether?

17. In how many different ways can a student answer a ten-question exam

    (a) if each question calls for a true-false response?

    (b) if each question calls for one of three possible responses?

18. In how many different ways can a student answer a ten-question multiple-choice exam, if the first five questions each have three possible responses and the last five questions each have four possible responses?

19. How many different patterns of heads and tails are possible

    (a) if one coin is tossed n times?

    (b) if n distinguishable coins (different sizes) are all tossed at the same time?

20. A die is a cube with faces numbered 1 through 6. How many different outcomes are possible

    (a) if one die is rolled n times?

    (b) if n distinguishable dice (different colors) are all rolled at the same time?

21. Consider all five-digit numbers that do not begin with zero (for example, 34037).

    (a) How many such numbers are there?

    (b) How many of them contain only odd digits (1, 3, 5, 7, 9)?

    (c) How many of them begin and end with an even digit?

    (d) How many of them contain the digit 5 at least once? (Hint: How many do not?)

22. Given the letters of the word LEXICOGRAPH, each written on a separate tile,

    (a) How many different four-letter words can be formed?

    (b) How many of them contain no vowels?

    (c) How many of them begin with a consonant and end with a vowel?

    (d) How many of them contain the letter X? (Hint: How many do not?)

23. We saw in Section 6.1 (Example 2) that a three-element set has eight subsets. Use the Counting Principle to prove that an n-element set A has exactly 2^n subsets. (Hint: A subset $X \subset A$ can be formed by making n decisions.)

24. We saw in Section 6.1 (Examples 3 and 4) that a two-set Venn diagram determines four component regions, and a three-set diagram determines eight regions. Use the Counting Principle to prove that an n-set Venn diagram determines 2^n component regions. (Hint: Each region can be characterized as an intersection.)

6.3 Permutations and Combinations

The solution of many counting problems depends on our being able to determine the number of ways in which k distinct elements can be chosen from a set of n elements ($k \le n$). In this section we derive some useful formulas for this purpose which are part of every mathematician's collection of counting techniques.

Permutations   In how many different ways can three elements be chosen from a set of five elements? The answer, as we shall see, depends on whether or not we want to take the *order of selection* into account. We begin by considering a case where order is important.

Example 1   A club has just been formed with five members. At their first meeting they decide to elect a slate of officers, consisting of a president, vice-president, and treasurer (all different). How many slates are possible?

Discussion   A slate of officers is determined by making three successive choices.

$$
\begin{array}{lll}
C_1\text{:} & \text{Choose president} & 5 \text{ ways} \\
C_2\text{:} & \text{Choose vice-president} & 4 \text{ ways} \\
C_3\text{:} & \text{Choose treasurer} & \underline{3 \text{ ways}} \\
 & \text{Total} \quad 5 \cdot 4 \cdot 3 = & 60 \text{ ways}
\end{array}
$$

The number of possible slates is the same as the number of different ways to make this sequence of choices, namely $5 \cdot 4 \cdot 3 = 60$. If we represent the five members of the club by the letters

$$A, B, C, D, E$$

then each slate of officers can be represented by a word made up of three distinct letters. The 60 possible outcomes are listed below.

ABC	ABD	ABE	ACD	ACE	ADE	BCD	BCE	BDE	CDE
ACB	ADB	AEB	ADC	AEC	AED	BDC	BEC	BED	CED
BAC	BAD	BAE	CAD	CAE	DAE	CBD	CBE	DBE	DCE
BCA	BDA	BEA	CDA	CEA	DEA	CDB	CEB	DEB	DEC
CAB	DAB	EAB	DAC	EAC	EAD	DBC	EBC	EBD	ECD
CBA	DBA	EBA	DCA	ECA	EDA	DCB	ECB	EDB	EDC

Notice that the outcome ABC (A is president, B is vice-president, C is treasurer) is not the same as the outcome BCA (B is president, C is vice-president, A is treasurer), even though the same three individuals are involved. In other words, we are interested not only in *which* three individuals are chosen, but also in the *specific order* in which they are chosen. Ordered arrangements of this type arise often in mathematical applications, and they are known by a special name.

Definition

Let n distinct objects be given. By a **permutation** of these objects **taken k at a time** we mean an ordered arrangement of k of the n objects, without repetitions. The number of such arrangements is denoted* by $P(n, k)$.

---

*Other common notations are $P_{n,k}$ and $_nP_k$.

Our list above shows all permutations of the five letters A, B, C, D, E, taken three at a time, and we see that

$$P(5, 3) = 5 \cdot 4 \cdot 3 = 60. \quad \diamond$$

More generally, let n distinct objects be given, and suppose $1 \le k \le n$. To form a permutation of these objects taken k at a time, we must fill in each of the blanks

$$\underline{\qquad} \quad \underline{\qquad} \quad \underline{\qquad} \quad \cdots \quad \underline{\qquad} \quad \underline{\qquad}$$
$$\;\; 1 \qquad\quad 2 \qquad\quad 3 \qquad\qquad k-1 \qquad k$$

with a different object. The first object can be chosen in n ways, the second in $n - 1$ ways, the third in $n - 2$ ways, and so forth. By the time we reach the kth blank we will have used $k - 1$ of the n objects, leaving $n - (k - 1) = n - k + 1$ ways to make the last choice. Hence, by the Counting Principle, the number of possible permutations is

$$n(n - 1)(n - 2) \cdots (n - k + 1).$$

Note that this "descending product" starts with n and contains exactly k factors.

The number of permutations of n objects taken k at a time is given by

$$P(n, k) = n(n - 1)(n - 2) \cdots (n - k + 1) \qquad (1)$$

where the product on the right contains exactly k factors.

For instance, using formula (1) we find that

$$P(10, 2) = 10 \cdot 9 = 90$$
$$P(10, 3) = 10 \cdot 9 \cdot 8 = 720$$
$$P(10, 4) = 10 \cdot 9 \cdot 8 \cdot 7 = 5040, \text{ and so on.}$$

Example 2   A pianist participating in a Chopin competition has decided to perform five of the fourteen Chopin waltzes. How many different programs are possible, consisting of five waltzes played in a certain order?

Solution   The pianist must select five different waltzes out of the fourteen and arrange them in a specific order. By formula (1), the number of such ordered arrangements (permutations) is

$$P(14, 5) = \underbrace{14 \cdot 13 \cdot 12 \cdot 11 \cdot 10}_{5 \text{ factors}} = 240{,}240. \quad \blacklozenge$$

When we speak simply of a **permutation of *n* objects,** without specifying the value of k, it is understood that $k = n$. That is, a permutation of *n* objects is an ordered arrangement containing all *n* objects, without repetitions. Letting $k = n$ in formula (1), we obtain the following formula.

The number of permutations of *n* objects (taken altogether) is given by

$$P(n, n) = n(n - 1)(n - 2) \cdots 3 \cdot 2 \cdot 1. \qquad (2)$$

Example 3   In how many different ways can the letters of the word BLANKET be rearranged?

Solution   Since all seven letters are different,* we want to find the number of permutations, or ordered arrangements, of seven distinct objects. By formula (2), this number is

$$P(7, 7) = 7 \cdot 6 \cdot 5 \cdot 4 \cdot 3 \cdot 2 \cdot 1 = 5040. \quad \blacklozenge$$

Some counting problems require both the Counting Principle *and* the permutation formulas (1) or (2), as the next example illustrates.

Example 4   In how many different ways can three men and three women be seated in a row,
(a) if the seating pattern does not matter?
(b) if men and women must take alternate seats?

Solution   (a) If the seating pattern does not matter then any ordered arrangement of the six individuals is acceptable. By formula (2), the number of such arrangements is

$$P(6, 6) = 6 \cdot 5 \cdot 4 \cdot 3 \cdot 2 \cdot 1 = 720.$$

(b) There are two possible alternating patterns of men and women, namely MWMWMW and WMWMWM. Once the pattern is chosen we must then fill the M positions with the three men and the W positions with the three women. This leads to the following counting scheme:

C_1:	Choose a pattern		2 ways
C_2:	Fill the M positions	$P(3, 3) = 3 \cdot 2 \cdot 1$	$= 6$ ways
C_3:	Fill the W positions	$P(3, 3) = 3 \cdot 2 \cdot 1$	$= 6$ ways
		Total $2 \cdot 6 \cdot 6$	$= 72$ ways

By the Counting Principle, the number of different alternating arrangements is therefore 72. $\blacklozenge$

---

*The case of repeated letters will be taken up in Section 6.4.

Combinations   We began our discussion of permutations by asking for the number of ways in which three elements can be chosen from a set of five elements. In Example 1, where the order of selection was important, we found that the answer was $P(5, 3) = 60$. We now consider the case in which order is not important.

Example 1,
Revisited

Consider the club described in Example 1, with five members A, B, C, D, E. Suppose they decide to appoint an executive committee consisting of three members having equal status. How many different committees are possible?

Discussion

If we choose the three members of the committee in order, one after the other, the outcome will be one of the 60 permutations listed in Example 1, which is reproduced below.

ABC	ABD	ABE	ACD	ACE	ADE	BCD	BCE	BDE	CDE
ACB	ADB	AEB	ADC	AEC	AED	BDC	BEC	BED	CED
BAC	BAD	BAE	CAD	CAE	DAE	CBD	CBE	DBE	DCE
BCA	BDA	BEA	CDA	CEA	DEA	CDB	CEB	DEB	DEC
CAB	DAB	EAB	DAC	EAC	EAD	DBC	EBC	EBD	ECD
CBA	DBA	EBA	DCA	ECA	EDA	DCB	ECB	EDB	EDC

{A, C, D}

Now, however, we are only interested in *which* three individuals are chosen, not in their specific order; hence, the permutations in any given column actually represent the same outcome. For instance, the permutations ACD, ADC, CAD, CDA, DAC, DCA (boxed) all represent the committee whose members are A, C, and D. This suggests that the committee should be represented by a *set,* or unordered grouping.*

Definition

Let *n* distinct objects be given. By a **combination** of these objects **taken k at a time** we mean a set containing *k* of the *n* given objects. The number of such sets is denoted† by $C(n, k)$.

From our list above, we see that there are ten different combinations of A, B, C, D, E, taken three at a time. To see why, notice that each combination corresponds to six permutations. For example, the combination {A, C, D} consists of the elements, A, C, and D, which can be permuted in

$$P(3, 3) = 3 \cdot 2 \cdot 1 = 6$$

---

*Recall from Section 6.1 that a set remains the same no matter how we list its elements; hence {A, C, D} = {A, D, C} = {C, A, D}, and so on.

†Other common notations for $C(n, k)$ are $\binom{n}{k}$, $C_{n, k}$ and $_nC_k$.

different ways. It follows that the number of combinations is one-sixth the number of permutations, or

$$C(5, 3) = \frac{P(5, 3)}{P(3, 3)} = \frac{5 \cdot 4 \cdot 3}{3 \cdot 2 \cdot 1} = \frac{60}{6} = 10. \quad \blacklozenge$$

In general, each combination of n objects taken k at a time can be ordered in $P(k, k)$ different ways. Hence, the number of combinations is equal to the number of permutations divided by $P(k, k)$; that is,

$$C(n, k) = \frac{P(n, k)}{P(k, k)}.$$

Using formula (1) to evaluate $P(n, k)$ and $P(k, k)$, we arrive at our next important counting formula.*

The number of combinations of n objects taken k at a time is given by

$$C(n, k) = \frac{n(n - 1)(n - 2) \cdots (n - k + 1)}{k(k - 1)(k - 2) \cdots \cdot 3 \cdot 2 \cdot 1}. \qquad (3)$$

Both the numerator and the denominator in (3) are "descending products" with k factors. Since $C(n, k)$ is always an integer, the denominator always cancels completely into the numerator. Examples:

$$C(10, 1) = \frac{10}{1} = 10$$

$$C(10, 2) = \frac{10 \cdot 9}{2 \cdot 1} = 45$$

$$C(10, 3) = \frac{10 \cdot 9 \cdot 8}{3 \cdot 2 \cdot 1} = 120.$$

Thus, in a group of 10 people we could form 10 one-member committees, 45 two-member committees, and 120 three-member committees.

Example 5   Suppose eight people raise their glasses in a toast. If every person clinks glasses with every other, how many clinks will be heard?

Solution   Each pair of individuals determines one clink. Moreover, X and Y produce the same clink as Y and X, so order does not matter. Thus, the number of clinks is the same as the number of combinations of 8 people taken 2 at a time, or

$$C(8, 2) = \frac{8 \cdot 7}{2 \cdot 1} = 28. \quad \blacklozenge$$

---

*This formula appears in the work of the Hindu mathematician Bhāskara (b. 1114) and the Frenchman Levi ben Gerson (1321).

Example 6   How many different 5-card poker hands are possible?

Solution   Each poker hand consists of 5 distinct cards chosen from a deck of 52 cards. Since the order of the five cards does not matter, we want the number of combinations of 52 things taken 5 at a time.

$$C(52, 5) = \frac{52 \cdot 51 \cdot 50 \cdot 49 \cdot 48}{5 \cdot 4 \cdot 3 \cdot 2 \cdot 1} = 2{,}598{,}960 \quad \blacklozenge$$

Example 7   A committee of four is chosen from a group consisting of six men and six women.
(a) How many different committees are possible?
(b) How many consist of two men and two women?

Solution   (a) Since the members of the committee all have the same status, the order in which they are selected does not matter. We therefore want the number of combinations of 12 people taken 4 at a time.

$$C(12, 4) = \frac{12 \cdot 11 \cdot 10 \cdot 9}{4 \cdot 3 \cdot 2 \cdot 1} = 495 \quad \text{All possibilities}$$

For part (b), we need the Counting Principle. A two-man, two-woman committee is obtained by first choosing two men (without regard to order), then choosing two women. The counting scheme is therefore

C_1:	Choose 2 men	$C(6, 2) = \dfrac{6 \cdot 5}{2 \cdot 1} = 15$ ways
C_2:	Choose 2 women	$C(6, 2) = \dfrac{6 \cdot 5}{2 \cdot 1} = 15$ ways
		Total $15 \cdot 15 = 225$ ways

Thus, out of the 495 possible committees, a total of 225 consist of two men and two women. $\blacklozenge$

In the next example we employ both permutations *and* combinations, together with the Counting Principle.

Example 8   How many five-letter words (meaningful or not) contain three different consonants and two different vowels?

Solution   The English alphabet contains 26 letters, of which 21 are consonants and 5 are vowels.* We must therefore decide (i) which three consonants to use; (ii) which two vowels to use; and (iii) how to arrange these five letters to form a word.

---

*The vowels are *a, e, i, o, u*. We ignore the fact that the letter *y* sometimes acts as a vowel.

C_1:	Choose 3 consonants	$C(21, 3) = \dfrac{21 \cdot 20 \cdot 19}{3 \cdot 2 \cdot 1} = 1330$ ways
C_2:	Choose 2 vowels	$C(5, 2) = \dfrac{5 \cdot 4}{2 \cdot 1} = 10$ ways
C_3:	Arrange them in order	$P(5, 5) = 5 \cdot 4 \cdot 3 \cdot 2 \cdot 1 = 120$ ways

Total $1330 \cdot 10 \cdot 120 = 1{,}596{,}000$ ways

By the Counting Principle, the number of possible words is therefore 1,596,000. ◆

Factorial Notation   We can express our counting formulas concisely by means of the following notation:

Definition

Let n be a positive integer. The symbol $n!$ (read n **factorial**) denotes the product of all positive integers less than or equal to n. That is,

$$n! = n(n - 1)(n - 2) \cdots 3 \cdot 2 \cdot 1.$$

In addition, we define

$$0! = 1.$$

The first few values of $n!$ are

$$0! = 1$$
$$1! = 1$$
$$2! = 2 \cdot 1 = 2$$
$$3! = 3 \cdot 2 \cdot 1 = 6$$
$$4! = 4 \cdot 3 \cdot 2 \cdot 1 = 24$$
$$5! = 5 \cdot 4 \cdot 3 \cdot 2 \cdot 1 = 120.$$

For larger values of n we can use the relation

$$n! = n \cdot (n - 1)!$$

which follows immediately from the definition of the factorial symbol. Thus,

$$6! = 6(5!) = 6 \cdot 120 = 720$$
$$7! = 7(6!) = 7 \cdot 720 = 5040$$
$$8! = 8(7!) = 8 \cdot 5040 = 40{,}320$$

and so forth. The factorial function grows very rapidly; in fact, 10! exceeds 3 million, and 13! exceeds 6 billion (which may account for the exclamation point)!

Permutation and combination numbers can be expressed very simply in terms of factorials. For example, observe that

$$P(5, 5) = 5 \cdot 4 \cdot 3 \cdot 2 \cdot 1 = 5!$$

$$P(8, 3) = 8 \cdot 7 \cdot 6 = 8 \cdot 7 \cdot 6 \cdot \frac{5 \cdot 4 \cdot 3 \cdot 2 \cdot 1}{5 \cdot 4 \cdot 3 \cdot 2 \cdot 1}$$

$$= \frac{8 \cdot 7 \cdot 6 \cdot 5 \cdot 4 \cdot 3 \cdot 2 \cdot 1}{5 \cdot 4 \cdot 3 \cdot 2 \cdot 1} = \frac{8!}{5!}$$

$$C(8, 3) = \frac{8 \cdot 7 \cdot 6}{3 \cdot 2 \cdot 1} = \frac{8 \cdot 7 \cdot 6}{3 \cdot 2 \cdot 1} \cdot \frac{5 \cdot 4 \cdot 3 \cdot 2 \cdot 1}{5 \cdot 4 \cdot 3 \cdot 2 \cdot 1}$$

$$= \frac{8 \cdot 7 \cdot 6 \cdot 5 \cdot 4 \cdot 3 \cdot 2 \cdot 1}{(3 \cdot 2 \cdot 1)(5 \cdot 4 \cdot 3 \cdot 2 \cdot 1)} = \frac{8!}{3! \, 5!}.$$

In the same way, we can derive the following general formulas, corresponding to our earlier formulas (1), (2), (3).

The number of permutations of n objects taken k at a time is given by

$$P(n, k) = \frac{n!}{(n - k)!}. \qquad (1a)$$

The number of permutations of n objects (taken altogether) is given by

$$P(n, n) = n! \qquad (2a)$$

The number of combinations of n objects taken k at a time is given by

$$C(n, k) = \frac{n!}{k!(n - k)!}. \qquad (3a)$$

In (1a) and (3a) it is understood that n and k are integers, with $0 \le k \le n$. We do not use (1a) or (3a) for purposes of computation, because our earlier formulas (1) and (3)—while more complicated in appearance—actually involve less arithmetic. The new formulas, however, have important theoretical uses. In particular, formula (3a) gives us the following important relations:

---

1. For any integer $n \ge 0$,

$$C(n, 0) = 1 \quad \text{and} \quad C(n, n) = 1.$$

2. For any integers n and k with $0 \le k \le n$,

$$C(n, k) = C(n, n - k).$$

For example:

$$C(8, 0) = \frac{8!}{0!(8 - 0)!} = \frac{8!}{0! \, 8!} = 1$$

$$C(8, 8) = \frac{8!}{8!(8 - 8)!} = \frac{8!}{8! \, 0!} = 1$$

$$C(8, 3) = \frac{8!}{3! \, 5!} = \frac{8!}{5! \, 3!} = C(8, 5).$$

The general proof of properties (1) and (2) is left for the reader. (See Exercises 31 and 32.)

6.3 Exercises

1. Evaluate:

   (a) $P(7, 5)$ (c) $P(6, 6)$ (e) $P(16, 3)$

   (b) $C(9, 2)$ (d) $C(8, 1)$ (f) $C(10, 5)$

2. Evaluate:

   (a) $P(15, 5)$ (c) $P(5, 1)$ (e) $P(8, 4)$

   (b) $C(15, 5)$ (d) $C(5, 5)$ (f) $C(13, 12)$

3. Given the objects a, b, c, d, e:

   (a) Use a tree diagram to list all twenty permutations of these objects taken two at a time.

   (b) List all ten combinations of these objects taken two at a time.

4. Given the objects x, y, z, w:

   (a) Use a tree diagram to list all 24 permutations of these objects taken three at a time.

   (b) List all four combinations of these objects taken three at a time.

5. Fifteen athletes have qualified for the Olympic broad jump competition. In how many possible ways can the gold, silver, and bronze medals be awarded?

6. How many different nautical signals, consisting of five flags strung vertically on a rope, can be made using nine flags of different colors?

7. Twenty medical school applicants are given a screening interview, the purpose of which is to narrow the field down to the top five candidates. How many different outcomes are possible

   (a) if the top five are ranked in order of preference?

   (b) if the top five are unranked?

8. A newly formed consumer action group has thirty members. In how many different ways can the group choose

   (a) a president, vice-president, secretary, and treasurer?

   (b) an executive council consisting of four members?

9. A bureau has eight drawers of the same size. In shipping the bureau to a new house, the movers took out the drawers and afterward replaced them in different positions, with the result that they no longer open and shut smoothly. The owner has been advised to try rearranging them until they all work. How many arrangements are possible?

10. In how many different ways can the letters of the word CREAM be rearranged?

11. Six couples are standing in a single-file line at a movie theater, in such a way that no couple is separated. How many such lineups are possible?

12. Four children, four teenagers, and four adults are to be arranged for a group photograph, with the children in the front row, the teenagers in the middle row, and the adults in the back row. How many different arrangements are possible?

13. Suppose ten points in the plane, no three of which lie in a straight line, are connected to each other by lines in all possible ways.

    (a) How many different lines must be drawn?

    (b) How many different triangles will be formed, having the given points as vertices?

14. Given twelve lines in the plane, what is the greatest possible number of intersection points they can determine?

15. In how many different ways can the Supreme Court render

    (a) a 5–to–4 decision?

    (b) a 6–to–3 decision?

    (c) a majority decision?

16. In Section 2.5 we discussed the solution of linear programs by the geometric method. How many basic points are possible, given a linear program with

    (a) two decision variables and five constraints?

    (b) three decision variables and seven constraints?

    (c) n decision variables and m constraints?

17. In a certain state assembly, 13 of of the 20 members come from urban districts, the remainder from rural districts. The assembly must appoint a committee of six to study energy problems.

    (a) How many different committees are possible?

    (b) How many are evenly balanced between urban and rural members?

18. A three-card hand is dealt from a standard deck of 52 cards.

    (a) How many different hands are possible?

    (b) How many contain only face cards (jack, queen, or king)?

    (c) How many contain exactly two face cards?

19. Sweet's Treats classifies its candies as creams (10 types), nuts (7 types), and chews (8 types). A customer has ordered an assortment of fifteen types, to contain six creams, four nuts, and five chews. How many different assortments are possible?

20. A woman must select two skirts, three blouses, and one pair of shoes to pack for a business trip. If she has four skirts, eight blouses, and five pairs of shoes to choose from, in how many different ways can she fill her suitcase?

21. A Chinese restaurant charges a standard price for its "family dinner," consisting of any four different dishes on the menu. The restaurant's advertisement claims that "over 300 different family dinners are possible." If this is true, what must be the least possible number of dishes listed on the menu?

22. A tenant's organization with 25 members must elect, according to its bylaws, (i) a policy board consisting of five members, and (ii) a negotiating team consisting of three of the five policy board members. How many different outcomes can the election have?

23. At the Bakersfield Film Festival, 14 American and 9 foreign films have been nominated for awards.

(a) In how many different ways can the judges award first, second, third, and fourth prize?

(b) What if it is determined beforehand that exactly two of the top four must be American films?

24. A violinist's repertoire contains ten classical pieces and eight modern pieces. How many different concerts can she arrange, consisting of five pieces—three classical and two modern—performed in a certain order?

25. Three Americans, four Russians, and two Chinese delegates are attending an international trade conference.

(a) In how many ways can these nine delegates line up for a group photo, with those of the same nationality standing together?

(b) What if the Russian and Chinese delegates refuse to stand next to each other?

26. A book dealer has seven different textbooks on French, five on German, and four on Russian. She wants to make a window display consisting of three books on each language, arranged on a shelf in such a way that books on the same language are together. How many different displays are possible?

27. Evaluate:

(a) $10!$ (b) $\dfrac{14!}{11!}$ (c) $\dfrac{8!}{7!}$ (d) $\dfrac{12!}{3! \cdot 9!}$

28. Evaluate:

(a) $(3!)!$ (b) $\dfrac{20!}{18! \cdot 2!}$ (c) $\dfrac{12!}{(6!)^2}$ (d) $\dfrac{7! + 8!}{6!}$

29. Derive formulas (1a) and (3a) in this section from (1) and (3).

30. Prove:

(a) $n! + (n + 1)! = n!(n + 2)$

(b) $\dfrac{(n + 1)!}{(n - 1)!} = n^2 + n$

(c) $\dfrac{1}{(m - 1)!n!} + \dfrac{1}{m!(n - 1)!} = \dfrac{m + n}{m!n!}$

31. Show that formula (3a) in this section gives

$$C(n, 0) = 1 \quad \text{and} \quad C(n, n) = 1$$

for any $n \geq 0$. Does this make sense in terms of combinations?

32. Use formula (3a) in this section to prove the relation

$$C(n, k) = C(n, n - k). \qquad (0 \leq k \leq n)$$

Can you justify this result intuitively, in terms of combinations?

Computer Projects   Write a BASIC program to compute the number of permutations and combinations of n objects taken k at a time. Use this program to answer the following questions:

(a) In the casino game of Keno, the house chooses twenty winning numbers from the integers 1 through 80. How many different outcomes are possible?

(b) In bridge, each player is dealt 13 cards from a standard deck of 52 cards. How many different bridge hands are possible? How many contain only red cards (hearts and diamonds)?

6.4 The Binomial Theorem

Combination numbers arise naturally not only in counting problems but also in algebra itself, where they are referred to as *binomial coefficients*. In this section we examine the Binomial Theorem—in which combination numbers make their algebraic debut—and also a generalization known as the Multinomial Theorem, which leads to some new combinatorial techniques.

Binomial Coefficients Expressions of the form $(x + y)^n$ arise routinely in all sorts of mathematical calculations. The reader has no doubt spent more than one evening expanding such expressions out in powers of x and y.

$$(x + y)^0 = 1$$
$$(x + y)^1 = x + y$$
$$(x + y)^2 = x^2 + 2xy + y^2$$
$$(x + y)^3 = x^3 + 3x^2y + 3xy^2 + y^3$$
$$(x + y)^4 = x^4 + 4x^3y + 6x^2y^2 + 4xy^3 + y^4$$

$$\cdots\cdots$$

In general, the expansion of $(x + y)^n$ is a sum of terms of the form

$$ax^{n-k}y^k \qquad (0 \leq k \leq n)$$

where a is an integer coefficient which depends on n and k. Remarkably, these coefficients turn out to be nothing other than the combination numbers we encountered in the preceding section.

Binomial Theorem*

The **coefficient** of $x^{n-k}y^k$ in the expansion of $(x + y)^n$ is given by

$$\binom{n}{k} = C(n, k) = \frac{n!}{k!(n - k)!}.$$

Thus,

$$(x + y)^n = \binom{n}{0}x^n + \binom{n}{1}x^{n-1}y + \binom{n}{2}x^{n-2}y^2 + \cdots + \binom{n}{n}y^n.$$

---

*This theorem was known to Arab mathematicians in the 13th century.

Observe that the terms of the expansion are arranged so that the powers of x decrease from n to 0, while the powers of y increase from 0 to n. In each term, the exponents add up to n.

Example 1   Expand $(x + y)^5$ in powers of x and y.

Solution   Applying the Binomial Theorem with $n = 5$, we obtain

$$(x + y)^5 = \binom{5}{0}x^5 + \binom{5}{1}x^4y + \binom{5}{2}x^3y^2 + \binom{5}{3}x^2y^3 + \binom{5}{4}xy^4 + \binom{5}{5}y^5$$

$$= x^5 + 5x^4y + 10x^3y^2 + 10x^2y^3 + 5xy^4 + y^5.$$

To see how combinations enter the picture, we write $(x + y)^5$ as a product of five factors.

$$(x + y)^5 = (x + y)(x + y)(x + y)(x + y)(x + y)$$

choose x or y
in each factor

We expand the right-hand side by choosing either x or y from each factor and forming the product, in every possible way. To obtain the coefficient of, say, the x^3y^2 term in the final expansion, we count the number of products—such as $xxxyy$, $xxyxy$, or $yxxxy$—that contain exactly two y's and three x's. Since the two y's can be chosen from the five factors in

$$\binom{5}{2} = C(5, 2) = 10$$

different ways, this is the desired coefficient.   ◆

Example 2   Expand $(2a - b)^4$ in powers of a and b.

Solution   Letting $n = 4$ in the Binomial Theorem, we have the expansion

$$(x + y)^4 = \binom{4}{0}x^4 + \binom{4}{1}x^3y + \binom{4}{2}x^2y^2 + \binom{4}{3}xy^3 + \binom{4}{4}y^4$$

$$= x^4 + 4x^3y + 6x^2y^2 + 4xy^3 + y^4.$$

If we now substitute $x = 2a$ and $y = -b$ in this expansion, we obtain

$$(2a - b)^4 = (2a)^4 + 4(2a)^3(-b) + 6(2a)^2(-b)^2 + 4(2a)(-b)^3 + (-b)^4$$

$$= 16a^4 - 32a^3b + 24a^2b^2 - 8ab^3 + b^4.   ◆$$

Pascal's Triangle   We can write the binomial coefficients $\binom{n}{k}$ in an array known as **Pascal's Triangle**.*

---

*Named after the French philosopher and mathematician Blaise Pascal (1623–62), although he was not the first to discover it.

$$\binom{4}{3} \quad \frac{4 \cdot 3 \cdot 2}{3 \cdot 2 \cdot 1}$$

$$
\begin{array}{c}
n = 0 \rightarrow \quad 1 \\
n = 1 \rightarrow \quad 1 \quad 1 \\
n = 2 \rightarrow \quad 1 \quad 2 \quad 1 \\
n = 3 \rightarrow \quad 1 \quad 3 \quad 3 \quad 1 \\
n = 4 \rightarrow \quad 1 \quad 4 \; \nabla \; 6 \quad 4 \quad 1 \\
n = 5 \rightarrow \quad 1 \quad 5 \quad 10 \quad 10 \quad 5 \quad 1
\end{array}
$$

$k = 0$, $k = 1$, $k = 2$, $k = 3$, $k = 4$, $k = 5$

Figure 1

This array is obtained from the table of binomial expansions by simply de-taching the coefficient of each term. Notice that the rows of the triangle are indexed by n and the diagonals by k, so that, for instance, the coefficient

$$\binom{5}{2} = \frac{5!}{2!3!} = 10$$

is found on row $n = 5$ and diagonal $k = 2$. Each row begins and ends with 1, since

$$\binom{n}{0} = 1 \quad \text{and} \quad \binom{n}{n} = 1$$

for any $n \geq 0$. More remarkable is the fact that each entry except the first and last in a given row is the sum of the two entries immediately above it. Thus, for example, the entry 10 (row $n = 5$, diagonal $k = 2$) is the sum of the en-tries 4 and 6 immediately above it; that is,

$$\binom{5}{2} = \binom{4}{1} + \binom{4}{2}$$

and more generally,

$$\binom{n}{k} = \binom{n-1}{k-1} + \binom{n-1}{k}. \tag{1}$$

This property (the proof of which is discussed in Exercise 8), enables us to generate the rest of the triangle very easily. For instance, we obtain row 6 ($n = 6$) by adding consecutive entries in row 5, then putting a "1" at both ends.

$$
\begin{array}{c}
n = 5 \rightarrow \quad 1 \quad\;\; 5 \quad\;\; 10 \quad\;\; 10 \quad\;\; 5 \quad\;\; 1 \\
\nabla \quad \nabla \quad \nabla \quad \nabla \quad \nabla \\
n = 6 \rightarrow \quad 1 \quad 6 \quad 15 \quad 20 \quad 15 \quad 6 \quad 1
\end{array}
$$

This immediately gives us the binomial expansion for $n = 6$.

$$(x + y)^6 = x^6 + 6x^5y + 15x^4y^2 + 20x^3y^3 + 15x^2y^4 + 6xy^5 + y^6$$

Multinomial Coefficients   The typical term in the expansion of $(x + y)^n$ can be written as

$$(\text{coefficient}) \cdot x^k y^l \qquad (k + l = n)$$

where the exponents k and l range over all nonnegative integers that add up to n. Similarly, the expansion of

$$(x + y + z)^n$$

in powers of x, y, and z will contain terms of the form

$$(\text{coefficient}) \cdot x^k y^l z^m \qquad (k + l + m = n)$$

where the integer exponents k, l, m take on all possible nonnegative values that add up to n. For example, the expansion of $(x + y + z)^5$ will include

$$\text{an } x^5 \text{ term} \quad (x^5 y^0 z^0)$$
$$\text{a } y^2 z^3 \text{ term} \quad (x^0 y^2 z^3)$$
$$\text{an } x^3 yz \text{ term} \quad (x^3 y^1 z^1)$$

and so forth, each with a certain integer coefficient.

In the general case we consider an expression of the form

$$(x_1 + x_2 + \cdots + x_r)^n$$

in which a sum of r terms, called a **multinomial,** is raised to an integer power. The expansion of this expression will contain terms of the form

$$(\text{coefficient}) \cdot x_1^{k_1} x_2^{k_2} \cdots x_r^{k_r}$$

where the exponents $k_1, \ldots, k_r$ range over all nonnegative integers that add up to n. The coefficient above is called a **multinomial coefficient.** The following theorem, which we state without proof, enables us to compute these coefficients—and therefore also the expansion of a multinomial power—very simply.

Multinomial Theorem*

Let $k_1, \ldots, k_r$ be nonnegative integers for which $k_1 + \cdots + k_r = n$. The **coefficient** of

$$x_1^{k_1} x_2^{k_2} \cdots x_r^{k_r}$$

in the expansion of $(x_1 + \cdots + x_r)^n$ is given by

$$\binom{n}{k_i, \ldots, k_r} = \frac{n!}{k_1! k_2! \cdots k_r!}.$$

---

*This result is credited to the German philosopher and mathematician Leibniz (1646–1716), one of the co-discoverers of the calculus.

Example 3 In the expansion of $(x + y + z)^5$ in powers of x, y, z, find the coefficient of
(a) x^5 (b) y^2z^3 (c) x^3yz.

Solution (a) The coefficient of x^5 is $\left(\begin{matrix} 5 \\ 5, 0, 0 \end{matrix}\right) = \dfrac{5!}{5!\,0!\,0!} = 1.$

(b) The coefficient of y^2z^3 is $\left(\begin{matrix} 5 \\ 0, 2, 3 \end{matrix}\right) = \dfrac{5!}{0!\,2!\,3!} = 10.$

(c) The coefficient of x^3yz is $\left(\begin{matrix} 5 \\ 3, 1, 1 \end{matrix}\right) = \dfrac{5!}{3!\,1!\,1!} = 20.$ ◆

Notice that when $k + \ell = n$ we have

$$\binom{n}{k} = \binom{n}{\ell} = \binom{n}{k, \ell} = \frac{n!}{k!\,\ell!}. \qquad (2)$$

It follows from (2) that the coefficient of $x^k y^\ell$ in the expansion of $(x + y)^n$ is the same whether we compute it by the Binomial Theorem or by the Multinomial Theorem (with $r = 2$). Thus, the Multinomial Theorem yields the Binomial Theorem as a special case.

Applications to Counting It is natural to wonder whether the multinomial coefficients

$$\left(\begin{matrix} n \\ k_1, \ldots, k_r \end{matrix}\right) = \frac{n!}{k_1!\,k_2! \cdots k_r!}$$

have a combinatorial significance similar to that of the binomial coefficients $\binom{n}{k}$. The answer is yes, as we see in the next example.

Example 4 In how many different ways can a painting contractor assign his twelve workers to three tasks, as follows: four to scrape, three to sand, and five to paint?

Solution We first choose the scrapers, then (from the remaining workers) the sanders, and finally the painters. The counting scheme is as follows:

C_1: Choose 4 scrapers $C(12, 4) = \dfrac{12!}{4!\,8!} = 495$ ways

C_2: Choose 3 sanders $C(8, 3) = \dfrac{8!}{3!\,5!} = 56$ ways

C_3: Choose 5 painters $C(5, 5) = \dfrac{5!}{5!\,0!} = 1$ way

Total $495 \cdot 56 \cdot 1 = 27{,}720$ ways

Observe that the answer can be expressed in factorial notation as

$$\frac{12!}{4!\,8!} \cdot \frac{8!}{3!\,5!} \cdot \frac{5!}{5!\,0!} = \frac{12!}{4!\,3!\,5!} = \left(\begin{matrix} 12 \\ 4, 3, 5 \end{matrix}\right).$$

Thus, the multinomial coefficient $\begin{pmatrix} 12 \\ 4, 3, 5 \end{pmatrix}$ represents the number of ways in which the set of twelve workers can be divided into three designated subsets, containing four, three, and five elements, respectively. ◆

Such a division of a set is called an **ordered partition.** As in our example, the number of ordered partitions of a set can always be computed using a multinomial coefficient.

Let S be a set containing n distinct objects. The number of ways to partition S into

a subset S_1 containing k_1 objects,

a subset S_2 containing k_2 objects,

$\cdots,$

a subset S_r containing k_r objects

is given by the multinomial coefficient

$$\begin{pmatrix} n \\ k_1, \ldots, k_r \end{pmatrix} = \frac{n!}{k_1! \, k_2! \cdots k_r!}. \qquad (3)$$

It is assumed here that the subsets $S_1, \ldots, S_r$ are pairwise disjoint (nonoverlapping) and their union is all of S, so that $k_1 + \cdots + k_r = n$.

Another way to think of an ordered partition is to say that we are *labeling* each of the objects in S with one of r types of labels—different colors, say—where we have available k_1 labels of type 1, k_2 labels of type 2, and so forth. The number of different labelings is then given by the multinomial coefficient (3).

Example 5   In how many different ways can an instructor assign one A, two B's, four C's, two D's, and one F to a class of ten students?

Solution   The class is to be partitioned into

a subset containing 1 member   (label "A"),
a subset containing 2 members   (label "B"),
a subset containing 4 members   (label "C"),
a subset containing 2 members   (label "D"),
and   a subset containing 1 member   (label "F").

By (3) above, the number of such partitions, or labelings, is

$$\begin{pmatrix} 10 \\ 1, 2, 4, 2, 1 \end{pmatrix} = \frac{10!}{1! \, 2! \, 4! \, 2! \, 1!} = 37,800. \quad ◆$$

An important application of formula (3) concerns permutations of n objects, some of which are indistinguishable from each other.

Example 6   In how many distinguishable ways can the seven letters of the word RE-TREAT be rearranged?

Solution   If the letters were all different, the answer would be $P(7, 7) = 7! = 5040$. The fact that we have repeated letters, however, means that some of these arrangements are indistinguishable from each other.

We therefore try another approach: The ''words'' we are trying to count can be obtained by filling in, or labeling, the blanks

$$\underline{\hspace{1cm}}\quad\underline{\hspace{1cm}}\quad\underline{\hspace{1cm}}\quad\underline{\hspace{1cm}}\quad\underline{\hspace{1cm}}\quad\underline{\hspace{1cm}}\quad\underline{\hspace{1cm}}$$
$$\;\;1\qquad\;2\qquad\;3\qquad\;4\qquad\;5\qquad\;6\qquad\;7$$

with the letters R, E, T and A, in such a way that exactly two R's, two E's, two T's and one A are used. According to (3) above, the number of different ways to do this is

$$\binom{7}{2, 2, 2, 1} = \frac{7!}{2!\,2!\,2!\,1!} = 630.$$

Note that in this example, S is the above set of seven distinct blanks, S_1 is the subset of blanks filled by the letter R, S_2 is the subset of blanks filled by the letter E, and so on. ◆

More generally, we have the following result:

Given n symbols, of which

k_1 are the same, of type 1,

k_2 are the same, of type 2,

$\cdots,$

k_r are the same, of type r,

the number of distinguishable permutations of these n symbols, taken altogether, is

$$\binom{n}{k_1, \ldots, k_r} = \frac{n!}{k_1!\,k_2!\cdots k_r!}. \tag{4}$$

Note that when all the symbols are different we have $k_1 = k_2 = \cdots = k_r = 1$, so that the formula above reduces to $n!$, as before.

6.4 Exercises   1. Evaluate:

(a) $\binom{6}{3}$ (c) $\binom{15}{12}$ (e) $\binom{100}{2}$

(b) $\binom{9}{5,\,4}$ (d) $\binom{6}{1,\,2,\,3}$ (f) $\binom{8}{1,\,3,\,3,\,1}$

2. Evaluate:

(a) $\binom{n}{1}$ (c) $\binom{n}{2}$ (e) $\binom{n}{n-3}$

(b) $\binom{n}{n,\,0,\,0}$ (d) $\binom{n+2}{1,\,n,\,1}$ (f) $\binom{n}{1,\,1,\,\ldots\,,\,1}$

3. Use the Binomial Theorem to expand $(x + y)^7$ in powers of x and y.

4. Use the Binomial Theorem to expand $(2a + 3b)^5$ in powers of a and b.

5. In the expansion of $(x + y)^{14}$ in powers of x and y, find the coefficient of

(a) x^{14} (b) $x^2 y^{12}$ (c) $x^7 y^7$

6. In the expansion of $\left(2x - \dfrac{1}{2}\right)^{12}$ in powers of x, find

(a) the coefficient of x^{10}

(b) the coefficient of x^5

(c) the constant term.

7. Extend Pascal's Triangle through the row $n = 10$.

8. Prove that

(i) $\binom{n-1}{k-1} = \dfrac{(n-1)!}{(k-1)!(n-k)!} = \dfrac{(n-1)!(k)}{k!(n-k)!}$

(ii) $\binom{n-1}{k} = \dfrac{(n-1)!}{k!(n-k-1)!} = \dfrac{(n-1)!(n-k)}{k!(n-k)!}$

Use equations (i) and (ii) to prove formula (1) in this section.

9. (a) Verify for $n \le 5$ that the entries in row n of Pascal's Triangle add up to 2^n; that is,

$$\binom{n}{0} + \binom{n}{1} + \cdots + \binom{n}{n} = 2^n.$$

(b) Using the Binomial Theorem, prove this identity for all $n \ge 0$.

(c) Can you justify this result in terms of combinations? (See Exercise 23 in Section 6.2)

10. Using the Binomial Theorem, prove the identity

$$\binom{n}{0} - \binom{n}{1} + \binom{n}{2} - \cdots \pm \binom{n}{n} = 0. \qquad (n > 0)$$

11. Prove:

$$\binom{n}{k} = \frac{n - k + 1}{k}\binom{n}{k-1}. \qquad (1 \le k \le n)$$

12. Notice that Pascal's Triangle is *symmetric:* A vertical line drawn through the middle of the array divides it into two halves which are mirror-images of each other. What property of the binomial coefficients $\binom{n}{k}$ accounts for this symmetry?

13. Use the Multinomial Theorem to expand $(x + y + z)^3$ in powers of x, y, z.

14. Use the Multinomial Theorem to expand $(u + 2v^2 + 3w^3)^2$ in powers of u, v, w.

15. In the expansion of $(x + y + z + w)^6$ in powers of x, y, z, w, find the coefficient of

    (a) y^6 (b) x^2w^4 (c) xy^2z^2w.

16. In the expansion of $(a + 3b + 2)^9$ in powers of a and b, find

    (a) the coefficient of a^6b^3

    (b) the coefficient of a^7

    (c) the constant term.

17. In how many ways can eight different toys be distributed equally among four children?

18. The late sheik Abu Dhabi willed that his fifteen villas are to be distributed as equally as possible among his four sons, except that his first-born, Emir, is to receive a double share. In how many different ways can the villas be distributed?

19. Express the following answers in terms of factorials:

    (a) In how many different ways can each of four players be dealt a 13-card bridge hand from a single deck?

    (b) In how many different ways can each of four players be dealt a 5-card poker hand from a single deck? (Hint: Let the dealer receive a fictitious "hand" consisting of the 32 cards left over.)

20. In how many different ways can fifteen people be divided into three groups of five

    (a) if the groups are distinguishable (for instance, each has a different task to perform)?

    (b) if the groups are indistinguishable (for instance, three discussion groups each preparing for the same examination)?

21. How many words—meaningful or not—can be made by rearranging the letters of the following words?

    (a) REENTER (b) TATTOO (c) REARRANGER

22. How many different nautical signals, consisting of eight flags strung vertically on a rope, can be made using

    (a) three red, two white, and three blue flags?

    (b) four black and four white flags?

    (c) two red, two green, two blue, and two yellow flags?

23. Out of a group of eleven foreign service officers, it has been determined that two will be sent to Brussels and three each to Tel Aviv, Addis Ababa, and Istanbul. How many possible assignments are there?

24. A molecular biologist has isolated a single strand of DNA, the chemical sub-stance that carries the genetic code. The strand is made up of eighteen nucleic acid molecules strung together in a specific sequence as in the figure below:

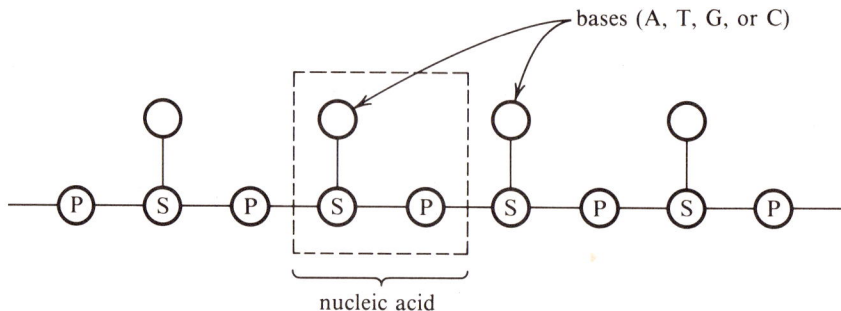

Each nucleic acid consists of one of four organic bases—adenine (A), thymine (T), guanine (G), or cytosine (C)—attached to a sugar-phosphate pair as shown in the figure. Thus, the strand is completely determined once the sequence of bases is known. By chemical analysis, the biologist finds that six of the bases are adenine, four are thymine, five guanine, and three cytosine. How many different strands are theoretically possible?

25. In the map below, point B is located four blocks east and three blocks north of point A.

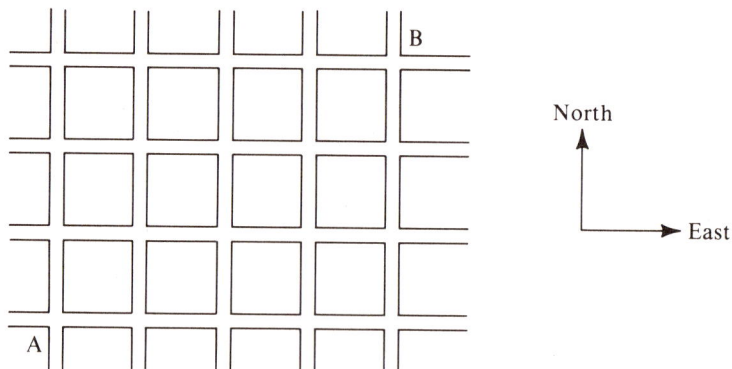

To go from A to B one must travel 4 blocks east and 3 blocks north, in some order (for example, north 1 block, then east 3 blocks, then north 2 blocks, then east 1 block).

(a) Explain how each such route can be represented by a seven-letter "word".

(b) Use formula (4) in this section to calculate the number of such routes pos-sible.

(c) Generalize to the case where B is m blocks east and n blocks north of A.

26. A coin is tossed eight times, and the outcome of each toss (heads or tails) is re-corded.

(a) Explain how each possible pattern of heads and tails can be represented by an eight-letter "word".

(b) Use formula (4) in this section to calculate the number of different ways in which three heads and five tails can occur.

(c) Generalize to find the number of ways k heads and ℓ tails can occur out of $n = k + \ell$ tosses.

Computer Projects

Write a BASIC program to compute multinomial coefficients. Use this program to answer the following questions:

(a) In how many different ways can each of four players be dealt a 13-card bridge hand from a single deck?

(b) In how many different ways can each of four players be dealt a 5-card poker hand from a single deck?

(c) How many words—meaningful or not—can be made by rearranging the letters of the following word?

ANTIDISESTABLISHMENTARIANISM

Chapter 6   Summary

Important Terms

6.1 set
element
set-builder $\{x \mid P(x)\}$
universal set U
empty set $\varnothing$
equality $A = B$
inclusion $A \subset B$
union $A \cup B$
intersection $A \cap B$
complement A^c

disjoint sets
Venn diagram

6.2 tree diagram
the counting principle

6.3 permutation
combination

6.4 binomial coefficient
multinomial coefficient
ordered partition of a set

Important Formulas

6.3 Permutations of n objects taken k at a time

$$P(n, k) = n(n - 1)(n - 2) \cdots (n - k + 1) = \frac{n!}{(n - k)!}$$

6.3 Permutations of n objects (taken altogether)

$$P(n, n) = n(n - 1)(n - 2) \cdots \cdots 3 \cdot 2 \cdot 1 = n!$$

6.3 Combinations of n objects taken k at a time

$$C(n, k) = \frac{n(n - 1)(n - 2) \cdots (n - k + 1)}{k(k - 1)(k - 2) \cdots \cdots 3 \cdot 2 \cdot 1} = \frac{n!}{k!(n - k)!}$$

Review Exercises

Consider the subsets A = {1, 2, 3}, B = {2, 3, 4, 5}, C = {4, 5, 6, 7, 8} *of the universe* U = {1, 2, 3, . . ., 9, 10}. *In Exercises 1–6, determine whether the given assertion is true or false.*

1. $8 \in A^c \cup B$

2. $5 \in A \cup (B \cap C)^c$

3. $A \cap B = \varnothing$

4. $C \subset A^c$

5. $A \cup B \cup C = U$

6. $A \subset \{x \mid x < 5\}$

7. Draw a Venn diagram showing typical sets A and B, and shade the region corresponding to $A \cap B^c$.

8. Draw a Venn diagram showing typical sets A, B, and C, and shade the region corresponding to $(A \cup B)^c \cap C$.

9. When a country music radio station surveyed 250 listeners at random it found that 83 were female, 96 were teenagers, and 67 fell into both categories (teenage girls). How many of those surveyed were nonteenage males?

10. A committee with three members, A, B, and C, must determine a regular time for its one-hour weekly meetings. The chairman has distributed a punch-out card showing all 40 possible one-hour time slots during the week, and has asked each member to punch out holes in those time slots when he or she *cannot* attend a meeting. The card returned by member A contained 27 holes, B's card contained 23 holes, and C's card contained 21 holes. Holding A and B's cards together against the light, the chairman counted 16 holes in common. Similarly, A and C's cards had 12 holes in common, and B and C's had 14 holes in common. There were 9 holes punched out on all three cards. Is there any time when all three members can meet?

11. A baseball team has eight pitchers, three catchers, and six outfielders. In how many different ways can the manager select a pitcher, catcher, left-fielder, center-fielder, and right-fielder?

12. A police artist makes a composite sketch of a crime suspect by asking witnesses to select one of ten noses, twelve pairs of eyes, eight mouths, and seven hairstyles. How many different faces are possible?

13. Evaluate:

    (a) $C(52, 4)$
    (b) $10!$
    (c) $\dfrac{18!}{15!}$
    (d) $P(14, 3)$

14. Evaluate:

    (a) $\begin{pmatrix} 8 \\ 5, 0, 3 \end{pmatrix}$
    (b) $\begin{pmatrix} 52 \\ 50 \end{pmatrix}$
    (c) $\begin{pmatrix} 6 \\ 0 \end{pmatrix}$
    (d) $\begin{pmatrix} 15 \\ 3, 4, 5, 3 \end{pmatrix}$

15. How many five-card hands dealt from a standard deck of 52 cards consist of three aces and two kings?

16. The chairman of a mathematics department with nine instructors must assign four instructors to teach algebra, three to teach statistics, and two to teach calculus. In how many different ways can these assignments be made?

17. Find the coefficient of xy^3z^4 in the expansion of $(x + y + z)^8$.

18. In how many distinguishable ways can the letters of the word SUCCESSES be rearranged?

7 Probability

From its shady origin around the gaming tables of 17th century Europe, the theory of probability has gradually progressed to a position of great prominence in modern science. Many of the greatest figures in mathematical history contributed to its development—Pascal, Bernoulli, Euler, Laplace, and Gauss, just to name a few—but the theory was rarely applied outside of gambling until the 19th century, when it began to be appreciated as a powerful new conceptual tool. Today, probabilistic methods permeate the physical and biological sciences, and through its applied arm—statistics—probability theory has found its way into every area of human endeavor where measurement is involved. In this chapter we examine the basic principles of the theory and explore some of its many applications.

7.1 The Meaning of Probability

Uncertainties are part of life. When we roll the dice or take an airplane flight or start a new business, to some extent we must simply hope for the best. These actions are examples of *random processes,* in the sense that the outcome in each case cannot be predicted with absolute certainty. Probability can be defined simply as the mathematical theory of random processes; in other words, the "science of uncertainty." It may seem paradoxical that mathematics, the most certain of the sciences, has something to say about the very nature of uncertainty itself. In this section we begin to investigate this curious relationship.

Sample Spaces and Events   To begin with, we need a mathematical model for the notion of a "random process." Underlying such a process is what we call an *experiment:* an action that has various possible outcomes. The first step is therefore to represent these outcomes by mathematical entities.

233

Definition

An experiment is an action having various possible outcomes.

A sample space for an experiment is a set **S**, the members of which represent the different possible outcomes. The members of *S* are called sample points.

The following examples illustrate the terminology.

Example 1   *Experiment:* A die is rolled, and the number showing on the top face is observed.

Discussion   We take as our sample space the set

$$S = \{1, 2, 3, 4, 5, 6\}$$

each member of which represents one possible outcome. (See Figure 1.)

Outcome = 4

Figure 1   ◇

Example 2   *Experiment:* A legal information service observes the number of telephone calls received between 9 and 10 am on a given weekday.

Discussion   Let us assume, for argument's sake, that a maximum of ten calls can be received and processed in any one-hour period. Then we can take as our sample space the set

$$S = \{0, 1, 2, 3, 4, 5, 6, 7, 8, 9, 10\}$$

where the sample point *n* represents the outcome "*n* calls are received" ($n = 0, 1, \ldots, 10$).   ◇

Example 3   *Experiment:* Two astronauts are chosen from a group of four candidates.

Discussion   If we denote the candidates by A, B, C, and D, then we can take as our sample space the set of all combinations of these four individuals taken two at a time.

$$S = \{\{A, B\}, \{A, C\}, \{A, D\}, \{B, C\}, \{B, D\}, \{C, D\}\}$$

(This assumes, of course, that the *order* in which the astronauts are selected is not important.) As this example illustrates, sample points do not have to be numbers; we can also use symbols, sequences or sets of symbols, and other things. ◆

When an experiment is about to be performed, we are often interested in whether or not a particular event will occur. Thus, in rolling a die (Example 1) we may want to know whether or not the die will come up even. For this event to occur, the outcome of the experiment must be either 2, 4, or 6. Hence, the event

"the die comes up even"

can be represented mathematically by the set

$$\{2, 4, 6\}$$

consisting of those outcomes that are "favorable" for the event. Listed below are several other events related to the experiment of Example 1, describing each event in words and showing the set of favorable outcomes.

Description of event	Set of favorable outcomes
1. The die comes up odd	$\{1, 3, 5\}$
2. The die comes up less than 5	$\{1, 2, 3, 4\}$
3. The die comes up 2	$\{2\}$
4. The die comes up 7	$\varnothing$
5. The die comes up less than 7	$\{1, 2, 3, 4, 5, 6\}$

Note that event (4) is *impossible* and therefore the set of favorable outcomes is empty, while event (5) is *certain* to occur and so the set of favorable outcomes is the entire sample space S. These examples suggest this definition:

Definition

An **event** is a subset E of the sample space S.

The **impossible** event is the subset $E = \varnothing$ (empty set).

The **certain** event is the subset $E = S$ (sample space).

Note that *any* subset of S represents an event, according to this definition. Thus, in rolling a die the subset $E = \{1, 2, 5\}$ represents an event, which we can describe in words as "the die comes up 1, 2, or 5."

Example 2, Revisited A legal information service observes the number of telephone calls received between 9 and 10 am on a given weekday.

Discussion We list below several events related to this experiment, representing each event as a subset of the sample space

$$S = \{0, 1, 2, \ldots, 10\}.$$

Description of event	Subset of S
1. Exactly 5 calls are received	{5}
2. At least 8 calls are received	{8, 9, 10}
3. Fewer than 5 calls are received	{0, 1, 2, 3, 4}
4. Between 4 and 6 calls are received	{4, 5, 6}
5. More than 12 calls are received	∅ ◇

The following example illustrates an important point, namely that there may be *more than one sample space* for a given experiment. How we choose to represent the outcomes depends on the events we want to describe.

Example 4 *Experiment:* A basketball player takes three free shots from the foul line.

Discussion We could use as a sample space the set

$$S_1 = \{0, 1, 2, 3\}$$

with the understanding that the sample point n represents the outcome "player makes exactly n baskets" ($n = 0, 1, 2, 3$). However, this sample space will be inadequate for some purposes. For instance, the event

"the player misses the second shot"

cannot be represented as a subset of S_1. To represent the possible outcomes in more detail, we can use a tree diagram as in Figure 2, where H represents a hit and M a miss.

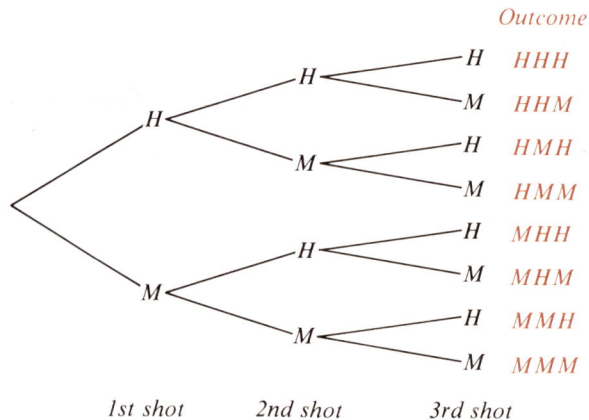

1st shot   2nd shot   3rd shot

Figure 2

Each path through this tree represents one possible outcome of the experiment. We thus arrive at a sample space containing eight sample points.

$$S_2 = \{HHH, HHM, HMH, HMM, MHH, MHM, MMH, MMM\}$$

In S_2 we can represent the event "the player misses the second shot" by the subset

$$E = \{HMH, HMM, MMH, MMM\}.$$

Both S_1 and S_2 are acceptable sample spaces for this experiment, but in S_2 the sample points provide more detail about the particular outcomes they represent. In most applications, therefore, we would prefer S_2 to S_1. ◆

Probability Spaces   Once we have chosen a sample space S for an experiment, the next question is: What do we mean by the *probability* of an event E? As it happens, this innocent-seeming question presents one of the most controversial problems in the philosophy of science! Among the various concepts of probability, the most widely used is the **relative frequency** interpretation.

$$\text{The probability } P(E) \text{ of an event } E \text{ represents} \tag{1}$$
the proportion of the time it tends to occur over the
long run.

According to this, when we say that a fair coin will come up heads "with probability 1/2," we mean that heads tends to come up half the time, out of many tosses. Notice that (1) gives us an *empirical* definition of probability: To compute $P(E)$ we repeat the underlying experiment many times, and let

$$P(E) = \frac{\text{number of times } E \text{ occurs}}{\text{number of trials of the experiment}}.$$

Example 1, Revisited   A die is rolled 500 times and the outcome of each trial is recorded. The results are shown in the following frequency table:

Outcome	Frequency
1	79
2	83
3	85
4	74
5	91
6	88
Total	500

$83+74+88 \quad \frac{245}{500}$

What is the empirical probability that this die will come up even?

Discussion   Out of the 500 trials, the die came up even (2, 4, or 6) a total of $83 + 74 + 88 = 245$ times. Hence,

$$P(\text{die comes up even}) = \frac{245}{500} = .49.$$

That is, the event "die comes up even" tends to occur 49% of the time, over the long run.   ◆

Question: Are 500 trials enough to define "the long run"? How do we know the observed frequencies were really typical and not just a fluke?

Answer: We don't. In general, the relative frequency interpretation of probability is only as good as the empirical data on which it depends. Hence, to get a better idea of the "true" probability that the die will come up even we might decide to roll it 5000 or even 10,000 times.

If we divide each frequency in our table by the total number of trials (500), we obtain the *relative frequency* table shown below.

Outcome	Relative frequency
1	.158
2	.166
3	.170
4	.148
5	.182
6	.176
Total	1.000

Thus, outcome 1 occurred 79 times out of 500 trials, giving a relative frequency

$$P(1) = \frac{79}{500} = .158.$$

These relative frequencies represent the empirical probabilities of each of the six possible outcomes. To compute the probability of a given event E, we can simply add the probabilities of the outcomes favorable for E. For instance,

$$P(\text{die comes up even}) = P(2) + P(4) + P(6)$$
$$= .166 + .148 + .176$$
$$= .490$$

as we already know, and

$$P(\text{die comes up} < 5) = P(1) + P(2) + P(3) + P(4)$$
$$= .158 + .166 + .170 + .148$$
$$= .642.$$

Thus, we can compute the probability of any event E, once we know the probabilities of each of the possible outcomes. This leads us to the following "abstract" notion of a probability space:

Definition

A **probability space** for an experiment is a sample space

$$S = \{a_1, a_2, \ldots, a_n\}$$

together with a rule which assigns to each sample point a_i a number p_i, satisfying

1. $0 \leq p_i \leq 1$ for $i = 1, 2, \ldots, n$
2. $p_1 + p_2 + \cdots + p_n = 1$.

In this definition, the number p_i represents the probability of the outcome a_i. Notice that nothing is said about *how* the values p_i are to be obtained; from the mathematical point of view they are regarded as "given." Thus, in our die-rolling experiment we might want to make the theoretical assumption that the die is perfectly "fair," that is, that all six outcomes are equally likely. This means we would expect each outcome to occur one sixth of the time, over the long run, as reflected in the probability space of the table below.

Outcome	Probability
1	$\frac{1}{6}$
2	$\frac{1}{6}$
3	$\frac{1}{6}$
4	$\frac{1}{6}$
5	$\frac{1}{6}$
6	$\frac{1}{6}$
Total	1

Note that here again, as in the relative frequency table, the probabilities p_i sum to 1. The question whether or not the probabilities in the above table are "correct" for a particular die can only be determined by actual experiment.

As we saw above, the probability of an event E can be computed by adding the probabilities of the outcomes favorable for E. This observation now takes the following form:

Definition

The **probability** of an event

$$E = \{a_i, a_j, \ldots\}$$

is defined by

$$P(E) = p_i + p_j + \cdots.$$

If $E = \emptyset$ we define $P(E) = 0$.

Note that the *impossible* event $\varnothing$ has probability 0, while the *certain* event

$$S = \{a_1, a_2, \ldots, a_n\}$$

has probability

$$P(S) = p_1 + p_2 + \cdots + p_n = 1.$$

Thus, the probability of an event E can never be less than 0 or greater than 1.

Let E be an event in a probability space S. Then

$$0 \leq P(E) \leq 1.$$

If E is impossible then $P(E) = 0$.
If E is certain to occur, then $P(E) = 1$.

Example 2, Revisited

A legal information service records the number of telephone calls received between 9 and 10 am on 250 successive weekdays. The results are shown in Table 1.

Calls	Frequency
0	28
1	42
2	59
3	47
4	28
5	19
6	12
7	7
8	5
9	2
10	1
Total	250

Table 1

Calls	Probability
0	.112
1	.168
2	.236
3	.188
4	.112
5	.076
6	.048
7	.028
8	.020
9	.008
10	.004
Total	1.000

Table 2

Use this empirical data to define a probability space.

Solution

Dividing each frequency by the number of observations (250), we obtain the empirical probability of each outcome (Table 2). Using these probabilities, we find that

$$P(\text{fewer than 5 calls}) = P(0) + P(1) + P(2) + P(3) + P(4)$$
$$= .112 + .168 + .236 + .188 + .112$$
$$= .816.$$

In other words, the service receives fewer than five calls 81.6% of the time. ◆

The probabilities associated with sample points can sometimes be deduced from the conditions of a problem, as the next example illustrates.

Example 5 A coin is loaded in such a way that heads comes up 50% more often than tails, over the long run. What is the probability that the coin will come up heads on the next toss?

Solution The experiment of tossing a coin results in one of the outcomes heads (H) or tails (T). Hence, our probability space is of the form

Outcome	Probability
H	p
T	q

where the probabilities p and q are unknown. Since heads comes up 50% more often than tails, it follows that the probability of heads is 50% greater than the probability of tails; that is,

$$p = q + .5q = 1.5q. \tag{2}$$

Also, from the second condition in the definition of a probability space we must have

$$p + q = 1. \tag{3}$$

Using (2) to eliminate p in equation (3), we obtain

$$1.5q + q = 1$$
$$2.5q = 1$$
$$q = \frac{1}{2.5} = .4.$$

By (3), therefore, the probability of heads on the next toss is

$$p = 1 - q = .6.$$

That is, for this particular coin, heads will come up 60% of the time, over many tosses. ◆

Example 6 Four biologists—Alvarez, Björnson, Cramer, and Duvchenko—are in the final running for the Nobel Prize. Insiders say that (i) Björnson and Duvchenko have equal chances of winning it, (ii) Alvarez's chances are twice as good as Björnson's, and (iii) Duvchenko's chances are twice as good as Cramer's. Find the probability that one of the Americans, either Alvarez or Cramer, will win the prize.

Solution   The experiment "the Nobel Prize is awarded" has four possible outcomes, which we denote by A, B, C, and D. Our probability space is therefore of the form

Outcome	Probability
A	p
B	q
C	r
D	s

where the probabilities p, q, r, s are unknown. Conditions (i), (ii) and (iii) in the problem translate into the equations

$$q = s \tag{4}$$
$$p = 2q \tag{5}$$
$$s = 2r. \tag{6}$$

Since the four probabilities must sum to 1, we also have

$$p + q + r + s = 1. \tag{7}$$

We solve equations (4)–(7) for the four unknowns p, q, r, s. For instance, (4)–(6) yield

$$s = 2r$$
$$q = s = 2r$$
$$p = 2q = 4r$$

and substituting these values into (7), we obtain

$$4r + 2r + r + 2r = 1$$
$$9r = 1$$
$$r = \frac{1}{9}$$

so that
$$p = \frac{4}{9}, q = \frac{2}{9}, r = \frac{1}{9}, s = \frac{2}{9}.$$

Hence:
$$P(\text{American wins prize}) = P(\{A, C\})$$
$$= P(A) + P(C)$$
$$= \frac{4}{9} + \frac{1}{9}$$
$$= \frac{5}{9}$$

or about a 55.6% chance. ◆

Remark to Reader   As the last example illustrates, we sometimes use probability theory in situations where the relative frequency interpretation does not apply. Since the experiment of Example 6—namely, awarding the

Nobel Prize in a certain specific year—cannot be repeated, it makes no sense to speak of the "proportion of the time" an event tends to occur. Here the probability of an event simply represents a subjective measure of likelihood, based on the information provided by our "insiders." We frequently encounter this subjective notion of probability; for instance, when we speak of the chances that a given candidate will win an election or the chances that life exists on other planets.

In some cases the notion of a **comparison class** is useful in determining probabilities. For example, to estimate the probability that a particular driver—say Louise, age 22—will have an auto accident in a given year, an insurance company might compare her with other female drivers in the 20–25 age group. If their records show that out of 15,000 drivers in this category 90 had an accident the previous year, it might be reasonable to assign the probability 90/15,000 = .006, or 0.6%, to the event that Louise will have an accident this year. Of course, the choice of comparison class is crucial. The fact that Louise is a stunt driver, for example, could raise this probability considerably.

Odds   Probabilities are often expressed in the language of "odds." For instance, we might say of a certain event E that *the odds are 3 to 2 in favor of E*. In terms of the relative frequency interpretation, this means that over the long run E will occur 3 times for every 2 times it does not occur. In other words, E will occur 3 times out of 5, over the long run, and therefore

$$P(E) = \frac{3}{5} = .6.$$

More generally:

Definition

The **odds in favor of an event** E **are** m **to** n if and only if

$$P(E) = \frac{m}{m + n}. \tag{8}$$

In this case we also say the **odds against** E **are** n **to** m.

Example 7   Find the probability of E, given that
(a) the odds in favor of E are 4 to 1
(b) the odds in favor of E are 5 to 3
(c) the odds against E are 13 to 7.

Solution   (a) Since the odds in favor of E are 4 to 1,

$$P(E) = \frac{4}{4 + 1} = \frac{4}{5} = .8.$$

(b) Odds in favor are 5 to 3, so

$$P(E) = \frac{5}{5 + 3} = \frac{5}{8} = .625.$$

(c) Since the odds against are 13 to 7, this means that the odds in *favor* are 7 to 13, and therefore

$$P(E) = \frac{7}{7 + 13} = \frac{7}{20} = .35.   \blacklozenge$$

We can also go in the other direction, from probability to odds, as shown by the next example.

Example 8   Find the odds in favor of E, given that
(a) $P(E) = 11/16$ (b) $P(E) = .76.$

Solution   (a) We write $P(E) = \frac{11}{16} = \frac{11}{11 + 5}.$

This has the form (8), with $m = 11$, $n = 5$. Thus, the odds in favor of E are 11 to 5.

(b) We write $P(E) = .76 = \frac{76}{100} = \frac{76}{76 + 24}.$

Hence by (8), the odds in favor of E are 76 to 24, or (in reduced form) 19 to 6. $\blacklozenge$

From the examples above we see that odds can be converted to probabilities, and vice-versa; the two concepts are interchangeable. It should be emphasized, however, that we are speaking here of what are called "true odds," not the "house odds" quoted in a casino. The latter refer to betting payoffs, which are discussed in Chapter 8.

7.1 Exercises   1. Fifteen auto engines are tested, and the number of defective units is recorded. Describe a sample space S for this experiment, and represent each of the following events as a subset of S:

(a) None of the engines is defective.

(b) At most two engines are defective.

(c) More than half the engines are defective.

2. In Example 3 in the text, represent each of the following events as a subset of the sample space S:

(a) Candidate A is one of those chosen.

(b) Candidates B and C are both chosen.

(c) Candidate D is not one of those chosen.

3. A coin is tossed twice. Describe a sample space S for this experiment, and represent each of the following events as a subset of S:

   (a) The coin comes up heads on the second toss.

   (b) The coin comes up heads at least once.

   (c) The coin comes up heads twice as many times as it comes up tails.

4. In Example 4 in the text, represent each of the following events as a subset of the sample space S_2:

   (a) The player misses the third shot.

   (b) The player makes exactly two baskets.

   (c) The player makes a basket on some shot but misses the next shot.

Consider an experiment with sample space $S = \{a, b, c, d\}$. In Exercises 5–8, determine whether the given assignment of values makes S a probability space.

5.

Outcome	Probability
a	.1
b	.2
c	.3
d	.4

6.

Outcome	Probability
a	$\frac{1}{5}$
b	$\frac{1}{4}$
c	$\frac{1}{3}$
d	$\frac{1}{2}$

7.

Outcome	Probability
a	.27
b	.53
c	$-.22$
d	.42

8.

Outcome	Probability
a	$\frac{3}{8}$
b	$\frac{1}{4}$
c	$\frac{5}{16}$
d	$\frac{1}{16}$

Consider the probability spaces (i), (ii), (iii) below, each based on a sample space $S = \{a, b, c, d, e\}$.

Outcome	Probability	Outcome	Probability	Outcome	Probability
a	$\frac{1}{2}$	a	$\frac{1}{5}$	a	.15
b	$\frac{1}{4}$	b	$\frac{1}{5}$	b	.20
c	$\frac{1}{8}$	c	$\frac{1}{5}$	c	.25
d	$\frac{1}{16}$	d	$\frac{1}{5}$	d	.30
e	$\frac{1}{16}$	e	$\frac{1}{5}$	e	.10
	(i)		(ii)		(iii)

In Exercises 9–12, calculate the probability of the given event E in each of these spaces.

9. $\{b, c\}$

10. $\{a, d, e\}$

11. $\{a, b, c, d, e\}$

12. $\{b, c, d, e\}$

13. Of the last 5862 sites drilled by an oil company, 927 yielded oil. What is the empirical probability that the next site drilled will yield oil?

14. A basketball player has made 271 baskets from the foul line, out of 359 attempts. What is the empirical probability that his next shot will go in?

15. A baker is trying to determine the number of chocolate cakes he is likely to sell on any given day, to help him decide how many to prepare. He keeps a record of the number requested on each of 300 working days, with these results:

Requests	Frequency
0	33
1	60
2	96
3	57
4	36
5	15
6	3
Total	300

(a) Use this frequency table to define a probability space.

(b) What is the probability that the number of cakes requested on a given day will be (i) exactly two? (ii) at most two? (iii) at least four?

(c) What is the maximum number of cakes he can prepare if he wants to be at least 50% sure of selling them all?

16. A stockbroker notices that her in-city mailings are rarely delivered within one day. Following up on 500 items mailed to find out when they arrived, she obtains the following data:

Delivery time (working days)	Number of items
1	109
2	159
3	136
4	65
5	31
Total	500

(a) Use this frequency table to define a probability space.

(b) What is the probability that a given item mailed on Monday will arrive (i) on Wednesday, Thursday, or Friday? (ii) by Thursday at the latest?

(c) By what day must she mail an item in order to be at least 80% sure it will arrive before the weekend?

17. Watching the cars at a certain intersection, an observer notices that they turn right 50% more often than they turn left, and 40% more often than they go straight. What is the probability that the next car approaching the intersection will turn left?

18. A die is loaded in such a way that (i) 1, 2, and 3 are equally likely; (ii) 4, 5, and 6 are equally likely; but (iii) 1 is 50% more likely than 6. What is the probability of rolling an even number with this die? ≈ .466

19. A lawyer informs her client that, in her judgment, a stiff sentence is twice as likely as a light one, and some sentence — either stiff or light — is twice as likely as no sentence at all. What is the probability of a stiff sentence? (Hint: There are three possible outcomes.)

20. Five Olympic swimmers — two from East Germany and three from the U.S. — are competing in the women's 200 m freestyle event. Three of the swimmers, one East German and two Americans, are all given exactly the same chance of winning. Of the two swimmers remaining, the East German is twice as likely to win as the American, but only half as likely to win as her East German teammate. What is the probability that the East German team will win the event?

21. Which of the following probabilities could be interpreted in terms of relative frequency? In terms of a comparison class?

    (a) The probability that a quarter, dropped by accident on a floor tiled with 6-inch squares, will land entirely on one tile (without touching a crack)

    (b) The probability that out of ten random people, at least two will have the same birthday

    (c) The probability that a heart bypass operation will be successful

    (d) The probability that pollution will eventually melt the polar ice caps.

22. Outfielder Larry Garfield has singled late in a World Series game, and the manager must decide whether Garfield should try to steal second base. The records show that Garfield has stolen successfully 248 times out of 371 career attempts, for a lifetime "stealing average" of .668. Is this a reasonable value to use for the probability of a successful steal? What factors might raise or lower the manager's estimate?

23. Find the probability of E, given that

    (a) the odds in favor of E are 7 to 1

    (b) the odds in favor of E are 1 to 1

    (c) the odds against E are 6 to 5

    (d) the odds against E are 1.5 to 1.

24. Complete the missing entries in the following table:

Event	Probability	Odds in favor	Odds against
A	.75	3 to 1	1 to 3
B	.85		
C			18 to 7
D	$\frac{4}{9}$		
E			12 to 13
F	$\frac{7}{12}$		
G		9 to 11	

7.2   The Classical Formula

By idealizing certain experiments it becomes possible to calculate probabilities *theoretically* as well as empirically, and thus to predict what the long-run relative frequencies will be. This "classical" concept of probability, which dates back to the 16th century, is still indispensable in analyzing games of chance and in the theory of random sampling.

Uniform Spaces   It is often intuitively plausible from the nature of an experiment that *all possible outcomes are equally likely*. In such a case the underlying probability space S is called a **uniform space.**

Example 1   *Experiment:* A fair die is rolled, and the number showing on the top face is observed.

Discussion   Here the word "fair" means that the die is perfectly symmetric (or very nearly so), and of constant density throughout. Under this assumption it is reasonable to expect each of the six faces to come up about one sixth of the time, over the long run; we would therefore assign probability 1/6 to each point in the sample space

$$S = \{1, 2, 3, 4, 5, 6\}. \quad \blacklozenge$$

In general:

If a uniform space S contains n sample points, then each has probability $1/n$.

Proof   If each of the sample points $a_1, a_2, \ldots, a_n$ has the same probability $p_i = p$, then

$$
\begin{aligned}
1 &= p_1 + p_2 + \cdots + p_n \\
&= p + p + \cdots + p \quad (n \text{ terms}) \\
&= np
\end{aligned}
$$

from which it follows that $p = 1/n$.

The phrases "fair die," "fair coin," "chosen at random," and so on, are used to signify that the underlying probability space is uniform. The early probabilists generally took uniformity for granted, on the grounds that two outcomes should be considered equally likely if no reason can be given why one rather than the other will occur. Today we are generally more cautious. For example, hospital records indicate that the probability of a random child being born male is about .514—not 1/2, as pure reason would suggest—and even gambling devices such as dice and roulette wheels have been shown to have slight but sometimes significant biases.*

The Classical Formula In a uniform space we can calculate probabilities by the following formula:†

The Classical Formula

Let S be a uniform space containing n sample points, and let E be an event containing k sample points. Then

$$P(E) = \frac{k}{n} = \frac{\text{number of favorable outcomes}}{\text{number of possible outcomes}}. \qquad (1)$$

To see why formula (1) works, let us return to the fair die of Example 1 and calculate the probability of the event "die comes up greater than 2".

$$P(\text{die comes up} > 2) = P(3) + P(4) + P(5) + P(6)$$
$$= \frac{1}{6} + \frac{1}{6} + \frac{1}{6} + \frac{1}{6}$$
$$= \frac{4}{6}$$

or in reduced form 2/3. Notice that 4/6 is the ratio of the number of favorable to the number of possible outcomes; more generally, if k out of n equally likely outcomes are favorable to an event E, then

$$P(E) = \underbrace{\frac{1}{n} + \frac{1}{n} + \cdots + \frac{1}{n}}_{k \text{ terms}} = \frac{k}{n}.$$

The classical formula thus enables us to calculate "theoretical" probabilities by simply counting sample points and forming a ratio.

Example 2   Baron Clouseau has 175 bottles in his wine cellar, classified as follows: 77 red Bordeaux, 39 white Bordeaux, 32 red Burgundies and 27 white Burgundies. As one of the Baron's guests enters the cellar to select a wine, his candle blows out, forcing him to choose a bottle at random. What is the probability that he will choose a white Bordeaux? A Burgundy? A red wine?

---

*In 1947 two Caltech students, A. R. Hibbs and R. Walford, found that one out of every four roulette wheels in the casinos they surveyed had a bias sufficient to cancel out the "house take."

†This formula, which was used by Cardano, Galileo, and others, is mentioned in Jacob Bernoulli's famous treatise, the *Ars Conjectandi,* published in 1713.

Solution   Since the choice is made at random, all 175 bottles are equally likely to be chosen. Hence, by the classical formula,

$$P(\text{white Bordeaux}) = \frac{\text{number of white Bordeaux}}{\text{number of bottles}} = \frac{39}{175} \doteq .223$$

$$P(\text{Burgundy}) = \frac{\text{number of Burgundies}}{\text{number of bottles}} = \frac{32 + 27}{175} = \frac{59}{175} \doteq .337$$

$$P(\text{red wine}) = \frac{\text{number of red wines}}{\text{number of bottles}} = \frac{77 + 32}{175} = \frac{109}{175} \doteq .623.$$

Notice that the probability of choosing a red wine (.623) is the same as the proportion of red wines in the total "population" (62.3%). ◆

In general, *proportions (or percentages) translate into probabilities* when we sample at random from a population.

Example 3   7.5% of the items produced by a certain machine are known to be defective. If an item is chosen at random from the machine's weekly output, what is the probability that it will be defective?

Solution   The weekly output contains an unspecified number (which we will call *n*) of items, all of which are equally likely to be chosen. Since 7.5% of these are defective, formula (1) gives

$$P(\text{item is defective}) = \frac{\text{number of defectives}}{\text{number of items}} = \frac{.075n}{n} = .075.$$

Thus the probability of a defective is the same as the proportion of defectives in the population sampled (regardless of the population size). ◆

Example 4   A charity has sold 2500 tickets in a raffle. One ticket, chosen at random, wins a free trip to Hawaii. What is the probability that Laurette will win the raffle
(a) if she buys one ticket?   (b) if she buys fifty tickets?

Solution   (a) Since the winning ticket is chosen at random, it is equally likely to be any of the 2500 tickets sold. Only one of these tickets is favorable for Laurette; hence, by the classical formula,

$$P(\text{Laurette wins raffle}) = \frac{1}{2500} = .0004 \ (= 0.04\%).$$

(b) If she buys fifty tickets then

$$P(\text{Laurette wins raffle}) = \frac{50}{2500} = \frac{1}{50} = .02 \ (= 2\%) \quad ◆$$

Games of Chance   Games involving cards and dice played a central role in the discovery of the principles of probability. In fact, the famous correspondence between Pascal and Fermat in 1654, which many historians mark as the origin of probability theory, dealt with two gambling problems posed to Pascal by the Chevalier de Méré, a noted dilettante in the French court. Since card and dice games will be used throughout this chapter to illustrate basic concepts, we give here a brief description of the underlying probability spaces.

Cards The standard deck consists of 52 cards, each having one of thirteen *values*

Ace (A), 2, 3, 4, . . ., 10, Jack (J), Queen (Q), King (K)

and one of four *suits*

spades (s), hearts (h), diamonds (d), clubs (c).

The values are usually ordered as listed (thus, a jack is "higher" than a nine, and so on); however, in some games aces are counted higher than kings. Jacks, queens and kings are called *face cards*. Spades and clubs are *black* cards, while hearts and diamonds are *red*.

Each card can be represented by one of 52 symbols, as shown in Table 1.

VALUES

		A	2	3	4	$\cdots$	10	J	Q	K
	s	A_s	2_s	3_s	4_s	$\cdots$	10_s	J_s	Q_s	K_s
	h	A_h	2_h	3_h	4_h	$\cdots$	10_h	J_h	Q_h	K_h
SUITS	d	A_d	2_d	3_d	4_d	$\cdots$	10_d	J_d	Q_d	K_d
	c	A_c	2_c	3_c	4_c	$\cdots$	10_c	J_c	Q_c	K_c

Table 1

When a card is dealt at random from a deck, all 52 outcomes are equally likely, and the classical formula applies.

Example 5 What is the probability that a card dealt at random will be a club? A face card? Both a club and a face card?

Solution Counting favorable and possible outcomes, we obtain

$$P(\text{club}) = \frac{\text{number of clubs}}{\text{number of cards possible}} = \frac{13}{52} = \frac{1}{4} = .25.$$

In terms of relative frequency, this means that out of many repeated trials, the card will be a club about 25% of the time. Likewise, we find that

$$P(\text{face card}) = \frac{12}{52} = \frac{3}{13} \doteq .231$$

$$P(\text{club and face card}) = \frac{3}{52} \doteq .058. \quad \blacklozenge$$

In most card games, not one but several cards are dealt at random. To analyze such games we need the counting techniques developed in Chapter 6.

Example 6* In blackjack (or "21"), two cards are dealt at random to the player. Face cards have a value of 10 points, and aces count as either 1 or 11, at the discretion of the player. What is the probability that the value of the player's original hand will total 21?

Solution A blackjack hand consists of two cards out of 52, with order unimportant. Thus there exist

$$C(52, 2) = \frac{52 \cdot 51}{2 \cdot 1} = 1326$$

possible hands, all equally likely. In order to total 21, one of the cards must be either a ten or a face card and the other must be an ace. Since a ten or face card can be chosen in 16 ways, and an ace in 4 ways, the Counting Principle gives a total of

$$16 \cdot 4 = 64$$

such hands. By the classical formula we conclude that

$$P(\text{hand totals 21}) = \frac{\text{number of favorable hands}}{\text{number of possible hands}} = \frac{64}{1326} \doteq .0483.$$

This assumes that the cards are dealt from a single deck. In some casinos two or more decks are used, which lowers the probability slightly. (See Exercise 15.) ◆

Example 7* In poker, a player is dealt five cards at random. What is the probability that the player will receive a "flush," that is, five cards all of the same suit?

Solution The number of possible 5-card hands is

$$C(52, 5) = \frac{52 \cdot 51 \cdot 50 \cdot 49 \cdot 48}{5 \cdot 4 \cdot 3 \cdot 2 \cdot 1} = 2,598,960$$

and if the cards are dealt at random all of these outcomes are equally likely. To determine the number of flushes we can use the following counting scheme:

C_1: Choose a suit   4 ways
C_2: Choose 5 cards of that suit $C(13, 5) = 1287$ ways
                                  Total $4 \cdot 1287 = 5148$ ways

Thus, by the classical formula we have

$$P(\text{flush}) = \frac{5148}{2,598,960} \doteq .002.$$

*These examples require counting techniques discussed in Section 6.3.

That is, out of many hands dealt we would expect to receive a flush about 0.2% of the time, or about one time in five hundred. ◆

Dice Certain games such as craps and backgammon depend on a roll of two dice simultaneously. In order to distinguish the dice physically, let us say one is white and the other black as in Figure 1.

Outcome = (5, 3)

Figure 1

The outcomes can be represented by pairs (i, j), where i shows on the white die and j on the black die. The 36 possible pairs are shown in Table 2.

		BLACK				
	1	2	3	4	5	6
1	(1, 1)	(1, 2)	(1, 3)	(1, 4)	(1, 5)	(1, 6)
2	(2, 1)	(2, 2)	(2, 3)	(2, 4)	(2, 5)	(2, 6)
3	(3, 1)	(3, 2)	(3, 3)	(3, 4)	(3, 5)	(3, 6)
4	(4, 1)	(4, 2)	(4, 3)	(4, 4)	(4, 5)	(4, 6)
5	(5, 1)	(5, 2)	(5, 3)	(5, 4)	(5, 5)	(5, 6)
6	(6, 1)	(6, 2)	(6, 3)	(6, 4)	(6, 5)	(6, 6)

WHITE (left of rows 3–4)

Table 2

These pairs comprise the sample space S for a roll of two dice. If the dice are fair we can regard S as a uniform space, in which each sample point (i, j) has probability 1/36.

Example 8 What are the possible sums that can be rolled with two fair dice? What is the probability of each sum?

Solution The sum can be as low as two ("snake eyes") and as high as twelve ("box cars"). For each sum we can count the favorable pairs (i, j) by referring to the table. For example, a sum of eight can occur in five ways—(2, 6), (3, 5), (4, 4), (5, 3), and (6, 2)—and hence

$$P(\text{sum} = 8) = \frac{\text{number of favorable outcomes}}{\text{number of possible outcomes}} = \frac{5}{36} \, (\doteq .139).$$

In other words, the odds against throwing an eight are 31 to 5. The other sums and their probabilities are shown in Table 3.

Sum	Probability
2	$\frac{1}{36}$
3	$\frac{2}{36}$
4	$\frac{3}{36}$
5	$\frac{4}{36}$
6	$\frac{5}{36}$
7	$\frac{6}{36}$
8	$\frac{5}{36}$
9	$\frac{4}{36}$
10	$\frac{3}{36}$
11	$\frac{2}{36}$
12	$\frac{1}{36}$

Table 3

Note that the most likely sum is seven, with probability $6/36 \doteq .167$. The numbers 2 through 12, representing the sum rolled, can be used as an alternative sample space. This space is not uniform, however, as Table 3 shows. ◆

Sampling* The classical formula can be applied to a wide variety of problems involving ordered and unordered samples. These more difficult applications generally require the use of our permutation and combination formulas.

Example 9 If five letters are typed at random, what is the probability that no letter will be repeated?

Solution By the Counting Principle, the number of possible "words" is

$$26^5 = 11,881,376.$$

Of these, the number containing no repetitions is given by

$$P(26, 5) = 26 \cdot 25 \cdot 24 \cdot 23 \cdot 22 = 7,893,600.$$

Thus,

$$P(\text{no repetition}) = \frac{7,893,600}{11,881,376} \doteq .664. \quad ◆$$

---

*These examples require counting techniques discussed in Section 6.3.

Example 10 Four of the paintings in Baron Clouseau's chateau are priceless originals, while the other twelve are merely good copies. A thief, untutored in such distinctions, plans to make off with five paintings from the chateau, chosen at random. What is the probability that he will take none of the originals? Exactly two of the originals?

Solution (a) Since there are 16 paintings altogether, and the thief is sampling 5 at random (without respect to order), the number of possible outcomes is

$$C(16, 5) = \frac{16 \cdot 15 \cdot 14 \cdot 13 \cdot 12}{5 \cdot 4 \cdot 3 \cdot 2 \cdot 1} = 4368$$

all of which are equally likely. Among these, the number of samples containing no original is

$$C(12, 5) = \frac{12 \cdot 11 \cdot 10 \cdot 9 \cdot 8}{5 \cdot 4 \cdot 3 \cdot 2 \cdot 1} = 792.$$

Hence,

$$P(\text{none are originals}) = \frac{792}{4368} \doteq .181.$$

(b) To determine the number of samples containing exactly two originals we use a counting scheme.

C_1: Choose 2 originals $\qquad C(4, 2) = \dfrac{4 \cdot 3}{2 \cdot 1} = \quad 6 \text{ ways}$

C_2: Choose 3 copies $\qquad C(12, 3) = \dfrac{12 \cdot 11 \cdot 10}{3 \cdot 2 \cdot 1} = 220 \text{ ways}$

$$\text{Total} \quad 6 \cdot 220 = 1320 \text{ ways}$$

By the Counting Principle, a total of 1320 samples contain exactly two originals, and therefore

$$P(\text{two are originals}) = \frac{1320}{4368} \doteq .302. \quad \blacklozenge$$

Note that we needed *both* the Counting Principle *and* the formula for combinations in order to find the number of favorable outcomes in part (b).

7.2 Exercises

1. If a fair die is rolled, what is the probability that it will come up
   (a) odd?
   (b) less than 3?
   (c) both odd and less than 3?
   (d) both odd and greater than 5?

2. If a card is drawn at random from a standard deck, what is the probability that it will be
   (a) a black card?
   (b) the jack of hearts?
   (c) higher than an eight (ace is low)?
   (d) neither black nor higher than an eight?

3. What is the probability that a random person's birthday will fall in March? That it will fall between May 15 and August 15, inclusive? (Assume all birthdays are equally likely.)

4. The American roulette wheel has 38 equal compartments, numbered 0, 00, and 1 through 36. A ball is spun counter to the wheel's motion and comes to rest in one of the compartments. What is the probability that the winning number will be

   (a) the number 29?

   (b) a number in the "second dozen" (13–24)?

   (c) an odd number in the second dozen?

5. If a two-digit number from 00 to 99 is generated at random by a computer, what is the probability that it will be

   (a) less than 47?   (b) evenly divisible by three?   (c) prime?

6. 250,000 tickets have been sold in a lottery. One ticket, chosen at random, will win first prize ($100,000), ten tickets will win second prize ($10,000), 50 tickets will win third prize ($2000), and 100 tickets will win fourth prize ($1000). What is the probability that a random ticketholder will win

   (a) $100,000?   (b) $2000 or more?   (c) nothing?

7. A large number of glass beads of uniform shape and size are mixed together to form an assortment. 34% of the beads are clear blue, 12% are cloudy blue, 10% are clear green, 18% are cloudy green, and 26% are clear yellow. If a bead is chosen from the assortment in the dark, what is the probability that it will be

   (a) cloudy green?   (b) clear?   (c) either blue or yellow?

8. The registered voters in Baker County are 65% white, 25% black, and 10% Hispanic. 60% of the white voters are male, compared with 48% of the black voters and 50% of the Hispanic voters. What is the probability that a juror, selected at random from the voting rolls, will be

   (a) a nonwhite?   (b) a Hispanic male?   (c) a female?

9. Two fair dice are rolled. What is the probability that

   (a) the sum rolled will be at least eight?

   (b) both dice will come up greater than 2?

   (c) both dice will come up the same?

10. A player rolls two fair dice repeatedly until either a sum of five occurs—called the player's "point"—in which case he wins, or a sum of seven occurs, in which case he loses.

    (a) Show that the game can end in any of 10 equally likely ways.

    (b) Find the probability that the player will win.

    (c) What if the player's "point" is four? Eight?

11. A fair coin is tossed three times. What is the probability that it will

    (a) come up heads three times?

    (b) come up heads exactly twice?

    (c) come up heads on some toss and tails on a later one?

12. Three fair dice are rolled. What is the probability that

    (a) they will all come up 6?

    (b) two will come up 6 and one will come up 5?

    (c) the sum rolled will be at least 17?

*13. Five distinct letters are chosen at random from the first ten letters of the alphabet (A–J) to form a five-letter "word." What is the probability that the word will

    (a) contain no vowels?

    (b) begin with a consonant and end with a vowel?

*14. A state lottery distributes 100,000 tickets, numbered 00000 to 99999. One of these five-digit numbers is then chosen at random, and prizes are awarded according to the number of matching initial digits, as follows:

    Ticket with all five winning digits pays $10,000
    Tickets matching first four digits pay only $1000
    Tickets matching first three digits pay only $100.

    What is the probability that a ticketholder will win

    (a) $10,000?   (b) $1000 or more?   (c) $100 or more?

†15. In Example 6 in the text, we found the probability that a random blackjack hand will total 21, if the cards are dealt from a single deck. What is the probability of receiving such a hand if the cards are dealt from

    (a) two decks, shuffled together?   (b) three decks?

†16. The premier of Liechtenstein must appoint three of the country's ten cabinet members to a highly sensitive NATO committee. Unknown to the premier, two Soviet agents have infiltrated the cabinet. If the three cabinet members are selected at random, what is the probability that neither of the Soviet agents will be chosen?

†17. If five cards are dealt at random, what is the probability that they will all have value nine or lower (ace is low)?

†18. A 13-card bridge hand is dealt at random. What is the probability that it will contain no diamonds? (Express the answer in terms of factorials, or use a calculator.)

†19. If three men and three women are seated randomly in a row, what is the probability that men and women will have alternate seats?

†20. A committee of four is chosen at random from a group consisting of six men and six women. What is the probability that the committee will contain

    (a) no men?   (b) two men and two women?

†21. In a simplified form of poker, a player is dealt three cards at random. What is the probability that the hand will be:

    (a) a "flush," that is, three cards of the same suit?

---

*These exercises require counting techniques discussed in Section 6.2.
†These exercises require counting techniques discussed in Section 6.3.

(b) "three-of-a-kind," that is, three cards of the same value?

(c) a "pair," that is, two cards of one value and one card of a different value?

†22. A simplified version of Keno is played as follows: On a special ticket the player marks any five numbers from 1 to 20, called the player's "spots." The house then chooses, at random, a set of five winning numbers from 1 to 20, and the player receives a payoff according to how many of his spots are winning numbers:

Winning spots	Payoff	Probability
0	0	.19369
1	0	.44021
2	$1	.29347
3	$5	.06772
4	$25	.00484
5	$2500	.00006

(a) Show that the probability that the house's "winning set" will contain exactly two of the player's spots is given by

$$\frac{C(5, 2) \cdot C(15, 3)}{C(20, 5)} = .29347.$$

(b) Calculate the other probabilities in the table above.

‡23. If the letters of the word TATTOO are randomly rearranged, what is the probability that

(a) they will still spell TATTOO?   (b) the new word will begin with T?

‡24. Three construction firms have submitted bids on six different government contracts. If each contract is awarded to one of the three firms at random, what is the probability that

(a) one firm will get all six?   (b) each firm will get two?

Simulation Repeated trials of an experiment can often be simulated using such devices as random number generators. For example, we can simulate five successive tosses of a fair coin by generating a random 5-digit binary number (sequence of 0's and 1's) and interpreting 1 as "heads" and 0 as "tails". Table II in the Appendix shows 250 such random numbers, generated by a computer.

25. Using these numbers, estimate empirically the probability of a fair coin coming up heads exactly three times in five tosses. How does your result compare with the theoretical value 5/16?

26. Using these numbers, estimate empirically the probability of obtaining a "run" of at least two heads in a row (such as THHTH or TTHHH) when a fair coin is tossed five times. How does your result compare with the theoretical value 19/32?

---

†This exercise requires counting techniques discussed in Section 6.3.
‡These exercises require counting techniques discussed in Section 6.4.

Computer Projects

A. The DICE program simulates repeated rolls of two fair dice by generating a sequence of random integers i and j from 1 to 6. The program counts the frequency of each sum $s = i + j$. Use this program to estimate empirically the probability of each sum, based on (a) 500 trials (b) 5000 trials. How do your results compare with the theoretical probabilities in Table 3?

B. Write a BASIC program to simulate repeated spins of a roulette wheel. (See Exercise 4.) Use this program to estimate empirically the probability of the events described in Exercise 4, based on (a) 500 trials (b) 5000 trials. How do your results compare with the theoretical probabilities?

7.3 Basic Laws of Probability

When events are logically related to one another, it is reasonable to expect their probabilities to be mathematically related in some way. In this section we develop a few simple rules which constantly come into play in probability calculations.

Events as Sets In Section 7.1 an *event* was defined as a subset E of the sample space S, which we think of as comprising those outcomes that are "favorable" for the event. In Figure 1 a Venn diagram represents this concept, where the sample space S is the universal set and $E \subset S$ is a typical event.

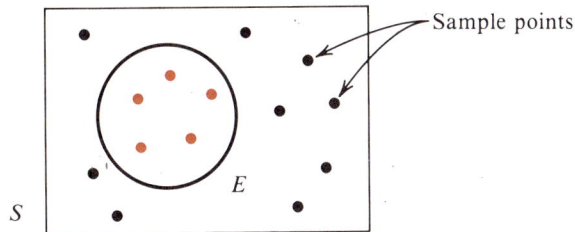

Figure 1

The probability of E is the sum of the probabilities assigned to the points in E. If S is uniform we can use the classical formula

$$P(E) = \frac{n(E)}{n(S)} = \frac{\text{number of points in } E}{\text{number of points in } S}. \tag{1}$$

where $n(X)$ denotes the number of elements in a set X. We shall assume uniformity at various points in this section, in order to simplify the proofs of certain laws. These laws are valid, however, in any probability space.

Operations on Events   Since events are sets, we can combine them using the operations of union, intersection, and complement (see Section 6.1), illustrated in Figure 2.

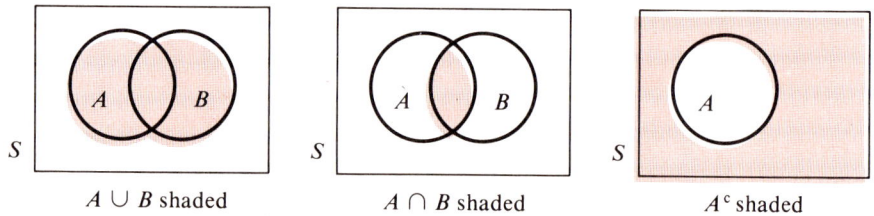

$A \cup B$ shaded $A \cap B$ shaded A^c shaded

Figure 2

From the figure we see that the union $A \cup B$ contains those outcomes favorable for either *A or B,* the intersection $A \cap B$ contains those outcomes favorable for both *A and B,* and the complement A^c contains those outcomes *un*favorable for *A.* Thus

$A \cup B$ represents the event   *"either A or B occurs (or both)"*

$A \cap B$ represents the event   *"both A and B occur"*

A^c represents the event   *"A does not occur."*

Remark to Reader   The union, intersection, and complement operations enter into many of our later formulas. The reader should therefore memorize the meaning of each of these symbols.

Example 1   In the experiment of rolling two fair dice, consider the following events:

$$A = \text{``sum rolled is eight''}$$
$$B = \text{``both dice come up greater than 2.''}$$

Express each of the events $A \cup B$, $A \cap B$, and A^c in words, and calculate its probability.

Solution $A \cup B$ represents   "sum = 8   *or*   both dice come up > 2"

$A \cap B$ represents   "sum = 8   *and*   both dice come up > 2"

A^c represents   "sum ≠ 8."

Figure 3 shows the sample space S for two dice, with the pairs (i, j) represented by points and the events A and B shown as subsets of S. (See Table 2, Section 7.2.) Counting the points in each region and using formula (1), we see that

$$P(A) = \frac{5}{36}, \quad P(B) = \frac{16}{36},$$

$$P(A \cup B) = \frac{18}{36}, \quad P(A \cap B) = \frac{3}{36}, \quad P(A^c) = \frac{31}{36}.$$

BLACK DIE

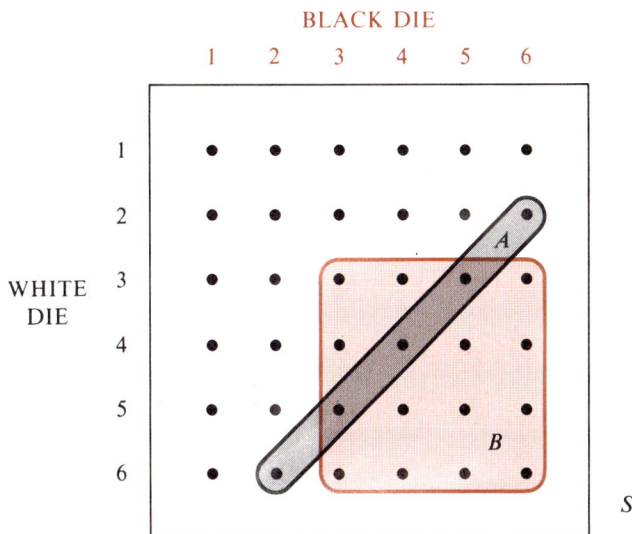

Figure 3

The Complement Rule If k out of n outcomes in a uniform space S are favorable to an event A, then the remaining $n - k$ outcomes are favorable to A^c. Thus,

$$P(A) + P(A^c) = \frac{k}{n} + \frac{n - k}{n} = \frac{n}{n} = 1.$$

Using this equation we can express $P(A^c)$ in terms of $P(A)$, or vice-versa.

Complement Rule

For any event A,

$$P(A^c) = 1 - P(A) \quad \text{and} \quad P(A) = 1 - P(A^c)$$

This is illustrated by Example 1 above (rolling two fair dice), where we found that

$$P(\text{sum} \neq 8) = P(A^c)$$
$$= 1 - P(A)$$
$$= 1 - \frac{5}{36} = \frac{31}{36}.$$

Example 2 The American roulette wheel has 38 equal compartments, numbered 0, 00, and 1 through 36. The numbers 0 and 00 are green, and the numbers 1–36 are evenly divided between red and black. If a player makes a $1 bet on red, what is the probability that he or she will lose?

Solution The ball can come to rest in any of 38 compartments, of which 18 (the red numbers) are favorable for the player. Hence

$$P(\text{player wins}) = \frac{18}{38} = \frac{9}{19}$$

so that

$$P(\text{player loses}) = 1 - P(\text{player wins})$$

$$= 1 - \frac{9}{19}$$

$$= \frac{10}{19} \ (\doteq .526).$$

Thus, the player can expect to lose about 52.6% of the time, over the long run. ◆

✶ When the probability of an event seems hard to calculate directly, a good strategy is to look at the complement—that is, calculate the probability that the event will *not* occur—and then apply the Complement Rule.

Example 3 If two fair dice are rolled, what is the probability that at least one of them will come up even?

Solution The sample space for rolling two dice is shown in Figure 3. Now consider the event

$$E = \text{``at least one die comes up even.''}$$

The favorable outcomes for E are represented by those pairs (i, j) in which either i or j (or both) is even. We could count these sample points directly in Figure 3, of course, but there is an easier way. Notice that the complement of E can be expressed in words as

$$E^c = \text{``neither die comes up even''}$$

which is the same as saying

$$E^c = \text{``both dice come up odd.''}$$

How many outcomes (i, j) are favorable for this event? Since there are three possible odd values for i (1, 3, or 5) and three for j, the Counting Principle gives us $3 \cdot 3 = 9$ such pairs. Hence,

$$P(E^c) = \frac{9}{36}$$

and therefore

$$P(E) = 1 - P(E^c)$$

$$= 1 - \frac{9}{36}$$

$$= \frac{27}{36} = \frac{3}{4} = .75. \quad ◆$$

The next example also involves the phrase "at least one," and again the Complement Rule simplifies the problem.

Example 4* What is the probability that a random poker hand (5 cards) will contain at least one diamond? *order not important*

Solution The number of possible 5-card hands is

$$C(52, 5) = \frac{52 \cdot 51 \cdot 50 \cdot 49 \cdot 48}{5 \cdot 4 \cdot 3 \cdot 2 \cdot 1} = 2{,}598{,}960.$$

If we introduce the event

$$D = \text{"hand contains at least one diamond"}$$

then the complement can be expressed as

$$D^c = \text{"hand contains no diamonds."}$$

This is the same as saying that the hand consists of five cards, all of which are nondiamonds. Since the deck contains $52 - 13 = 39$ nondiamonds, the number of such hands is

$$C(39, 5) = \frac{39 \cdot 38 \cdot 37 \cdot 36 \cdot 35}{5 \cdot 4 \cdot 3 \cdot 2 \cdot 1} = 575{,}757.$$

Hence

$$P(D^c) = \frac{575{,}757}{2{,}598{,}960} \doteq .222$$

and the Complement Rule gives

$$P(D) = 1 - P(D^c)$$
$$\doteq 1 - .222$$
$$= .778.$$

In other words, out of many hands dealt about 77.8% will contain at least one diamond. ◆

The Union Rule Consider two events A and B in a uniform space S, as shown in Figure 4. The number of points in the union $A \cup B$, denoted by

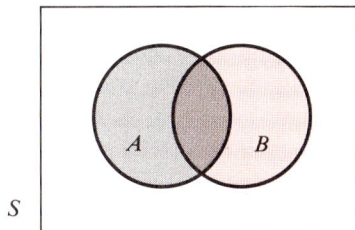

Figure 4

---

*This example requires counting techniques discussed in Section 6.3.

$n(A \cup B)$, will be less than the sum $n(A) + n(B)$ if the events overlap, as in the figure. The reason for this is that points common to A and B are counted *twice* in the sum—once as elements of A and once as elements of B. If we correct for this by subtracting the number of common points from the sum, we obtain the identity

$$n(A \cup B) = n(A) + n(B) - n(A \cap B)$$

which holds for any sets A and B. Dividing this equation by $n(S)$ and applying formula (1), we arrive at an important law of probability.

Union Rule

For any events A and B,

$$P(A \cup B) = P(A) + P(B) - P(A \cap B).$$

Thus, in Example 1 we found that when two fair dice are rolled, the events

$$A = \text{``sum is 8''}$$
$$B = \text{``both dice come up} > 2\text{''} \quad \text{and}$$
$$A \cap B = \text{``sum is 8 \textit{and} both dice come up} > 2\text{''}$$

have respective probabilities 5/36, 16/36, and 3/36. The Union Rule now gives

$$P(\text{sum is 8 \textit{or} both come up} > 2) = P(A \cup B)$$
$$= P(A) + P(B) - P(A \cap B)$$
$$= \frac{5}{36} + \frac{16}{36} - \frac{3}{36}$$
$$= \frac{18}{36}$$

as we found earlier by actually counting sample points.

Example 5   In order to be accepted by the army a volunteer must pass both a physical and a written exam. Among recent volunteers 17% have failed the physical exam, 26% have failed the written exam, and 8% have failed both. What is the probability that a random volunteer will be accepted?

Solution   Consider the events

$$A = \text{``volunteer fails the physical exam''}$$
$$B = \text{``volunteer fails the written exam.''}$$

Using the recent volunteers as a comparison class, we see that

$$P(A) = .17, \quad P(B) = .26, \quad P(A \cap B) = .08.$$

A volunteer is rejected if he or she fails *either* exam. Hence, by the Union Rule,

$$P(\text{volunteer rejected}) = P(A \cup B)$$
$$= P(A) + P(B) - P(A \cap B)$$
$$= .17 + .26 - .08$$
$$= .35.$$

The Complement Rule now gives

$$P(\text{volunteer accepted}) = 1 - .35 = .65. \quad \blacklozenge$$

The Union Rule can also be used to give us a lower bound for $P(A \cap B)$ when $P(A)$ and $P(B)$ are known. This is illustrated in the next example.

Example 6   Suppose events A and B have probabilities .70 and .75, respectively. Show that the probability is at least .45 that *both* events will occur.

Solution   By the Union Rule, we have

$$P(A \cup B) = P(A) + P(B) - P(A \cap B)$$
$$= .70 + .75 - P(A \cap B)$$
$$= 1.45 - P(A \cap B).$$

But $P(A \cup B)$ cannot be greater than 1; hence

$$1.45 - P(A \cap B) \leq 1$$

and therefore $P(A \cap B) \geq 1.45 - 1 = .45$

as claimed. $\blacklozenge$

Example 7   An oil company has undertaken exploratory drilling in northern Canada. Preliminary studies indicate a 90% chance of finding oil and an 85% chance of finding natural gas. What can be concluded about the probability of finding both?

Solution   By the Union Rule,

$$P(\text{oil} \cup \text{gas}) = P(\text{oil}) + P(\text{gas}) - P(\text{oil} \cap \text{gas}).$$

Since the term on the left cannot exceed 1, we have

$$P(\text{oil}) + P(\text{gas}) - P(\text{oil} \cap \text{gas}) \leq 1.$$

After rearranging terms as in the preceding example, this becomes

$$P(\text{oil} \cap \text{gas}) \geq P(\text{oil}) + P(\text{gas}) - 1$$
$$= .90 + .85 - 1$$
$$= .75.$$

Thus, there is at least a 75% chance of finding both oil *and* gas. In order to determine this probability exactly we would have to be given more information. ◆

Mutually Exclusive Events   An important special case of the Union Rule arises when we consider events A and B which "exclude" each other, in the sense that they cannot both occur at the same time.

Definition

Events A and B are called **mutually exclusive** if the event "A and B" is impossible; that is,

$$A \cap B = \emptyset.$$

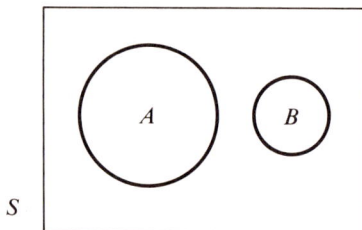

A and B are mutually exclusive

Figure 5

Figure 5 shows two mutually exclusive events A and B as subsets of a sample space S. Since no outcome is favorable for both A and B, the sets are disjoint. When this is the case we have $P(A \cap B) = P(\emptyset) = 0$, and therefore the Union Rule takes the following special form:

Special Union Rule

If A and B are mutually exclusive events, then

$$P(A \cup B) = P(A) + P(B).$$

This rule is one of the most often applied principles of probability.

Example 8   Swimmers A and B are entered in a race, along with several others. The odds in favor of A winning the race are 1 to 3, while the odds in favor of B are 1 to 5. What is the probability that either A or B will win the race?

Solution   Converting odds to probabilities, we have

$$P(\text{A wins}) = \frac{1}{1+3} = \frac{1}{4}$$

$$\text{and } P(\text{B wins}) = \frac{1}{1+5} = \frac{1}{6}.$$

Since they cannot both win the race, these two events are mutually exclusive. Hence, by the Special Union Rule,

$$P(\text{A wins} \cup \text{B wins}) = P(\text{A wins}) + P(\text{B wins})$$

$$= \frac{1}{4} + \frac{1}{6}$$

$$= \frac{5}{12} \ (\doteq .417). \quad \blacklozenge$$

Example 9*   If two cards are dealt at random from a standard deck, what is the probability that they will be either two aces or two face cards?

Solution   The events "two aces are dealt" and "two face cards are dealt" are mutually exclusive, since aces are not face cards. Hence,

$$P(2 \text{ aces} \cup 2 \text{ face cards}) = P(2 \text{ aces}) + P(2 \text{ face cards})$$

$$= \frac{C(4, 2)}{C(52, 2)} + \frac{C(12, 2)}{C(52, 2)}$$

$$= \frac{6}{1326} + \frac{66}{1326}$$

$$= \frac{72}{1326} \ (\doteq .0543). \quad \blacklozenge$$

The Special Union Rule can be generalized to three or more events. Specifically, events $A_1, A_2, \ldots, A_n$ are called mutually exclusive if no two of them can occur at the same time. In this case the Special Union Rule becomes

$$P(A_1 \cup \cdots \cup A_n) = P(A_1) + \cdots + P(A_n).$$

This formula enables us to calculate the probability of an event by breaking it down into separate cases.

Example 10   If two fair dice are rolled, what is the probability that the sum will be at least 10?

Solution   The probabilities of the various sums were calculated in Section 7.2. We divide the event "sum is at least 10" into mutually exclusive cases and apply the Special Union Rule.

---

*This example requires counting techniques discussed in Section 6.3.

5,5/4,6/6,4
6+5/5,6
6+6

$$P(\text{sum is at least 10}) = P(\text{sum is 10 or more})$$
$$= P(\text{sum} = 10 \cup \text{sum} = 11 \cup \text{sum} = 12)$$
$$= P(\text{sum} = 10) + P(\text{sum} = 11) + P(\text{sum} = 12)$$
$$= \frac{3}{36} + \frac{2}{36} + \frac{1}{36} \quad\quad 6 \; + 6$$
$$= \frac{6}{36} = \frac{1}{6} \; (\doteq .167) \quad \blacklozenge$$

We can analyze any event of the form "sum is at least k" or "sum is at most k" in a similar manner. However, it is sometimes easier to use the Complement Rule to reduce the number of cases. For example, the event "sum is at least four" leads to nine cases.

$$P(\text{sum is at least 4}) = P(\text{sum is 4 or more})$$
$$= P(\text{sum} = 4) + P(\text{sum} = 5) + \cdots + P(\text{sum} = 12)$$

Its complement, however, leads to only two cases.

$$P(\text{sum is less than 4}) = P(\text{sum is 3 or less})$$
$$= P(\text{sum} = 2) + P(\text{sum} = 3)$$
$$= \frac{1}{36} + \frac{2}{36} = \frac{3}{36}$$

Hence, by the Complement Rule,

$$P(\text{sum is at least 4}) = 1 - \frac{3}{36} = \frac{33}{36} \; (\doteq .917).$$

In addition to the Complement and Union Rules there is also an Intersection Rule, which we shall see in Section 7.5. A summary of these rules can be found at the end of this chapter.

7.3 Exercises

1. In the experiment of observing the economy for one year, consider the events

$$A = \text{"inflation increases"}$$
$$B = \text{"unemployment increases."}$$

Symbolize each of the following events in terms of A and B using set operations, and illustrate with a Venn diagram:

(a) Inflation does not increase.

(b) Both inflation and unemployment increase.

(c) Inflation increases but not unemployment.

(d) Neither inflation nor unemployment increases.

2. In the experiment of marketing three new products, consider the events

$$A = \text{"Product A is successful"}$$
$$B = \text{"Product B is successful"}$$
$$C = \text{"Product C is successful."}$$

Symbolize each of the following events in terms of A, B, and C using set operations, and illustrate with a Venn diagram:

(a) All three products are successful.

(b) None of the three products is successful.

(c) Only Product B is successful.

(d) At least two of the three products are successful.

3. In the experiment of drawing a card at random from a standard deck, consider the events

$$A = \text{``card is red''}$$
$$B = \text{``card is higher than a five (ace is low).''}$$

Express each of the following events in words, and calculate its probability:

(a) $A \cap B$ (b) $A \cup B$ (c) $A \cap B^c$ (d) $A^c \cap B^c$

4. In the experiment of generating a three-digit number from 000 to 999 at random, consider the events

$$A = \text{``number is less than 400''}$$
$$B = \text{``number is even''}$$
$$C = \text{``number is at least 200.''}$$

Express each of the following events in words, and calculate its probability:

(a) A^c (b) $A \cap C$ (c) $(A \cup B)^c$ (d) $A \cap B \cap C$

5. Suppose the probability that A will occur is .53, the probability that B will occur is .72, and the probability that both will occur is .48. What is the probability that

(a) A will not occur?   (c) neither A nor B will occur?

(b) either A or B will occur?

6. Suppose A, B, and C are mutually exclusive events with respective probabilities 1/5, 1/4, and 1/2. What is the probability that

(a) one of the three events will occur?

(b) none of the three events will occur?

(c) more than one of the three events will occur?

In Exercises 7–12, use the Complement Rule.

7. If a fair coin is tossed three times, what is the probability that it will come up heads at least once?

8. If three fair dice are rolled, what is the probability that at least one will come up 6?

* 9. If five letters (not necessarily distinct) are typed at random, what is the probability that at least one will be a vowel?

*10. If four hats are returned to four women at random, what is the probability that someone will receive the wrong hat?

_____

*These exercises require counting techniques discussed in Section 6.2.

†11. A shipment of twenty refrigerators contains five defective units. If three units are chosen from the shipment at random, what is the probability that at least one defective unit will be among them?

†12. What is the probability that a random blackjack hand (two cards) will contain at least one face card?

In Exercises 13–18, use the Union Rule.

13. In a large sample of families it was found that 86% of the husbands and 63% of the wives were employed, while in 54% of the cases both were employed. Assuming the sample is representative, what is the probability that a random family will have

    (a) at least one spouse employed?

    (b) neither spouse employed?

14. The percentage of voters supporting Senator Foghorn went from 46% on August 1 to 52% on September 1, although only 38% supported him in both polls. If a voter is chosen at random, what is the probability that the voter supported Foghorn

    (a) either on August 1 or September 1?

    (b) at neither time?

15. What is the probability that the winning number in roulette will be either odd (from 1 to 35) or in the second dozen (13–24)? (See Exercise 4 in Section 7.2.)

16. What is the probability that a card drawn at random from a standard deck will be either a club or a face card?

17. A matchmaker figures the odds are 11 to 1 that Norman will like Sylvia, and 7 to 5 that she will like him. Show that the odds are at least even (probability .5) that they will both like each other.

18. Two mechanical systems must both function properly in order for a Venus probe to land successfully. NASA officials estimate that under actual landing conditions the chances are 80% and 75% that the respective systems will work. What can be concluded about the chances of a successful landing?

In Exercises 19–24, use the Special Union Rule for mutually exclusive events.

19. A racetrack is paying 3 to 1 on Affirmation and 4 to 1 on Brobdingnag in the third race. If these payoffs represent the true odds against these horses, what is the probability that one of the two will win the race?

20. Three New York critics are watching a preview of a trendy new French film. The film's producer rates the chances of a favorable review at 15% for Alexis, 20% for Bruce, and 25% for Clive. However, no two of the critics ever like the same film. What is the probability that all three reviews will be unfavorable?

21. A punctilious commuter observes that the 5:30 train is usually anywhere from one to ten minutes late. Keeping records over a long period, she obtains the following probability table:

---

†These exercises require counting techniques discussed in Section 6.3.

Minutes late	Probability
0	.03
1	.07
2	.12
3	.15
4	.17
5	.16
6	.13
7	.09
8	.05
9	.02
10	.01

What is the probability that on a random day, the train will be

(a) at least five minutes late?

(b) at least two minutes late?

(c) at most seven minutes late?

22. In the simplified version of Keno described in Exercise 22 of Section 7.2, what is the probability that the player's payoff will be

(a) zero?   (b) at most $5?   (c) at least $5?

†23. If five people are chosen at random from a group of four men and six women, what is the probability that a majority of those selected will be women?

†24. If two marbles are chosen at random from an urn containing ten white marbles and ten black marbles, what is the probability that they will both be the same color?

25. Draw a Venn diagram showing three events A, B, and C in a uniform space S (see Figure 3 in Section 6.1).

(a) Explain why points in $A \cap B^c \cap C$ are counted twice in the sum $n(A) + n(B) + n(C)$, while points in $A \cap B \cap C$ are counted three times.

(b) Prove the *Union Rule for Three Events*.

$$P(A \cup B \cup C) = P(A) + P(B) + P(C)$$
$$-P(A \cap B) - P(A \cap C) - P(B \cap C)$$
$$+P(A \cap B \cap C)$$

Difference of Sets   The **difference** of sets A and B, written $A - B$, is defined by

$$A - B = A \cap B^c = \{x \mid x \in A \text{ but } x \notin B\}.$$

If A and B are events, $A - B$ represents the event "A occurs but B does not."

26. Draw a Venn diagram for two events A and B, with $A - B$ shaded. Use the Special Union Rule to prove that

$$P(A - B) = P(A) - P(A \cap B).$$

†These exercises require counting techniques discussed in Section 6.3.

27. In Exercise 13, what is the probability that a random family will have
    (a) only the husband employed?
    (b) exactly one spouse employed?

28. In Exercise 14, what is the probability that a random voter
    (a) supported Senator Foghorn on August 1 but not on September 1?
    (b) changed his or her opinion either way during August?

7.4 Conditional Probability

The probability of an event A is, in a sense, a measure of our degree of ignorance concerning the outcome. Thus, if we have some information about the outcome—like a stock market investor who has picked up an inside tip—it may alter the probability of A. In this section we see how to calculate such *conditional* probabilities.

Reduced Sample Spaces   Let S be a uniform space representing the possible outcomes of an experiment, and let A be an event in S, as in Figure 1a. Suppose that the experiment has been performed, but we have not been told the outcome; and suppose we are interested in whether or not the event A occurred.

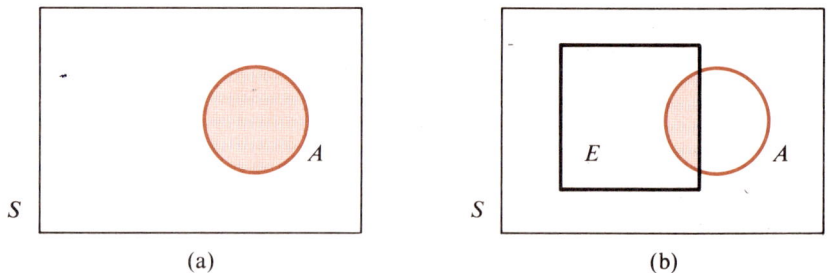

(a)   (b)

Figure 1

Case 1: We are given no information at all. In this case we are in the same position as before the experiment was performed; namely, all outcomes in S are equally likely. By the classical formula, the probability that A occurred is therefore

$$P(A) = \frac{n(A)}{n(S)} = \frac{\text{number of points in } A}{\text{number of points in } S}.$$

Case 2: We are given the information that a certain event E occurred, as illustrated in Figure 1b. In this case, the only possible outcomes are those in E, so that E becomes in effect a *reduced sample space*. The outcomes in E that are favorable to A are those in the intersection $E \cap A$, which is shaded in Figure 1b. Thus, applying the classical formula to the new sample space E, we find that the *probability that A occurred, given the fact that E occurred,* is

$$P(A|E) = \frac{n(E \cap A)}{n(E)} = \frac{\text{number of points in } E \text{ and } A}{\text{number of points in } E}. \tag{1}$$

Example 1 Two fair dice are rolled. What is the probability that the sum rolled was seven, if we are told that both dice came up less than 5?

Solution Figure 2 shows the sample space for two dice, together with the events

$$A = \text{``sum rolled is seven''}$$

and $\quad E = \text{``both dice come up less than 5.''}$

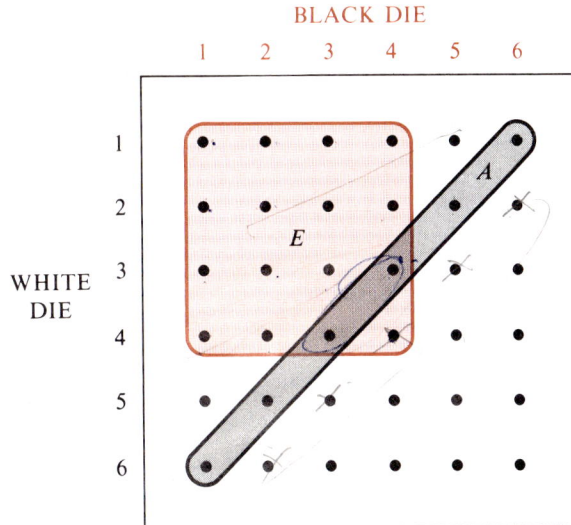

Figure 2

Applying formula (1), we find that

$$P(A|E) = \frac{n(E \cap A)}{n(E)} = \frac{2}{16} = .125.$$

In other words, the fact that both dice came up less than 5 implies that only 16 points (i, j) are possible, instead of the usual 36. Since only two of these points are favorable to A—the pairs $(3, 4)$ and $(4, 3)$—the probability of A, given E, is 2/16 or .125. Note that if we ignore E we have

$$P(A) = \frac{6}{36} \doteq .167.$$

Thus, the probability that the sum was seven *decreased* from 16.7% to 12.5% because of the information that both dice came up less than 5. ◆

Conditional Probability Probabilities of the form (1) are called *conditional* probabilities. Notice that

$$\frac{n(E \cap A)}{n(E)} = \frac{n(E \cap A)/n(S)}{n(E)/n(S)} = \frac{P(E \cap A)}{P(E)}$$

so the conditional probability $P(A|E)$ can be defined as a ratio of ordinary probabilities.

Definition

Let A and E be events in a probability space S, with $P(E) \neq 0$. The **conditional probability of A given E** is defined by

$$P(A|E) = \frac{P(E \cap A)}{P(E)}. \tag{2}$$

Definition (2) applies in any probability space. In a uniform space, however, we can also use the equivalent formula (1).

Example 2 In a pharmaceutical research study, half of a group of 500 patients suffering from chronic headaches were given a new painkilling drug; the other half were given a placebo. The results are shown below.

	Felt better	Felt no better	Total
Given drug	183	67	250
Given placebo	117	133	250
Total	300	200	500

(a) What is the probability that a random patient felt better, given that he or she received the drug?

(b) What is the probability that a random patient received the placebo, given that he or she felt no better?

Solution (a) Since 250 patients received the drug, 183 of whom reported an improvement, formula (1) yields

$$P(\text{better}|\text{drug}) = \frac{n(\text{drug} \cap \text{better})}{n(\text{drug})}$$

$$= \frac{183}{250} = .732.$$

That is, 73.2% of those given the drug felt better.

(b) Of those given the drug, 67 reported no improvement, compared ...
133 of those given the placebo. Thus,

$$P(\text{placebo}|\text{no better}) = \frac{n(\text{no better} \cap \text{placebo})}{n(\text{no better})}$$

$$= \frac{133}{200} = .665.$$

That is, 66.5% of those who felt no better were given the placebo. ◆

Example 3 In a large survey of its customers, a department store finds that 19% make major purchases, 45% have a charge account, and 12% fall into both categories. What is the probability that
(a) a customer who makes a major purchase has a charge account?
(b) a customer with a charge account will make a major purchase?

Solution Assuming the sample surveyed was representative, the figures indicate that for a random customer,

$$P(\text{major purchase}) = .19$$
$$P(\text{charge account}) = .45$$
$$P(\text{major purchase} \cap \text{charge account}) = .12.$$

In part (a) we want to know the conditional probability that a random customer has a charge account, *given* the fact that the customer makes a major purchase. By formula (2), this conditional probability is

$$P(\text{charge account}|\text{major purchase}) = \frac{.12}{.19} \doteq .632.$$

In other words, among those customers who make a major purchase 63.2% have a charge account, compared to 45% of the customers at large. (b) Here we want to know the conditional probability that a random customer will make a major purchase, given the fact that the customer has a charge account. Again using formula (2), we obtain

$$P(\text{major purchase}|\text{charge account}) = \frac{.12}{.45} \doteq .267$$

that is, 26.7% of the customers with a charge account make a major purchase, compared to 19% of the customers at large. ◆

Note that conditional probabilities are easily disguised in ordinary discourse. For instance, the phrase "customer who makes a major purchase" gives us information about our "random" customer, but in an indirect way.

Example 4 An insurance company has found that in any given year, 8.4% of its policyholders will have at least one accident, and 1.4% will have at least two accidents. What is the probability that a policyholder who has already had one accident this year will have another one?

Solution   The data indicate that for a random policyholder,

$$P(\text{at least 1 accident}) = .084$$
$$P(\text{at least 2 accidents}) = .014.$$

We are given the fact that the policyholder has already had one accident; that is, the event "at least 1 accident" has already occurred. Hence, the probability that the policyholder will have another accident during the year is given by

$$P(\text{at least 2}|\text{at least 1}) = \frac{P(\text{at least 1} \cap \text{at least 2})}{P(\text{at least 1})}. \qquad (3)$$

But the event "at least 1 ∩ at least 2" is the same as the event "at least 2," because anyone who has two or more accidents during the year will necessarily have one or more. Thus, (3) becomes

$$P(\text{at least 2}|\text{at least 1}) = \frac{P(\text{at least 2})}{P(\text{at least 1})}$$

$$= \frac{.014}{.084} \doteq .167.$$

Thus, the chance of another accident is about 16.7%.   ◆

Applications   Conditional probabilities arise very naturally in connection with "multi-stage" experiments, since additional information becomes available as each stage is completed. Typically in such cases we do not employ formula (2); instead we use the available information to reduce the sample space, then count favorable and possible outcomes and apply the classical formula. Games of chance provide abundant illustrations of this technique.

Example 5   A blackjack player receives two cards from the dealer, who also gives herself two cards, one face up and one face down. The player can request an additional card as long as the total value of his hand is less than 21. If the new card gives him a total greater than 21 ("bust"), he loses the game. Suppose the player holds the jack of clubs and the four of hearts, and the dealer's exposed card is the six of diamonds.

(a) What is the probability that the player will bust if he "hits," that is, takes another card?

(b) Suppose he hits and receives the two of spades. What is the probability that he will bust if he hits again?

Solution   (a) The situation is illustrated in Figure 3a. Recall* that face cards are worth 10 points and aces count as either 1 or 11, at the player's discretion. Thus, the player's initial hand is worth 14 points, and in order to bust he must draw an eight or higher, that is, 8, 9, 10, J, Q, or K.

---

*See Example 6 of Section 7.2.

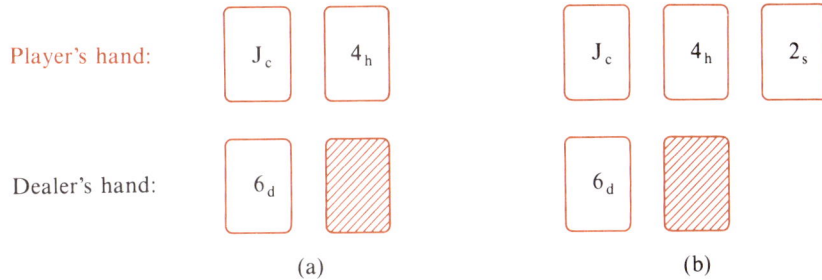

Figure 3

If the player now hits, the card he receives could be anything except J_c, 4_h, or 6_d, since these cards are already out. Thus there are $52 - 3 = 49$ possible outcomes, all equally likely.* Of these 49 possible outcomes, 23 lead to a bust—namely, any eight through king, except the jack of clubs. Hence

$$P(\text{player busts}) = \frac{23}{49} \doteq .469.$$

(b) Suppose the player draws the two of spades, producing the situation shown in Figure 3b. Since his hand is now worth 16 points, he will bust if he draws any card higher than a five, namely, 6, 7, 8, 9, 10, J, Q, or K. Moreover, two of these cards (J_c and 6_d) are already out. Thus, of the $52 - 4 = 48$ possible outcomes if he hits again, $32 - 2 = 30$ lead to a bust, and therefore

$$P(\text{player busts}) = \frac{30}{48} = .625. \quad \blacklozenge$$

We remark that both (a) and (b) are *conditional* probabilities, since in each case we are given the information that certain events have already occurred. The next example is similar, except that combinations are involved.

Example 6† You and an opposing player are each dealt a five-card poker hand from a standard deck. If your hand contains two aces, what is the probability that your opponent has at least one ace?

Solution The simplest approach here is to look at the "conditional complement" of the event in question; that is, to calculate

$$P(\text{your opponent has } no \text{ aces} | \text{you have 2 aces})$$

and then subtract from 1 (see Exercise 19).

---

*Note that the dealer's second card, since we know nothing about it, must be treated as though it were still in the deck!

†This example requires counting techniques discussed in Section 6.3.

Since you hold five known cards of which two are aces, your opponent's hand is chosen from the remaining 47 cards, of which two are aces. The number of possible hands is thus $C(47, 5)$, and the number with no aces is $C(45, 5)$, so that

$$P(\text{opponent has no aces}|\text{you have 2}) = \frac{C(45, 5)}{C(47, 5)} \doteq .796$$

and therefore

$$P(\text{opponent has an ace}|\text{you have 2}) \doteq 1 - .796 = .204. \quad \blacklozenge$$

Notice that in the situation described in Example 6, the probability of the event "opponent has an ace" depends on who is computing it and what information is available to him. From *your* point of view this probability is approximately .204. However, from your *opponent's* point of view it is either 0 or 1, since he knows for certain whether or not he holds an ace. A third observer, who can see neither your hand nor your opponent's, would come up with still a different answer! Since information alters probabilities, it is not surprising that different individuals, possessing different information about an experiment, might arrive at different estimates for the probability of an event, each of which is "correct" from the observer's point of view.

7.4 Exercises

1. Suppose the probability that A will occur is .53, the probability that B will occur is .72, and the probability that they will both occur is .48. Calculate

   (a) the conditional probability of A, given B

   (b) the conditional probability of B, given A.

2. Suppose that A and B are mutually exclusive events with respective probabilities 1/3 and 1/4. Calculate

   (a) the conditional probability of A, given B

   (b) the conditional probability of B, given A.

3. Two fair dice are rolled. What is the probability that the sum rolled is at least eight, given

   (a) no information at all?

   (b) that at least one die comes up 5?

   (c) that both dice come up the same?

4. A card is drawn at random from a standard deck. What is the probability that the card is higher than a seven (ace is low), given

   (a) no information at all?

   (b) that the card is not an ace?

   (c) that the card is a face card?

5. A fair coin is tossed three times. What is the probability that it comes up heads at least once, given

   (a) no information at all?

(b) that all three tosses come up the same?

(c) that it comes up heads at most once?

*6. Two letters, not necessarily distinct, are typed at random. What is the probability that they are both vowels, given

(a) no information at all?

(b) that they are different?

(c) that at least one is a vowel?

7. Show that the probability that a random card is an ace is not affected by the information that it is a club.

8. Show that the probability that a random number from 00 to 99 is less than 30 is not affected by the information that it is even.

9. In a sociological survey, 1200 voters were interviewed and rated as either "liberal" or "conservative." The respondents were also classified by age, with the following results:

		Liberal	Conservative	Total
	Under 30	195	162	357
Age	30–50	219	321	540
	Over 50	110	193	303
	Total	524	676	1200

If this study is representative of the public at large, what is the probability that

(a) a random voter 30 or older will be conservative?

(b) a random voter who is liberal will be 50 or younger?

10. In the Framingham study,† a cardiovascular research project, a group of 828 men in their thirties were classified according to serum cholesterol level (in milligrams per deciliter). These cases were followed up many years later to determine how many developed coronary disease. The results are shown below.

		Coronary disease	No coronary disease	Total
	Under 200	22	293	315
	200–219	12	116	128
Cholesterol level	220–239	29	140	169
	240–259	19	78	97
	Over 259	35	84	119
	Total	117	711	828

If this data is representative of men in their thirties, what is the probability that

(a) a man in his thirties with a cholesterol level of 237 will develop coronary disease in later life?

---

*This exercise requires counting techniques discussed in Section 6.2.

†From William B. Kannel, M.D., "Recent Findings of the Framingham Study," *Resident & Staff Physician*, January 1978, p. 57. Copyright by The Resident, Inc.

(b) a man who develops coronary disease had a cholesterol level of at least 240 when he was in his thirties?

11. Among army volunteers it is found that 17% fail the physical exam, 26% fail the written exam, and 8% fail both. What is the probability that a volunteer who fails the physical exam will also fail the written exam?

12. The percentage of voters supporting Senator Foghorn went from 46% on August 1 to 52% on September 1, although only 38% supported him in both polls. What is the probability that a random voter who supported him on August 1 still supported him on September 1?

13. U.S. population life tables indicate that the probability that a white female will survive from birth to age 60 is .864, while the probability that she will survive to age 70 is .733. What is the probability that a white female already 60 years of age will survive to age 70?

14. Track oddsmakers say that Brobdingnag is a 3-to-1 favorite to win the Kentucky Derby and a 5-to-2 favorite to win the Triple Crown (the Kentucky Derby and two other races). If they are right, what is the probability that if the horse wins the Derby he will go on to win the Triple Crown?

15. A poker player holds a 5, 6, 7, 8, and jack. The hands of the other players are unknown. If she discards the jack and takes another card at random, what is the probability that she will have a "straight," a hand with values 4 through 8 or 5 through 9?

16. A blackjack player (see Example 5 in this section) decides to "stand" with a seven and a ten; that is, he declines to take another card. The dealer's exposed card is a nine, and when she turns over her other card it turns out to be a five, which means she must draw a third card for herself. From the player's viewpoint what is the probability that when the dealer takes the third card

(a) she will bust, that is, go over 21 points?

(b) she will beat the player's hand, that is, have a total of 18, 19, 20, or 21?

†17. A poker player holds a 3, 5, 9, and two queens. If he discards the 3, 5, and 9 and draws three new cards at random, what is the probability that he will pick up at least one more queen?

†18. A Scrabble player has a Q among the seven letter-tiles in her tray, but no U (or blank tile, which may function as a U) to go with it. She observes that (i) there are 100 letter-tiles altogether, of which four are U's and two are blanks; (ii) 49 tiles have already been played on the board, including one U and one blank; (iii) seven tiles are in her tray, as described above, and seven are in her opponent's tray, which is not visible. The remaining 37 tiles are in a bag. If the player puts three of her letters on the board and chooses three new ones from the bag at random, what is the probability that she will pick a U or a blank?

19. Prove: If A and E are events with $P(E) \neq 0$, then

$$P(A^c|E) = 1 - P(A|E).$$

(See Exercise 26 in Section 7.3.)

---

†These exercises require counting techniques discussed in Section 6.3.

20. Let A and E be events with $P(E) \neq 0$. Prove:

    (a) If $E \subset A$ then $P(A|E) = 1$.

    (b) If E and A are mutually exclusive then $P(A|E) = 0$.

21. An argument similar to the following actually occurred in a murder trial:

    Prosecutor: Ladies and gentlemen of the jury, you have heard the witnesses testify that the murderer drove a blue GM sedan whose license plate began with Z. Out of 735,000 cars registered in this state, there are only five which (a) fit the description given, and (b) could have been at the scene of the crime. More-over, Snavely—the defendant—owns one of those five cars! Mathematically, the odds are therefore overwhelming that Snavely is the murderer.

    As Snavely's attorney, how would you defend your client against this "mathematical" attack?

22. In a certain inner-city high school, 4000 students were asked (i) whether they were regular alcohol users, (ii) whether they were regular marijuana users, and (iii) whether they had ever tried "hard" drugs such as amphetamines, barbiturates, or heroin. Their responses are summarized in the Venn diagram below:

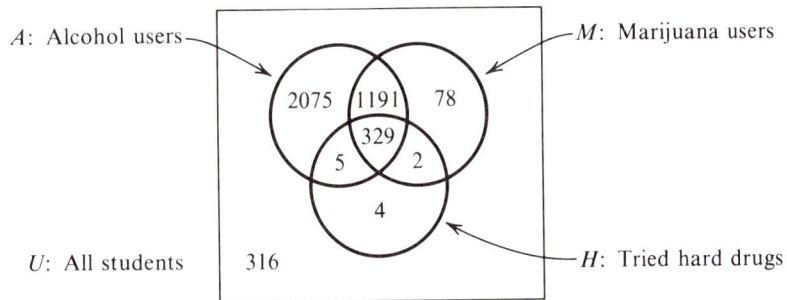

A marijuana advocate claimed that these data disproved the theory that marijuana use leads to hard drugs, since an even higher percentage of those trying hard drugs were alcohol users (98.2%) than marijuana users (97.4%). Use the data to show that, in fact, marijuana users are more than twice as likely as alcohol users to try hard drugs.

7.5 Stochastic Processes

Many experiments are performed in a sequence of stages, each of which may have various possible outcomes. Such a multi-stage experiment is called a **stochastic process,** from the Greek word *stochos* for "guess." In this section we see how to analyze such processes, which arise in many applications, and also encounter the important notion of *independence*.

The Intersection Rule   The conditional probability of an event B, given an event A, was defined in Section 7.4 by the formula

$$P(B|A) = \frac{P(A \cap B)}{P(A)}.$$

If we multiply by $P(A)$, we obtain a simple formula for the probability of the intersection of two events.

Intersection Rule

For any events A and B,

$$P(A \cap B) = P(A) \cdot P(B|A).$$

In words: The probability that two events will *both* occur is equal to the probability of the first, times the probability of the second given the first. For three events the corresponding formula is

$$P(A \cap B \cap C) = P(A) \cdot P(B|A) \cdot P(C|A \cap B)$$

which follows from the definition of the right-hand factors. We assume, of course, that the conditional probabilities in question are all defined.

Example 1   A prisoner of war is considering an escape attempt. He knows that 40% of those who try never make it past the barbed wire, and 20% of the rest are caught before reaching the border. What is the probability that he will succeed in reaching the border?

Solution   From the given information we see that

$$P(\text{get past wire}) = 1 - .40 = .60$$

and $P(\text{reach border} \mid \text{get past wire}) = 1 - .20 = .80.$

Hence, the probability that he will *both* get past the wire (W) *and* reach the border (B) is given by

$$P(W \cap B) = P(W) \cdot P(B|W)$$
$$= (.60)(.80)$$
$$= .48.$$

This means that 48% of all escape attempts are successful.   ◆

Stochastic Diagrams   We can illustrate the stochastic process of Example 1 using a tree diagram, as in Figure 1. Stage 1 of the experiment consists in the prisoner's trying to get past the barbed wire. This leads to the two alternatives shown, with probabilities as labeled on each branch. If the prisoner does get past the barbed wire, then Stage 2 of the experiment—

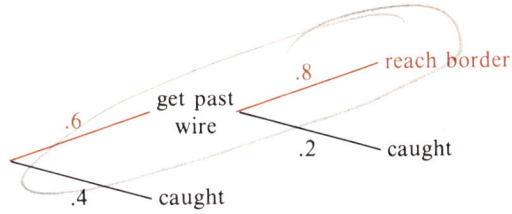

Figure 1

trying to reach the border—again produces two alternatives, with conditional probabilities as indicated in the figure.

A tree diagram of this sort, in which each branch is labeled with an appropriate probability or conditional probability, is called a **stochastic diagram.** As we have seen, the Intersection Rule implies that

$$P(\text{get past wire} \cap \text{reach border}) = (.6)(.8) = .48.$$

Notice that this is just the product of those probabilities on the uppermost branch of the tree. In general, the Intersection Rule gives us the following useful principle:

> In a stochastic diagram, the probability that all the events along a particular branch will occur is equal to the product of the probabilities on that branch.

Thus, for example, from Figure 1 we see that the probability that the prisoner will get past the barbed wire but will *not* reach the border is equal to (.6)(.2) = .12.

Example 2   Two cards are dealt at random from a standard deck. What is the probability that they will both be the same color (red or black)?

Solution   We can solve this problem using combinations, but now we consider another approach. Instead of viewing the cards as an unordered sample, we imagine that they are dealt one after the other without replacement.* This leads to the stochastic diagram shown in Figure 2. Since the first card is dealt from 52 cards of which 26 are red, the probability that it will be red is 26/52, or 1/2. If it *is* red, then the second card will be dealt from 51 cards of which 25 are red; hence the conditional probability of getting a second red card is 25/51. In the same way we arrive at the other probabilities shown in the figure.

The event "both cards are the same color" corresponds to two mutually exclusive branches in the diagram, namely red-red and black-black. Thus,

---

*It is intuitively clear that the answer will be the same whether the cards are dealt in order or simultaneously.

by the Special Union Rule,

$$P(\text{both same color}) = P(\text{both red} \cup \text{both black})$$
$$= P(\text{both red}) + P(\text{both black})$$
$$= \frac{26}{52} \cdot \frac{25}{51} + \frac{26}{52} \cdot \frac{25}{51}$$
$$= \frac{1300}{2652} \doteq .490.$$

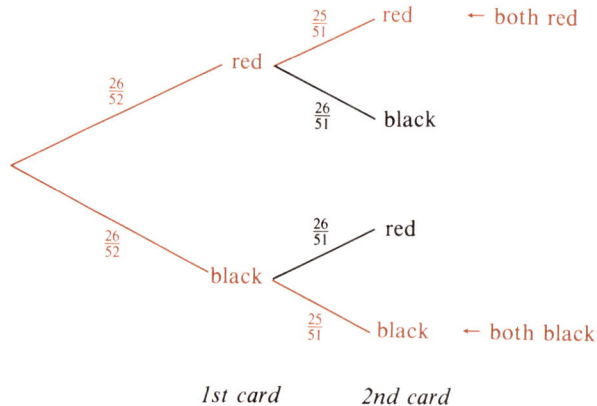

Figure 2

That is, the cards will be the same color about 49% of the time. ◆

The next example is a special case of the "division problem" or "problem of points," one of two questions which the Chevalier de Méré posed to Pascal in 1654.*

Example 3   Two players, A and B, are playing a best-of-seven series of games. They are forced to cancel the series after B has won two games, A one game. Assuming the players are equally skilled, how should the prize money be divided between them?

Solution   To answer the question we must first determine the probability that each player would win the series if it were continued. The prize money should then be divided between A and B in the same ratio as their respective probabilities of winning.

It is intuitively clear that since the players are equally skilled, and player B is ahead 2 games to 1 at the end of the third game, player B would have a greater chance of winning the series if it were continued. Now, a

*The general problem was known much earlier, at least as early as 1380. For an interesting account, see O. Ore, "Pascal and the Invention of Probability Theory," *American Mathematical Monthly*, May 1960.

"best-of-seven" series is determined as soon as one player or the other has won a total of four games. Thus, in the situation described, player A needs three more games to win the series, while player B needs only two. The possible outcomes for the remainder of the series (starting with Game 4) are represented in the stochastic diagram of Figure 3. The series terminates as soon as A wins three games or B wins two, whichever happens first. Since the players are equally skilled, we give them each the same probability of winning any particular game, namely 1/2. From the diagram we see that

$$P(A \text{ wins series}) = P(AAA) + P(AABA) + P(ABAA) + P(BAAA)$$
$$= \left(\frac{1}{2}\right)^3 + \left(\frac{1}{2}\right)^4 + \left(\frac{1}{2}\right)^4 + \left(\frac{1}{2}\right)^4 = \frac{5}{16}$$

and therefore
$$P(B \text{ wins series}) = 1 - \frac{5}{16} = \frac{11}{16}.$$

Thus we should allot 5/16 of the prize money to player A and 11/16 to player B.

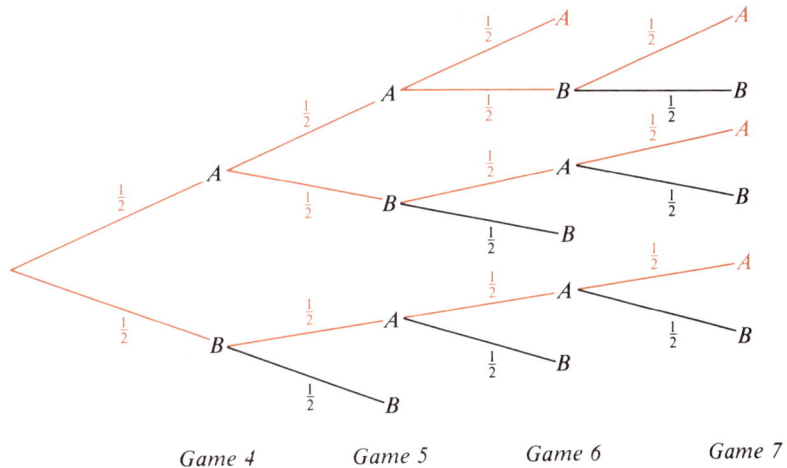

Game 4   Game 5   Game 6   Game 7

Figure 3

Remark to Reader Note that although there are ten paths through the tree of which four lead to a win for player A, we cannot conclude that A has probability 4/10 of winning the series. The reason is that *not all paths through the tree are equally likely*. For instance, $P(AAA) = 1/8$, but $P(BB) = 1/4$.

Independent Events We often encounter events A and B which are independent, in the sense that neither one affects the probability of the other. This means that the probability of B is not changed by the information that A has occurred, and vice-versa. Using conditional probability we can make this idea precise as follows:

Definition

Events A and B are called **independent events** if

$$P(B|A) = P(B) \quad \text{and} \quad P(A|B) = P(A). \tag{1}$$

We also call A and B independent when either of the conditional probabilities in (1) is undefined, that is, when $P(A) = 0$ or $P(B) = 0$.

We illustrate this concept using the familiar experiment of rolling two dice.

Example 4   In the experiment of rolling two fair dice (one white and one black), show that the events

$$A = \text{``white die comes up less than 4''}$$
$$B = \text{``black die comes up greater than 4''}$$

are independent.

Solution   It is intuitively obvious that what happens to the white die cannot affect what happens to the black die, and vice-versa, so the events A and B should be independent. To verify this, we represent A and B as subsets of the sample space S for two dice, shown in Figure 4. Counting sample points, we see that

$$P(A) = \frac{n(A)}{n(S)} = \frac{18}{36} = \frac{1}{2}$$

$$P(B) = \frac{n(B)}{n(S)} = \frac{12}{36} = \frac{1}{3}$$

$$P(B|A) = \frac{n(A \cap B)}{n(A)} = \frac{6}{18} = \frac{1}{3}$$

$$P(A|B) = \frac{n(B \cap A)}{n(B)} = \frac{6}{12} = \frac{1}{2}.$$

Hence $\quad P(B|A) = \dfrac{1}{3} = P(B) \quad \text{and} \quad P(A|B) = \dfrac{1}{2} = P(A)$

so that formula (1) is satisfied and the events are indeed independent.   ◆

Remark   This example illustrates the distinction between *independent* events and *mutually exclusive* events. The events A and B in Example 4 are independent, as we have seen, but they are not mutually exclusive because it is possible for both of them to occur at the same time. It can be shown (see Exercise 25) that independent events A and B are *never* mutually exclusive, except when $P(A) = 0$ or $P(B) = 0$.

We began this section with the Intersection Rule

$$P(A \cap B) = P(A) \cdot P(B|A)$$

BLACK DIE

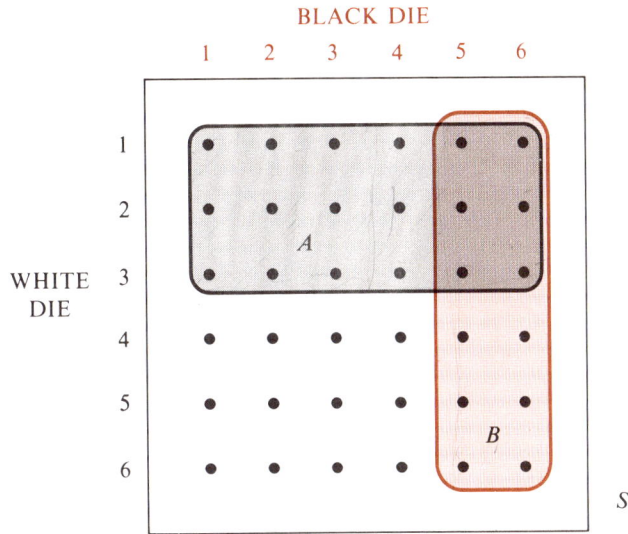

Figure 4

which holds for any events A and B. When A and B are independent, this equation simplifies to

$$P(A \cap B) = P(A) \cdot P(B). \tag{2}$$

Conversely, if equation (2) holds then

$$P(B|A) = \frac{P(A \cap B)}{P(A)} = \frac{P(A) \cdot P(B)}{P(A)} = P(B)$$

$$\text{and } P(A|B) = \frac{P(B \cap A)}{P(B)} = \frac{P(A) \cdot P(B)}{P(B)} = P(A).$$

In other words, equation (2) is a necessary and sufficient condition for two events A and B to be independent.

Special Intersection Rule

If A and B are independent events, then

$$P(A \cap B) = P(A) \cdot P(B).$$

Conversely, if this equation holds then A and B are independent.

Example 5 In the experiment of drawing a card at random from a standard deck, determine whether the events

$$A = \text{``card is an ace''}$$
$$B = \text{``card is black''}$$

are independent.

Solution The experiment has 52 equally likely outcomes. By the classical formula,

$$P(A) = \frac{\text{number of aces}}{\text{number of cards}} = \frac{4}{52} = \frac{1}{13}$$

$$P(B) = \frac{\text{number of black cards}}{\text{number of cards}} = \frac{26}{52} = \frac{1}{2}$$

$$P(A \cap B) = \frac{\text{number of black aces}}{\text{number of cards}} = \frac{2}{52} = \frac{1}{26}$$

Hence,

$$P(A \cap B) = \frac{1}{26} = \frac{1}{13} \cdot \frac{1}{2} = P(A) \cdot P(B).$$

Since the Special Intersection Rule is satisfied, the events are independent; thus, neither one affects the probability of the other. ◆

Example 6 Two independent mechanical systems must both function properly in order for a Venus probe to land successfully. NASA officials estimate that under actual landing conditions the chances are 80% and 75% that the respective systems will work. What is the probability of a successful landing?

Solution Since the systems are independent, a malfunction in one does not affect the probability of a malfunction in the other. Hence, the events

$$F_1 = \text{``system 1 functions properly''}$$
$$F_2 = \text{``system 2 functions properly''}$$

are independent. By the Special Intersection Rule,

$$P(\text{successful landing}) = P(F_1 \cap F_2)$$
$$= P(F_1) \cdot P(F_2)$$
$$= (.80)(.75)$$
$$= .60.$$

Thus, there is a 60% chance of a successful landing. ◆

The Special Intersection Rule can be generalized to three or more events. We say that $A_1, \ldots, A_n$ are independent if the probability of each one is unaffected by information concerning the others. In this case the Special Intersection Rule takes the form

$$P(A_1 \cap A_2 \cap \cdots \cap A_n) = P(A_1) \cdot P(A_2) \cdots \cdots P(A_n).$$

In applying the rule, remember that intersection means "and". In words, the Special Intersection Rule therefore says that the probability that a number of independent events will *all* occur is equal to the product of their separate probabilities.

Example 7 Three marksmen, A, B, and C, have respective probabilities 80%, 70%, and 50% of hitting a stationary target at 500 feet. If all three fire at the target independently, what is the probability that the target will be hit?

Solution We want to know

$$P(\text{at least one hit}) = P(A \text{ hits} \cup B \text{ hits} \cup C \text{ hits}).$$

We cannot use the Special Union Rule, however, because these events are not mutually exclusive. Instead, we look at the complementary event

$$P(\text{all three miss}) = P(A \text{ misses} \cap B \text{ misses} \cap C \text{ misses}).$$

Since they are firing independently, the probability that any one of them will miss is unaffected by what the other two do. Hence we can apply the Special Intersection Rule:

$$P(\text{all three miss}) = P(A \text{ misses}) \cdot P(B \text{ misses}) \cdot P(C \text{ misses})$$
$$= (.20)(.30)(.50)$$
$$= .03$$

and therefore $\quad P(\text{at least one hit}) = 1 - .03 = .97.$

That is, there is a 97% chance that the target will be hit. ◆

An important special case of independence arises whenever an experiment is repeated a number of times, as in the next example.

Example 8 A player rolls two fair dice repeatedly until either a sum of five occurs, in which case she wins, or a sum of seven occurs, in which case she loses. What is the probability that
(a) she will win on the third roll?
(b) she will have to roll more than three times?

Solution (a) On each roll we have*

$$P(\text{sum} = 5) = \frac{4}{36}, \qquad P(\text{sum} = 7) = \frac{6}{36}, \qquad P(\text{other}) = \frac{26}{36}.$$

In order to win on the third roll, the player must throw a sum other than 5 or 7 on the first and second rolls, and a sum of 5 on the third roll. Since these rolls are independent,

$$P(\text{win on 3rd roll}) = \frac{26}{36} \cdot \frac{26}{36} \cdot \frac{4}{36} \doteq .0580.$$

(b) The event "player rolls more than three times" is the same as the event "first three rolls produce sums other than 5 or 7." Thus,

---

*See Table 3 in Section 7.2.

$$P(\text{roll more than 3 times}) = \frac{26}{36} \cdot \frac{26}{36} \cdot \frac{26}{36} \doteq .377. \quad \blacklozenge$$

We shall return to the topic of "repeated independent trials" of an experiment in Section 8.2. Notice that the independence condition embodies a basic assumption, well justified by experience: *On each roll of the dice or turn of the roulette wheel, the probabilities remain the same,* no matter what has happened earlier. This runs counter to the superstitions of many gamblers, who feel that the dice are "hotter" at some times than others, or that odd numbers are "overdue" after a run of evens; but the steady profits of gambling casinos should provide abundant evidence that dice and roulette wheels possess "neither conscience nor memory."

7.5 Exercises

1. Let A and B be events with $P(A) = .6$ and $P(B|A) = .3$. What is the probability that both A and B will occur?

2. Let A, B, and C be events with $P(A) = 3/4$, $P(B|A) = 2/3$, and $P(C|A \cap B) = 1/2$. What is the probability that all three events A, B, and C will occur?

3. Engines on the Generic Motors production line are inspected for defects at two points. The first inspection catches 70% of the defective units sent through; of those that get by, 80% are caught by the second inspection. What is the probability that a defective unit will evade both inspections?

4. Sixty percent of the graduates of Plimpton Law School pass the bar exam on the first try. Of those who fail the first time 50% pass on the second try, and among those who fail twice 40% pass on the third try. What is the probability that a random graduate will pass the exam within three tries?

5. A basketball player hits her first free shot 80% of the time. When this occurs her confidence increases, raising her accuracy on the second shot to 90%; however, when she misses the first shot she is demoralized, and her accuracy on the second shot goes down to 70%. If she takes two shots, what is the probability that she will make

   (a) both of them?   (b) at least one of them?

6. A lawyer tells her client that in his forthcoming trial he has a 70% chance of conviction and a 30% chance of acquittal. If he is convicted, she feels that the odds are 3 to 2 he will be acquitted when the verdict is appealed to a higher court. What is the probability that her client will be acquitted either way?

7. From an audience of 15 boys and 10 girls two children are chosen at random to assist a magician. What is the probability that

   (a) two girls will be chosen?

   (b) a boy will be chosen first and a girl second?

   (c) a girl will be chosen second?

8. A jeweler has accidentally dropped two diamonds into a tray containing three zircons of identical shape and size. He must examine the five gems one by one until both diamonds are found. What is the probability that he will examine

   (a) no more than two of the gems?   (b) no more than three of the gems?

9. Two cards are dealt at random from a standard deck, one after the other without replacement. What is the probability that neither one will be a club?

10. From an urn containing 10 red, 10 white, and 10 blue marbles, three are chosen at random, one after the other without replacement. What is the probability that they will be red, white, and blue, in that order?

11. Team A has won two games and team B has won none in a best-of-seven series. What is the probability that team A will go on to win the series

   (a) if the teams are of equal strength?

   (b) if team A has a 60% chance of winning each game?

12. In the map below, point B is located three blocks east and two blocks north of point A.

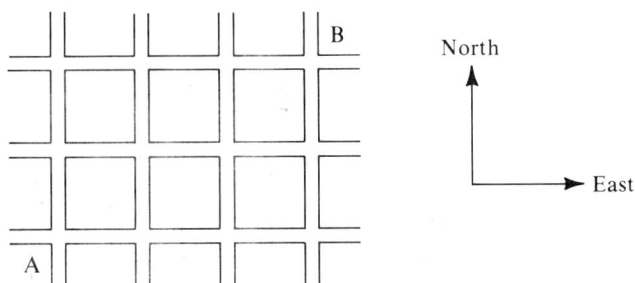

To go from A to B a cyclist must travel 3 blocks east and 2 blocks north, in some order (for instance, north 1 block, then east 3 blocks, then north 1 block).

   (a) Use a tree diagram to represent all possible routes from A to B.

   (b) Suppose the cyclist chooses at random between north and east whenever he has a choice. What is the probability that he will make exactly one turn during his trip? Exactly two turns?

13. Suppose the probability that A will occur is .8, the probability that B will occur is .5, and the probability that they will both occur is .4. Determine whether A and B are independent.

14. A geneticist observes that among plants of a certain species 12% are dwarf size, 8% lack pigmentation, and 6% fall into both categories. Are the two traits independent?

15. In the experiment of tossing a fair coin twice, determine whether the events

$$A = \text{``first toss comes up heads''}$$
$$B = \text{``both tosses come up the same''}$$

are independent.

16. In the experiment of drawing two cards at random from a standard deck, one after the other without replacement, determine whether the events

$$A = \text{``first card is red''}$$
$$B = \text{``second card is red''}$$

are independent.

17. Let A and B be events with $P(A) = .65$, $P(B) = .30$. Find $P(A \cap B)$ and $P(A \cup B)$, assuming that

(a) A and B are mutually exclusive

(b) A and B are independent.

18. Let A and B be events with $P(A) = .85$, $P(B) = .90$.

(a) What is the smallest possible value for $P(A \cap B)$?

(b) Is it possible for A and B to be mutually exclusive?

(c) What is the value of $P(A \cap B)$ if A and B are independent?

19. A manufacturer estimates that there is an 8% chance of a materials shortage and a 10% chance of labor unrest during the course of a certain project the company is planning to undertake. If these events are independent, what is the probability that neither problem will arise?

20. The probability that a female over 60 years of age will develop breast cancer is 4%, and the probability that she will develop diabetes is 2%. Assuming these events are independent, what is the probability that a female over 60 will develop (a) both diseases? (b) at least one of the two diseases?

21. Three teams are attempting to climb different faces of Mt. Everest. If their chances of success are 40%, 45%, and 50%, respectively, what is the probability that at least one team will succeed?

22. Two intelligence operatives, A and B, are captured by the Albanian secret police and separately subjected to rigorous interrogation. The chances are 10% and 20%, respectively, that the agents will crack, revealing the nature of their mission. What is the probability that at least one of the two will crack?

23. The first three batters in a baseball team's lineup have averages .250, .240, and .300, respectively. Assuming that a player's average represents his probability of getting a hit, what is the probability that in the first inning

(a) all three will get a hit?

(b) exactly one of the three will get a hit?

24. In their past encounters, chess grandmaster P. Lipner has defeated J. Traub 40% of the time, lost 10% of the time, and drawn 50% of the time. If they play three games, what is the probability that

(a) all three games will be drawn?

(b) Lipner will win one game and draw two?

25. Prove: If events A and B are both mutually exclusive and independent, then $P(A) = 0$ or $P(B) = 0$.

26. Using Exercise 26 of Section 7.3, prove: If A and B are independent events, then A and B^c are also independent.

27. In the game of craps, a player rolls two fair dice. If the sum is 7 or 11 ("naturals") he wins; if it is 2, 3, or 12 ("craps") he loses. In the other cases (4, 5, 6, 8, 9, or 10), the sum rolled becomes the player's "point." He continues to roll until he either rolls his point again, in which case he wins, or he rolls a 7, in which case he loses.

(a) Given each of the six "point" values, find the probability that the player will go on to win. (See Exercise 10 in Section 7.2.)

(b) Use a stochastic diagram to find the probability of "making a pass" in craps, that is, winning the game.

Probability of Distinct Outcomes *Suppose an experiment with n equally likely outcomes is performed k times in succession. By the Intersection Rule, the probability that all k outcomes will be different is given by*

$$\frac{n}{n} \cdot \frac{n-1}{n} \cdot \frac{n-2}{n} \cdot \ldots \cdot \frac{n-k+1}{n} = P(n, k)/n^k.$$

Exercises 28–30 use this formula.

28. Three fair dice are rolled. What is the probability that they will all come up different?

29. Five letters are typed at random. What is the probability that no letter will be repeated?

30. *The Birthday Problem*   Using a hand calculator, show that in a group of 23 random people the chances are greater than 50% that at least two have exactly the same birthday. (Assume all birthdays are equally likely.) $\left(\sim P(\text{DIFF. BIRTHDAY})\right)$

$$\frac{P(365, 23)}{365^{23}}$$

Computer Projects

A. The SERIES program simulates 2000 trials of a complete best-of-seven series of games between two teams A and B, where team A has a prescribed probability p of winning each game. For several different values of p, use this program to estimate empirically the probability that team A will win the series. How do your results compare with the theoretical probabilities?

B. Write a BASIC program to simulate repeated trials of the game of craps (see Exercise 27). Use this program to estimate empirically the probability of "making a pass," based on (a) 100 trials (b) 1000 trials. How do your results compare with the theoretical probability?

Chapter 7   Summary

Important Terms

7.1 experiment
   sample space S
   event E
   impossible event
   certain event
   relative frequency
   probability space
   probability $P(E)$
   odds in favor of E
   odds against E

7.2 uniform space
   the classical formula

7.3 mutually exclusive events

7.4 conditional probability $P(A|E)$

7.5 stochastic diagram
   independent events

Important Formulas

7.3 Complement Rule

$$P(A^c) = 1 - P(A) \text{ and } P(A) = 1 - P(A^c)$$

Union Rule

$$P(A \cup B) = P(A) + P(B) - P(A \cap B)$$

Special Union Rule

For mutually exclusive events A and B,
$$P(A \cup B) = P(A) + P(B)$$

7.5 Intersection Rule

$$P(A \cap B) = P(A) \cdot P(B|A)$$

Special Intersection Rule

For independent events A and B,
$$P(A \cap B) = P(A) \cdot P(B)$$

Review Exercises

1. The frequency table below shows the distribution of letter-grades awarded over the last few semesters in Professor Gorki's Russian literature course.

Grade	Frequency
A	36
B	63
C	129
D	45
F	27
Total	300

Based on this empirical data, what is the probability that a random student will earn

(a) an A?   (b) at least a B?   (c) at most a C?   (d) a passing grade?

2. A major TV network has narrowed its search for a new programming director down to three candidates: Anderson, Bertolucci, and Claypool. Bertolucci has been told that her chances of being selected are twice as good as Anderson's but only half as good as Claypool's. What is the probability that Bertolucci will get the job?

3. If the odds in favor of an event E are 18 to 7, what is the probability of E?

4. If an event E has probability .4, what are the odds against E?

5. If a person is chosen at random, what is the probability that he or she was born on a Tuesday?

6. If two fair dice are rolled, what is the probability that the sum rolled will be less than five?

7. If six cards are dealt at random from a standard deck, what is the probability that exactly half of them will be red?

8. Adriana and Barbara are fifth-grade classmates and inseparable friends. If their class of twenty students is divided at random into two new classes of ten students each, what is the probability that they will remain together?

9. Suppose the probability that A will occur is .2, the probability that B will occur is .5, and the probability that they will both occur is .1. What is the probability that

   (a) either A or B will occur?

   (b) neither A nor B will occur?

   (c) B will occur, given that A occurs?

10. In a certain town, 40% of the population reads the morning *Times* and 32% reads the afternoon *Herald*. Moreover, 18% of the population reads both papers.

    (a) What percentage of the population reads at least one of the two papers?

    (b) If someone reads the *Times*, what is the probability that he or she also reads the *Herald?*

11. Let A, B, and C be mutually exclusive events with probabilities .07, .23, and .12, respectively. What is the probability that none of these three events will occur?

12. A clothes dryer contains twelve matching gray socks (six identical pairs) and eight matching brown socks. If two socks are taken out at random, what is the probability of obtaining a matching pair?

13. A poker player holds four diamonds and a club. The hands of the other players are unknown. If he discards the club and takes another card at random, what is the probability that he will end up with a "flush," that is, that he will pick up a fifth diamond?

14. Of the students entering a certain medical school 36% will fail out during the first year and 48% will complete the full four-year program. If an entering student survives the first year, what is the probability that he or she will complete the program?

15. An encyclopedia salesman has found that he has a 20% chance of getting his foot in the door when he calls at any given house. When he does get his foot in the door he has a 50% chance of making a sale, but when he does not he has only a 10% chance of making a sale. What is the probability that he will make a sale at his next stop?

16. Diplomatic experts say there is a 75% chance that Country A will impose a naval blockade of its territorial waters against Country B's fishing boats. If this occurs, there is an 80% chance that Country B will defy the blockade, in which case the chance of a full-scale war will be 90%. What is the probability that this chain of events will in fact occur?

17. Two swimmers, A and B, are independently attempting to swim from Cuba to Florida. If their respective chances of success are 20% and 30%, what is the probability that at least one of the two will succeed?

18. An oil exploration firm is currently drilling in three different countries, A, B, and C, where geologists estimate the chances of a major find at 50%, 60%, and 70%, respectively. What is the probability that at least one of the sites will yield a major find?

Further Topics in Probability

We now turn to some more specialized aspects of probability theory, which utilize the concepts developed in the preceding chapter.

8.1 Bayes' Formula

Much of science is based on *inductive* rather than deductive inference. That is, instead of reasoning from hypotheses to conclusions we work backwards, assigning probabilities to various hypotheses on the basis of known conclusions. A fundamental tool in the theory of inductive inference is Bayes' Formula, which has widespread applications in statistics. But before presenting this formula, we consider the following example.

Example 1   Each bottle of Solar Cola is filled and capped by one of three machines in the company's plant in Decatur, Georgia. Twenty percent of the total output goes through Machine 1, which has a 0.1% defective rate—that is, one out of every 1000 bottles that go through it is improperly filled or sealed. Machine 2 handles 30% of the output, with a 0.6% defective rate, and Machine 3 processes the rest, with a 0.4% defective rate. If a man in Duluth buys a bottle of Solar Cola, what is the probability that it will be defective?

Solution   We can use a stochastic diagram to solve this problem. From the Figure 1 we see that the event "bottle is defective" can be broken down into three mutually exclusive cases, represented by the three different branches ending in "defective." The probability of each case is found by multiplying the probabilities along the corresponding branch of the tree. Hence,

$$P(\text{defective}) = (.20)(.001) + (.30)(.006) + (.50)(.004) \qquad (1)$$
$$= .004.$$

In other words, there is a .4% chance that the bottle will be defective.

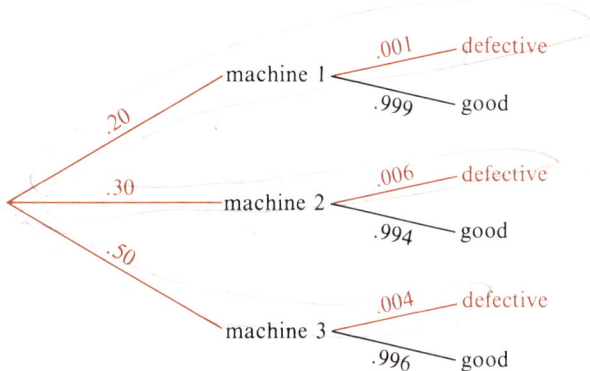

Figure 1

Now let M_1, M_2, and M_3 be the following events:

$$M_1 = \text{``bottle went through machine 1''}$$
$$M_2 = \text{``bottle went through machine 2''}$$
$$M_3 = \text{``bottle went through machine 3''}$$

and let D be the event "bottle is defective." Then $P(M_1) = .20$, and $P(D|M_1) = .001$. (Recall that "$P(D|M_1)$" means "the probability that the bottle is defective, given that it went through machine 1.") In the same way, $P(M_2) = .30$, $P(D|M_2) = .006$, $P(M_3) = .50$, and $P(D|M_3) = .004$. We can therefore write equation (1) as

$$P(D) = P(M_1)P(D|M_1) + P(M_2)P(D|M_2) + P(M_3)P(D|M_3). \quad \Diamond$$

It turns out that this formula for the probability of D is true for *any* event D, as long as one of the events M_1, M_2, M_3 *must* occur, and as long as only one of these events *can* occur. To be precise, we make the following statement:

Definition

Events $A_1, \ldots, A_n$ form a **complete set of alternatives** if

1. they are mutually exclusive; that is, $A_i \cap A_j = \emptyset$ for $i \neq j$
2. they are exhaustive; that is, $A_1 \cup \cdots \cup A_n = S$.

In set-theoretic terms, this means that $A_1, \ldots, A_n$ partition the sample space S into nonoverlapping subsets. Figure 2a (next page) shows a partition for the case $n = 3$. The effect of such a partition on an arbitrary event E is to

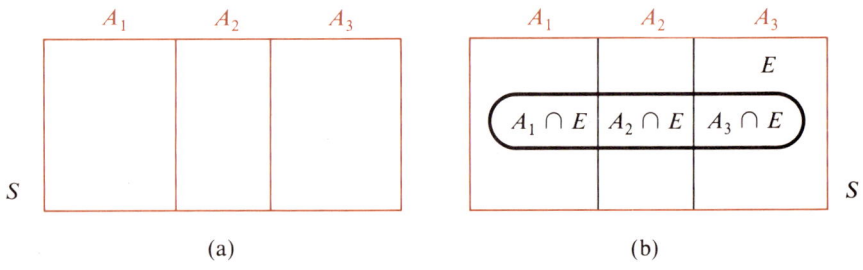

Figure 2

divide it into mutually exclusive subsets, as Figure 2b shows.

$$E = (A_1 \cap E) \cup (A_2 \cap E) \cup \cdots \cup (A_n \cap E)$$

By the Special Union Rule, we have

$$P(E) = P(A_1 \cap E) + P(A_2 \cap E) + \cdots + P(A_n \cap E).$$

Applying the Intersection Rule to each term on the right, we obtain

$$P(E) = P(A_1)P(E|A_1) + P(A_2)P(E|A_2) + \cdots + P(A_n)P(E|A_n). \tag{2}$$

Formula (2) is sometimes called the law of complete probability.

Bayes' Formula   Given a complete set of alternatives $A_1, \ldots, A_n$, we can think of each A_i as a "hypothesis" which has a certain probability of being true. Once an event E has occurred, we can ask the question, How does this new "observed evidence" alter the probability of A_i? In other words, what is the probability of A_i, given the information that E has occurred? This is, of course, exactly what is denoted by the conditional probability

$$P(A_i|E) = \frac{P(E \cap A_i)}{P(E)} = \frac{P(A_i \cap E)}{P(E)}. \tag{3}$$

If we apply the Intersection Rule and formula (2) to the right-hand side of equation (3), we immediately obtain our main result.

Bayes' Formula*

Let $A_1, \ldots, A_n$ be a complete set of alternatives. Then for any event E with $P(E) \neq 0$

$$P(A_i|E) = \frac{P(A_i)P(E|A_i)}{P(A_1)P(E|A_1) + \cdots + P(A_n)P(E|A_n)}. \tag{4}$$

---

*Thomas Bayes (1702–61) was an English clergyman and mathematician. This theorem appeared in a famous paper entitled "An Essay Toward Solving a Problem in the Doctrine of Chances," published in 1763.

Although the notation is a bit cumbersome, the significance of the formula can be summarized simply as follows: If we know the probability of E under each "hypothesis" A_i, then we can calculate the probability of each A_i, given E.

Bayes' Formula is useful when n, the number of alternatives, is large. For small values of n we can use the simpler relation (3), together with a stochastic diagram.

Example 1, Continued

A woman in Cleveland complains to the manager of her market that a bottle of Solar Cola which she bought the day before tasted completely flat, due to improper sealing. What is the probability that her bottle went through machine 1?

Solution

Here we are *given* that the bottle is defective, and we want to know the probability that it went through machine 1; thus, we are looking for the "inverse" probability

$$P(\text{machine 1}|\text{defective}) = \frac{P(\text{machine 1} \cap \text{defective})}{P(\text{defective})}.$$

The right-hand side of this equation is easily evaluated using the stochastic diagram of Figure 1. Indeed, from the diagram we see that

$$P(\text{machine 1} \cap \text{defective}) = (.20)(.001) = .0002$$

and as we saw earlier,

$$P(\text{defective}) = (.20)(.001) + (.30)(.006) + (.50)(.004) = .004.$$

Hence

$$P(\text{machine 1}|\text{defective}) = \frac{.0002}{.004} = \frac{1}{20} = .05.$$

In the same way, we find that

$$P(\text{machine 2}|\text{defective}) = \frac{.0018}{.004} = \frac{9}{20} = .45$$

and

$$P(\text{machine 3}|\text{defective}) = \frac{.002}{.004} = \frac{1}{2} = .5.$$

Thus, if we know that a bottle is defective the chance that it went through machine 1 *decreases* from 20% to 5%, the chance that it went through machine 2 increases from 30% to 45%, and the chance that it went through machine 3 remains the same at 50%. In Bayesian terms, the probability of each "hypothesis" M_1, M_2, M_3 has been altered by the "observed evidence" that D occurred. ◆

Bayes' Formula—and the diagram method, which is based on it— sometimes lead to surprising answers, as in the next example.

Example 2   *Bertrand's Box Paradox* Three identical boxes each contain two coins. In Box 1 both coins are gold, in Box 2 one coin is gold and one is silver, and in Box 3 both are silver. One of the boxes is chosen at random, and a coin is taken out of it at random. If the coin is gold, what is the probability that the other coin in the box is also gold?

Solution   The situation is illustrated in Figure 3. One might be tempted to reason this way: Since a gold coin was taken out, the box must be either Box 1 or Box 2. In one case the remaining coin will be gold and in the other case it will be silver. Thus the answer should be 1/2.

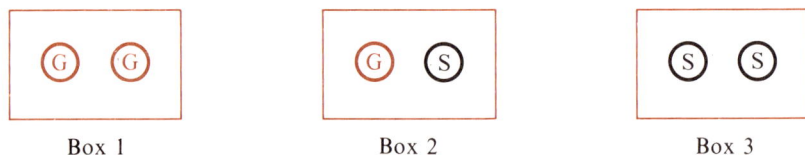

Figure 3

To see that this is fallacious, we represent the experiment by a stochastic diagram as in Figure 4. Each box has the same probability of being chosen, namely 1/3. If Box 1 is chosen we are certain of drawing a gold coin; if Box 2 is chosen we have equal chances of drawing a gold or a silver coin; and if Box 3 is chosen we are certain of drawing a silver coin.

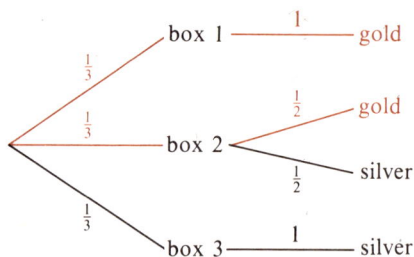

Figure 4

Now, if we know that the coin drawn was gold, then the other coin will be gold just in case Box 1 was the one chosen. What we want is therefore the "inverse" probability

$$P(\text{box 1}|\text{gold}) = \frac{P(\text{box 1} \cap \text{gold})}{P(\text{gold})}.$$

Since two branches of the tree end in "gold," we see from Figure 4 that

$$P(\text{box 1}|\text{gold}) = \frac{(1/3)(1)}{(1/3)(1) + (1/3)(1/2)} = \frac{1/3}{3/6} = 2/3.$$

Thus the probability is 2/3 that the other coin is gold, not 1/2 as we might have believed.* ◆

Reliability of Tests Bayes' Formula has important applications to the statistical evaluation of tests, particularly in medicine. In one typical situation we want to know how accurately a particular diagnostic test predicts a certain condition, that is, to what extent a positive test result can be believed.

Example 3 Approximately 8% of all black Americans are carriers of sickle cell anemia. A new test to detect the trait was administered to a large group of subjects known to be carriers and another group known to be noncarriers. The test picked up 97% of the known carriers and correctly eliminated 95% of the known noncarriers. If a random black American is tested and the test result is positive (indicating the presence of the trait), what is the probability that the person really is a carrier?

Solution We take as our sample space the set of all black Americans, and consider the alternatives ''person is a carrier'' and ''person is a noncarrier,'' as in Figure 5. The probability that a random black American is a carrier is equal to the proportion of carriers in the population, namely .08. Therefore, the probability that he or she is a noncarrier is .92. Since the test was positive for 97% of the known carriers, the conditional probability that someone will have a positive test result, given that he or she is a carrier, is .97 (and therefore the conditional probability of a negative test result is .03). Similarly, the conditional probability of a *negative* test result, given that someone is a *noncarrier*, is .95 (and therefore the conditional probability of a positive test result is .05).†

We thus arrive at the stochastic diagram of Figure 5. What we want to know is *the probability that someone is a carrier, given that he or she has a*

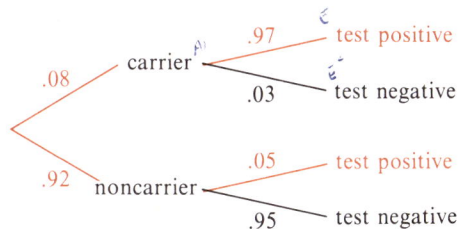

Figure 5

---

positive test result; that is,

$$P(\text{carrier}|\text{test positive}) = \frac{P(\text{carrier} \cap \text{test positive})}{P(\text{test positive})}.$$

Since two branches lead to "test positive," we have

$$P(\text{carrier}|\text{test positive}) = \frac{(.08)(.97)}{(.08)(.97) + (.92)(.05)}$$

$$= \frac{.0776}{.0776 + .0460}$$

$$= \frac{.0776}{.1236} = \frac{776}{1236} \doteq .628.$$

In other words, a positive test result is correct only 62.8% of the time; in 37.2% of the cases it is a false alarm!   ◆

The reader may be surprised by this answer, having expected a much higher reliability. In general, *the rarer a condition is among the population, the less reliable a diagnostic test will be.* In other words, as the percentage of actual carriers decreases, the percentage of "false alarms" increases. As the preceding example shows, our intuition can be very misleading in problems of statistical inference.

8.1 Exercises   *In Exercises 1–4 find the indicated probability using the stochastic diagram shown below.*

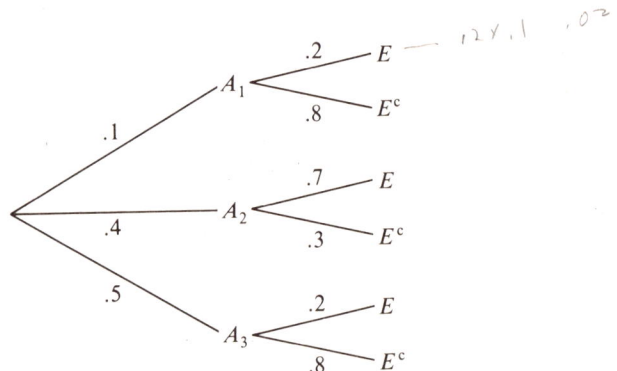

1. $P(E)$ 2. $P(A_1|E)$ 3. $P(A_2|E)$ 4. $P(A_2|E^c)$

5. Three urns each contain ten marbles. In Urn 1, five marbles are black and five are white. In Urn 2, two are black and eight are white, and in Urn 3, eight are black and two are white. An urn is chosen at random and a marble is taken out of it at random. If the marble is black, what is the probability that Urn 3 was chosen?

6. An oil exploration firm has found in the past that only 10% of the sites drilled yield oil. The firm is considering using a new prediction method based on infra-

red studies. When tested on a large number of random drilling sites, the new method correctly predicted oil in 82% of the sites where oil was subsequently found, and it correctly predicted no oil in 74% of the sites which subsequently turned out to be dry. If the new method predicts oil at a new site, what is the probability that the prediction will be correct?

7. An airline has hired a psychologist to determine whether prospective pilots will be able to react well in emergencies. The airline currently employs a large number of pilots, of whom 85% are known to react well. Each pilot was interviewed anonymously by the psychologist and given either a "steady" or "unsteady" rating. Among the pilots known to react well, 92% were given a "steady" rating, compared with only 12% of the pilots known to react poorly. If a random applicant is interviewed and given a "steady" rating, what is the probability that he or she will react well in emergencies?

8. An auto insurance company classifies its policyholders as either good risks (class G) or bad risks (class B). Among their current policyholders 95% are class G and 5% are class B. The probability of having an accident in a given year is .01 for class G drivers and .06 for class B drivers. If a policyholder reports that he or she has just had an accident, what is the probability that he or she is a class B driver?

9. A TV consumer reporter finds that 20% of the consumer complaints she investigates are satisfactorily resolved at the first stage following an inquiry to the company in question. The remaining 80% are presented on television. Of these, 70% are satisfactorily resolved after the complaint has been presented on television, and 30% remain unresolved. Of those complaints which are eventually resolved, what percentage are presented on television?

10. A mail order business has an 18% response rate when it uses its prime mailing list, purchased from a market research firm, but the response rate on its random mailings is only 9%. If 40% of the items mailed go to people on the prime list, what percentage of those responding come from the prime list?

11. Half the students in a certain class read the text after each class meeting. Among those who do, the percentage who receive an A is twice as high as among those who do not. If a student gets an A, what is the probability that he or she read the text regularly? (Hint: Let p be the probability that a nonreader will get an A.)

12. A sales representative finds that 45% of the buyers she sees are male. Moreover, her success rate with the male buyers is three times that with the female buyers. If a meeting results in a sale, what is the probability that the buyer was male? (Hint: Let p be her success rate with female buyers.)

13. A box contains three coins, one two-headed, one two-tailed and one fair. A coin is chosen at random and tossed twice. If it comes up heads both times, what is the probability that the coin is two-headed?

14. A gambler has one pair of fair dice and one pair of loaded dice, but he has forgotten which is which. With the loaded dice the probability of rolling a sum of seven is 1/5. If he chooses one of the pair at random and rolls a sum of seven, what is the probability that the dice are loaded?

15. A blood bank uses a screening test to determine whether potential donors have hepatitis. When the disease is present the test is positive 80% of the time, and when it is not the test is negative 70% of the time. If 0.5% of the potential donors

have the disease, and the blood bank accepts only those whose tests are negative, what percentage of the donors accepted will have hepatitis?

16. A physician knows that her patient's unusual symptoms necessarily imply one and only one of three diseases, A, B, or C. Her research shows that these symptoms occur among 80% of those with disease A, 90% of those with disease B, and 95% of those with disease C. If these diseases are known to occur in 0.2%, 0.4%, and 0.1% of the population, respectively, what is the most probable diagnosis?

17. One percent of the inhabitants of a certain island suffer from a congenital disease, the symptoms of which only appear in adulthood. Immunologists searching for a "marker" associated with the disease have noticed a heightened incidence of two antigens, A and B, in the blood of patients who have the disease. Specifically, among those with the disease 60% have antigen A, 70% have antigen B, and 50% have both; while for the nondiseased population the corresponding rates of occurrence are 12%, 15%, and 9%, respectively. What is the probability that a child carries the disease, if his or her blood contains

(a) both antigens?   (b) antigen A but not B?

18. A do-it-yourself pregnancy test known as Ova II was evaluated in 1976 by the Center for Disease Control in Atlanta.* When administered to women known to be pregnant the test was positive 50% of the time, while it was negative among 41% of the women known not to be pregnant. Show that according to these findings, a positive test result actually *lowers* the probability that a random woman is pregnant! (Hint: Let p be the probability of pregnancy.)

8.2 Bernoulli Trials†

It often happens that an experiment is performed not once but repeatedly, and we are interested in the frequency of occurrence of a particular event E. In this section we derive a simple and useful formula for the probability that E will occur any specified number of times.

Repeated Independent Trials The relative frequency interpretation of probability, which we encountered in Section 7.1, is based on the notion of an experiment as a *repeatable action* which has various possible outcomes. According to this interpretation, the probability of an event E predicts the proportion of the time E will occur over the long run.

Suppose for example that two fair dice are rolled, and on each roll we are interested only in whether or not a sum of seven is obtained. Since

$$P(\text{sum} = 7) = \frac{6}{36} = \frac{1}{6}$$

---

*L. D. Baker, et al., "Evaluation of a 'do-it-yourself' pregnancy test," *American Journal of Public Health* 66(1976): p. 166.

†This section requires the formula for combinations in Section 6.3.

we would expect this event to occur about one-sixth of the time, over many trials. Thus, out of 3000 rolls we would expect to obtain a seven approximately 500 times. But what if we obtained no sevens at all? Or obtained a seven on every roll? Such an occurrence is unlikely, of course, but it *could* happen. In general, if we record the number of occurrences of an event E when an experiment is repeated n times, the result might be any value k from 0 to n—assuming, of course, that the event E is neither certain nor impossible, that is, that $0 < P(E) < 1$. Our question now becomes: For a given value of k, what is the *probability* that an event E will occur exactly k times out of n repeated trials?

Bernoulli Processes Since we are only interested in whether or not E occurs on each trial, we can simplify the problem by calling an occurrence of E a "success" and an occurrence of E^c a "failure." This leads to the following definition:

Definition

A **Bernoulli process (or sequence of Bernoulli trials)*** is a sequence of n trials of an experiment, with the following characteristics:

1. Each trial has only two possible outcomes, denoted by S (success) and F (failure).
2. The trials are independent.
3. The probability of success is the same for each trial.

It is customary to denote the probability of success by p and the probability of failure by q, where $p + q = 1$.

Example 1 A card is drawn at random from a standard deck. If it is a club, the outcome is considered a "success," otherwise it is considered a "failure." After each draw the card is replaced and the deck reshuffled. This is repeated three times.

Discussion The three drawings represent $n = 3$ Bernoulli trials, where for each trial we have

$$p = P(S) = \frac{13}{52} = .25$$

$$q = P(F) = \frac{39}{52} = .75.$$

---

*Named after the Swiss mathematician Jacob Bernoulli (1654–1705), whose work the *Ars Conjectandi* (pub. 1713) was a landmark in probability theory. The Bernoulli family produced eight outstanding mathematicians over a period of three generations.

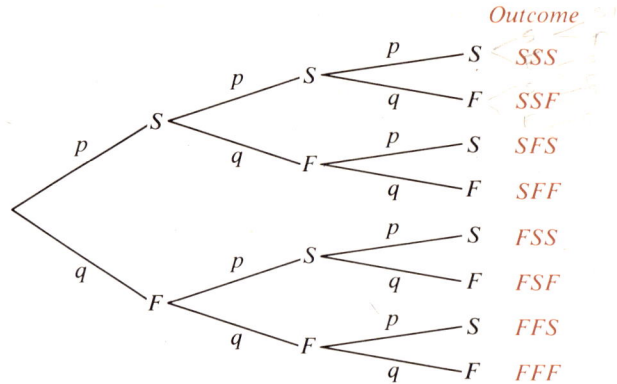

Figure 1

This process can be represented by a stochastic diagram, as in Figure 1.

Although there are $2^3 = 8$ possible outcomes, represented by the various paths through the tree, they are not all equally likely. For example, the probability of drawing three clubs is given by

$$P(SSS) = p \cdot p \cdot p = p^3 \doteq .0156$$

while the probability of drawing three nonclubs is

$$P(FFF) = q \cdot q \cdot q = q^3 \doteq .4219.$$

To find the probability of drawing *exactly one* club we add the probabilities of the three favorable branches.

$$
\begin{aligned}
P(SFF) + P(FSF) + P(FFS) &= pqq + qpq + qqp \\
&= pq^2 + pq^2 + pq^2 \\
&= 3pq^2 \doteq .4219
\end{aligned}
$$

Similarly, the probability of drawing exactly two clubs is given by

$$
\begin{aligned}
P(SSF) + P(SFS) + P(FSS) &= ppq + pqp + qpp \\
&= p^2q + p^2q + p^2q \\
&= 3p^2q \doteq .1406.
\end{aligned}
$$

Hence, for each value $k = 0, 1, 2, 3$, we can compute the probability of drawing exactly k clubs out of $n = 3$ trials. ◇

**Example 1,
Continued**   Suppose that $n = 5$ cards are drawn, replacing and reshuffling after each draw as in Example 1. What is the probability that exactly two of the cards will be clubs?

Discussion   If we extend the tree diagram of Figure 1 by adding two more levels of branching, the favorable outcomes are represented by the following five-letter "words," each of which contains the letter S exactly twice:

$$SSFFF \quad SFFSF \quad FSSFF \quad FSFFS \quad FFSFS$$
$$SFSFF \quad SFFFS \quad FSFSF \quad FFSSF \quad FFFSS$$

There are ten such words, because the two positions that the S's occupy can be chosen in

$$C(5, 2) = \frac{5 \cdot 4}{2 \cdot 1} = 10$$

different ways. Moreover, *all ten favorable outcomes have the same probability*, namely p^2q^3. For instance,

$$P(SFFSF) = p \cdot q \cdot q \cdot p \cdot q \qquad \text{(by independence)}$$
$$= p^2q^3$$

and similarly for the others. Putting these two facts together, we obtain

$$P(2 \text{ clubs in 5 draws}) = P(SSFFF) + \cdots + P(FFFSS)$$
$$= \underbrace{p^2q^3 + \cdots + p^2q^3}_{10 \text{ terms}}$$
$$= 10p^2q^3 \doteq .264.$$

Thus, there is a 26.4% chance of obtaining exactly two clubs in five draws. ◆

If we generalize the reasoning above we arrive at the following formula for the probability of obtaining exactly k successes in a sequence of n Bernoulli trials:

Bernoulli's Formula

If p is the probability of success and $q = 1 - p$ the probability of failure for a single Bernoulli trial, then

$$P(k \textbf{ successes in } n \textbf{ trials}) = C(n, k)p^kq^{n-k}. \tag{1}$$

Here k may take any integer value from 0 to n; in particular, since $C(n, 0) = C(n, n) = 1$, we have as special cases the formulas

$$P(0 \text{ successes in } n \text{ trials}) = C(n, 0)p^0q^n = q^n \tag{2}$$
$$P(n \text{ successes in } n \text{ trials}) = C(n, n)p^nq^0 = p^n. \tag{3}$$

The reader should observe that Bernoulli's Formula agrees with what we found in the 3-trial card-drawing problem of Example 1; that is,

$$P(0 \text{ clubs in 3 draws}) = C(3, 0)p^0q^3 = q^3$$
$$P(1 \text{ club in 3 draws}) = C(3, 1)p^1q^2 = 3pq^2$$
$$P(2 \text{ clubs in 3 draws}) = C(3, 2)p^2q^1 = 3p^2q$$
$$P(3 \text{ clubs in 3 draws}) = C(3, 3)p^3q^0 = p^3$$

where $p = .25$, $q = .75$.

Applications   We can bring Bernoulli's Formula to bear in a wide variety of situations—generally speaking, whenever an experiment is repeated a number of times under identical conditions.

Example 2   What is the probability that out of ten tosses, a fair coin will come up heads
(a) exactly five times?
(b) either four, five, or six times?

Solution   (a) If we define heads as a "success" and tails as a "failure," then on each toss $p = P(H) = 1/2$ and $q = P(T) = 1/2$. Hence, by formula (1),

$$P(\text{5 heads in 10 tosses}) = C(10, 5)p^5q^5 = 252\left(\frac{1}{2}\right)^5\left(\frac{1}{2}\right)^5 = \frac{252}{1024} \doteq .246.$$

In other words, there is only a 24.6% chance that a fair coin will come up heads *exactly* half the time, out of ten tosses.

(b) Using formula (1) again, we find that

$$P(\text{4 heads in 10 tosses}) = C(10, 4)p^4q^6 = 210\left(\frac{1}{2}\right)^4\left(\frac{1}{2}\right)^6 = \frac{210}{1024} \doteq .205$$

$$P(\text{6 heads in 10 tosses}) = C(10, 6)p^6q^4 = 210\left(\frac{1}{2}\right)^6\left(\frac{1}{2}\right)^4 = \frac{210}{1024} \doteq .205.$$

Hence, dividing the event "4, 5, or 6 heads" into mutually exclusive cases we have

$$\begin{aligned}
P(\text{4, 5, or 6 heads}) &= P(\text{4 heads}) + P(\text{5 heads}) + P(\text{6 heads}) \\
&\doteq .205 + .246 + .205 \\
&= .656.
\end{aligned}$$

That is, out of ten tosses the chances are pretty good (65.6%) that heads will come up between forty and sixty percent of the time.   ◆

In an earlier section we discussed the "problem of points," one of two problems posed to Pascal by the Chevalier de Méré.* The second problem posed by the Chevalier was the following "dice problem":

Example 3   What is the probability that
(a) at least one six will appear if a fair die is rolled four times?
(b) at least one double-six will appear if two fair dice are rolled 24 times?

Solution   (a) If we define a "success" to mean rolling a six with a single die, then the four rolls represent $n = 4$ Bernoulli trials, in which $p = P(S) = 1/6$ and $q = P(F) = 5/6$. By formula (2), the probability of *not* rolling a six is given by

---

*See Example 3 in Section 7.5.

$$P(0 \text{ successes in 4 trials}) = \left(\frac{5}{6}\right)^4 \doteq .482.$$

Hence, by the Complement Rule,

$$P(\text{at least one six}) \doteq 1 - .482 = .518.$$

(b) If we define a "success" to mean rolling a double-six with two dice, then we have $n = 24$ Bernoulli trials, in which $p = P(S) = 1/36$ and $q = P(F) = 35/36$. Using formula (2) once again, the probability of *not* rolling a double-six in 24 tries is given by

$$P(0 \text{ successes in 24 trials}) = \left(\frac{35}{36}\right)^{24} \doteq .509$$

and therefore

$$P(\text{at least one double-six}) \doteq 1 - .509 = .491. \quad \blacklozenge$$

Actually, de Méré calculated the correct answer himself—being a fairly good amateur mathematician—but he was troubled by the fact that it violated a well-established gambling rule. This rule seemed to imply that if the player in (a) has better than even chances, so should the player in (b). Obviously, the "well-established rule" was incorrect.

The next example illustrates the role of Bernoulli's Formula in the theory of statistical sampling.

Example 4   Sixty percent of the voters in a certain district favor a proposed mass transit project. If seven voters are sampled at random, what is the probability that a majority of them will favor the project?

Solution   Each time a voter is questioned we declare the outcome to be a "success" if the voter favors the project and a "failure" otherwise. Since the population is large compared to the sample size, the probability of success on each trial remains approximately the same, namely $p = .60$. Thus, the probability that a majority of the seven voters will favor the project is given by

$$P(\text{at least 4 successes in 7 trials})$$
$$= P(4 \text{ succ.}) + P(5 \text{ succ.}) + P(6 \text{ succ.}) + P(7 \text{ succ.})$$
$$= C(7, 4)p^4q^3 + C(7, 5)p^5q^2 + C(7, 6)p^6q^1 + C(7, 7)p^7q^0$$
$$= 35(.6)^4(.4)^3 + 21(.6)^5(.4)^2 + 7(.6)^6(.4) + (.6)^7$$
$$\doteq .290 + .261 + .131 + .028$$
$$= .710.$$

Note that there is a fairly good chance (approximately 29%) that a majority of those sampled will *not* favor the project, and therefore that this small sample will give a misleading picture of the population at large. $\blacklozenge$

Example 5 When a large number of transport helicopters were tested under desert conditions, 10% failed due to mechanical malfunctions. A planned desert rescue mission will require nine working helicopters to be successful. The commander in charge of the operation is sending ten helicopters, on the theory that he can expect one of these to fail and he will still have enough to complete the mission. How big a risk is he taking?

Solution On the basis of the test data, we can assume that each helicopter has a 10% chance of malfunction, independent of what happens to the others. Thus, the ten helicopters can be viewed as a sequence of $n = 10$ Bernoulli trials, where the single-trial probability of success (no malfunction) is $p = .9$, and the probability of failure (malfunction) is $q = .1$. For the mission to succeed, at least nine helicopters must function properly. Using Bernoulli's Formula, we find that

$$P(\text{at least 9 successes in 10 trials})$$
$$= P(9 \text{ successes}) + P(10 \text{ successes})$$
$$= C(10, 9)p^9 q^1 + C(10, 10)p^{10}q^0$$
$$= 10(.9)^9(.1)^1 + (.9)^{10}$$
$$\doteq .387 + .349$$
$$= .736.$$

By the Complement Rule, the commander is taking a 26.4% chance of ruining the mission by sending only ten helicopters! ◆

We conclude this section with an application of Bernoulli's Formula in ballistics, where the number of trials (n) is the unknown.

Example 6 A new type of ground-to-air missile has been shown to have a 20% chance of hitting a moving target at a standard speed and distance. How many such missiles must be fired in order to be at least 90% sure of a hit?

Solution We regard the firing of each missile as a Bernoulli trial, in which the probability of success (a hit) is $p = .20$ and the probability of failure is $q = .80$. If n such missiles are fired, then according to formula (2) the probability that they will *all miss* is given by

$$P(0 \text{ successes in } n \text{ trials}) = (.8)^n.$$

Thus, by the Complement Rule,

$$P(\text{at least one hit}) = P(\text{at least 1 success in } n \text{ trials})$$
$$= 1 - (.8)^n.$$

We therefore want to know the smallest value of n for which

$$1 - (.8)^n \geq .9.$$

Trying $n = 1, 2, 3, \ldots$, we find that

$$1 - (.8)^{10} \doteq .893$$
$$1 - (.8)^{11} \doteq .914.$$

Thus $n = 11$ is the least number of missiles for which we can be 90% sure of a hit.* ◆

Bernoulli trials arise in a great many applications. More on this topic can be found in Sections 11.2 and 11.3, in the discussion of the *binomial distribution*.

8.2 Exercises

1. Consider a sequence of $n = 6$ Bernoulli trials, where $p = .7$ and $q = .3$ are the single-trial probabilities of success and failure, respectively. What is the probability of obtaining

   (a) exactly 3 successes?   (c) 6 successes?

   (b) at least 5 successes?   (d) at least one failure?

2. Consider a sequence of $n = 10$ Bernoulli trials, where $p = 1/3$ and $q = 2/3$ are the single-trial probabilities of success and failure, respectively. What is the probability of obtaining

   (a) no successes?   (c) exactly 2 successes?

   (b) at least one success?   (d) at most 2 successes?

3. A fair die is rolled three times. What is the probability that the number 4 will come up

   (a) zero times?   (b) once?   (c) twice?   (d) three times?

4. A coin is weighted in such a way that the chance of heads on each toss is 30%. If this coin is tossed four times, what is the probability of obtaining

   (a) four tails?   (d) three heads and one tail?

   (b) one head and three tails?   (e) four heads?

   (c) two heads and two tails?

5. The American roulette wheel has 38 equal compartments, numbered 0, 00, and 1 through 36. The numbers 0 and 00 are green, and the numbers 1–36 are evenly divided between red and black. If a player bets on red four times in a row, what is the probability that

   (a) she will win all four bets?

   (b) she will win at least three out of the four?

6. A basketball player has a consistent 80% average shooting from the foul line. If he takes five shots, what is the probability that he will make

   (a) exactly four of them?   (b) at least four of them?

---

*The equation $1 - (.8)^n = .9$ has an exact (noninteger) solution n, which can be found using logarithms.

7. The recovery rate for a certain disease is 75%; that is, on the average, 75% of those with the disease recover. Given twelve patients with the disease, what is the probability that exactly 75% of them will recover?

8. Suppose that 40% of the registered voters in a certain county are women. If twelve jurors are selected at random, what is the probability that fewer than 20% of those chosen will be women?

9. A machine produces 5% defectives. If ten items are chosen at random from its output, what is the probability that at most one will be defective?

10. A multiple-choice exam contains ten questions each worth ten points. Each question has five possible answers, only one of which is correct. If a student guesses the answer to each question at random, what is the probability that his or her score will be at most 20%?

11. A husband and wife are both carriers of a recessive gene for sickle-cell anemia. Although they are not affected by the disease, the probability is 1/4 that any particular offspring will have the disease. If they plan to have four children, what is the probability that all four will be healthy?

12. Physicist Freeman Dyson, describing the British bombing missions during World War II, writes:*

The normal tour of duty for a crew in a regular squadron was thirty missions. The loss-rate during the middle years of the war [the probability of being shot down during a mission] averaged about four per cent. This meant that a crewman had three chances in ten of completing a normal tour.

Show how he arrived at this figure.

13. What is the probability of obtaining a sum of seven at least once in six rolls of two fair dice?

14. What is the probability that a .250 hitter will get at least one hit in four times at bat?

15. The famous English diarist Samuel Pepys posed the following question to Isaac Newton in 1693:†

A has six dice in a box, with which he is to fling [at least] one six. B has in another box twelve dice, with which he is to fling [at least] two sixes. C has in another box eighteen dice, with which he is to fling [at least] three sixes. Question: Whether B and C have not as easy a taske as A, at even luck?

Newton replied that "an easy computation" showed A to have the advantage. Can you supply Newton's reasoning?

16. A district attorney has the names of ten individuals all of whom witnessed a recent crime. She also knows that in general only 20% of all witnesses can be convinced to testify. If she needs the testimony of at least two witnesses to get a conviction, what are her chances of success?

---

*Freeman Dyson, "Reflections: Disturbing the Universe—I," *The New Yorker,* 6 August 1979, p. 46.
†E. D. Schell, "Samuel Pepys, Isaac Newton and Probability, *The American Statistician,* October 1960, pp. 27–30.

17. A workshop has seven lathes and ten full-time employees who work independently of one another. If each worker uses a lathe 60% of the time, what is the probability that there will not be enough lathes available for all who need them, at any given moment?

18. A restaurant finds that there is a 20% chance that someone with a reservation will fail to show up. If it books twelve reservations for ten available tables, what is the probability that everyone who shows up will be seated?

19. A sales representative has a 10% chance of making a sale at any given house. How many stops must she make to be 50% sure of making at least one sale?

20. What is the least number of fair coins that must be thrown in order to be 99% sure that at least one will come up heads?

21. Two evenly matched teams, A and B, are playing a best-of-seven series. What is the probability that team A will win the series in exactly seven games? (Hint: Team A must win exactly three of the first six games, and must also win the seventh game.)

22. Fermat solved the "problem of points" as follows:* Suppose player A needs three more games to win a series and player B needs two more, where the players are equally skilled. Then the series will be decided one way or the other in four more games, some of which may be "fictional"—for instance, if the outcome is *BABB*, then the series is actually decided on the third game. Fermat therefore reasoned that the probability that A will win the series is the same as *the probability that A will win at least three out of the next four games.*

    (a) Use Bernoulli's Formula to evaluate this probability.

    (b) Use the same method to solve Exercise 11(b) in Section 7.5.

23. A pharmaceutical house claims that its new drug is 90% effective in curing certain disease. It is known that 50% of those with the disease recover without any treatment. To test the manufacturer's claim, the drug is administered to ten patients with the disease, and it is agreed that the drug will be approved provided that at least eight of them recover.

    (a) What is the probability that the drug will fail the test, if in fact the manufacturer's claim is true?

    (b) What is the probability that the drug will pass the test, if in fact it is worthless?

24. In a test of extrasensory perception, the "agent" concentrates on a card chosen at random from a full deck, and the "percipient" attempts to guess its suit (spade, etc.). If the percipient guesses correctly at least four times out of five tries, he or she is considered to be "clairvoyant."

    (a) What is the probability that someone guessing completely at random will show clairvoyance?

    (b) What is the probability that out of 100 subjects, each guessing at random, at least one will show clairvoyance?

---

*See Example 3 in Section 7.5.

25. Generalize the argument used in the text for the 5-trial card-drawing problem to obtain a complete proof of Bernoulli's Formula.

26. Prove: In the binomial expansion of $(p + q)^n$, the term in which p appears to the power k represents the probability of k successes in n Bernoulli trials, where $p = P(S)$ and $q = P(F)$.

Computer Projects

A. For a Bernoulli process with given n, p, q, the BINOM program computes the theoretical probability of obtaining between k_1 and k_2 successes, inclusive. Use this program to confirm your answer in Exercises 1–24.

B. Suppose a roulette player repeatedly bets a fixed amount on red (see Exercise 5). Use the BINOM program to determine the probability that he will at least break even after 50 rolls (that is, that red will come up at least 25 times).

C. The TRIALS program simulates a Bernoulli process with given n, p, q. The program repeats the process 2000 times and counts the frequency of k successes for $k = 0, 1, \ldots, n$. Use this program to estimate empirically the probability of each value k. How do your results compare with the theoretical probabilities?

8.3 Expected Value

Throughout this chapter we have been concerned with the calculation of various probabilities. In the analysis of games of chance, however, it is not enough to know the probabilities of winning and losing; we also need the notion of *expected value,* which comes into play whenever we have numerical values—monetary gains or losses, for instance—associated with the outcomes of an experiment.

Random Variables   A **random variable** is a numerical quantity X whose value depends on the outcome of a chance experiment.

Example 1   One thousand tickets have been sold in a lottery. One ticket, selected at random, will win the first prize of $500. Ten tickets will win second prizes of $50 each, and 20 tickets will win third prizes of $25 each.

Discussion   The payoff for a particular ticketholder will depend on the outcome of the drawing. That is, if we let

$$X = \text{amount won}$$

(in dollars), then X is a random variable which may take the values 0, 25, 50, or 500. Since only one ticket out of 1000 gives a payoff of $500 we have

$$P(X = 500) = \frac{1}{1000} = .001.$$

Similarly,
$$P(X = 50) = \frac{10}{1000} = .010$$

$$P(X = 25) = \frac{20}{1000} = .020$$

and, since $1000 - 31 = 969$ tickets yield no payoff,

$$P(X = 0) = \frac{969}{1000} = .969.$$

We can summarize these probabilities as in the table below, which is called the **probability distribution** for X.

Value	Probability
0	.969
25	.020
50	.010
500	.001

Notice that the four probabilities sum to 1, since all possible values of X are represented. ◇

Expected Value   For any random variable X, we introduce the following notion:

Definition

Suppose X is a random variable taking k distinct values $x_1, x_2, \ldots, x_k$ with respective probabilities $p_1, p_2, \ldots, p_k$. The **expected value*** of X is given by

$$E(X) = x_1 p_1 + x_2 p_2 + \cdots + x_k p_k. \tag{1}$$

Thus, $E(X)$ is a "weighted sum" of the values of X, in which each value is weighted—that is, multiplied—by its probability of occurrence.

Example 1, Continued   In the lottery of Example 1, what is the ticketholder's *expected* payoff?

Solution   Using the distribution in the table, we see that
$$\begin{aligned}
\text{expected payoff} = E(X) &= x_1 p_1 + \cdots + x_k p_k. \\
&= 0(.969) + 25(.020) + 50(.010) + 500(.001) \\
&= 0 + .5 + .5 + .5 \\
&= 1.5.
\end{aligned}$$

---

*Also called the **mathematical expectation** or **mean value** of X. This concept was introduced by the great Dutch scientist Christiaan Huygens in his treatise *De Ratiociniis in Ludo Aleae* (1657).

Thus, a ticketholder should "expect" to win $1.50 in such a lottery. This will never actually happen, of course; what it predicts is the *long-run average* of X. In other words, someone who enters such a lottery day after day can expect to win $1.50 per day, on the average. If a lottery ticket costs $1.75, then for each trial we have

$$\text{expected net gain} = \text{expected payoff} - \text{cost of ticket}$$
$$= 1.50 - 1.75$$
$$= -.25.$$

Thus, a person who buys a lottery ticket every day should expect to *lose 25¢ a day*, over the long run. This interpretation of E(X) is valid for any random variable X. ◆

The expected value E(X) predicts the *average* of the values X will take over many trials.

To see why this is so, let the possible values of X be denoted by $x_1, \ldots, x_k$, with corresponding probabilities $p_1, \ldots, p_k$. Suppose the underlying experiment is repeated n times and the values of X are recorded on a list. If the numbers $x_1, \ldots, x_k$ occur in this list $n_1, \ldots, n_k$ times, respectively, then the relative frequency interpretation predicts that we will have

$$n_1/n \doteq p_1, \quad \ldots, \quad n_k/n \doteq p_k$$

provided n is large enough. Now, the *average* value of X over the n trials is equal to

$$\frac{x_1 n_1 + x_2 n_2 + \cdots + x_k n_k}{n}$$

which can be rewritten as

$$x_1\left(\frac{n_1}{n}\right) + \cdots + x_k\left(\frac{n_k}{n}\right).$$

Thus, when n is large the average value of X should be closely approximated by

$$x_1 p_1 + \cdots + x_k p_k$$

which is just the expected value E(X).

Applications In calculating the expected value of a random variable X, we always proceed as follows:

1. Determine the possible values of X.
2. Calculate the probability of each value.
3. Combine the results using formula (1).

Example 2   A fair die is rolled. What is the expected outcome?

Solution   If we let X be the number rolled, then X takes each of the values 1, 2, . . ., 6 with the same probability 1/6. Hence X has the *uniform* probability distribution shown below:

Value	Probability
1	$\frac{1}{6}$
2	$\frac{1}{6}$
3	$\frac{1}{6}$
4	$\frac{1}{6}$
5	$\frac{1}{6}$
6	$\frac{1}{6}$

Thus,
$$E(X) = 1\left(\frac{1}{6}\right) + 2\left(\frac{1}{6}\right) + \cdots + 6\left(\frac{1}{6}\right)$$
$$= \frac{21}{6} = 3.5.$$

In other words, the average of the numbers rolled over many trials will tend toward 3.5.   ◆

Example 3   An insurance company charges a $155 premium for a one-year $30,000 life insurance policy for a woman 25 years of age. Mortality tables indicate that a 25-year-old woman has a 99.8% chance of surviving one more year. What is the company's expected net gain on such a policy?

Solution   We let
$$X = \text{company's net gain}$$
(in dollars). If the woman lives through the year the company keeps the premium, for a net gain of $155. If she dies, the company keeps the premium but must pay out $30,000 in benefits, for a net gain of
$$155 - 30{,}000 = -29{,}845$$
(i.e., a *loss* of $29,845). The probability distribution for X is:

Value	Probability
155	.998
$-29{,}845$	.002

The company's expected net gain is therefore
$$E(X) = 155(.998) + (-29{,}845)(.002)$$
$$= 154.69 - 59.69$$
$$= 95.$$

That is, on the average the company will gain $95 from such a policy.   ◆

Example 4　A new janitor has been told that two of the five keys he carries will open the mail room, but he does not know which they are. If he tries the keys one by one at random until one works, how many keys must he expect to try?

Solution　We let

$$X = \text{number of keys tried}$$

which has the possible values 1, 2, 3, and 4 (by the fourth trial he is certain of success). To find the probability distribution of X, we represent the experiment by the stochastic diagram in Figure 1, where W represents "key works" and F represents "key fails to work." From the diagram we see that

$$P(X = 1) = P(W) = \frac{2}{5}$$

$$P(X = 2) = P(FW) = \frac{3}{5} \cdot \frac{2}{4} = \frac{3}{10}$$

$$P(X = 3) = P(FFW) = \frac{3}{5} \cdot \frac{2}{4} \cdot \frac{2}{3} = \frac{1}{5}$$

$$P(X = 4) = P(FFFW) = \frac{3}{5} \cdot \frac{2}{4} \cdot \frac{1}{3} \cdot \frac{2}{2} = \frac{1}{10}.$$

Hence X has the distribution shown below:

Value	Probability
1	$\frac{2}{5}$
2	$\frac{3}{10}$
3	$\frac{1}{5}$
4	$\frac{1}{10}$

The expected number of trials is thus given by

$$E(X) = 1\left(\frac{2}{5}\right) + 2\left(\frac{3}{10}\right) + 3\left(\frac{1}{5}\right) + 4\left(\frac{1}{10}\right)$$

$$= \frac{20}{10} = 2.$$

In other words, he will have to try two keys, on the average, in order to find one that works.

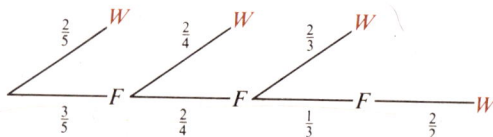

Figure 1

Games of Chance　Like the other central concepts of probability theory, the notion of expected value first arose in the context of gambling. In the next example we encounter some of the familiar terminology connected with casino games, and begin to see why, in the words of the old English proverb, "The best throw of the dice is to throw them away."

Example 5 One of the bets available at the craps table is the "field" bet, in which the player wagers a certain amount that the next roll of the dice will produce a sum other than 5, 6, 7, or 8. The "house odds" are 2 to 1 if a sum of 2 or 12 occurs, with an even payoff (1 to 1) for the other five winning sums. What is the player's expected net gain on a $1 bet?

Solution We begin by letting

$$X = \text{player's net gain}$$

(in dollars). The "house odds" refer to the payoff ratio in case the player wins; thus, on a $1 bet the player wins $2 (and his $1 bet is returned) for a sum of 2 or 12, wins $1 for a sum of 3, 4, 9, 10, or 11, and loses $1 for a sum of 5, 6, 7, or 8. Using Table 3 in Section 7.2, we arrive at the following probability distribution for X:

Value	Probability
−1	20/36
1	14/36
2	2/36

The player's expected net gain is thus

$$E(X) = (-1)\left(\frac{20}{36}\right) + 1\left(\frac{14}{36}\right) + 2\left(\frac{2}{36}\right)$$

$$= \frac{-2}{36} \doteq -.056.$$

That is, the player should expect a *net loss* of about 5.6¢ per game, over the long run. A series of, say, 500 such bets should therefore leave the player poorer by about (.056)(500) = $28.

Note that the player's expected loss (and therefore the casino's expected gain) is 5.6% of the amount wagered ($1). This percentage, known as the "house percentage" or "house take," remains the same no matter how much the player bets, since the payoffs are all proportional to the amount wagered. Thus, on a $10 "field" bet the player should expect to lose (.056)(10) = .56, or 56¢. (This is reasonable if we think of a $10 bet as ten simultaneous $1 bets.) ◆

In general, the **value** of a game of chance to the player is defined by

$$V = E(X) = \text{player's expected net gain.}$$

If $V > 0$ the player has the advantage, since he or she stands to come out ahead in the long run. If $V < 0$ the house has the advantage, and if $V = 0$ the game is **fair;** that is, both sides expect to break even over the long run. What we have seen for the "field" bet in craps holds for every wager available in a casino; namely, *the house always has the advantage.** Thus, while the house percentage may vary from one game to another, the player is almost certain to lose if he or she plays long enough.

Example 6   A casino has an urn containing ten marbles, of which five are red and five are white. A player draws two marbles at random from the urn. If they are both the same color he wins $5; otherwise he loses $5. What is the value of this game to the player?

Solution   The experiment of drawing two marbles from the urn, one after another, is represented in the stochastic diagram of Figure 2 where R represents red and W represents white. The player's net gain X (in dollars) is determined by the outcome. Thus,

$$
\begin{aligned}
P(X = 5) &= P(\text{both same color}) \\
&= P(RR) + P(WW) \\
&= \frac{5}{10} \cdot \frac{4}{9} + \frac{5}{10} \cdot \frac{4}{9} \\
&= \frac{40}{90} = \frac{4}{9}
\end{aligned}
$$

and

$$
\begin{aligned}
P(X = -5) &= P(\text{not both same color}) \\
&= P(RW) + P(WR) \\
&= \frac{5}{10} \cdot \frac{5}{9} + \frac{5}{10} \cdot \frac{5}{9} \\
&= \frac{50}{90} = \frac{5}{9}.
\end{aligned}
$$

Hence, the value of the game to the player is given by

$$
\begin{aligned}
E(X) &= 5\left(\frac{4}{9}\right) + (-5)\left(\frac{5}{9}\right) \\
&= -\frac{5}{9} \doteq -.56.
\end{aligned}
$$

That is, the player stands to lose about 56¢ per game, on the average. Since the player is betting $5, the house percentage on this game is

$$
\frac{\text{expected loss}}{\text{amount wagered}} \doteq \frac{.56}{5.00} \doteq .11 = 11\%
$$

---

*The only exception to the rule is blackjack, in which a "system" player can gain a slight advantage over the house. Casinos are always on the lookout for system players, however, and have various ways of frustrating them.

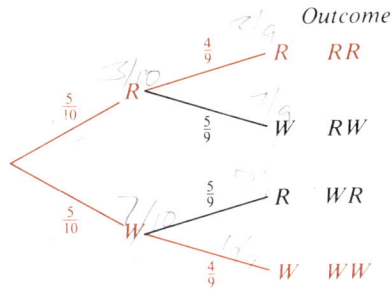

Figure 2

In other words, this game is even worse for the player than the "field" bet in Example 5. ◆

Utility and Expected Gain   As we have seen, lotteries and casino games have a negative value to the player—the player can expect to lose money in the long run. Why then do people participate in such games?

Often such behavior is simply irrational, reflecting a lack of understanding of the objective probabilities involved. But there is also another element, namely that *the utility which a person assigns to a possible gain or loss may not equal its monetary value*. Almost everyone agrees that "the second $1000 gained is worth less than the first"—in other words, that a gain of $2000 is not quite twice as valuable as a gain of $1000. Conversely, a *loss* of $2000 is *more* than twice as painful as a loss of $1000. Thus, given a choice between (a) receiving an outright gift of $10,000, or (b) receiving $20,000 if a fair coin comes up heads, and nothing otherwise, most of us would choose (a), even though both offers present exactly the same expected gain. Likewise, we would prefer a certain loss of $10 to a 0.1% risk of losing $10,000; indeed, were it not for this fact the insurance industry would cease to exist.

The average person who buys a lottery ticket or pays an occasional visit to a casino probably feels that the most likely outcome—a small net loss—is more than compensated for by the possibility, however remote, of a big win. Similarly, the individual who buys an insurance policy is not interested in the average return on such an investment, but rather on being protected against the worst possible case. In both cases the player's expected net gain is negative, but his expected utility may be positive.

There have been various attempts to measure subjective utilities, that is, to determine how much a particular gain or loss is "really worth" to a given individual. The interested reader is referred to Howard Raiffa, *Decision Analysis* (Reading, Mass.: Addison-Wesley, 1968), in which both the practical and theoretical aspects of utility theory are discussed in detail.

8.3 Exercises

1. A baker finds that the number of chocolate cakes requested on any given day is a random variable with the following distribution:

Number of requests	Probability
0	.11
1	.20
2	.32
3	.19
4	.12
5	.05
6	.01

How many requests should the baker expect, on the average?

2. A stockbroker observes that the delivery time for her in-city mailings is a random variable with the following distribution:

Delivery time (working days)	Probability
1	.218
2	.318
3	.272
4	.130
5	.062

What is the expected delivery time for such a mailing?

3. 250,000 tickets have been sold in a lottery. One ticket, chosen at random, will win first prize ($100,000), ten tickets will win second prize ($10,000), 50 tickets will win third prize ($2000), and 100 tickets will win fourth prize ($1000). How much can a ticketholder expect to win?

4. A bank teller's cash drawer contains 100 singles ($1 bills), 50 fives, 50 tens, and 25 twenties. What is the expected value of a bill chosen at random from the drawer?

5. An oil company has found that only 10% of the sites drilled yield oil. When oil is found, a site brings in $75 million in revenues, on the average. In either case, however, the company loses its drilling costs of $5 million. What is the expected net profit from a new site?

6. An investor has bought a tract of land for $50,000. There is a 20% chance that an airport will be built nearby, increasing the value of the property to $250,000. However, if the airport commission decides against the move, the value of the land will drop to $35,000. What is the investor's expected net gain?

7. An insurance company charges a $250 premium for a one-year homeowner's insurance policy. The company's statistics indicate that 1.2% of their policy-

holders file a claim each year, and the average amount paid on a claim is $12,000. How much should the company expect to make on such a policy?

8. A government agency lends an individual $25,000 to start a new business. The loan is to be repaid in one year at 10% interest. However, there is a 15% chance that the borrower will default, resulting in a $25,000 loss for the agency. What is the agency's expected net gain on such a loan?

9. A basketball player hits her first free shot 80% of the time. When this occurs her confidence increases, raising her accuracy on the second shot to 90%. However, when she misses the first shot she is demoralized, and her accuracy on the second shot goes down to 70%. How many baskets can she expect to make, out of two free shots?

10. A jeweler has accidentally dropped two diamonds into a tray containing three zircons of identical shape and size. She must examine the five gems one by one until both diamonds are identified. What is the expected number of gems she will have to examine?

11. A player draws a card at random from a standard deck. If it is an ace she wins $5; if it is a spade but not an ace she wins $1; and if it is neither a spade nor an ace she loses $1. What is her expected net gain?

12. A player tosses a fair coin repeatedly until either one head or three tails have occurred. If heads comes up on the first, second or third toss the player wins $1, $2, or $4, respectively; if three tails come up she wins nothing. What is the value of this game to the player?

13. An urn contains ten marbles, of which three are red and seven are white. A player chooses two marbles at random from the urn. If they are both the same color he wins $5; otherwise he loses $5. Which side has the advantage in this game?

14. A player repeatedly draws a card at random from a standard deck, replacing the card after each draw, until either an ace appears—in which case she wins $3— or a face-card appears, in which case she loses $1. Is this a fair game?

15. The American roulette wheel has 38 compartments, numbered 0, 00, and 1 through 36. The numbers 0 and 00 are green, and the numbers 1–36 are evenly divided between red and black. Find a player's expected net gain if he bets $1 on red, which pays even odds (1 to 1). What is the house percentage?

16. Betting a "seven" in craps is the wager that a sum of seven will occur on the next roll of the dice. The house pays 4 to 1 if the player wins. Find the player's expected net gain on a $1 bet. What is the house percentage?

17. In the simplified version of Keno described in Exercise 22 of Section 7.2, find the expected net gain of a player who buys a ticket for $1.20. What is the house percentage?

18. A roulette player decides she will put $5 on red (see Exercise 15). If she loses she will quit, but if she wins she will "let it ride," that is, put her winnings plus the original bet down on red again. No matter what happens on the second roll she will quit. What is her expected net gain using this strategy?

Sums of Random Variables *If $X_1, \ldots, X_n$ are random variables associated with an experiment, we can form their sum by adding together their values for any outcome. An important result* states that*

$$E(X_1 + \cdots + X_n) = E(X_1) + \cdots + E(X_n),$$

the expected value of the sum is the sum of the expected values. *This is intuitively plausible if we think of $E(X)$ as the long-run average of X. Exercises 19–22 use this formula.*

19. In the lottery of Example 1 in the text, what is the expected net gain of a person who buys *two* tickets at $1.75 each? Three tickets?

20. In Example 2 in the text we found the expected outcome when one fair die is rolled. What sum should be expected, on the average, when *two* fair dice are rolled? Three fair dice?

21. A roulette player (see Exercise 15) puts $1 on red, which pays even odds, and $1 on the "second dozen" (numbers 13–24), which pays 2 to 1. Find his expected net gain.

22. The "field" bet in craps was described in Example 5 in this section. Another bet is the "big six," which is the wager that a sum of six will occur before a sum of seven, when the dice are rolled repeatedly. This bet pays even odds if the player wins. What is the expected net gain of a player who puts $1 on the "field" and $1 on the "big six"?

23. Consider a game in which the player either wins $1 or loses $1. If the probability of winning is p, what is the value of the game to the player? For what value of p will the "house take" by 10%?

24. A player bets $125 against $100 that an event A will occur (thus, if A occurs he wins $100, but if it does not he loses $125). What is the player's expected net gain, if the true odds in favor of A are 3 to 2? 5 to 4? 8 to 7?

25. In Exercise 11(a) of Section 7.5, how many more games can the series be expected to last?

26. In Exercise 12 of Section 7.5, how many turns can the bicyclist expect to make?

Computer Projects

A. The DICE program simulates repeated rolls of two fair dice. Use this program to estimate empirically the average net gain on a $1 "field" bet (see Example 5 in this section), based on (a) 500 trials (b) 5000 trials. How do your results compare with the theoretical expected net gain?

B. The SERIES program simulates 2000 trials of a complete best-of-seven series of games between two teams A and B, where team A has a prescribed probability p of winning each game. For a given value of p, use this program to estimate empirically the average number of games in such a series. How does your result compare with the theoretical expected number of games?

---

*See, for example, William Feller, *An Introduction to Probability Theory and Its Applications*, vol. 1, 3rd ed. (New York: Wiley, 1968), p. 222.

C. Write a BASIC program to simulate 5000 rolls of a roulette wheel. Use this program to estimate empirically the average net gain on a $1 "second dozen" bet (numbers 13–24), which pays 2 to 1. How does your result compare with the theoretical expected net gain?

8.4 Decision Theory

In earlier chapters we have seen how mathematical tools such as linear programming and matrix algebra can help us make optimal decisions in a wide variety of practical problems. In this section we shall see how the ideas of probability theory can be used to compare the different options available to a decision-maker.

Decisions Under Uncertainty Making a decision amounts to choosing between various possible courses of action. When we can predict the outcome of each action exactly we say we are operating under *conditions of certainty*. In such a case, the best decision is simply the one that produces the most desirable outcome, for example, the greatest profit or the least cost.

More often than not, however, we must operate under *conditions of uncertainty,* where the outcome of each action depends in part on factors beyond our control. To model such a situation, we assume the decision-maker has available a number of possible actions, and that one of various possible states of nature will occur, independent of the action taken. The consequences of each action, under each state of nature, can be summarized in a **payoff matrix,** the entries of which show the decision-maker's net gain.

Example 1 An inventor has patented a new, fuel-efficient carburetor. She would like to produce the device herself, but this is a costly gamble: if the device is eventually adopted by the major automakers she will make a $2 million profit, but if not she will lose $500,000. A company has offered to finance the project, for 70% of the profits; in that case, the inventor would either make $600,000 or break even. A second company has offered to buy the patent outright for $100,000. What is the inventor's best course of action?

Discussion The investor can either (1) produce the device herself; (2) retain a 30% interest; or (3) sell the patent. Her payoff, or net gain, will depend on whether or not the device is adopted, as summarized in the following table, called a *payoff matrix:*

	State of nature	
Action	Adopted	Not adopted
	Net gain (in thousands)	
1. Produce device	$2000	− $500
2. Retain 30%	$600	0
3. Sell patent	$100	$100

If the inventor *knew* her device would be adopted she would obviously do best to produce it herself (payoff = 2000). Likewise, if she knew it would *not* be adopted her best course of action would be to sell the patent (payoff = 100). Unfortunately, her decision must be made under conditions of uncertainty as to the true (future) state of nature. In such a situation there are several different criteria for selecting the "best" course of action:

    1. **The maximin criterion:** Choose the action that *maxi*mizes the *min*imum possible payoff.

In our example, action 3 has a minimum possible payoff of 100, while actions 1 and 2 have minimum payoffs of − 500 and 0, respectively. By the maximin criterion, therefore, the inventor should choose action 3 (selling the patent), since this option offers the least risk, that is, the greatest payoff assuming the worst will happen. Clearly, this criterion reflects a conservative or "pessimistic" decision strategy.

    2. **The maximax criterion:** Choose the action that *maxi*mizes the *max*imum possible payoff.

In other words, assume the best will happen and compare the various options accordingly. In our example, the maximum possible payoffs for actions 1, 2, and 3 are 2000, 600, and 100, respectively. Using the maximax criterion, therefore, the inventor would choose action 1 (producing the device herself), since this option offers the greatest opportunity, that is, the greatest payoff under the best of circumstances. The maximax criterion thus reflects an adventurous or "optimistic" strategy. ◆

When the "payoffs" represent losses or costs, rather than gains, the roles of maximum and minimum are interchanged. Thus, the conservative strategy would be to *min*imize the *max*imum possible loss (minimax), while the adventurous strategy would be to *min*imize the *min*imum possible loss (minimin). Alternatively, losses can be expressed as *negative* payoffs, as we have done here.

Bayes' Criterion   Since the state of nature is determined by external factors, a rational decision strategy should take all the available information into account. Specifically, if we know the *probability* of each state of nature, we can compare our options using the notion of expected value. This leads to a third criterion for selecting the "best" course of action:

    3. **Bayes' criterion*** Choose the action that maximizes the *expected* payoff.

For instance, suppose the inventor in Example 1 has estimated the chances at 25% that her device will be adopted by the major automakers. Then her expected payoffs for actions 1, 2, and 3 are given by:

---

*Also called the **expected payoff** criterion.

$$\bar{E}_1 = 2000(.25) + (-500)(.75) = 125$$
$$E_2 = 600(.25) + 0(.75) = 150$$
$$E_3 = 100(.25) + 100(.75) = 100$$

Thus, Bayes' criterion would lead her to select action 2 (retaining a 30% interest), since this action offers the greatest expected payoff ($150,000).

Notice that our three decision criteria—maximin, maximax and Bayes'—each lead to a different course of action. In other words, there is no single answer to the question which course of action is "best". We shall return to this question shortly, but first let us look at another decision problem.

Example 2   A Houston-based firm puts out oil fires on short notice anywhere in the world. A number of trained crews are kept on call for this purpose, each crew receiving a salary of $10,000 a week. The number of calls has been observed to vary between two and six per week, according to the following distribution:

Number of calls	Probability
2	.1
3	.3
4	.3
5	.2
6	.1

The firm charges $50,000 for its services, and each crew can handle one call per week. Since the crews must be paid whether they are used or not, the firm would like to determine the optimal number of crews to have on hand.

Discussion   The firm's weekly net profit will depend on both the action taken (number of crews available) and the state of nature (number of calls received), as indicated in the following payoff matrix:

Number of crews	Number of calls per week				
	2	3	4	5	6
	(Net profit per week in thousands)				
2	80	80	80	80	80
3	70	120	120	120	120
4	60	110	160	160	160
5	50	100	150	200	200
6	40	90	140	190	240

For instance, if the firm has 4 crews on hand and 3 calls are received, then

$$\text{profit} = \text{revenue} - \text{cost}$$
$$= 3(50,000) - 4(10,000)$$
$$= 150,000 - 40,000$$
$$= 110,000.$$

Important Note that with 4 crews on hand, the profit is the same ($160,000) whether 4, 5, or 6 calls come in, since only 4 calls can be accepted.

Using the table and the given probability distribution, we can compute the *expected* profit for each course of action, as summarized below:

Number of crews	Expected profit (in thousands)
2	80
3	115
4	135
5	140
6	135

For instance, with 4 crews on hand we find that

$$\text{expected profit} = 60(.1) + 110(.3) + 160(.6) = 135$$

so that, *on the average,* the firm will make $135,000 per week by keeping 4 crews available. Since the maximum expected profit ($140,000/wk) occurs when 5 crews are kept on hand, this would be the best course of action according to Bayes' criterion (a somewhat surprising result, perhaps, since the expected number of calls per week is only 3.9).

Note that according to the maximin criterion we should keep only 2 crews on hand, since this gives the greatest payoff under the *worst* of circumstances (when only 2 calls are received). Similarly, using the maximax criterion we would decide to keep 6 crews on hand at all times. ◆

Utility and Risk Aversion   Which criterion should a "rational" decision-maker employ in a given situation? When relatively small amounts are at stake, Bayes' criterion makes the most sense, since we know that it gives the greatest average payoff over the long run. In Example 2, for instance, the fire-fighting outfit is not especially concerned with its profit for any one week, but rather with the long term profit if it *repeatedly* follows a particular course of action. In Example 1, however, we could argue that the situation is different: Here the decision-maker must decide how best to exploit her new invention—a nonrepeatable decision in which the stakes are considerable. If the inventor needs $80,000 immediately in order to buy a house, she may decide to take the conservative course (action 3), which assures her of $100,000 no matter what happens. On the other hand, if she is on the verge of bankruptcy and needs $800,000 to survive, she may well decide she has no choice but to "go for broke" and gamble on the large payoff (action 1). These decisions are not at all "irrational"; they merely reflect the different *utility* of money in different situations.

In the case of a private or public corporation with large assets, it is obviously in the long-term best interests of the institution to use Bayes' criterion in making investment decisions. Some projects will fail and some will succeed, but *overall* the net profit will be maximized. Some interesting studies have shown, however, that corporate managers tend to show excessive *risk aversion* in their decisions, preferring safe courses of action to those promising the greatest expected gain.* The reason for this phenomenon lies in the psychology of the individual decision-maker, who worries—often justifiably—that his or her performance will be evaluated on the basis of the success or failure of a few key decisions, rather than on their inherent "rationality." This may explain why, the more bureaucratic such institutions become, the more conservative their decisions tend to be—with the long-term result that millions of dollars in potential profits are lost!

Decision Trees Often a decision-maker must play a cat-and-mouse game against nature: The decision-maker makes a move, then nature responds, creating a new situation calling for a new decision, and so on. Such sequential decision problems can become quite involved, but with the aid of a device known as a **decision tree** they can be systematically analyzed.

Example 3 An aircraft company is bidding on a contract worth $50 million. One of the company's directors wants to build a $10 million prototype immediately, since this will give the company a 50% chance of winning the contract at the preliminary stage of the bidding. Another director wants to submit the bid without the prototype, which will give the company only a 20% chance of success. If this first-stage bid is rejected, she argues, the company will still retain the option of building a rush prototype for $15 million, which will restore a 50% chance of success in the final stage of bidding. What decision strategy will maximize the company's expected net profit?

Discussion The decision tree for this problem is shown in Figure 1. The points where branching occurs are called **nodes.** Those represented by small squares are called **choice-nodes,** representing points where a decision must be made. Those represented by small circles are called **chance-nodes,** representing points where nature takes one of several states. The tree branches from left to right, each branch finally terminating in an **end-node** (heavy dot). The initial choice-node A represents the decision whether or not to build the prototype immediately. The decision to build it leads to the chance-node B, which branches according to whether the bid is then accepted (probability = .5) or rejected (probability = .5). If the bid is accepted, the company's payoff (net profit, in millions) will be $50 - 10 = 40$, as indicated on the corresponding end-node. If the bid is rejected, the company will lose the cost of the prototype, for a payoff of -10. On the other hand, if the prototype is not built immediately we arrive at chance-node C, which branches according to whether the bid is accepted (probability = .2) or rejected (probability = .8). If the bid is accepted the payoff will be 50; if not, we must decide (node D) whether to build the rush prototype or drop out of the bidding and take a

*See Ralph O. Swalm, "Utility Theory—Insights into Risk Taking," *Harvard Business Review,* November–December 1966, pp. 123–36.

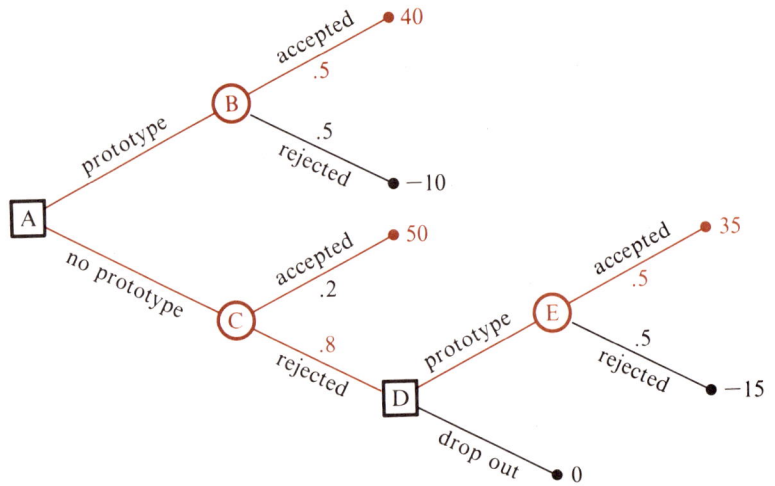

Figure 1

payoff of 0. If the rush prototype is built (node E), the bid will either be accepted at the final stage (payoff = 50 − 15 = 35) or rejected (payoff = −15).

Thus, Figure 1 represents the entire sequential process described in the problem. Note that each end-node is labeled with a numerical payoff denoting the company's net profit if that branch of the tree is followed. In order to compare the different strategies available, we work *backward* through the tree, assigning a value to each node according to the following rules:

Value of a Node

1. The value of an end-node is the corresponding payoff.
2. The value of a chance-node is the *expected* value of the nodes to which it branches.
3. The value of a choice-node is the *maximum* value of the nodes to which it branches.

In our example, this computation proceeds as follows (see Figure 1; the reader may find it helpful to label each node with its value as it is computed):

Value of node E This is a chance-node, so according to rule 2 its value is given by

$$35(.5) + (-15)(.5) = 10.$$

This represents the expected payoff at this point.

Value of node D This is a choice-node, so we compare the values of our available options. Since building the rush prototype (node E) has value 10, while dropping out of the bidding has value 0, we would choose to build the prototype. Hence, in accordance with rule 3 we assign value 10 to node D.

Value of node C Since node D has value 10, the value of the chance-node C is given by

$$50(.2) + 10(.8) = 18$$

in accordance with rule 2.

Value of node B This is a chance-node, so by rule 2 its value is

$$40(.5) + (-10)(.5) = 15.$$

Value of node A Comparing our options at the choice-node A, we see that building the prototype (node B) has value 15, while not building it (node C) has value 18. Thus we would choose *not* to build the prototype, and by rule 3 we assign value 18 to node A.

From this analysis we conclude that the company's maximum expected profit is $18 million, and its optimal decision strategy is to *submit the bid without the prototype, and build a rush prototype later if necessary.* ◆

The definition of value used above is based on Bayes' criterion, and therefore the sort of analysis we have just completed is called a **Bayesian analysis.** It provides the most rational decision method for an institution which faces many situations of this type, and wishes to maximize its long-term profits.

Other Applications There are other problems in which probability theory can be applied to the decision-making process. In the next example, not only the payoffs but also the probabilities of the "states of nature" depend on the course of action taken.

Example 4 The Atlas Insurance Company is considering hiring a number of private investigators to look for a $50,000 necklace which was recently stolen from one of the company's policyholders. Experience has shown that such investigators have a 20% chance of success, working alone. Assuming the investigators work independently of each other and that each investigator costs the company $1,000, how many should be hired to minimize the company's expected loss?

Solution If the company hires no investigators, it will have to reimburse the policyholder $50,000. Obviously, then, it does not pay to hire more than 50 investigators at $1,000 each, since in this case the company is *certain* to lose $50,000 (and still may have to reimburse the policyholder!)

Suppose the company hires n investigators ($0 \le n \le 50$). Since they work independently of each other, the probability that none of them will recover the necklace is $(.8)^n$, and in this case the company's net loss will be

$$50,000 + 1000n.$$

The probability that the necklace *will* be recovered is therefore

$$1 - (.8)^n$$

and in this case the company's net loss will be $1000n$. Thus, the *expected* net loss, using n investigators, is given by

$$E_n = (50{,}000 + 1000n)(.8)^n + (1000n)[1 - (.8)^n]$$

which simplifies to

$$E_n = 50{,}000(.8)^n + 1000n. \tag{1}$$

Trying successive values $n = 0, 1, 2, \ldots$ in formula (1), we obtain

$$E_0 = 50{,}000(.8)^0 + 1000(0) = 50{,}000$$
$$E_1 = 50{,}000(.8)^1 + 1000(1) = 41{,}000$$
$$E_2 = 50{,}000(.8)^2 + 1000(2) = 34{,}000$$
$$E_3 = 50{,}000(.8)^3 + 1000(3) = 28{,}600$$
$$\vdots$$

Continuing in this way we find that E_n decreases until $n = 11$ and then begins increasing again. Thus, the company's minimum expected loss is

$$E_{11} = 50{,}000(.8)^{11} + 1000(11) \doteq \$15{,}294.97$$

attained by hiring $n = 11$ investigators.   ◆

8.4 Exercises

1. Consider the hypothetical payoff matrix shown below.

	State of nature			
Action	s_1	s_2	s_3	s_4
a_1	20	25	−10	0
a_2	30	15	30	−20
a_3	40	0	50	0

Suppose the probabilities of the four states are .3, .4, .2, and .1, respectively. Determine the best courses of action using the maximin, maximax, and Bayes' criteria.

2. Consider the hypothetical payoff matrix shown below.

	State of nature		
Action	s_1	s_2	s_3
a_1	10	15	25
a_2	−20	30	0
a_3	25	10	−10
a_4	−10	15	45

Suppose the probabilities of the three states are .2, .7, and .1, respectively. Determine the best courses of action using the maximin, maximax, and Bayes' criteria.

3. A carnival operator is worried about the possibility of rain over the weekend. If he cancels the carnival he will lose $2000 in promotional expenses. If he sets up the equipment he will make $12,000 if the weather holds out but he will lose $8000 if it rains. Assume he applies Bayes' criterion.

   (a) What should he do if the chance of rain is 65%?

   (b) How great must the chance of rain be to make canceling worthwhile?

4. An insurance company sells single-flight airplane accident policies in major airports. A $10,000 policy costs $10, and a $20,000 policy costs $15. Let p be the probability of an airplane accident.

   (a) If $p = .0003$, which of the two policies is better for the company?

   (b) If the company's expected gain is the same for both policies, what is the value of p?

5. The owner of a newsstand gets a few requests each day for out-of-town papers. One of them, the Passaic *Sun-Times*, costs the dealer 40¢ a copy and sells for 90¢. Unsold copies cannot be returned and thus represent a net loss for the dealer. The number of daily requests for the *Sun-Times* has been observed to vary as follows:

Number requested	Probability
0	.1
1	.2
2	.3
3	.3
4	.1

How many papers should the dealer stock to maximize his long-term profit?

6. A fish dealer observes that the daily demand for fresh salmon varies between 5 and 9 pounds, as follows:

Number of lbs requested	Probability
5	.1
6	.2
7	.4
8	.2
9	.1

The salmon, which sells for $6 per lb, must be ordered by the dealer the day before at a cost of $3 per lb. The dealer can get $2 per lb for day-old salmon from a local cat food manufacturer. Thus, unsold fish represents a $1 per lb loss for the dealer. How many pounds should he order each day to maximize his long-term profit?

In Exercises 7 and 8, find the value of each node in the decision tree and determine the expected payoff if the decision-maker follows an optimal strategy.

7.

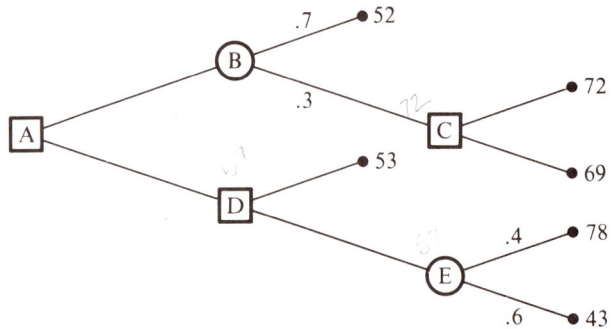

8.

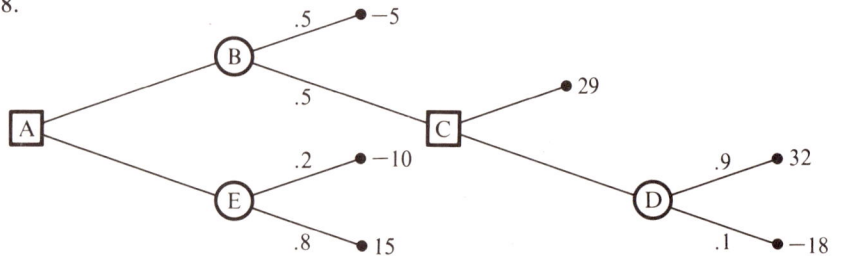

In Exercises 9–14, use a decision tree to determine the optimal decision strategy.

9. A special interest group is trying to decide whether or not to hire a lobbyist, at a cost of $250,000, to help pass a key piece of legislation now before Congress. Hiring the lobbyist will increase the bill's chance of passage from 20% to 30%, and passage of the bill is worth $3 million in revenue to the group. What is the group's best course of action, in terms of expected net gain?

10. The director of the San Francisco Opera must decide whether or not to hire a famous tenor for the new season. If he does not hire the tenor he expects a net profit for the company of $4 million. If he does hire the tenor the company's net profit will increase by $750,000. However, in this case there is a 50% chance that the company's star soprano will become temperamental and threaten to quit. If so, the director must either give her a $1 million increase in salary (reducing the net profit by that amount) or else let her go, in which case he expects a net profit of $3,800,000. What decision strategy will maximize the company's expected net profit?

11. Elaine often has to park downtown for an hour. She can pay 85¢ at a parking lot or look for a space on the street, which she finds 60% of the time. If she finds a space she can either put 25¢ in the meter or take a 5% risk of getting a $10 ticket. If she doesn't find a space, she must go back to the parking lot, having wasted 75¢ worth of her time. What decision strategy will minimize her expected loss?

12. A homeowner is asking $100,000 for his house, which he must sell within two weeks. A buyer has made what she claims is a "final offer" of $80,000. The seller can either accept it or make a counteroffer of $90,000. If he makes the counteroffer, the chances are even that it will be accepted. If it is not, there is a 20% chance that the buyer will come back with a compromise offer of $85,000—which the seller is prepared to accept—and an 80% chance that she will drop out of the bidding, in which case the seller will have to accept a lower offer of $75,000 from another party. What should the seller do to maximize his expected gain?

13. A woman is considering whether or not to file a civil suit against a local roofing company. If she files the suit there is a 50% chance the company will offer her $2000 to settle her claim out of court. If she refuses the company's offer (or if no offer is made) the case will go to trial, in which case she will stand a 70% chance of winning her full $4000 claim. However, if she loses in court she must pay $3000 in legal fees. What decision strategy will maximize her expected gain?

14. An oil company is trying to decide whether to do shallow or deep drilling at a new site. Shallow drilling costs $5 million, with a 12% chance of finding oil, while deep drilling costs $15 million, with a 23% chance of finding oil. If shallow drilling is unsuccessful the company can undertake deep drilling at an extra cost of $15 million, but the chance of finding oil will be only 12.5%. Alternatively, it can abandon the project (taking a loss of $5 million). If oil is found at a site the company will realize $90 million in revenues. What decision strategy will maximize the company's expected net profit?

15. A television network has exclusive rights to cover a transatlantic balloon race, using airborne cameras mounted in separate helicopters. Each airborne camera costs the network $8000 and has probability .95 of functioning properly. If all the available cameras malfunction the network will lose an estimated $4 million in revenues. How many cameras should the network use in order to minimize its expected loss?

16. A player pays a $1 entrance fee and simultaneously rolls n fair dice. If no die comes up six she wins n dollars, but if a six appears she wins nothing. What value of n will give her the greatest advantage?

17. A player is dealt n cards, face down, from a well-shuffled deck ($1 \leq n \leq 52$). If the ace of spades is among them he loses n dollars. If not, he wins n dollars. What value of n will give him the greatest advantage?

18. In Exercise 17, suppose the player loses n dollars if *any* ace is among the cards received, and wins n dollars otherwise. What value of n will give him the greatest advantage?

Value of Perfect Information   *Consider the payoff matrix shown below, where states* s_1, s_2, *and* s_3 *have respective probabilities .5, .3, and .2.*

| | State of nature | | | Expected |
Action	s_1	s_2	s_3	payoff
a_1	$4000	$7000	−$1000	$3900
a_2	$5000	$4000	$3000	$4300
a_3	$6000	−$2000	$4000	$3200

Using Bayes' criterion the best action is a_2, with an expected payoff of $4300. But suppose the decision-maker had perfect information; that is, he knew in advance which state of nature would occur, and acted accordingly to maximize his payoff. His expected gain would then be

$$6000(.5) + 7000(.3) + 4000(.2) = \$5900$$

because the maximum payoffs for states s_1, s_2, and s_3 are $6000, $7000, and $4000, respectively. The difference

$$5900 - 4300 = \$1600$$

*thus represents the **value of perfect information**. This is the maximum amount the decision-maker should be willing to pay for advice, market forecasts, astrological predictions, or other decision aids. Find the value of perfect information in each of the following decision problems:*

19. Exercise 1   20. Exercise 2   21. Exercise 3(a)   22. Exercise 5

Chapter 8   Summary

Important Terms

8.1 complete set of alternatives
law of complete probability
Bayes' Formula

8.2 Bernoulli process

8.3 expected value

8.4 payoff matrix
maximin criterion
maximax criterion
Bayes' criterion
decision tree

Important Formulas

8.2 Bernoulli's Formula

$$P(k \text{ successes in } n \text{ trials}) = C(n, k)p^k q^{n-k}$$

8.3 Expected Value

$$E(X) = x_1 p_1 + x_2 p_2 + \cdots + x_k p_k$$

Review Exercises

1. A certain disease is known to occur among 5% of the population. Among those who have the disease, 80% carry a blood antigen known as the X-factor, whereas only 10% of those who do not have the disease carry this factor. If a random member of the population is found to carry the X-factor, what is the probability that he or she has the disease?

2. Of the voters in Congressman Flipside's district 60% are white, 15% are black, and 25% are Hispanic. The congressman is supported by 55% of the white voters, 80% of the black voters, and 60% of the Hispanic voters. What percentage of his support comes from the black community?

3. Assuming that a newborn child is equally likely to be male or female, what is the probability that a woman with four children has

(a) two boys and two girls?

(b) at least one child of each sex?

4. Two evenly matched teams, A and B, are playing a best-of-seven series. What is the probability that the series will last the full seven games; in other words, that Team A will win exactly three of the first six games?

5. Each year the tax board in a certain state subjects 2% of all individual income tax returns to a random audit. What is the probability that a given taxpayer

(a) will not be audited within the next ten years?

(b) will be audited at most once in the next ten years?

6. A television network is starting eight new situation comedies this season. If past experience shows that on the average only one out of every five shows is successful, what is the probability that

(a) at least one of the new sitcoms will succeed?

(b) at least two will succeed?

7. An investment analyst has determined that during the next year the value of a certain stock will decrease by 10% with probability .1, remain the same with probability .2, increase by 10% with probability .4, and increase by 20% with probability .3. What is the expected net gain of an investor who buys $1000 worth of the stock?

8. A company which sells solar heating units has found that on any given day there is a 70% chance of not selling any units, a 15% chance of selling one unit, a 10% chance of selling two units, and a 5% chance of selling three units. If the company makes $1000 profit on each unit sold, what will the average daily profit be?

9. A player rolls two fair dice. If neither die comes up six she loses $1, if exactly one comes up six she wins $1, and if both come up six she wins $10. What is her expected net gain?

10. At the end of years 1, 2, and 3 of a four-year graduate program students must pass an examination to determine whether they are allowed to continue. Fifty percent of the first-year students pass the exam at the end of year 1. Of these, 80% pass the exam at the end of year 2, and of these 90% pass the exam at the end of year 3. On the average, how many years can a new student expect to spend in the program?

11. A textbook publisher estimates that the first-year demand for a new calculus text (to the nearest 5000) will have the following probability distribution:

Number ordered	Probability
5000	.2
10,000	.5
15,000	.2
20,000	.1

The publisher must decide whether to run 5, 10, 15, or 20 thousand copies in the first printing. Each copy printed costs the publisher $12, and the book sells for

$25. Moreover, each unsold copy costs an additional $4 in inventory expenses. What size printing run will maximize the publisher's expected net profit?

12. Each morning a baker makes a certain number of chocolate layer cakes to be sold during the day. The number of daily requests for this type of cake has been observed to vary as follows:

Number requested	Probability
0	.2
1	.4
2	.3
3	.1

If each cake sold yields a $4 profit and each cake not sold results in a $1 loss, how many cakes should the baker prepare in order to maximize his long-term profit?

13. If it rains during the next two weeks a farmer will make $100,000 profit, but if it does not he will lose $80,000. The chance of rain is estimated at 50%, but a cloud-seeding operation, which costs $15,000, would increase the chance of rain to 60%. Is it worth it, in terms of expected net gain?

14. A quiz show contestant has already won $4200. He can either keep the money or go for a "jackpot" question, which he feels 50% sure of answering correctly. If he answers it correctly he wins an additional $5000, but if not he will lose the money already won. If he answers the jackpot question correctly he can either keep his winnings or risk them on a "superjackpot" question with a $20,000 additional payoff, which he feels 25% sure of answering correctly. What is his optimal decision strategy, in terms of expected net gain?

9

Markov Chains

Although mathematics tends to get divided into separate areas, the truth is that almost every part of mathematics makes contact in some way with every other part. This is especially the case where real-world applications are involved. In this chapter we examine a very useful theory which depends on the interaction of probability and matrix algebra.

9.1 Introduction to Markov Chains

In the preceding chapter we encountered the Bernoulli-trials model, in which an experiment with two outcomes is repeated a certain number of times. In this section the Bernoulli-trials model is generalized to include processes in which the outcome of one trial affects the outcome of the next.

Markov Chains

A stochastic process made up of an infinite sequence of trials, numbered $n = 0, 1, 2, \ldots$ is called a **Markov chain*** if the following two conditions hold:

1. Each trial has the same set of possible outcomes.
2. If the outcome of a given trial is i, there is a fixed probability p_{ij} that the outcome of the next trial will be j.

---

*Named after the Russian mathematician A. A. Markov (1856–1922), who developed much of the modern theory of stochastic processes.

The possible outcomes are called **states,** and the numbers p_{ij} are called **transition probabilities.**

In order to understand this terminology it is important to have in mind a certain "picture" of a Markov chain process. Imagine a system which has, at each moment $n = 0, 1, 2, \ldots$, one of a finite number of possible states. Instead of saying "the outcome of the nth trial is i," we say that "the system is in state i at time n." Condition (2) above can be paraphrased as follows:

> If the system is in state i at a given time n, there is a fixed probability p_{ij} that it will be in state j at time $n + 1$.

In other words, p_{ij} represents *the probability of going from state i to state j in one step,* or unit time period. The fact that p_{ij} does not depend on n means that we can calculate the probability of the next state of the system, if we know the current state.

Example 1 Basketball players 1 and 2 are taking turns shooting baskets from the foul line. Each player continues to shoot until he misses, at which point the other player takes over. Suppose player 1 is an 80% shooter and player 2 is a 70% shooter. Represent this process as a Markov chain.

Discussion We can assume the first shot is taken at time $n = 0$, and subsequent shots are taken at regular unit intervals. Thus, the second shot occurs at time 1, the third at time 2, and so on. Then at each time $n = 0, 1, 2, \ldots$, exactly one of the players is shooting, so that the "system" is in one of two states:

> 1 : Player 1 is shooting;
> 2 : Player 2 is shooting.

Suppose now that player 1 is shooting at a given time n. If he makes the shot (probability = .8) he will be shooting again at time $n + 1$. If not (probability = .2), player 2 will be shooting at time $n + 1$. Similarly, if player 2 is shooting at a given time, the probability that he will be shooting next is .7, and the probability that player 1 will be shooting next is .3.   ◇

As Example 1 shows, *we can calculate the probability of the next state of the system if we know the current state.* This makes the process a Markov chain.

Figure 1, called a **transition diagram,** shows the states of the system in Example 1 together with the transition probabilities from one state to another.

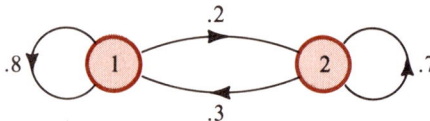

Figure 1

For example, the arrow from 1 to 2, labeled .2, indicates that the probability p_{12} of going from state 1 to state 2 in one step—in other words, the probability that player 2 will be shooting next if player 1 is shooting now—is .2. Similarly, $p_{11} = .8$, $p_{21} = .3$, and $p_{22} = .7$.

The Transition Matrix

If a Markov chain has states 1, 2, . . ., r, the transition probabilities p_{ij} form an $r \times r$ matrix $P = [p_{ij}]$ called the **transition matrix.**

For the two-state Markov chain of Example 1, we have the 2×2 transition matrix:

$$\begin{array}{c} \\ \text{Current} \\ \text{state} \end{array} \begin{array}{cc} & \text{Next state} \\ & \begin{array}{cc} 1 & 2 \end{array} \\ \begin{array}{c} 1 \\ 2 \end{array} & \begin{bmatrix} .8 & .2 \\ .3 & .7 \end{bmatrix} = P \end{array}$$

The entries .8 and .2 in the first row represent the transition probabilities from state 1 to states 1 and 2, respectively. Since from state 1 we *must* go to one and only one of the two states, these probabilities necessarily sum to 1. Likewise, the second-row entries .3 and .7, representing the transition probabilities from state 2 to states 1 and 2, also sum to 1. In general, the ith row of transition matrix P contains the transition probabilities from the "current" state i to each of the possible "next" states 1, 2, . . ., r, and therefore:

The transition matrix P for a Markov chain is always a **stochastic matrix;** that is, it has the following properties:

1. All entries are nonnegative.
2. The entries in each row sum to 1.

(Notice, however, that the entries in a given *column* need not sum to 1.)

The transition matrix P plays an important role in the theory of Markov chains, since—as we shall see shortly—it enables us to use matrix methods to analyze stochastic processes. Before pursuing this idea, however, let us look at another example.

Example 2   During each month that a healthy person remains in Grosso Maxo, he runs a 5% risk of contracting an acute infection caused by a local parasite. Once contracted, the disease never completely leaves one's system, but remains present in a dormant state and periodically flares up in acute form. If the disease is acute in a given month there is a 60% chance it will still be acute the next month, while if it is dormant in a given month there is only a 10% chance it will be acute the next month. Represent this process as a Markov chain and form the transition matrix.

Solution   Taking one month as the time unit, we see that at each time $n = 0, 1, 2, . . .,$ a person is in one of the following three possible states:

$$H = \text{healthy}$$
$$A = \text{acute state}$$
$$D = \text{dormant state}$$

The transition diagram from one month to the next is shown in Figure 2.

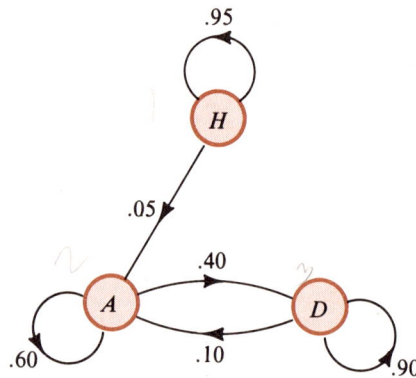

Figure 2

As the diagram shows, there is a 5% chance of going from state H to state A in one step (in a given one-month period) and thus a 95% chance of remaining in state H. Once in state A (or state D) there is no returning to state H. From state A there is a 60% chance of returning to state A, and thus a 40% chance of going to state D; while from state D there is a 10% chance of going to state A, and thus a 90% chance of remaining in state D. The transition matrix for this Markov chain is given below.

$$
\begin{array}{c}
\text{Current} \\
\text{state}
\end{array}
\begin{array}{cc}
 & \begin{array}{ccc}
\text{Next state} \\
H \quad\ A \quad\ D
\end{array} \\
\begin{array}{c} H \\ A \\ D \end{array} &
\begin{bmatrix}
.95 & .05 & 0 \\
0 & .6 & .4 \\
0 & .1 & .9
\end{bmatrix}
\end{array} = P
$$

Observe once again that the transition matrix is stochastic; that is, all rows sum to 1.   ◇

The State Vector   Our main interest in Markov chains centers on this question: *What is the probability that the system will be in a given state at a given time?*

If we let $x_i^{(n)}$ denote the probability that the system is in state i at time n, then the row vector

$$X^{(n)} = [x_1^{(n)} x_2^{(n)} \cdots x_r^{(n)}]$$

is called the **state vector** of the Markov chain at time n. Its entries represent the probabilities of the respective states $1, 2, \ldots, r$ after n steps of the process, and therefore they always sum to 1.

Example 1,
Continued

Suppose players 1 and 2 toss a fair coin to decide who shoots first. What is the probability that the second shot will be taken by player 1? By player 2?

Discussion

Since each player has probability 1/2 of taking the first shot (time 0), the **initial state vector** is

$$X^{(0)} = [.5 \quad .5].$$

That is, the system is equally likely to be in state 1 or state 2 at time 0. To find the corresponding probabilities at the time of the second shot (time 1), we represent the first two stages of the process—namely, the coin toss and the first shot—as in Figure 3.

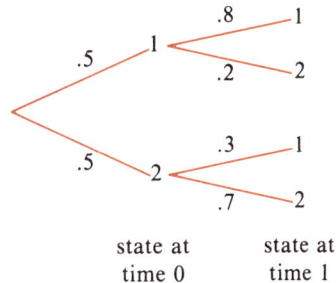

Figure 3

The probabilities .8, .2, .3, and .7 are just the entries of the transition matrix

$$P = \begin{matrix} 1 \\ 2 \end{matrix} \begin{bmatrix} .8 & .2 \\ .3 & .7 \end{bmatrix}.$$

Since two branches lead to state 1, we see that the probability that the system will be in state 1 at time 1 is given by

$$(.5)(.8) + (.5)(.3) = .40 + .15 = .55$$

and the probability that it will be in state 2 is

$$(.5)(.2) + (.5)(.7) = .10 + .35 = .45.$$

Thus, the state vector at time 1 is

$$X^{(1)} = \overset{1 \qquad 2}{[.55 \quad .45]}.$$

That is, there is a 55% chance that the second shot will be taken by player 1, and a 45% chance that it will be taken by player 2.   ◆

The computation performed above can be expressed very simply in terms of matrix multiplication.*

$$\underset{\substack{\text{State vector} \\ \text{at time 0}}}{[.5 \quad .5]} \times \underset{\substack{\text{Transition} \\ \text{matrix}}}{\begin{bmatrix} .8 & .2 \\ .3 & .7 \end{bmatrix}} = \underset{\substack{\text{State vector} \\ \text{at time 1}}}{[.55 \quad .45]}$$

In symbols, $X^{(0)}P = X^{(1)}.$

More generally, the following result holds:

Let P be the transition matrix for a Markov chain. If the state vector at a given time is X, then the state vector one step later is XP. In symbols,

$$X^{(n)}P = X^{(n+1)}. \tag{1}$$

Thus, continuing with Example 1, we can compute the following state vectors:

$$\underset{\substack{\text{State vector} \\ \text{at time 1}}}{[.55 \quad .45]} \begin{bmatrix} .8 & .2 \\ .3 & .7 \end{bmatrix} = \underset{\substack{\text{State vector} \\ \text{at time 2}}}{[.575 \quad .425]}$$

$$\underset{\substack{\text{State vector} \\ \text{at time 2}}}{[.575 \quad .425]} \begin{bmatrix} .8 & .2 \\ .3 & .7 \end{bmatrix} = \underset{\substack{\text{State vector} \\ \text{at time 3}}}{[.5875 \quad .4125]}$$

$$\underset{\substack{\text{State vector} \\ \text{at time 3}}}{[.5875 \quad .4125]} \begin{bmatrix} .8 & .2 \\ .3 & .7 \end{bmatrix} \doteq \underset{\substack{\text{State vector} \\ \text{at time 4}}}{[.5938 \quad .4062]}$$

In other words, there is a 59.38% chance that the system will be in state 1 at time 4; that is, that the fifth shot will be taken by player 1. In the next section we shall see that the state vector approaches an "equilibrium" value

$$X = [.6 \quad .4]$$

---

*The definition of matrix multiplication can be found in Section 4.2.

as the process continues; thus, player 1 will take about 60% of the shots over the long run.

Example 2, Continued If a healthy person arrives in Grosso Maxo for a four-month stay, what is the probability that he will leave with an acute parasitic infection? A dormant infection?

Solution At each time $n = 0, 1, 2, \ldots$ (months after arrival), the person is in one of the states

$$H = \text{healthy}$$
$$A = \text{acute state}$$
$$D = \text{dormant state.}$$

We saw earlier that the month-to-month transition matrix for this Markov chain is

$$
P = \begin{array}{c} \\ H \\ A \\ D \end{array}
\begin{array}{ccc} H & A & D \\ \left[\begin{array}{ccc} .95 & .05 & 0 \\ 0 & .6 & .4 \\ 0 & .1 & .9 \end{array}\right] \end{array}.
$$

Since the person is known to be healthy (in state H) at the time of arrival, the initial state vector is

$$
X^{(0)} = \begin{array}{ccc} H & A & D \\ \left[\begin{array}{ccc} 1 & 0 & 0 \end{array}\right] \end{array}.
$$

That is, the initial probabilities of states H, A, and D are 1, 0, and 0, respectively. To find the state vector at time 4 (four months after arrival) we apply formula (1) repeatedly.

$$
\begin{bmatrix} 1 & 0 & 0 \end{bmatrix}
\begin{bmatrix} .95 & .05 & 0 \\ 0 & .6 & .4 \\ 0 & .1 & .9 \end{bmatrix}
= \begin{bmatrix} .95 & .05 & 0 \end{bmatrix}
\qquad \text{(time 1)}
$$

$$
\begin{bmatrix} .95 & .05 & 0 \end{bmatrix}
\begin{bmatrix} .95 & .05 & 0 \\ 0 & .6 & .4 \\ 0 & .1 & .9 \end{bmatrix}
= \begin{bmatrix} .9025 & .0775 & .0200 \end{bmatrix}
\qquad \text{(time 2)}
$$

$$
\begin{bmatrix} .9025 & .0775 & .0200 \end{bmatrix}
\begin{bmatrix} .95 & .05 & 0 \\ 0 & .6 & .4 \\ 0 & .1 & .9 \end{bmatrix}
= \begin{bmatrix} .8574 & .0936 & .0490 \end{bmatrix}
\qquad \text{(time 3)}
$$

$$
\begin{bmatrix} .8574 & .0936 & .0490 \end{bmatrix}
\begin{bmatrix} .95 & .05 & 0 \\ 0 & .6 & .4 \\ 0 & .1 & .9 \end{bmatrix}
\doteq \begin{bmatrix} .8145 & .1039 & .0815 \end{bmatrix}
\qquad \text{(time 4)}
$$

That is, an initially healthy person has an 81.45% chance of still being healthy after four months, a 10.39% chance of having an acute infection, and an 8.15% chance of having a dormant infection.   ◆

Applications   There are many applications of Markov chain models. One class of problems concerns *population shifts*, or the movement of individuals from one classification to another.

Example 3   TV channels 1 and 2 are competing for the evening news audience in Shreveport. After changing its format to attract more viewers, channel 2 conducted a survey which showed that over a six-week period 20% of those watching channel 1 switched to channel 2, while only 10% of those watching channel 2 switched to channel 1. If channel 2 had 25% of the audience at the time of the format change, and this pattern of viewer shift continues, what share of the audience will channel 2 have after six weeks? After twelve weeks? After eighteen weeks?

Solution   We let $n = 0$ denote the time of the format change, with each additional time unit representing a six-week period. At each time $n = 0, 1, 2, . . .,$ a random viewer will be in one of the states

$$1 = \text{watches channel 1 news}$$
$$2 = \text{watches channel 2 news.}$$

The transition diagram for this Markov chain is shown in Figure 4.

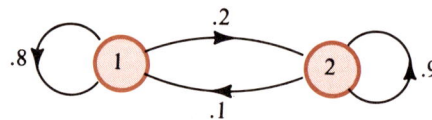

Figure 4

That is, someone watching channel 1 has a 20% chance of switching to channel 2 in any given six-week period, and therefore an 80% chance of remaining with channel 1; while someone watching channel 2 has a 10% chance of switching to channel 1, and thus a 90% chance of remaining with channel 2. These probabilities give us the 2×2 transition matrix P.

$$
\begin{array}{c}
\text{Current} \\
\text{state}
\end{array}
\begin{array}{cc}
 & \text{Next state} \\
 & \begin{array}{cc} 1 & \quad 2 \end{array} \\
\begin{array}{c} 1 \\ 2 \end{array} &
\begin{bmatrix} .8 & .2 \\ .1 & .9 \end{bmatrix} = P
\end{array}
$$

The initial state vector

$$X^{(0)} = \begin{bmatrix} 1 & \quad 2 \\ .75 & .25 \end{bmatrix}$$

represents the respective audience shares at the time of the format change. Thus, there is a 25% chance that a random viewer is watching channel 2 initially, and therefore a 75% chance he or she is watching channel 1. To find the audience shares at times $n = 1, 2,$ and 3, we apply formula (1).

$$[.75 \quad .25]\begin{bmatrix} .8 & .2 \\ .1 & .9 \end{bmatrix} = [.625 \quad .375] \qquad \text{(6 weeks)}$$

$$[.625 \quad .375]\begin{bmatrix} .8 & .2 \\ .1 & .9 \end{bmatrix} = [.5375 \quad .4625] \qquad \text{(12 weeks)}$$

$$[.5375 \quad .4625]\begin{bmatrix} .8 & .2 \\ .1 & .9 \end{bmatrix} \doteq [.4763 \quad .5237] \qquad \text{(18 weeks)}$$

Thus, channel 2 will have 37.5% of the audience after 6 weeks, 46.25% after 12 weeks, and 52.37% after 18 weeks. In the next section we shall see that, provided this pattern of viewer shift continues, channel 2 will eventually capture two thirds of the audience. ◆

9.1 Exercises

1. Draw a transition diagram for the Markov chain with states 1, 2 and transition matrix:

(a) $\begin{bmatrix} .5 & .5 \\ .6 & .4 \end{bmatrix}$
(b) $\begin{bmatrix} \frac{2}{3} & \frac{1}{3} \\ \frac{1}{4} & \frac{3}{4} \end{bmatrix}$
(c) $\begin{bmatrix} .2 & .8 \\ 0 & 1 \end{bmatrix}$

2. Draw a transition diagram for the Markov chain with states 1, 2, 3 and transition matrix:

(a) $\begin{bmatrix} 0 & .6 & .4 \\ .7 & 0 & .3 \\ .4 & .4 & .2 \end{bmatrix}$
(b) $\begin{bmatrix} 1 & 0 & 0 \\ \frac{1}{3} & \frac{1}{3} & \frac{1}{3} \\ 1 & 0 & 0 \end{bmatrix}$
(c) $\begin{bmatrix} 0 & .5 & .5 \\ 0 & 0 & 1 \\ 1 & 0 & 0 \end{bmatrix}$.

3. Find the transition matrix P for the Markov chain with transition diagram:

(a)

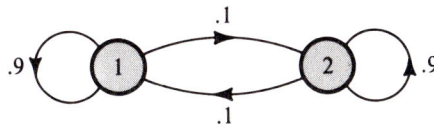

(b)

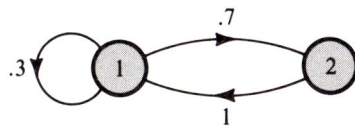

4. Find the transition matrix P for the Markov chain with transition diagram:

(a)

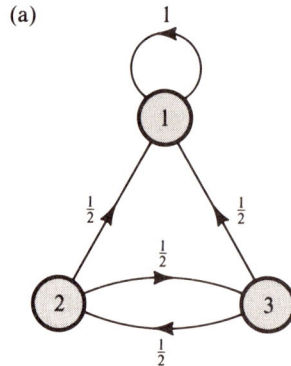

(b)

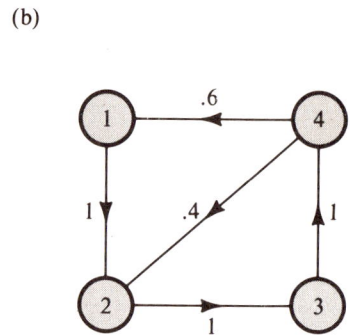

5. A two-state Markov chain has transition matrix

$$P = \begin{bmatrix} .6 & .4 \\ .1 & .9 \end{bmatrix}.$$

Find the state vector after three steps (that is, at time 3), if initially the system is

(a) known to be in state 1;   (b) equally likely to be in state 1 or 2.

6. A three-state Markov chain has transition matrix

$$P = \begin{bmatrix} 0 & 1 & 0 \\ 0 & 0 & 1 \\ \frac{1}{3} & 0 & \frac{2}{3} \end{bmatrix}.$$

Find the state vector after three steps (that is, at time 3), if initially the system is

(a) known to be in state 1;   (b) equally likely to be in state 1, 2, or 3.

7. An actor has found from experience that his performance on a given night affects his performance the following night. A good performance is followed by another good performance 90% of the time, but a bad performance is followed by a good one only 70% of the time. If he performs well on Monday, what is the probability that he will perform well the following Friday?

8. An author has been told that retyping her manuscript over and over to eliminate errors will yield diminishing returns. If a particular word contains an error there is a 48% chance it will be spotted and corrected on the copy; however, if the word is correct to begin with there is a 2% chance it will be typed incorrectly on the copy. If a word is correct initially, what is the probability that it will still be correct after the manuscript has been retyped three times?

9. The sales manager for a metropolitan newspaper has observed an alarming trend: 15% of the subscribers at the beginning of any given month fail to renew their subscriptions the following month, while only 5% of all nonsubscribers become subscribers during the same period. If this trend continues, what will happen to the newspaper's current market share of 60% in another three months?

10. There are some recent signs that Booneville's high unemployment rate may be turning around at last. A three-month study revealed that 45% of those unemployed on January 1 were employed by April 1, while only 5% of those employed

initially became unemployed during this period. If this trend continues, what will happen to the town's current 20% unemployment rate in another six months? In nine months?

11. Forty percent of the population of Xanadu lives in the cities, with the rest occupying the rural areas. Suppose that, in any given year, 10% of the rural population moves to the cities, while 5% of the city dwellers move to the rural areas. How long will it be before a majority of the population lives in the cities?

12. A sample of US car owners were asked whether their current car was of American or foreign make, and what make car they would buy next. Four out of five owners of foreign cars said they would buy them again the next time, while two out of five owners of American makes said they would switch to a foreign car the next time. If foreign cars currently have a 20% share of the market, and car owners make a new purchase every five years on the average, then how long will it take before foreign cars capture half the US market?

13. A mouse is placed in a box with three connecting compartments, as shown here.

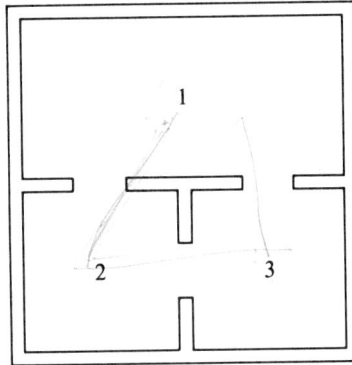

In each unit time period the mouse chooses at random one of the doors of the compartment it is occupying, and passes through it into an adjoining compartment. If the mouse starts in compartment 1 at time 0, what is the probability that it will be there at time 4?

14. A runner proceeds from one town to another in the network shown below, always choosing at random among the roads available.

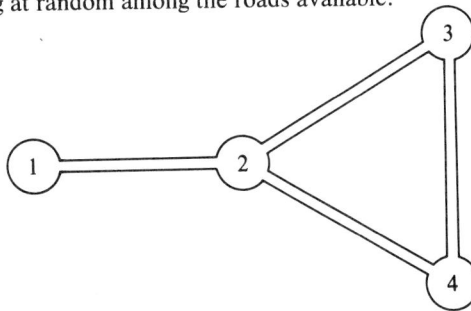

Thus, from town 1 she must go to town 2, but from town 2 she is equally likely to go to town 1, 3, or 4. If she starts at town 1 and covers one road per unit time, what is the probability that she will be in town 1 at time 4?

15. In a sociological study of class mobility* it was found that the children of upper-class parents were divided evenly between the upper and middle classes, while the children of lower-class parents were divided evenly between the lower and middle classes. Among children of middle-class parents, 10% moved to the upper class, 70% remained in the middle class, and 20% moved to the lower class. Assuming the same birthrate for all three groups, what will the social distribution be in two generations if it is currently 10% upper class, 40% middle class and 50% lower class?

16. A political scientist observes that among children of registered Democrats, 70% become Democrats, 10% become Republicans, and 20% register as Independents. For children of Republicans the distribution is 10% Democrats, 60% Republicans, and 30% Independents, and for children of Independents it is 10% Democrats, 10% Republicans, and 80% Independents. If the current voter registration is 40% Democratic, 40% Republicans, and 20% Independent, and the birthrate is the same for all three groups, what will the voter distribution be two generations from now?

17. A toll gate at the Braxton Bridge passes one car through in each unit time period (provided some car is in line). The probability that 0, 1, or 2 more cars may join the line during this interval is .2, .6, and .2, respectively; however, once four cars are in line new arrivals always choose another gate. Let the state of the system be the number of cars in line at the end of a given time period, that is, immediately before the lead car is passed through. Represent this process as a five-state Markov chain.

18. An auto dealer always has one or two Bracers on hand in his showroom at the beginning of each week. He knows from experience that the probability of being able to sell 0, 1, or 2 of these cars during a given week (assuming they are in the showroom) is .7, .2, and .1, respectively. If by the end of the week he has sold all the units on hand he orders two more for delivery Monday morning. If not, he does not place any order. Let the state of the system be the number of cars on hand at the end of a given week, immediately before ordering replacements. Represent this process as a three-state Markov chain.

Computer Projects   The MARKOV program computes the successive state vectors of a Markov chain, given the transition matrix and initial state vector. Use this program to confirm your answer in Exercises 5–16.

9.2 Regular Markov Chains

Many Markov chains tend to "stabilize" over the long run, in the sense that the state vector approaches a limiting or equilibrium value as the process continues. For such processes we can predict the proportion of the time the system will be in each state, over the long run.

---

*For the origin of this application see S. J. Prais, "Measuring Social Mobility," *Journal of the Royal Statistical Society* 118 (1955), pp. 56–66.

The Equilibrium Vector Let us look once again at the basket-shooting problem described in the preceding section.

Example 1 Players 1 and 2 take turns shooting baskets from the foul line. Each player continues to shoot until he misses, at which point the other player takes over. Suppose player 1 is an 80% shooter and player 2 is a 70% shooter. Over the long run, what percentage of the shots will be taken by Player 1?

Discussion We saw earlier that the transition matrix for this process is

$$
\begin{array}{c}
\text{Player shooting next} \\
\begin{array}{cc} 1 & 2 \end{array} \\
\begin{array}{c} \text{Player} \\ \text{shooting now} \end{array}
\begin{array}{c} 1 \\ 2 \end{array}
\begin{bmatrix} .8 & .2 \\ .3 & .7 \end{bmatrix} = P
\end{array}
$$

Moreover, if we know the initial state vector $X^{(0)}$, we can calculate the state vector at any later time by the iterative procedure of Section 9.1.

$$X^{(0)}P = X^{(1)}$$
$$X^{(1)}P = X^{(2)}$$
$$X^{(2)}P = X^{(3)}$$

and so forth. The table below shows the result of such a calculation, rounded off to four decimal places. In the first column we assume that player 1 shoots first, in the second that player 2 shoots first, and in the third that they toss a fair coin to start the process.

	Player 1 starts		Player 2 starts		Toss coin	
time 0:	[1	0]	[0	1]	[.5	.5]
time 1:	[.8	.2]	[.3	.7]	[.55	.45]
time 2:	[.70	.30]	[.45	.55]	[.575	.425]
time 3:	[.650	.350]	[.525	.475]	[.5875	.4125]
time 4:	[.6250	.3750]	[.5625	.4375]	[.5938	.4062]
time 5:	[.6125	.3875]	[.5813	.4187]	[.5969	.4031]
time 6:	[.6063	.3937]	[.5906	.4094]	[.5984	.4016]
time 7:	[.6031	.3969]	[.5953	.4047]	[.5992	.4008]
time 8:	[.6016	.3984]	[.5977	.4023]	[.5996	.4004]
time 9:	[.6008	.3992]	[.5988	.4012]	[.5998	.4002]
time 10:	[.6004	.3996]	[.5994	.4006]	[.5999	.4001]
time 11:	[.6002	.3998]	[.5997	.4003]	[.6000	.4000]
time 12:	[.6001	.3999]	[.5999	.4001]	[.6000	.4000]
time 13:	[.6000	.4000]	[.5999	.4001]	[.6000	.4000]
time 14:	[.6000	.4000]	[.6000	.4000]	[.6000	.4000]
time 15:	[.6000	.4000]	[.6000	.4000]	[.6000	.4000]

From the table we see that in all three cases, the state vectors converge toward the same value

$$X = [.6 \quad .4]$$

as the process continues. Once the state vector has reached this "stable" value X it will not change, for if the state vector at a given time is X, the state vector one step later is

$$
\begin{aligned}
XP &= [.6 \quad .4]\begin{bmatrix} .8 & .2 \\ .3 & .7 \end{bmatrix} \\
&= [(.6)(.8) + (.4)(.3) \quad (.6)(.2) + (.4)(.7)] \\
&= [.48 + .12 \quad .12 + .28] \\
&= [.6 \quad .4] = X
\end{aligned}
$$

In other words, *as the process continues, the probability that player 1 is shooting tends to stabilize at .6, no matter who went first.* It follows that over the long run, about 60% of the shots will be taken by player 1 and 40% by player 2. ◆

   This vector X is called the *equilibrium vector* (or *stable state vector*) for the Markov chain.

The **equilibrium vector** for an r-state Markov chain has the form

$$X = [x_1 \, x_2 \, \cdots \, x_r]$$

where $x_1, x_2, \ldots, x_r$ represent the long-run probabilities of states 1, 2, $\ldots, r$, respectively. These probabilities always sum to 1.

Regular Markov Chains   Two important questions may have already occurred to the reader, namely:

   1. Does every Markov chain approach equilibrium?
   2. If an equilibrium vector exists, how do we find it?

It is not hard to show that the answer to the first question is negative.* There are Markov chains whose state vectors fail to stabilize no matter how long the process goes on, and there are others whose state vectors stabilize to different values X, depending on the initial state of the system. The next theorem, however, shows that for a wide class of Markov chains, the equilibrium vector always exists and is unique.

---

*See Exercises 3 and 4.

Theorem 1

Let P be the transition matrix for a Markov chain, and suppose that some power of P contains all positive entries. Then regardless of the initial state of the system, the state vectors

$$X^{(0)}, X^{(1)}, X^{(2)}, \ldots$$

will approach a unique equilibrium vector X, which satisfies the equation

$$XP = X.$$

A Markov chain is said to be **regular** if it satisfies the above condition; that is, if at least one of the matrices

$$P, P^2, P^3, P^4, \ldots$$

contains all positive entries. It can be shown* that if this occurs at all, it must occur among the first $(r - 1)^2 + 1$ powers of P, where r is the number of states in the Markov chain. Thus, for instance, for a three-state chain we need only check the first five powers of P to see if the chain is regular.

For a regular Markov chain, Theorem 1 not only guarantees the existence of an equilibrium vector X but also enables us to find it. Thus, the Markov chain of Example 1 is regular, since

$$P = \begin{bmatrix} .8 & .2 \\ .3 & .7 \end{bmatrix}$$

already contains all positive entries. By Theorem 1, its equilibrium vector

$$X = [x_1 \quad x_2]$$

satisfies the matrix equation $XP = X$, or

$$[x_1 \quad x_2]\begin{bmatrix} .8 & .2 \\ .3 & .7 \end{bmatrix} = [x_1 \quad x_2].$$

Multiplying out the left-hand side, we obtain

$$[.8x_1 + .3x_2 \quad .2x_1 + .7x_2] = [x_1 \quad x_2]$$

which is equivalent to the system

$$.8x_1 + .3x_2 = x_1$$
$$.2x_1 + .7x_2 = x_2$$

or, transposing,

$$-.2x_1 + .3x_2 = 0 \tag{1a}$$
$$.2x_1 - .3x_2 = 0. \tag{1b}$$

---

*See Daniel P. Maki and Maynard Thompson, *Mathematical Models and Applications* (Englewood Cliffs, N.J.: Prentice-Hall, 1973), p. 101. See also op. cit., pp. 92–95 for a proof of Theorem 1.

This is a 2 × 2 linear system, which we would normally solve for the unknowns x_1 and x_2. Unfortunately this system is redundant, since equation (1b) follows from (1a). We must therefore introduce the additional fact that the equilibrium probabilities sum to 1.

$$x_1 + x_2 = 1 \qquad\qquad (1c)$$

Equations (1a)–(1c) can now be solved either by substitution (see Section 1.2) or Gauss-Jordan elimination (see Sections 3.1–3.2). Applying the substitution method, for example, we first obtain

$$x_2 = \frac{2}{3}x_1$$

using either (1a) or (1b). When substituted into (1c) this gives

$$x_1 + \frac{2}{3}x_1 = 1$$

$$\frac{5}{3}x_1 = 1$$

$$x_1 = \frac{3}{5}, \, x_2 = \frac{2}{3}x_1 = \frac{2}{5}.$$

The equilibrium vector is therefore

$$X = \begin{bmatrix} \frac{3}{5} & \frac{2}{5} \end{bmatrix} = \begin{bmatrix} .6 & .4 \end{bmatrix}$$

which agrees with the empirical result we found earlier.

Example 2   Consider the disease model described in Example 2 of Section 9.1, with the states H (healthy), A (acute infection) and D (dormant infection), and the following transition matrix:

$$
\begin{array}{cc}
 & \text{State next month} \\
 & \begin{array}{ccc} H & A & D \end{array}
\end{array}
$$

$$
\text{Current state} \quad
\begin{array}{c} H \\ A \\ D \end{array}
\begin{bmatrix}
.95 & .05 & 0 \\
0 & .6 & .4 \\
0 & .1 & .9
\end{bmatrix} = P
$$

Is this a regular Markov chain?

Solution   Since P contains several zero entries, we must first try to determine whether some higher power of P contains all positive entries. We can simplify the calculation by letting the symbol $\star$ represent a positive entry. Thus:

$$
P = \begin{bmatrix}
\star & \star & 0 \\
0 & \star & \star \\
0 & \star & \star
\end{bmatrix}
$$

$$P^2 = P \times P = \begin{bmatrix} \star & \star & 0 \\ 0 & \star & \star \\ 0 & \star & \star \end{bmatrix} \begin{bmatrix} \star & \star & 0 \\ 0 & \star & \star \\ 0 & \star & \star \end{bmatrix} = \begin{bmatrix} \star & \star & \star \\ 0 & \star & \star \\ 0 & \star & \star \end{bmatrix}$$

$$P^3 = P^2 \times P = \begin{bmatrix} \star & \star & \star \\ 0 & \star & \star \\ 0 & \star & \star \end{bmatrix} \begin{bmatrix} \star & \star & 0 \\ 0 & \star & \star \\ 0 & \star & \star \end{bmatrix} = \begin{bmatrix} \star & \star & \star \\ 0 & \star & \star \\ 0 & \star & \star \end{bmatrix}$$

Since P^2 amd P^3 have the same form, it is clear that this pattern will persist for higher powers. (In any case, we need only check P^4 and P^5, in view of the remark following Theorem 1.) Thus, every power P^n contains zero entries, and the Markov chain is not regular. Theorem 1 therefore gives us no information on whether or not an equilibrium vector exists (see Exercises 19–20). ◆

Example 3 Consider the viewer-shift model described in Example 3 of Section 9.1, in which a random viewer is watching one of two TV channels at the beginning of each six-week period. The transition matrix is

$$\begin{array}{c} \text{Next channel} \\ \begin{array}{cc} 1 & 2 \end{array} \\ \begin{array}{c} \text{Current} \\ \text{channel} \end{array} \begin{array}{c} 1 \\ 2 \end{array} \begin{bmatrix} .8 & .2 \\ .1 & .9 \end{bmatrix}. \end{array}$$

What proportion of the audience will each channel have over the long run?

Solution Since P has all positive entries, this process is a regular two-state Markov chain. The equilibrium vector

$$X = [x_1 \quad x_2]$$

satisfies $XP = X$, which leads to the system of equations

$$.8x_1 + .1x_2 = x_1$$
$$.2x_1 + .9x_2 = x_2.$$

Transposing terms and adding as before the condition that the variables sum to 1, we arrive at the (redundant) system

$$-.2x_1 + .1x_2 = 0$$
$$.2x_1 - .1x_2 = 0$$
$$x_1 + x_2 = 1$$

which has the solution $x_1 = 1/3$, $x_2 = 2/3$. Thus, channel 1 will have one-third of the audience over the long run, while channel 2 will have two-thirds. ◆

Example 4   Three players are tossing a frisbee. Player 1 always throws to player 2, who in turn always throws to player 3. However, player 3 is equally likely to throw to player 1 or player 2. Over the long run, what proportion of the throws will be made by each player?

Solution   The transition matrix for this three-state Markov chain is

$$
\begin{array}{c}
\\
\text{Player throwing now}
\end{array}
\begin{array}{c}
\text{Player throwing next} \\
\begin{array}{ccc} 1 & 2 & 3 \end{array} \\
\begin{array}{c} 1 \\ 2 \\ 3 \end{array}
\begin{bmatrix}
0 & 1 & 0 \\
0 & 0 & 1 \\
\frac{1}{2} & \frac{1}{2} & 0
\end{bmatrix} = P.
\end{array}
$$

To test for regularity, we write the matrix in the form

$$
P = \begin{bmatrix}
0 & \star & 0 \\
0 & 0 & \star \\
\star & \star & 0
\end{bmatrix}
$$

where the symbol $\star$ denotes a positive entry, and calculate successive powers of P as in Example 2. In this way we find that

$$
P^2 = \begin{bmatrix}
0 & 0 & \star \\
\star & \star & 0 \\
0 & \star & \star
\end{bmatrix}, \qquad
P^3 = \begin{bmatrix}
\star & \star & 0 \\
0 & \star & \star \\
\star & \star & \star
\end{bmatrix},
$$

$$
P^4 = \begin{bmatrix}
0 & \star & \star \\
\star & \star & \star \\
\star & \star & \star
\end{bmatrix}, \qquad
P^5 = \begin{bmatrix}
\star & \star & \star \\
\star & \star & \star \\
\star & \star & \star
\end{bmatrix}.
$$

Since P^5 contains all positive entries, we know by Theorem 1 that the Markov chain has an equilibrium vector

$$
X = [x_1 \quad x_2 \quad x_3].
$$

The condition $XP = X$ becomes

$$
[x_1 \quad x_2 \quad x_3]
\begin{bmatrix}
0 & 1 & 0 \\
0 & 0 & 1 \\
\frac{1}{2} & \frac{1}{2} & 0
\end{bmatrix} = [x_1 \quad x_2 \quad x_3]
$$

which is equivalent to the following three equations.

$$
\frac{1}{2}x_3 = x_1
$$

$$
x_1 \quad + \frac{1}{2}x_3 = x_2
$$

$$
x_2 \quad = x_3
$$

Adding the restriction that the equilibrium probabilities sum to 1, we therefore arrive at the linear system below.

$$-x_1 \qquad + \frac{1}{2}x_3 = 0$$

$$x_1 - x_2 + \frac{1}{2}x_3 = 0$$

$$x_2 - x_3 = 0$$

$$x_1 + x_2 + x_3 = 1$$

Solving this system, we find that $x_1 = 1/5$, $x_2 = 2/5$, and $x_3 = 2/5$. This means that over the long run player 1 will throw 20% of the time, while players 2 and 3 will each throw 40% of the time. ◆

Rates of Convergence Theorem 1 assures us that for a regular Markov chain the state vectors will eventually converge to an equilibrium vector X. The *rate* of convergence, which is often of special interest, depends on both the initial state vector $X^{(0)}$ and the transition matrix P.

Example 5 A patient is either sick (state S) or well (state W) at the beginning of each day, her state varying according to the transition matrix:

$$\begin{array}{c} \text{Next state} \\ \begin{array}{cc} S & W \end{array} \end{array}$$

$$\begin{array}{cc} \text{Current} \\ \text{state} \end{array} \quad \begin{array}{c} S \\ W \end{array} \begin{bmatrix} .997 & .003 \\ .001 & .999 \end{bmatrix} = P.$$

Discussion We leave it to the reader to verify that the equilibrium vector for this regular Markov chain is

$$\begin{array}{cc} S & W \end{array}$$

$$X = [.25 \quad .75].$$

Thus, there is a 75% chance that the patient will be well in the long run, regardless of her initial state. However, if she is sick initially she is likely to remain sick for quite some time. For instance, the first few state vectors are

$$\begin{array}{ll}
[1 \quad 0] & \text{(time 0)} \\
[.997 \quad .003] & \text{(time 1)} \\
[.994 \quad .006] & \text{(time 2)} \\
[.991 \quad .009] & \text{(time 3)} \\
[.988 \quad .012] & \text{(time 4)} \\
[.985 \quad .015] & \text{(time 5).}
\end{array}$$

If we continue the list with the aid of a computer, we obtain

$$\vdots$$

$$[.752 \quad .248] \qquad \text{(time 100)}$$

$$\vdots$$

$$[.585 \quad .415] \qquad \text{(time 200)}$$

$$\vdots$$

and so forth. Thus, even after 200 days—more than six months—the patient is more likely to be sick than well!　◆

9.2 Exercises

1. In the Markov chain diagrams below, each arrow represents a positive transition probability. Determine whether the given two-state chain is regular.

(a)

(b)

2. In the Markov chain diagrams below, each arrow represents a positive transition probability. Determine whether the given three-state chain is regular.

(a)

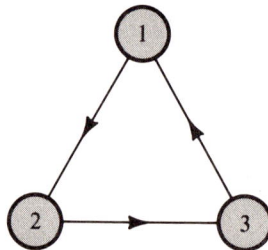

(b)

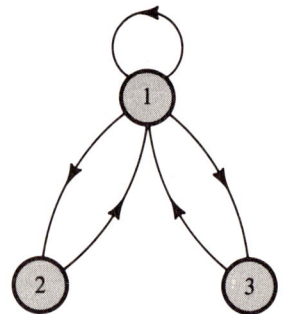

3. Consider the two-state Markov chain with transition diagram

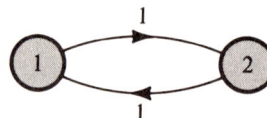

(a) Assume the system is initially in state 1, and find the state vectors at time 0, 1, 2, and 3.

(b) Do part (a) under the assumption that the system is initially in state 2.

(c) Does this Markov chain approach equilibrium?

4. Consider the three-state Markov chain with transition diagram

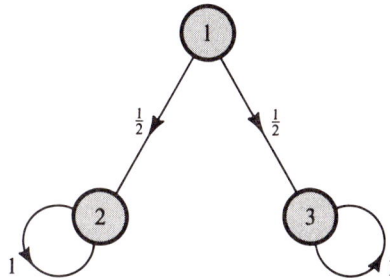

(a) Assume the system is initially in state 1, and find the state vectors at times 0, 1, 2, and 3.

(b) Do part (a) under the assumption that the system is initially in state 2. In state 3.

(c) Does this Markov chain approach equilibrium?

5. In Exercise 5 of Section 9.1, show that the Markov chain is regular and find the equilibrium vector.

6. In Exercise 6 of Section 9.1, show that the Markov chain is regular and find the equilibrium vector.

7. In Exercise 7 of Section 9.1, what proportion of the actor's performances will be good, over the long run?

8. In Exercise 8 of Section 9.1, what proportion of the words will contain an error, after many retypings?

9. In Exercise 9 of Section 9.1, what share of the market will the newspaper eventually have?

10. In Exercise 10 of Section 9.1, what unemployment rate will Booneville eventually have?

11. In Exercise 11 of Section 9.1, what proportion of the population of Xanadu will eventually live in the cities?

12. In Exercise 12 of Section 9.1, what share of the market will foreign cars eventually capture?

13. In Exercise 13 of Section 9.1, what is the long-run probability that the mouse will be in compartment 1?

14. In Exercise 14 of Section 9.1, what is the long-run probability that the runner will be in town 1?

15. In Exercise 15 of Section 9.1, what will the social distribution be after many generations?

16. In Exercise 16 of Section 9.1, what will the voter distribution be after many generations?

17. In Exercise 17 of Section 9.1, what will be the average number of cars in line?

18. In Exercise 18 of Section 9.1, how many cars will the dealer have on hand at the end of an average week?

19. Consider the disease model (Example 2 in this section) applied to a person who is known to be infected initially—that is, the person is either in state A or state D.

   (a) Show that under this assumption, the model can be simplified to obtain a regular two-state Markov chain.

   (b) Show that over the long run the infection will be acute 20% of the time and dormant 80% of the time.

20. Consider the disease model (Example 2 in this section) applied to a person who is known to be healthy initially (state H).

   (a) Show that the probability that he will still be healthy after n months is $(.95)^n$.

   (b) Conclude that the long-run probability of state H is zero.

   (c) Using part (b) and the result of Exercise 19, show that the disease model has an equilibrium vector.

Computer Projects

A. The MARKOV program computes the successive state vectors of a Markov chain, given the transition matrix and initial state vector. Use this program to confirm your value for the equilibrium vector in Exercises 5–16. Try several different choices of the initial state vector.

B. The PIVOT program performs a pivot transformation on a designated nonzero entry in a tableau. This program can be used to solve for the equilibrium vector in Exercises 5–18.

C. Write a BASIC program to determine whether or not a given 5-state Markov chain is regular.

D. Use the PIVOT program to find the equilibrium vector for the regular 5-state Markov chain with the following transition matrix:

$$
\begin{array}{c}
\text{Current} \\
\text{state}
\end{array}
\quad
\begin{array}{cc}
 & \text{Next state} \\
\begin{array}{c}
1 \\
2 \\
3 \\
4 \\
5
\end{array}
&
\begin{array}{ccccc}
1 & 2 & 3 & 4 & 5 \\
\begin{bmatrix}
0 & .1 & .8 & .1 & 0 \\
.2 & 0 & .6 & .1 & .1 \\
.1 & .4 & 0 & .4 & .1 \\
.1 & .1 & .6 & 0 & .2 \\
0 & .1 & .8 & .1 & 0
\end{bmatrix}
\end{array}
\end{array}
$$

9.3 Absorbing Markov Chains

We now turn to another class of Markov chains which arise in many applications. Like regular chains, these processes exhibit stable behavior over the long run, but here the "equilibrium" vector depends on the initial state of the system.

Absorbing States   A state in a Markov chain is called an **absorbing state** if it is impossible to leave it once it is entered.

An **absorbing Markov chain** is one in which:

1. there exists at least one absorbing state, and

2. from every state it is possible to reach an absorbing state in a finite number of steps.

Those states which are not absorbing are called **transient states.**

Example 1   The laundry-detergent market is divided among four brands, A, B, C, and D, all of which advertise that "Once you've tried ours, you'll never switch!" For brands A and C this claim is true. However, only 40% of those using brand B at a given time will be using it one month later; 30% will switch to A, 10% to C, and 20% to D. And everyone who tries brand D picks a different brand the following month, with 20% switching to A, 50% to B, and 30% to C. Represent this process as a Markov chain.

Discussion   The month-to-month transition diagram is shown in Figure 1, where the states correspond to the four soap brands.

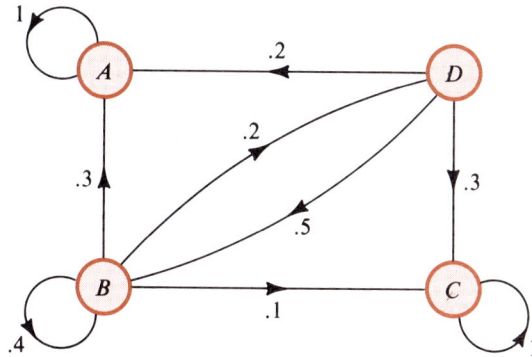

Figure 1

For example, if a person is currently using brand B, the probability is .3 that he will be using brand A one month later, .4 that he will still be using brand B, and so on. Observe that *states A and C are absorbing:* Once the process enters either of these states it remains there forever. Moreover, from states B and D it is possible to reach an absorbing state (in one step, in fact). This process is therefore an absorbing Markov chain. The transition matrix is

$$
\begin{array}{c}
\\
\text{Current} \\
\text{state}
\end{array}
\begin{array}{c}
\\
\text{A} \\
\text{B} \\
\text{C} \\
\text{D}
\end{array}
\begin{array}{cccc}
\multicolumn{4}{c}{\text{State next month}} \\
\text{A} & \text{B} & \text{C} & \text{D} \\
\left[\begin{array}{cccc}
1 & 0 & 0 & 0 \\
.3 & .4 & .1 & .2 \\
0 & 0 & 1 & 0 \\
.2 & .5 & .3 & 0
\end{array}\right]
\end{array} = P.
$$

◇

Notice that the rows corresponding to the absorbing states A and C have a special form. All entries are zero except for a 1 along the main diagonal. In general, for an absorbing state i we have $p_{ii} = 1$ and $p_{ij} = 0$ for $j \neq 1$, which just says that from state i we are certain to return to state i.

Long-Run Stability It is clear that a Markov chain with more than one absorbing state cannot have a unique equilibrium vector, and therefore such a chain can never be regular. In our example, for instance, if the system starts out in state A it will always have state vector

$$\begin{array}{cccc} A & B & C & D \end{array}$$
$$[1 \quad 0 \quad 0 \quad 0]$$

while if it starts in state C it will always have state vector

$$\begin{array}{cccc} A & B & C & D \end{array}$$
$$[0 \quad 0 \quad 1 \quad 0]$$

no matter how long the process goes on. What happens if the process starts out in one of the transient states B or D? The table below, obtained with the aid of a computer, shows the state vectors at one-month intervals for each of these starting states (figures have been rounded off to two places):

	Initial state B				Initial state D			
time 0:	[0	1	0	0]	[0	0	0	1]
time 1:	[.3	.4	.1	.2]	[.2	.5	.3	0]
time 2:	[.46	.26	.20	.08]	[.35	.20	.35	.10]
time 3:	[.55	.14	.25	.05]	[.43	.13	.40	.04]
time 4:	[.61	.08	.28	.03]	[.48	.07	.43	.03]
time 5:	[.64	.05	.30	.02]	[.50	.04	.44	.01]
time 6:	[.66	.03	.31	.01]	[.52	.02	.45	.01]
time 7:	[.67	.02	.31	.01]	[.53	.01	.45	.00]
time 8:	[.67	.01	.32	.00]	[.53	.01	.46	.00]
time 9:	[.68	.01	.32	.00]	[.54	.00	.46	.00]
time 10:	[.68	.00	.32	.00]	[.54	.00	.46	.00]

In both cases we see that the state vector has stabilized (to two places) by time 10, but *the equilibrium probabilities depend on the initial state*. Thus, if a person uses brand B initially there is a 68% chance he will be using brand A after 10 months and a 32% chance he will be using brand C; while for those who use brand D initially the corresponding figures are 54% and 46%.

Note that there is virtually *no* chance that the person will be using brand B or D after 10 months. In other words, as the process continues the system is increasingly certain to be in an absorbing state. This can be proved to hold in general.

In an absorbing Markov chain the long-run probability of every transient state is zero, regardless of the initial state. Hence, the long-run probability that the system will reach an absorbing state is 1.

When the system starts out in an absorbing state, the long-run behavior of the process is completely determined. When it starts in a transient state, however, two questions can be raised:

1. On the average, how long will the process take to reach an absorbing state?

2. What is the probability that the system will end up in a particular absorbing state?

We shall soon see that the transition matrix P holds the answer to both these important questions.

The Fundamental Matrix   If we rearrange the four states in Example 1 so that the absorbing states come first, the transition matrix becomes

$$
\begin{array}{c}
 \\
\text{Current}\\
\text{state}
\end{array}
\begin{array}{c}
\\
\\
A\\
C\\
B\\
D
\end{array}
\begin{array}{cc}
\multicolumn{2}{c}{\text{State next month}}\\
\begin{array}{cccc}
A & C & B & D
\end{array}\\
\left[\begin{array}{cc|cc}
1 & 0 & 0 & 0\\
0 & 1 & 0 & 0\\
\hline
.3 & .1 & .4 & .2\\
.2 & .3 & .5 & 0
\end{array}\right] = P^{\star}.
\end{array}
$$

This is called the **canonical form** of the transition matrix. Note that if we draw two lines as indicated, separating the absorbing states from the transient states, we partition $P^{\star}$ into four submatrices of dimension 2×2. The matrix in the upper left is just the identity matrix I, while the matrix in the upper right is the zero matrix O.

More generally, an absorbing chain with r absorbing states and s transient states will have a canonical transition matrix of the form

$$
P^{\star} = \begin{array}{c}
r\,\{\\
s\,\{
\end{array}
\overbrace{\left[\begin{array}{c|c}
I & O\\
\hline
R & Q
\end{array}\right]}^{\displaystyle r \quad s}
\tag{1}
$$

where I is an $r \times r$ identity matrix, O is an $r \times s$ zero matrix, and R and Q have dimensions $s \times r$ and $s \times s$, respectively. The matrix

$$
N = (I - Q)^{-1}
\tag{2}
$$

where I is an $s \times s$ identity matrix, is called the **fundamental matrix** for the absorbing chain.

Example 1,
Continued

Find the fundamental matrix for the soap-shift problem.

Solution The lower-right submatrix of $P^\star$ is $\qquad Q = \begin{bmatrix} .4 & .2 \\ .5 & 0 \end{bmatrix}$.

Hence, $\qquad I - Q = \begin{bmatrix} 1 & 0 \\ 0 & 1 \end{bmatrix} - \begin{bmatrix} .4 & .2 \\ .5 & 0 \end{bmatrix} = \begin{bmatrix} .6 & -.2 \\ -.5 & 1 \end{bmatrix}$

Inverting this matrix by the method of Section 4.3 we obtain

$$N = (I - Q)^{-1} = \begin{bmatrix} 2 & .4 \\ 1 & 1.2 \end{bmatrix}.$$

The reader can verify this result by direct matrix multiplication.   ◇

Absorption Times and Probabilities   After the transition matrix is written in the canonical form $P^\star$, and the fundamental matrix N is calculated using formula (2), we can find the answers to the two questions raised earlier. If U denotes an s-termed column vector all of whose entries are 1, then the vector

$$T = NU \tag{3}$$

is called the **vector of absorption times.** The matrix

$$A = NR \tag{4}$$

is called the **matrix of absorption probabilities.** With the interpretation given in the following theorem, the entries of T and A answer our questions about absorption times and probabilities.

Theorem 2*

Let $\qquad P^\star = \left[\begin{array}{c|c} I & O \\ \hline R & Q \end{array} \right]$

be the canonical transition matrix for an absorbing Markov chain, and let

$$N = (I - Q)^{-1}$$
$$T = NU$$
$$A = NR.$$

Then:

1. The entry in row i of T represents the expected number of steps before absorption, if the system starts in the transient state i.

2. The entry in row i, column j of A represents the probability that the system will end up in the absorbing state j if it starts in the transient state i.

---

*For a proof see J. G. Kemeny and J. L. Snell, *Finite Markov Chains* (Princeton, N.J.: Van Nostrand, 1960), Ch. 3.

Example 1, Continued

Find the expected absorption times and absorption probabilities for the soap-shift problem.

Solution

Recall the canonical transition matrix

$$P^\star = \begin{array}{c} \\ A \\ C \\ B \\ D \end{array} \begin{array}{cccc} A & C & B & D \\ \left[\begin{array}{cc|cc} 1 & 0 & 0 & 0 \\ 0 & 1 & 0 & 0 \\ \hline .3 & .1 & .4 & .2 \\ .2 & .3 & .5 & 0 \end{array}\right] \end{array} = \left[\begin{array}{c|c} I & O \\ \hline R & Q \end{array}\right]$$

and the fundamental matrix

$$N = (I - Q)^{-1} = \begin{bmatrix} 2 & .4 \\ 1 & 1.2 \end{bmatrix}$$

that we computed earlier. Formula (3) now yields

$$T = NU = \begin{bmatrix} 2 & .4 \\ 1 & 1.2 \end{bmatrix}\begin{bmatrix} 1 \\ 1 \end{bmatrix} = \begin{array}{c} B \\ D \end{array}\begin{bmatrix} 2.4 \\ 2.2 \end{bmatrix}.$$

The rows of T correspond to the transient states B and D. The entries tell us the expected absorption times for each transient starting state. Thus, if a person uses brand B initially it will take 2.4 months, on the average, before he reaches one of the absorbing brands A or C, and if he uses brand D initially it will take 2.2 months, on the average.

From formula (4) we find that

$$A = NR = \begin{bmatrix} 2 & .4 \\ 1 & 1.2 \end{bmatrix}\begin{bmatrix} .3 & .1 \\ .2 & .3 \end{bmatrix} = \begin{array}{c} B \\ D \end{array}\begin{array}{cc} A & C \\ \begin{bmatrix} .68 & .32 \\ .54 & .46 \end{bmatrix} \end{array}.$$

Here the rows correspond to the transient states B and D, and the columns correspond to the absorbing states A and C. The entries represent the absorption probabilities for each transient starting state. Thus, someone using brand B initially has a 68% chance of eventually ending up using brand A, and a 32% chance of ending up with brand C; while for those who use brand D initially 54% will end up with brand A and 46% with brand C. Our earlier computer data confirms these results. ◆

Example 2

To obtain a Ph.D. in mathematics from Burkly University, a graduate student must go through four successive states:

1. Has not yet passed the qualifying exams.
2. Has passed the exams, but does not yet have a dissertation advisor.
3. Has a dissertation advisor but does not yet have the degree.
4. Has received the degree.

If we add the state

0. Has dropped out of the program

then the year-to-year transition matrix is found to be:

$$
\begin{array}{c}
\\
\text{Current} \\
\text{state}
\end{array}
\begin{array}{c}
\\ \\ \\
\end{array}
\begin{array}{c}
 & \multicolumn{5}{c}{\text{State next year}} \\
 & 0 & 1 & 2 & 3 & 4 \\
0 \\ 1 \\ 2 \\ 3 \\ 4
\end{array}
\left[
\begin{array}{ccccc}
1 & 0 & 0 & 0 & 0 \\
.4 & .2 & .4 & 0 & 0 \\
.2 & 0 & .6 & .2 & 0 \\
.1 & 0 & 0 & .8 & .1 \\
0 & 0 & 0 & 0 & 1
\end{array}
\right] = P
$$

For example, a student who has passed the exams but does not have an advisor has a 20% chance of dropping out within the year, a 60% chance of remaining in the same state, and a 20% chance of finding an advisor. Analyze this process as an absorbing Markov chain.

Solution States 0 and 4 are absorbing and states 1, 2, and 3 are transient. To write the transition matrix in canonical form we rearrange the states so that the absorbing states come first.

$$
P^\star =
\begin{array}{c}
0 \\ 4 \\ 1 \\ 2 \\ 3
\end{array}
\left[
\begin{array}{cc|ccc}
1 & 0 & 0 & 0 & 0 \\
0 & 1 & 0 & 0 & 0 \\
\hline
.4 & 0 & .2 & .4 & 0 \\
.2 & 0 & 0 & .6 & .2 \\
.1 & .1 & 0 & 0 & .8
\end{array}
\right]
=
\left[
\begin{array}{c|c}
I & O \\
\hline
R & Q
\end{array}
\right]
$$

Next, we compute

$$
I - Q =
\begin{bmatrix}
1 & 0 & 0 \\
0 & 1 & 0 \\
0 & 0 & 1
\end{bmatrix}
-
\begin{bmatrix}
.2 & .4 & 0 \\
0 & .6 & .2 \\
0 & 0 & .8
\end{bmatrix}
=
\begin{bmatrix}
.8 & -.4 & 0 \\
0 & .4 & -.2 \\
0 & 0 & .2
\end{bmatrix}
$$

and invert this matrix by the method of Section 4.3 to obtain the fundamental matrix

$$
N = (I - Q)^{-1} =
\begin{bmatrix}
1.25 & 1.25 & 1.25 \\
0 & 2.5 & 2.5 \\
0 & 0 & 5
\end{bmatrix}
$$

The absorption-time vector is therefore given by

$$
T = NU =
\begin{bmatrix}
1.25 & 1.25 & 1.25 \\
0 & 2.5 & 2.5 \\
0 & 0 & 5
\end{bmatrix}
\begin{bmatrix}
1 \\ 1 \\ 1
\end{bmatrix}
=
\begin{array}{c}
1 \\ 2 \\ 3
\end{array}
\begin{bmatrix}
3.75 \\ 5 \\ 5
\end{bmatrix}
$$

In other words, a new student (state 1) will spend 3.75 years in the program, on the average, before either dropping out or receiving the degree (we don't know which), while a student in state 2 or 3 will spend 5 years, on the average.*

The matrix of absorption probabilities is

$$
A = NR = \begin{bmatrix} 1.25 & 1.25 & 1.25 \\ 0 & 2.5 & 2.5 \\ 0 & 0 & 5 \end{bmatrix} \begin{bmatrix} .4 & 0 \\ .2 & 0 \\ .1 & .1 \end{bmatrix} = \begin{matrix} 1 \\ 2 \\ 3 \end{matrix} \begin{bmatrix} .875 & .125 \\ .75 & .25 \\ .5 & .5 \end{bmatrix}
$$

Hence, a new student (state 1) has an 87.5% chance of dropping out and only a 12.5% chance of ever obtaining the degree; while for students in state 2 or 3 the success rates are 25% and 50%, respectively. ◆

The theory of absorbing chains can be applied to Bernoulli processes in several interesting ways. The next example illustrates the use of absorbing chains in the analysis of *success runs*.

Example 3   A fair coin is tossed repeatedly until three heads in a row have occurred. How long will this take, on the average?

Solution   We say that the system is in state i at a given time ($i = 0, 1, 2,$ or 3) if the last toss completes a run of i heads; that is, i is the largest number such that the preceding i tosses all came up heads. (For instance, after the sequence THHT the system is in state 0, after THHTH it is in state 1, and so on). Since the process ends when we obtain three heads in a row, we make state 3 absorbing. The transition matrix for this process is shown below.

$$
\begin{matrix} & & \text{State after next toss} \\ & & 0 \quad\ 1 \quad\ 2 \quad\ 3 \end{matrix}
$$

$$
\begin{matrix} \text{Current} \\ \text{state} \end{matrix} \quad \begin{matrix} 0 \\ 1 \\ 2 \\ 3 \end{matrix} \begin{bmatrix} \frac{1}{2} & \frac{1}{2} & 0 & 0 \\ \frac{1}{2} & 0 & \frac{1}{2} & 0 \\ \frac{1}{2} & 0 & 0 & \frac{1}{2} \\ 0 & 0 & 0 & 1 \end{bmatrix} = P
$$

For instance, from state 2 we will either go to state 3 (if the next toss is heads) or to state 0 (if the next toss is tails). If we list the absorbing state 3 before the transient states 0, 1, and 2, we obtain the canonical form

$$
P^\star = \begin{matrix} 3 \\ 0 \\ 1 \\ 2 \end{matrix} \begin{matrix} 3\quad\ 0\quad\ 1\quad\ 2 \end{matrix} \left[\begin{array}{c|ccc} 1 & 0 & 0 & 0 \\ \hline 0 & \frac{1}{2} & \frac{1}{2} & 0 \\ 0 & \frac{1}{2} & 0 & \frac{1}{2} \\ \frac{1}{2} & \frac{1}{2} & 0 & 0 \end{array} \right] = \left[\begin{array}{c|c} I & O \\ \hline R & Q \end{array} \right]
$$

---

*It can be shown that the average new student who *does* get the degree will spend 8.75 years in the program.

The fundamental matrix is found to be

$$N = (I - Q)^{-1} = \begin{bmatrix} \frac{1}{2} & -\frac{1}{2} & 0 \\ -\frac{1}{2} & 1 & -\frac{1}{2} \\ -\frac{1}{2} & 0 & 1 \end{bmatrix}^{-1} = \begin{bmatrix} 8 & 4 & 2 \\ 6 & 4 & 2 \\ 4 & 2 & 2 \end{bmatrix}$$

The absorption-time vector is therefore

$$T = NU = \begin{bmatrix} 8 & 4 & 2 \\ 6 & 4 & 2 \\ 4 & 2 & 2 \end{bmatrix} \begin{bmatrix} 1 \\ 1 \\ 1 \end{bmatrix} = \begin{matrix} 0 \\ 1 \\ 2 \end{matrix} \begin{bmatrix} 14 \\ 12 \\ 8 \end{bmatrix}$$

Starting in state 0 we see that the expected absorption time is 14 steps—it will take an average of 14 tosses to get a run of three heads in a row.* ◆

Absorbing chains also have important applications in the theory of gambling—specifically, to the problem known as *gambler's ruin,* in which a player wants to know the probability of going broke. In the next section this topic is considered in detail (See also Exercises 13 and 14).

9.3 Exercises

1. In the Markov chain diagrams below, each arrow represents a positive transition probability. Determine whether the given three-state chain is absorbing.

(a)

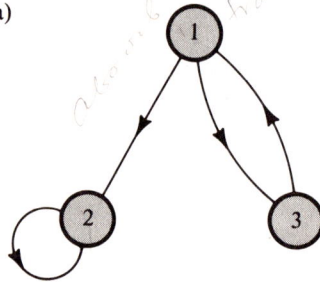

(b)

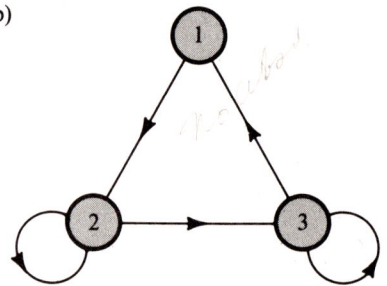

2. In the Markov chain diagrams below, each arrow represents a positive transition probability. Determine whether the given four-state chain is absorbing.

(a)

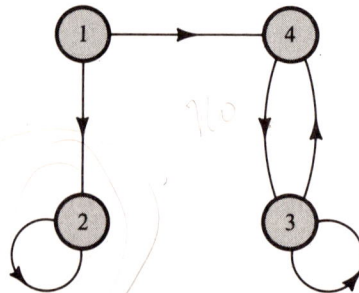

(b)

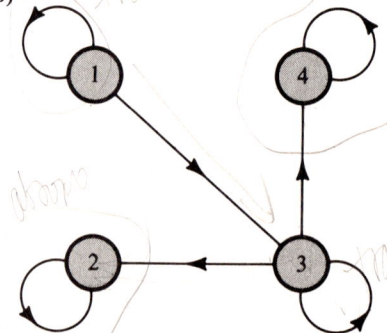

---

*The matrix of absorption probabilities is of no interest here, since the chain has only one absorbing state. The reader should verify that $A = NR = U$.

3. Consider the absorbing chain with transition diagram

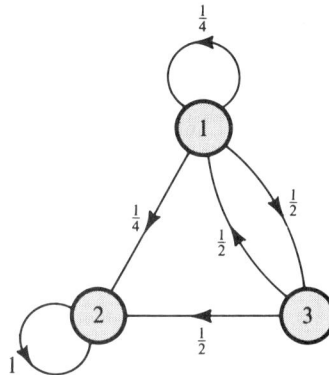

(a) How long will it take, on the average, to reach absorption if the system starts in state 1? In state 3?

(b) What is the probability that the system will end up in state 2, if it starts in state 1? If it starts in state 3?

4. Consider the absorbing chain with transition diagram

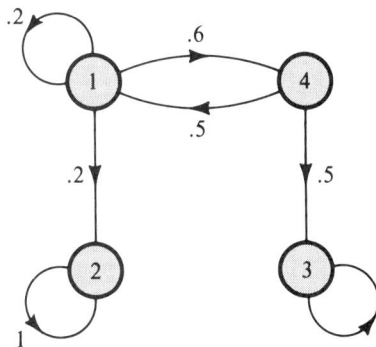

(a) How long will it take, on the average, to reach absorption if the system starts in state 1? In state 4?

(b) What is the probability that the system will end up in state 3, if it starts in state 1? If it starts in state 4?

5. Diplomatic personnel from an eastern bloc country X are randomly reassigned each year as follows: Of those stationed at home (state X), 60% remain there the next year, 30% are transferred to another eastern bloc country (state Y), and 10% are assigned to the west (state W). After one year outside the country all personnel must return home. Suppose that one of country X's diplomats currently stationed at home intends to defect as soon as he is assigned to the west. How long should he expect to wait?

6. A mouse is placed in a box with four compartments, as shown below:

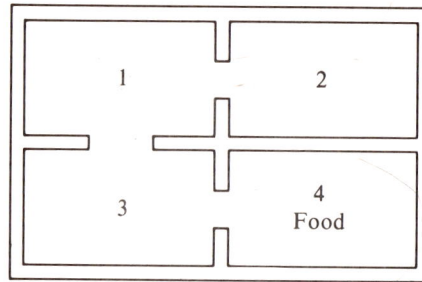

In each unit time period the mouse goes through one of the available doors, at random, until it reaches food in compartment 4. How long will this take, on the average, if the mouse starts in compartment 1? compartment 2? compartment 3?

7. A certain childhood disease follows a week-to-week transition pattern: If the patient is sick but has no fever (state S), he has a 40% chance of being in the same state the following week, 20% of developing a fever (state F), and 40% of recovering (state R). If he develops a fever, the chances are 60% that he will still have it the following week, 20% that he will revert to state S, and 20% that he will die (state D). States R and D are absorbing. If a patient is now in state S,

   (a) how many weeks will the disease last, on the average?

   (b) what is the probability that he will eventually recover?

8. Angela is a slightly stronger tennis player than Brigit. When the ball is in Angela's court (state A) she returns it to Brigit 80% of the time; the remaining 20% of the time she misses the shot and Brigit wins the point (state B★). When the ball is in Brigit's court (state B) she returns it 75% of the time; the remaining 25% of the time Angela wins the point (state A★).

   (a) How long will their volley last, on the average, if the ball starts out in Angela's court? In Brigit's court?

   (b) What is the probability that Angela will eventually win the point, if the ball starts out in her court? In Brigit's court?

9. A panel of three judges must reach a unanimous decision either to approve or reject a wiretap request. If the number voting for approval on a particular ballot is 0 or 3, the matter is settled. However, if the vote is split (either 1 to 2 or 2 to 1) there is a 20% chance that the dissenter will give in to the majority on the next ballot, a 10% chance he will win exactly one of the others over to his position, and a 70% chance the vote will remain the same on the next ballot. If the initial vote is 1 for approval and 2 against,

   (a) how many more ballots will be needed, on the average?

   (b) what is the probability that the wiretap will eventually be approved?

10. An urn initially contains three marbles, two black and one white. At each step a marble is drawn from the urn at random and replaced by one of the opposite color. The process ends as soon as all three marbles are the same color.

   (a) How many steps will this take, on the average?

(b) What is the probability that all three marbles will eventually be black?

11. A fair coin is tossed repeatedly until three heads have occurred (not necessarily together). How long will this take, on the average?

12. A player rolls two fair dice repeatedly until a sum of seven has occurred twice in a row. How many rolls will this take, on the average?

13. A player with $1 agrees to a series of games against an opponent who has $3. In each game the player has even chances of winning $1 from his opponent or losing $1 to him, and the series continues until either the player or his opponent is broke. Thus, the amount held by the player at any point has the possible values 0, 1, 2, 3, or 4, with states 0 and 4 absorbing.

(a) How many games will it take, on the average, before one of them goes broke?

(b) What is the probability that the player will go broke first?

(c) Answer parts (a) and (b) under the assumption that both the player and his opponent have $2 initially.

14. Work Exercise 13 under the assumption that in each game, the player has probability 3/4 of winning $1 from his opponent.

Computer Projects

A. The MARKOV program computes the successive state vectors of a Markov chain, given the transition matrix and initial state vector. Use this program to confirm the absorption probabilities found in Exercises 3–4 and 7–14.

B. Most versions of BASIC include matrix operations. Write a BASIC program to compute the expected absorption times and absorption probabilities for any given 5-state absorbing chain with two absorbing states and three transient states.

C. Write a BASIC program which simulates the Markov process of Example 2 in this Section (Burkly Ph.D. program). Starting in state 1, the program should randomly generate a sequence of successive states, in accordance with the given transition probabilities, until an absorbing state (0 or 4) is reached. Use this program to confirm the expected absorption times given in the text.

9.4 Gambler's Ruin

In Chapter 8 we saw how to use the concept of expected value to analyze games of chance. Even when a game is favorable, however, there is something else to worry about. The player may go broke before the "law of averages" has a chance to work in his favor. In this section we consider the classical problem known as *gambler's ruin*, first posed by Huygens and solved in its general form by Jacob Bernoulli.

The Absorbing Chain Model Consider a series of games between a player and his opponent, each of whom starts out with a known amount of money. The stakes in each game are assumed to be some fixed unit amount, like $1 or $5, and the series continues until either the player or his opponent goes broke. Specifically, it is assumed that:

1. In each game, the player either wins 1 unit with probability p, or loses 1 unit with probability $q = 1 - p$.
2. Initially the player has A units and his opponent has B units, where $A + B = N$.
3. The series ends when the player has either lost A units or won B units, whichever occurs first.

If we let the state of the system be the number of units held by the player at a given time, this process defines an absorbing Markov chain with states $0, 1, 2, \ldots, N$. The absorbing states are 0 (player is broke) and N (opponent is broke). From each transient state x ($0 < x < N$) the system goes in one step to either state $x + 1$ or $x - 1$, depending on whether the player wins (probability $= p$) or loses (probability $= q$).

In principle we could apply the method of Section 9.3 to determine the absorption probabilities and expected absorption times for this Markov chain. When N is large, though, this is clearly impractical since it would require inverting an $(N - 1) \times (N - 1)$ matrix! Fortunately there is an alternative approach, which we now describe.

Probabilities of Ruin   For each state $x = 0, 1, 2, \ldots, N$, we define

$$p_x = P(\text{player goes broke} | \text{current state is } x).$$

That is, if the player has x units at a certain point, then p_x represents his probability of eventual ruin and $1 - p_x$ the probability of his opponent's ruin. Clearly

$$p_0 = 1, \qquad p_N = 0 \tag{1}$$

because in state 0 the player is ruined, and in state N his opponent is ruined. If the player is in any other state x, then after one more game he will either be in state $x + 1$ or $x - 1$, as shown in Figure 1.

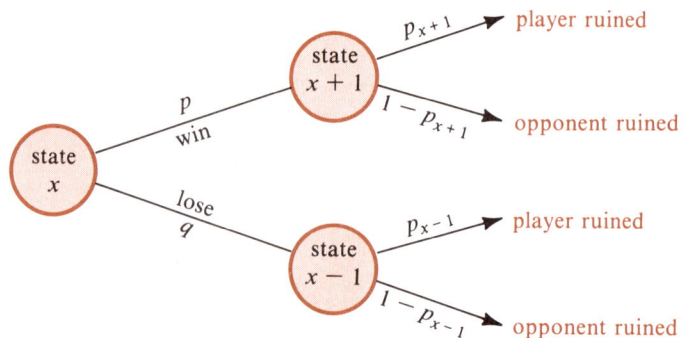

Figure 1

From state $x + 1$ the player's probability of going on to ruin is p_{x+1}, while from state $x - 1$ it is p_{x-1}, as indicated in the diagram. Thus, the probability of ruin from state x is given by

$$p_x = p \cdot p_{x+1} + q \cdot p_{x-1}. \tag{2}$$

Equation (2) is a type of condition known as a *second-order linear difference equation*. Together with the boundary conditions (1), this equation can be solved explicitly for p_x in terms of x.* When $p = q = 1/2$ the solution turns out to be

$$p_x = \frac{N - x}{N} \tag{3a}$$

while for $p \neq q$ the solution is

$$p_x = \frac{1 - r^{N-x}}{1 - r^N} \tag{3b}$$

where $r = p/q$. Substituting the initial state A for x, and recalling that $A + B = N$, we arrive at our main result:

Theorem 1

Under the assumptions (1)–(3) above, the probability of ruin for the player is given by

$$\frac{B}{A + B} \tag{4a}$$

if the game is fair ($p = q = 1/2$), and

$$\frac{1 - r^B}{1 - r^{A+B}} \tag{4b}$$

if the game is unfair, where $r = p/q \neq 1$.

Formula (4a) is intuitively reasonable. It states that in the case of a fair game the player's probability of ruin is equal to the proportion of their combined assets held by his opponent. Thus, if they play for $1 stakes and start out with $A = \$20$ and $B = \$30$, respectively, the player's probability of ruin is

$$\frac{30}{20 + 30} = \frac{3}{5}$$

which makes sense in view of the fact that the player's opponent holds three-fifths of the money.

---

*See Samuel Goldberg, *Introduction to Difference Equations* (New York: Wiley, 1958). Note also Exercises 13–14.

Applications Formula (4a) shows that casinos could make money even if their games were fair, since the assets of the house *(B)* far exceed those of the player *(A)*, and most players are unable to quit while they are ahead. The fact that casino games invariably favor the house only increases the player's probability of ruin—sometimes dramatically.

Example 1 A player with $20 decides to make a series of $1 bets on red in roulette until she either goes broke or wins $20. What is her probability of ruin?

Solution Since 18 of the 38 numbers on the roulette wheel are red,* the player's probability of winning each game is $p = 18/38$ and her probability of losing is $q = 20/38$. Hence

$$r = \frac{p}{q} = \frac{18/38}{20/38} = .9.$$

Since the player will quit if she wins $20, we can regard this amount as the house's total assets, so that

$$A = B = 20.$$

Formula (4b) then gives her probability of ruin as

$$\frac{1 - r^B}{1 - r^{A+B}} = \frac{1 - (.9)^{20}}{1 - (.9)^{40}} \doteq .892.$$

That is, she has an 89.2% chance of going broke! Note that if she makes $5 bets instead of $1 bets on each play, then initially both she and the house have $20/5 = 4$ units, so in this case her probability of ruin is

$$\frac{1 - r^B}{1 - r^{A+B}} = \frac{1 - (.9)^4}{1 - (.9)^8} \doteq .604.$$

In fact, her best strategy is to make a single $20 bet, which reduces her probability of ruin to $q = 20/38 \doteq .526$. In general, *a series of small bets is advantageous to the house,* since it gives the "law of averages" a chance to work in the house's favor. This is why casinos place a limit on the amount one can bet at any one time.   ◇

Example 2 The probability of winning ("making a pass") in craps† is 244/495, or approximately .493. If a compulsive gambler with $1000 makes a series of $100 bets on the pass line, how much must the house have to be 90% sure that he will eventually go broke?

---

*See Example 2 in Section 7.3.
†See Exercise 27 in Section 7.5.

Solution Since $100 changes hands on each play, we take this amount as our unit. Thus, the gambler starts out with $A = 1000/100 = 10$ units and the house with an unknown number B. Since $p = 244/495$ and $q = 1 - p = 251/495$, we compute

$$r = \frac{p}{q} = \frac{244/495}{251/495} = \frac{244}{251} \doteq .972.$$

For the gambler's probability of ruin to be 90%, formula (4b) shows that we must have

$$\frac{1 - r^B}{1 - r^{A+B}} = .90. \tag{5}$$

We can solve this equation for r^B, as follows:

$$1 - r^B = .9(1 - r^{A+B})$$
$$r^B - .9r^{A+B} = 1 - .9 = .1$$
$$r^B(1 - .9r^A) = .1$$

and therefore

$$r^B = \frac{.1}{1 - .9r^A} = \frac{1}{10 - 9r^A}. \tag{6}$$

Substituting $r \doteq .972$ and $A = 10$ into (6), we obtain

$$(.972)^B = \frac{1}{10 - 9(.972)^{10}} \doteq .310.$$

Trying successive values of B (or using logarithms), we find that B lies between 41 and 42. Thus, with only 42 units ($4200), the house can be 90% certain that the gambler will eventually be ruined, if he continues to play. ◆

Duration of Play The moral of our analysis is that "a fool and his money are soon parted." Moreover, we can even estimate *how* soon: Using difference equations it can be shown that the expected absorption time for the gambler's ruin is just

$$AB \tag{7a}$$

if the game is fair ($p = q = 1/2$), and

$$\frac{1 + r}{1 - r}\left[A - Nr^B\left(\frac{1 - r^A}{1 - r^N}\right)\right] \tag{7b}$$

if the game is unfair, where $r = p/q \neq 1$ and $N = A + B$.

Example 1, Revisited A player with $20 decides to make a series of bets on red in roulette until she either goes broke or wins $20. How long can she expect to play if she bets $1 at a time? $5 at a time?

Solution We saw earlier that $r = p/q = .9$. If the player bets $1 at a time then $A = B = 20$, and formula (7b) gives an expected absorption time of

$$\frac{1 + .9}{1 - .9}\left[20 - 40(.9)^{20}\left(\frac{1 - (.9)^{20}}{1 - (.9)^{40}}\right)\right] \doteq 297.6.$$

In other words, she can expect to play about 300 times, on the average. On the other hand, if she makes $5 bets then $A = B = 4$, and the expected absorption time is

$$\frac{1 + .9}{1 - .9}\left[4 - 8(.9)^4\left(\frac{1 - (.9)^4}{1 - (.9)^8}\right)\right] \doteq 15.8.$$

That is, on the average it will take about 16 plays for her to either win $20 or go broke. If an evening's entertainment is her goal, therefore, she might want to place smaller bets—or better still, leave the casino and take in a movie! ◆

9.4 Exercises

1. Work Exercise 13 of Section 9.3 using formulas (4a) and (7a) of this section.

2. Work Exercise 14 of Section 9.3 using formulas (4b) and (7b) of this section.

3. Player A has $15 and player B has $10. If they play a series of fair games at $1 a game,

   (a) what is the probability that player A will go broke before player B?

   (b) how many games will it take, on the average, for one of the two players to go broke?

4. Work Exercise 3 under the assumption that the players bet $5 a game.

5. When Abbie and Brenda play table tennis Abbie wins 60% of the time. If they bet $1 a game, what is the probability that Abbie will be able to win $5 from Brenda before losing $5 herself?

6. In Exercise 5, what is the probability that Abbie will be able to win $20 from Brenda before losing $5 herself?

7. Consider a game with an even payoff, in which the player has probability $p = .45$ of winning. What is the probability that he will win $20 before losing $50, if he bets $1 at a time? $5 at a time?

8. Work Exercise 7 under the assumption that the game is fair; that is, $p = q = 1/2$.

9. A fair coin is tossed repeatedly until the number of heads is 10 more than the number of tails, or vice-versa. How many tosses will this take, on the average? (Hint: Consider players H and T who bet $1 on the outcome of each toss.)

10. A marksman hits the target 75% of the time. His score at any point is computed by subtracting the number of misses from the number of hits. If his score is currently 5, what is the probability that it will reach 10 before it reaches 4? (Hint: Consider players H and M who bet $1 on the outcome of each shot.)

11. Consider the case of gambler's ruin in which the game is unfavorable to the player; that is, $r = p/q < 1$. Prove that the player's probability of ruin is at least $1 - r^B$, regardless of the amount A the player starts out with.

12. Suppose a player makes a series of one-unit bets on red in roulette until she either breaks the house or goes broke herself. Using Exercise 11, show that the house needs only 44 units in order to be 99% sure of ruining the player, no matter how much she has.

13. Show by direct substitution that formula (3a) in this section gives a solution of the difference equation (2) for $p = q = 1/2$, and satisfies the boundary conditions (1).

14. Show by direct substitution that formula (3b) in this section gives a solution of the difference equation (2) for $p \neq q$, and satisfies the boundary conditions (1).

Computer Projects   The RUIN program simulates a series of games played for unit stakes between a player with A units and an opponent with B units. The series ends when either the player or his opponent is ruined. This program can be used to illustrate the probabilities of ruin given in the text.

9.5 Genetics

The principles of probability have important applications to genetics, the science of heredity. In this section we outline the Mendelian theory of inheritance and show how Markov chains and other stochastic models can be applied.

The Mendelian Theory   In a famous series of experiments on garden peas, the Moravian monk Gergor Mendel (1822–84) arrived at several basic principles governing the inheritance of characteristics among living organisms. His account was published in 1866 but it was ignored by the scientific community of his day. In 1900—sixteen years after Mendel's death—his laws were rediscovered by de Vries, Correns and Tschermak.

In Mendel's theory, the basic unit of heredity is the **gene.** An inherited trait is determined by the combination of two genes, one from each parental reproductive cell. The offspring therefore carries two genes for each trait. For example, in garden peas one gene-pair determines whether the plant is tall or dwarfed, another determines whether the seeds are yellow or green, and so on. The alternative possible forms of a particular gene are called **alleles.** For those traits studied by Mendel only two alleles are possible, one of which is **dominant,** the other **recessive.** If we denote these alleles by A and a, respectively, we arrive at three possible **genotypes** for a given individual:

> AA : Both genes are A
>
> Aa : One gene is A, the other is a
>
> aa : Both genes are a

We refer to these genotypes as **pure dominant, hybrid,** and **pure recessive,** respectively. In the hybrid case (Aa) the **phenotype,** or displayed trait, is determined by the dominant allele A. For example, in garden peas the allele for yellow seed color (A) is dominant, while that for green seed color (a) is recessive. Plants with genotype AA or Aa will therefore have yellow seeds, while those of genotype aa will have green seeds.

Mendel's first law, called the **law of segregation,** states that exactly one gene from each parental gene-pair is transmitted to the offspring during reproduction; moreover, it is purely a matter of chance which gene is transmitted. Let us consider now what happens when we cross two individuals of known genotype.

Case 1: Both parents pure.   For example, if both are pure dominant (AA), the offspring must receive the A gene from both parents and therefore will again have genotype AA. In symbols:

$$AA \times AA \rightarrow AA$$

By similar reasoning we have:

$$aa \times aa \rightarrow aa$$
$$AA \times aa \rightarrow Aa$$

Case 2: One parent pure, the other hybrid.   If one parent is pure dominant (AA), say, and the other is hybrid (Aa), the offspring must receive the A gene from the AA parent but is equally likely to receive the A or a gene from the Aa parent. Thus, the offspring will have genotype AA or Aa, each with probability 1/2. Schematically:

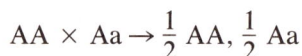

$$AA \times Aa \rightarrow \frac{1}{2} AA, \frac{1}{2} Aa$$

In other words, out of many such offspring we would expect about half to be pure dominant and half hybrid. In the same way we arrive at

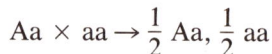

$$Aa \times aa \rightarrow \frac{1}{2} Aa, \frac{1}{2} aa$$

That is, the offspring of a hybrid and a pure recessive will be hybrid with probability 1/2 and pure recessive with probability 1/2.

Case 3: Both parents hybrid (Aa).   In this case the offspring is equally likely to receive the A or a gene from the father and, independently, the A or a gene from the mother. Thus, the probability that the offspring will receive two A genes—that is, that it will be pure dominant—is (1/2) (1/2) = 1/4. Similarly, the probability that it will be pure recessive is 1/4. To be hybrid the offspring must receive either an A gene from the father and an a gene from the mother, or vice versa. Since these cases are exclusive, the probability of a hybrid offspring is (1/2) (1/2) + (1/2) (1/2) = 1/2. Thus we have

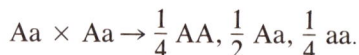

$$Aa \times Aa \rightarrow \frac{1}{4} AA, \frac{1}{2} Aa, \frac{1}{4} aa.$$

This means that, on the average, out of four offspring of hybrid parents one will be pure dominant, two will be hybrid and one will be pure recessive. Thus, three out of four will show the dominant trait.

For later reference these results are summarized in Table 1 below, where D, H, and R represent the pure dominant, hybrid, and pure recessive genotypes, respectively.

Genotypes of parents	Genotype probabilities for offspring		
	D	H	R
D × D	1	0	0
R × R	0	0	1
D × R	0	1	0
D × H	$\frac{1}{2}$	$\frac{1}{2}$	0
H × R	0	$\frac{1}{2}$	$\frac{1}{2}$
H × H	$\frac{1}{4}$	$\frac{1}{2}$	$\frac{1}{4}$

Table 1

Note that these are the only six combinations possible, since, for example, H × D represents the same cross as D × H.

Markov Chain Models In controlled breeding experiments we are interested in what happens to the genotype of the offspring after several generations. Markov chains are often useful in problems of this type.

Example 1 Albinism is a recessive trait determined by a single gene-pair. Suppose an albino mates with an individual who does not carry the recessive allele, then their first offspring mates with another noncarrier, and so on indefinitely. How many generations will it take, on the average, before the recessive allele is eliminated in the firstborn offspring?

Solution We denote the recessive allele by a, the dominant (normal) allele by A. Thus, albinos have genotype R = aa, carriers have genotype H = Aa, and noncarriers have genotype D = AA. Let the state of the system be the genotype of the firstborn offspring in the nth generation ($n = 0, 1, 2, \ldots$). The transition matrix is

$$\begin{array}{c} \text{Genotype in next generation} \\ \begin{array}{ccc} D & H & R \end{array} \end{array}$$

$$\text{Current genotype} \quad \begin{array}{c} D \\ H \\ R \end{array} \begin{bmatrix} 1 & 0 & 0 \\ \frac{1}{2} & \frac{1}{2} & 0 \\ 0 & 1 & 0 \end{bmatrix} = P$$

obtained by crossing each "current genotype" with the genotype D of a noncarrier, according to Table 1. This is an absorbing Markov chain with a single absorbing state D. The matrix P is already in canonical form (see Section 9.3), so we compute

$$I - Q = \begin{bmatrix} 1 & 0 \\ 0 & 1 \end{bmatrix} - \begin{bmatrix} \frac{1}{2} & 0 \\ 1 & 0 \end{bmatrix} = \begin{bmatrix} \frac{1}{2} & 0 \\ -1 & 1 \end{bmatrix}$$

and invert to obtain the fundamental matrix

$$N = (I - Q)^{-1} = \begin{bmatrix} 2 & 0 \\ 2 & 1 \end{bmatrix}.$$

The vector of expected absorption times is therefore

$$T = NU = \begin{bmatrix} 2 & 0 \\ 2 & 1 \end{bmatrix}\begin{bmatrix} 1 \\ 1 \end{bmatrix} = \begin{matrix} H \\ R \end{matrix}\begin{bmatrix} 2 \\ 3 \end{bmatrix}.$$

Since the expected absorption time starting from state R is 3 steps, we conclude that it will take three generations, on the average, for an albino's first-born descendants to become noncarriers.   ◆

Other Applications   In addition to Markov chains we can apply Bayes' formula, Bernoulli trials and other stochastic models to problems in genetics.

Example 2   Ethyl's brother has fibrocystic disease of the pancreas, a recessive condition determined by a single gene-pair. However, neither Ethyl nor her parents have the disease. What is the probability that Ethyl carries the recessive allele?

Solution   Since Ethyl's brother has the disease his genotype is R = aa, which means that both parents must carry the recessive allele (a). Since they do not have the disease, it follows that they both have genotype H = Aa.

For the offspring of hybrid parents, Table 1 shows that $P(D) = 1/4$, $P(H) = 1/2$ and $P(R) = 1/4$. In Ethyl's case, however, we can rule out genotype R since we know that she does not have the disease. Thus, what we want to know is *the conditional probability that her genotype is H, given that it is not R*. In symbols, this becomes

$$P(H|R^c) = \frac{P(H \cap R^c)}{P(R^c)}$$

$$= \frac{P(H)}{1 - P(R)}$$

$$= \frac{\frac{1}{2}}{1 - \frac{1}{4}} = \frac{2}{3}$$

That is, the chances are 2 in 3 that Ethyl carries the recessive allele.   ◆

The Hardy-Weinberg Principle   What happens to a recessive trait after many generations of random mating in a large population? It might seem at first glance that such a trait should gradually disappear, according to the Mendelian theory. Yet we observe many recessive characteristics in both animal and plant populations, and these traits seem to persist in fairly stable proportions from one generation to the next. To many biologists in the early 1900s, this paradox cast serious doubt on the Mendelian theory of inheritance. Then, in 1908, the English mathematician G. H. Hardy—and independently, the German physician W. Weinberg—produced a remarkably simple argument* to show that the disappearance of recessive traits is not a consequence of the Mendelian theory at all.

Consider a trait determined by a single gene-pair with two alleles, A (dominant) and a (recessive). Assume that the genotype proportions in the initial population are known. Let

$$p_0 = \text{proportion of pure dominants (AA)}$$
$$q_0 = \text{proportion of hybrids (Aa)}$$
$$r_0 = \text{proportion of pure recessives (aa)}.$$

We assume that these proportions are the same for both males and females, and that genotype has no influence on the choice of a mate or the average number of offspring produced. Under these conditions, we can calculate the probability u_0 that a random member of the first generation of offspring will receive an A allele from a given parent (e.g., the mother). The calculation is as follows:

$$u_0 = P(\text{offspring receives A})$$
$$= P(\text{parent is AA})(1) + P(\text{parent is Aa})\left(\frac{1}{2}\right)$$
$$= (p_0)(1) + (q_0)\left(\frac{1}{2}\right)$$
$$= p_0 + \frac{1}{2}q_0$$

Similarly,

$$v_0 = P(\text{offspring receives a})$$
$$= (q_0)\left(\frac{1}{2}\right) + (r_0)(1)$$
$$= \frac{1}{2}q_0 + r_0$$

(Note that $u_0 + v_0 = p_0 + q_0 + r_0 = 1$). To be pure dominant (AA) an offspring must receive an A allele from *both* parents, independently. Hence

$$P(\text{offspring is AA}) = u_0 \cdot u_0 = u_0^2.$$

Similarly

$$P(\text{offspring is aa}) = v_0 \cdot v_0 = v_0^2.$$

To be hybrid (Aa), an offspring must receive either the A allele from the mother and the a from the father, or vice versa. Since these cases are mutually exclusive, we have

$$P(\text{offspring is Aa}) = u_0 v_0 + v_0 u_0 = 2u_0 v_0.$$

---

*G. H. Hardy, "Mendelian Proportions in a Fixed Population," *Science,* N.S. vol 28 (1908), pp. 49–50.

If the population is large, these probabilities should approximate the actual genotype proportions in the first generation. That is, among offspring of the initial population we should have

$$p_1 = \text{proportion of AA} = u_0^2$$
$$q_1 = \text{proportion of Aa} = 2u_0v_0 \qquad (1)$$
$$r_1 = \text{proportion of aa} = v_0^2$$

where

$$u_0 = p_0 + \frac{1}{2}q_0$$
$$v_0 = \frac{1}{2}q_0 + r_0.$$

Equations (1) can be applied from each generation to the next. Thus, in the *second* generation of offspring we should expect

$$p_2 = \text{proportion of AA} = u_1^2$$
$$q_2 = \text{proportion of Aa} = 2u_1v_1 \qquad (2)$$
$$r_2 = \text{proportion of aa} = v_1^2$$

where

$$u_1 = p_1 + \frac{1}{2}q_1$$
$$v_1 = \frac{1}{2}q_1 + r_1.$$

But now, using the relations (1) we find that

$$u_1 = p_1 + \frac{1}{2}q_1 = u_0^2 + u_0v_0 = u_0(u_0 + v_0) = u_0$$

and similarly $v_1 = v_0$. Thus, equations (2) lead to the following unexpected conclusion:

$$p_2 = u_1^2 = u_0^2 = p_1$$
$$q_2 = 2u_1v_1 = 2u_0v_0 = q_1$$
$$r_2 = v_1^2 = v_0^2 = r_1$$

That is, *the genotype proportions in the second generation are the same as those in the first.* This result is summarized as follows:

Hardy-Weinberg Principle

In a large population with random mating, the genotype distribution will be theoretically stable after one generation.

For instance, if the initial population is 30% AA, 60% Aa and 10% aa, then $p_0 = .3$, $q_0 = .6$ and $r_0 = .1$, so that

$$u_0 = p_0 + \frac{1}{2}q_0 = .6, \qquad v_0 = \frac{1}{2}q_0 + r_0 = .4,$$

$$p_1 = u_0^2 = .36$$
$$q_1 = 2u_0v_0 = .48$$
$$r_1 = v_0^2 = .16.$$

That is, in the first generation of offspring—and forever after—the genotype proportions will be stable at 36% AA, 48% Aa and 16% aa. In each generation after the first, therefore, we can expect 16% of the population to show the recessive trait!

Note that for the genotype distribution to be stable we must have $p = u^2$, $q = 2uv$, $r = v^2$, where $u + v = 1$. If we know either p or r, we can solve for u and v and thus determine the distribution.

Example 3   A recessive condition known as phenylketonuria affects one child in 40,000. What proportion of the population are carriers (genotype Aa)?

Solution   Assuming the genotype distribution is stable, we must have

$$p = \text{proportion of AA} = u^2$$
$$q = \text{proportion of Aa} = 2uv$$
$$r = \text{proportion of aa} = v^2$$

where $u + v = 1$. The proportion of pure recessives is

$$r = v^2 = \frac{1}{40,000} = .000025.$$

Hence
$$v = \sqrt{r} = .005, \qquad u = 1 - v = .995$$

and therefore
$$q = 2uv = 2(.995)(.005) = .00995.$$

That is, about 1% of the population carry the recessive allele.   ◆

Remarks   The Hardy-Weinberg principle is only an approximation to reality, since it assumes that the genotype proportions in each generation will be exactly equal to their expected values, and that mating is completely random. We have also ignored the effects of migration, evolutionary selection, differential birth rates and genetic mutation, all of which affect the genotype distribution.

The Mendelian theory has been refined in a number of ways as a result of recent discoveries in molecular biology. For example, many genes are now known to have more than two alleles, and in such cases the dominance relations between them can be rather complicated. For that matter, even in the classical two-allele case it is by no means always true that one allele is dominant over the other. In snapdragons, for example, the three genotypes

AA, Aa and aa for blossom color produce three distinct phenotypes, or traits: AA plants have red blossoms, Aa pink and aa white.

It should also be noted that the "traits" we are interested in—even those as simple as eye-color—are generally determined not by a single gene-pair but by the interaction of many gene-pairs, which may or may not be inherited independently of each other. Moreover, an individual's phenotype is determined not only by genetic factors but also by the environment—that is, by "nurture" as well as "nature." For a simple introduction to the subject, the reader is referred to A. E. H. Emery, *Heredity, Disease and Man: Genetics in Medicine* (University of California Press, 1968).

9.5 Exercises

Exercises 1–4 refer to a trait determined by a single gene-pair, where one allele is dominant and the other is recessive.

1. A hybrid mates with a pure dominant. If they produce two offspring, what is the probability that

   (a) neither will be pure dominant?   (b) exactly one will be pure dominant?

2. Two hybrids produce a litter of four offspring. What is the probability that

   (a) none of the four will be pure recessive?

   (b) exactly one of the four will be pure recessive?

3. Sixty percent of the females in a certain population are pure dominant, 30% are hybrid, and 10% are pure recessive. If a pure recessive male mates with a female chosen at random, and their first offspring has the dominant trait, what is the probability that the female chosen was pure dominant?

4. In Example 2 in the text it was shown that Ethyl has a 2/3 chance of being a carrier, 1/3 of being a noncarrier. Suppose she marries a healthy man known to be a carrier. If their first offspring is healthy, what is the probability that Ethyl is a noncarrier?

5. Consider a population of snapdragons in which 20% have red blossoms (AA), 40% have pink blossoms (Aa) and 40% have white blossoms (aa). A plant is chosen at random and crossed with a pink-blossomed plant to produce a single offspring, which is in turn crossed with a pink-blossomed plant to produce a single offspring, and so on.

   (a) Represent this process as a three-state Markov chain, where the state at time n is the genotype of the nth offspring.

   (b) What is the probability that the second offspring will have red blossoms? Pink? White?

   (c) What is the long-run probability of each genotype?

6. In Exercise 5, suppose a snapdragon with pink blossoms is self-pollinated to produce a single offspring, which is in turn self-pollinated to produce a single offspring, and so on.

   (a) Represent this process as a three-state Markov chain.

   (b) How many generations will it take, on the average, before a pure strain (red or white) is obtained?

7. Consider a large population in which 70% are pure dominant, 20% are hybrid and 10% are pure recessive. Assuming random mating, what will the genotype distribution be in the first generation of offspring? The second generation?

8. Consider a large population in which 50% are pure dominant and 50% are hybrid. Assuming random mating, what will the genotype distribution be in the first generation of offspring? The second generation?

9. In parts of Africa, 4% of the population has sickle-cell anemia, a recessive blood disease determined by a single gene-pair. Assuming the genotype distribution is stable and mating is random, what proportion of the population are carriers (Aa)?

10. Consider a large population of rabbits in which mating is random. Suppose that 1% of the rabbits are albinos, and the genotype distribution is stable.

    (a) What is the probability that a random nonalbino rabbit carries the recessive allele?

    (b) If two random nonalbinos produce an offspring, what is the probability that it will be an albino?

11. Let p, q, and r be the proportion of AA, Aa, and aa genotypes, respectively, in a given population. Prove that under random mating, the next generation will have the same genotype distribution if and only if $q^2 = 4pr$.

12. Suppose a population contains the same proportion of pure dominants as pure recessives. Prove that under random mating, the next generation will contain 25% pure dominants, 50% hybrids and 25% pure recessives.

Exercises 13 and 14 depend on Example 2 in the text:

13. In the family tree below, males and females are represented by squares and circles, respectively.

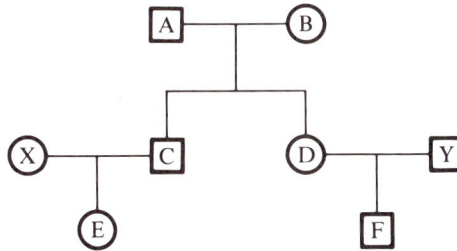

    Thus, A and B have a son C and daughter D who are married to X and Y, respectively. Note that E and F are first cousins. Suppose that C has a certain recessive trait, but D does not have it and both X and Y are known to be noncarriers.

    (a) What is the probability that E is a carrier? That F is a carrier?

    (b) In a first-cousin marriage between E and F, what is the probability that their first offspring will have the recessive trait?

14. Rework Exercise 13 under the following assumptions: (i) Neither A nor B has the recessive trait, but they are both known to be carriers, (ii) neither C nor D has the recessive trait, and (iii) both X and Y are known to be noncarriers.

Chapter 9   Summary

Important Terms

9.1 Markov chain
   transition probability
   transition diagram
   transition matrix
   state vector $X^{(n)}$

9.2 equilibrium vector
   regular Markov chain

9.3 absorbing state
   absorbing Markov chain
   transient state
   canonical transition matrix $P^{\star}$
   fundamental matrix N
   vector of expected absorption times T
   matrix of absorption probabilities A

Review Exercises

In Exercises 1–4, find the transition matrix P for the Markov chain with the given transition diagram, and determine whether the chain is regular, absorbing, or neither.

1.

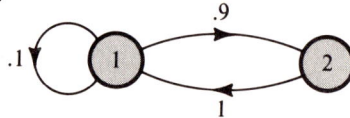

2.

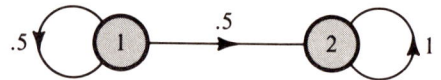

3.

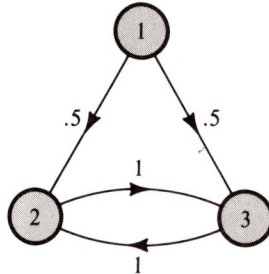

4.

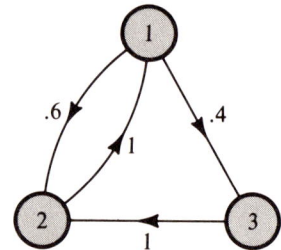

5. A three-state Markov chain has the following transition matrix:

$$
\begin{array}{c}
 \\
\text{Current} \\
\text{state}
\end{array}
\begin{array}{c}
\text{Next state} \\
\begin{array}{ccc}
1 & 2 & 3
\end{array} \\
\begin{array}{c}
1 \\
2 \\
3
\end{array}
\begin{bmatrix}
0 & 0 & 1 \\
.5 & .5 & 0 \\
.4 & .2 & .4
\end{bmatrix}
\end{array}
$$

   (a) Find the state vector after three steps (at time 3) if the system is initially in state 1.

   (b) Show that this chain is regular, and find the equilibrium vector.

6. A four-state absorbing Markov chain has the following transition matrix:

$$\begin{array}{c} \\ \text{Current} \\ \text{state} \end{array} \begin{array}{c} \\ 1 \\ 2 \\ 3 \\ 4 \end{array} \begin{array}{c} \text{Next state} \\ \begin{array}{cccc} 1 & 2 & 3 & 4 \end{array} \\ \begin{bmatrix} .2 & .4 & 0 & .4 \\ 0 & 1 & 0 & 0 \\ 0 & 0 & 1 & 0 \\ .4 & .2 & .1 & .3 \end{bmatrix} \end{array}$$

(a) How long will it take, on the average, to reach absorption if the system starts in state 1? In state 4?

(b) What is the probability that the system will end up in state 2, if it starts in state 1? If it starts in state 4?

7. Suppose that 52% of the children of doctors become doctors, while 2% of the children of nondoctors become doctors. Assuming the same birthrate for doctors and nondoctors, what proportion of the population will be doctors after many generations?

8. A man finds that whenever he calls the telephone company with a service complaint he is always put on hold. If he is on hold at any given time there is an 80% chance he will still be on hold one minute later, and a 20% chance that he will be speaking to a service representative. If he is speaking to a service representative, there is a 50% chance he will still be speaking to him or her one minute later, a 30% chance he will be back on hold, and a 20% chance he will be off the phone. On the average, how long can he expect to be on the phone?

9. Alfie and Ben repeatedly toss a fair coin. If it comes up heads Alfie wins a penny from Ben; otherwise Ben wins a penny from Alfie. If they start out with 25¢ and 40¢, respectively,

(a) what is the probability that Alfie will go broke before Ben?

(b) how many tosses will it take, on the average, before one of them goes broke?

10. Player A beats Player B at tennis 55% of the time. If they bet $5 a game, what is the probability that Player A will win $20 from Player B before losing $15 himself?

11. Albinism is a recessive trait determined by a single gene-pair. Consider a mating between two nonalbinos, both of whom carry the recessive allele (that is, both are of hybrid genotype). If they produce three offspring, what is the probability that

(a) none will be albinos (pure recessive)?

(b) all three will be noncarriers (pure dominant)?

12. Suppose that 9% of the individuals in a large population have a certain recessive trait. Assuming the genotype distribution is stable, what proportion of the population are carriers (Aa)?

10 Game Theory

Many human activities, such as warfare, politics, and business, involve competition between individuals or groups with opposing interests. In a series of papers written in the 1920s and 1930s, the brilliant Hungarian-born mathematician John von Neumann* laid the groundwork for the mathematical theory of conflict situations known as *game theory*. In this chapter we shall see how probability, matrix algebra, and linear programming can be used to find optimal strategies for the players in a certain class of games.

10.1 Strictly Determined Games

For our purposes a game can be defined as a competition involving two or more parties, or **players,** each of whom has available a number of possible courses of action, or **moves.** The **payoff** to each player depends, in general, not only on his own move but also those of the other players. We assume that each player must make his move without knowing what his opponents will do, and that the goal of each player is to maximize his own payoff.

Matrix Games The simplest class of games, called **matrix games** (or **two-person zero-sum games**), are those in which (i) there are only two players, and (ii) for each choice of moves their respective payoffs sum to zero. This means that whatever one player wins the other loses; that is, the outcome of each play will be a payment of a certain amount from one player to the other.

---

*Von Neumann (1903–57) did pioneer work in many areas of mathematics, physics, and computer science. Together with the economist Oskar Morgenstern he wrote the classic *Theory of Games and Economic Behavior* (Princeton University Press, 1944 and 1980.)

If we denote the players by R (the "row-player") and C (the "column-player"), then such a game can be characterized by a **payoff matrix,** the entries of which represent the row-player's gain (or, equivalently, the column-player's loss).

Example 1 *Two-finger Morra* Two players, R and C, each simultaneously show one or two fingers. If the sum of the numbers shown is even, R wins that many dollars from C. If the sum is odd, C wins that many dollars from R.

Discussion Here each player has two possible moves, namely, "show one finger" or "show two fingers." The payoff matrix is shown below.

$$
\begin{array}{c}
\text{Player C} \\
\begin{array}{cc} 1 & 2 \end{array} \\
\text{Player R} \begin{array}{c} 1 \\ 2 \end{array}
\begin{bmatrix} 2 & -3 \\ -3 & 4 \end{bmatrix}
\end{array}
$$

The entries represent the payoffs from the *row*-player's point of view, positive entries denoting a gain and negative entries a loss. Thus, if they both show one finger then R's payoff is 2 (R wins \$2 from C), while if R shows one finger and C shows two then R's payoff is -3 (R loses \$3 to C). Notice that the two players have directly opposing interests. The row-player wants to *maximize* the payoff entry, while the column-player wants to *minimize* it. Although it is not obvious, we shall see in the next section that this game is favorable to the column-player; that is, player C has an advantage if the game is played repeatedly. ◆

In general, any $m \times n$ matrix $A = [a_{ij}]$ can be used to define a matrix game. We simply let the rows of A represent player R's possible moves, the columns player C's possible moves, and the entries a_{ij} the payoffs from player C to player R.

Example 2 Discuss possible moves in the following 3×4 matrix game:

$$
\begin{array}{c}
\text{Player C} \\
\begin{array}{cccc} 1 & 2 & 3 & 4 \end{array} \\
\text{Player R} \begin{array}{c} 1 \\ 2 \\ 3 \end{array}
\begin{bmatrix} 0 & -4 & 2 & -1 \\ -2 & 5 & 6 & 0 \\ 3 & 0 & 1 & 4 \end{bmatrix}
\end{array}
$$

Discussion Player R has three possible moves (rows) and player C has four possible moves (columns). As before, the entries represent payoffs from player C to player R. Thus, if R chooses row 2 and C chooses column 3 then R will win 6 units from C, but if C chooses column 1 instead then C will win 2 units from R. ◆

Reduced Form   If we look at the game of Example 2 from player R's view-point there appears to be no clear superiority of any one move over another. For instance, comparing the first and second rows we see that move 1 will be better than move 2 if player C chooses column 1, but worse if he chooses column 2, 3, or 4. Comparing player C's four options, on the other hand, we see that *move 2 is definitely preferable to move 3*. The reason is that each entry in column 2 is less than or equal to the corresponding entry in column 3; so player C will always lose a smaller amount playing column 2 than he will playing column 3, *no matter what move the row-player makes*. Player C, being rational, should therefore eliminate column 3 from consideration. This gives us the reduced game shown below:

$$\begin{array}{c} & 1 & 2 & & 4 \\ 1 & \begin{bmatrix} 0 & -4 & . & -1 \\ -2 & 5 & \vdots & 0 \\ 3 & 0 & \vdots & 4 \end{bmatrix} \\ 2 \\ 3 \end{array}$$

Player R, of course, is no dummy either. Realizing that her opponent will not play column 3 and comparing her options once again, she now sees that move 3 is preferable to move 1, since each entry in row 3 is now greater than or equal to the corresponding entry in row 1. That is, she will always do better playing row 3 than playing row 1, no matter what player C does. We can thus eliminate row 1 from consideration, reducing the game still further.

$$\begin{array}{c} & 1 & 2 & & 4 \\ & \begin{bmatrix} \cdots & \vdots & \cdots \\ -2 & 5 & \vdots & 0 \\ 3 & 0 & \vdots & 4 \end{bmatrix} \\ 2 \\ 3 \end{array}$$

Comparing the columns in this reduced matrix, player C now observes that move 1 is preferable to move 4, so that column 4 can be deleted. We therefore arrive at the 2×2 matrix game

$$\begin{array}{c} & 1 & 2 \\ & \begin{bmatrix} \cdots & \vdots & \vdots \\ -2 & 5 & \vdots & \vdots \\ 3 & 0 & \vdots & \vdots \end{bmatrix} \\ 2 \\ 3 \end{array}$$

This is called the **reduced form** of the original game. Note that neither of player R's remaining moves (or player C's) is clearly preferable to the other, so the game cannot be simplified any further.

In general, to find the reduced form of a matrix game we begin by comparing the row-player's possible moves. Move i will be preferable to move j whenever the entries in row i are uniformly greater than or equal to those in row j. In this case, row j can be deleted from the payoff matrix. For the column-player, move i will be preferable to move j whenever the entries in column i are uniformly less than or equal to those in column j. In this case, column j can be deleted. By successively eliminating the interior moves of each player in turn, we finally arrive at a matrix game in reduced form.

Example 3 Find the reduced form of the matrix game

$$
\begin{array}{c}
 & \text{Player C} \\
 & \begin{array}{ccc} 1 & 2 & 3 \end{array} \\
\text{Player R} \;
\begin{array}{c} 1 \\ 2 \\ 3 \end{array}
\begin{bmatrix} 2 & 1 & 3 \\ 1 & -2 & 3 \\ 4 & 0 & -1 \end{bmatrix}
\end{array}
$$

Solution Comparing rows we see that move 1 is preferable to move 2 for player R. We therefore delete row 2.

$$
\begin{array}{c}
\begin{array}{ccc} 1 & 2 & 3 \end{array} \\
\begin{array}{c} 1 \\ \\ 3 \end{array}
\begin{bmatrix} 2 & 1 & 3 \\ \cdots\cdots\cdots \\ 4 & 0 & -1 \end{bmatrix}
\end{array}
$$

Comparing columns, we now see that move 2 is preferable to move 1 for player C, so column 1 can be deleted.

$$
\begin{array}{c}
\begin{array}{cc} 2 & 3 \end{array} \\
\begin{array}{c} 1 \\ \\ 3 \end{array}
\begin{bmatrix} \vdots & 1 & 3 \\ \cdots\cdots\cdots \\ \vdots & 0 & -1 \end{bmatrix}
\end{array}
$$

Move 1 is now preferable to move 3 for the row-player, so the matrix becomes

$$
\begin{array}{c}
\begin{array}{cc} 2 & 3 \end{array} \\
\begin{array}{c} 1 \\ \\ \end{array}
\begin{bmatrix} \vdots & 1 & 3 \\ \cdots\cdots\cdots \\ \cdots\cdots\cdots \end{bmatrix}
\end{array}
$$

Player C will now certainly prefer move 2 to move 3, so the reduced form of the game is

$$
\begin{array}{c}
\begin{array}{c} 2 \end{array} \\
\begin{array}{c} 1 \\ \\ \end{array}
\begin{bmatrix} \vdots & 1 & \vdots \\ \cdots\cdots\cdots \\ \cdots\cdots\cdots \end{bmatrix}
\end{array}
$$

Thus, assuming both players are rational, the game is completely solved. *Player R should play row 1 and player C should play column 2.* This will give player R a payoff of 1 unit per game. ◆

Minimax Optimal Moves   If a matrix game reduces to a single row and column, as in Example 3, the optimal moves of the players are completely determined. When the reduced game is not 1×1, however, we are still left with the problem of finding the best move for each player. Since the row-player has no way of anticipating his opponent's move, he might be justified in taking the pessimistic view that *no matter what he does, the column-player will make the best possible countermove*. His safest course is therefore to compare his available moves by looking at the *smallest* payoff he might receive with each one. Likewise, the column-player's safest course is to look at the *largest* payoff he might have to make for each possible move. This leads to the following definition:

Definition

The **minimax optimal moves** in a matrix game are determined as follows:

Row-player: Calculate the minimum entry in each row, and choose the row for which this value is greatest.

Column-player: Calculate the maximum entry in each column, and choose the column for which this value is least.

Example 4   Find the minimax optimal moves for each player in the following matrix game:

$$
\begin{array}{c}
 & & \text{Player C} \\
 & & \begin{array}{ccc} 1 & 2 & 3 \end{array} \\
\text{Player R} \quad \begin{array}{c} 1 \\ 2 \\ 3 \end{array} & &
\begin{bmatrix} 1 & 8 & -2 \\ 3 & 6 & 4 \\ 2 & -5 & 7 \end{bmatrix}
\end{array}
$$

Discussion   This game is in reduced form, since neither player has any move which is uniformly inferior to another. In the table below are recorded the *minimum* entry in each row and the *maximum* entry in each column.

		Player C			
		1	2	3	Min
Player R	1	1	8	−2	−2
	2	3★	6	4	3 ← best move
	3	2	−5	7	−5
Max		3	8	7	

↑
best move

The minimum entries in rows 1, 2, and 3 are −2, 3, and −5, respectively. These represent the worst payoffs player R could receive for each of his possible moves. Thus, his best move is to choose row 2, which guarantees him a payoff of at least 3 units no matter what player C does. Similarly, player C's best move is to choose column 1, which minimizes the maximum possible payoff to player R. The minimax optimal moves are therefore row 2 for player R, column 1 for player C. ◆

Notice that when the players choose their moves in this way, player R's minimum possible gain is the same as player C's maximum possible loss — namely, 3 units. This is because the entry in row 2, column 1 of the payoff matrix (starred) is both a *minimum* in its row and a *maximum* in its column. As a result, the minimax optimal solution is *stable* in the sense that each player is making the best countermove to his opponent's move. That is, neither player can do any better by unilaterally changing to another move.

Saddle Points An entry in a payoff matrix which is both a minimum in its row and a maximum in its column is called a **saddle point.** If a matrix game has a saddle point a the game is said to be **strictly determined,** and the number a is called the **value** of the game.* It is not difficult to establish the following result:

If a matrix game is strictly determined, then the minimax optimal moves determine a saddle point. This solution is stable, in the sense that neither player can benefit by changing to another move.

To find out if a given matrix game is strictly determined, we can find the minimax optimal moves and see if they determine a saddle point.

Example 5 Consider the matrix game

Player C

$$
\begin{array}{c}
 & 1 \quad\quad 2 \quad\quad 3 \\
\text{Player R} \quad \begin{array}{c} 1 \\ 2 \end{array}
\begin{bmatrix}
3 & -1 & 0 \\
-2 & 1 & -1
\end{bmatrix}
\end{array}
$$

Discussion This is a 2×3 game in reduced form. The row minima and column maxima are shown below.

		Player C				
		1	2	3	Min	
Player R	1	3	−1	0★	−1	← best move
	2	−2	1	−1	−2	
	Max	3	1	0		

↑
best move

---

*At any two saddle points the entries will be the same (see Exercise 20), so no ambiguity can arise.

The minimax optimal moves are row 1 for player R, column 3 for player C. This determines the entry 0 in row 1, column 3 (starred). However, this entry is *not* a saddle point since it is not the minimum value in its row; hence this game is not strictly determined. Note that the minimax optimal solution is unstable: If player C knows that player R intends to choose row 1 he can benefit by switching to column 2! ◆

It may happen that a player has more than one minimax optimal move. If the game is strictly determined, however, *any* pair of minimax optimal moves will determine a saddle point, and the entries at all such points will be the same.

Example 6 Find the optimal moves in the matrix game

$$
\begin{array}{c}
& \begin{array}{ccc} \text{Player C} \\ 1 & 2 & 3 \end{array} \\
\text{Player R} \begin{array}{c} 1 \\ 2 \\ 3 \\ 4 \end{array}
\left[\begin{array}{ccc}
3 & 2 & 5 \\
-1 & 0 & 6 \\
4 & 2 & 4 \\
5 & 1 & 0
\end{array}\right]
\end{array}
$$

Discussion The row minima and column maxima are shown below.

	Player C 1	2	3	Min	
1	3	2★	5	2	←
2	−1	0	6	−1	best moves
3	4	2★	4	2	←
4	5	1	0	0	
Max	5	2	6		

Player R (left label).

↑
best move

The minimax optimal moves are row 1 *or* row 3 for player R, and column 2 for player C. There are thus two entries which correspond to optimal moves (starred). Both of these entries are saddle points, so the game is strictly determined with value 2. Note that it does not matter which of his optimal moves player R makes, since his payoff is the same in either case. ◆

Applications In practical conflict situations between rational opponents, it is often difficult to specify numerical "payoffs" in a meaningful way. When this can be done, however, the concepts of game theory can be put into play.

Example 7 Two competing companies, R and C, are each planning to locate a store in one of the three towns shown in Figure 1. The population distribution for the three towns is indicated in the figure. If the companies locate in different towns, the store closer to a given town will get all that town's business. However, if they both locate in the same town, company R, which is better known, will get 60% of the total business of the three towns, company C the remaining 40%. Where should each company locate?

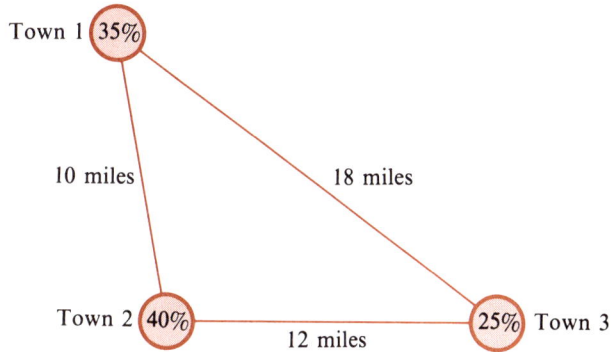

Town 1 35%

10 miles   18 miles

Town 2 40%   25% Town 3
12 miles

Figure 1

Solution   Since each company can locate in town 1, 2, or 3, the players R and C each have three possible moves. If we let company R's payoff be the percentage of the total population whose business it can expect to get, then the payoff matrix is as shown below:

Company C

	1	2	3
1	60	35	75
Company R 2	65	60	75
3	25	25	60

Thus, if R and C both locate in town 1 then R will receive 60% of the total business of the three towns, but if R locates in town 1 and C in town 3, then R will get the combined business of towns 1 and 2 (75%).*

The row minima and column maxima are calculated below.

Company C

	1	2	3	Min
1	60	35	75	35
Company R 2	65	60★	75	60 ← best move
3	25	25	60	25
Max	65	60	75	

↑
best move

*Strictly speaking, this is not a zero-sum game but a *constant-sum* game, since the payoffs of the two players always add up to 100%. Since one player's loss is the other's gain, however, the theory still applies.

The minimax optimal solution is for *both companies to locate in town 2*. Since the entry in row 2, column 2 is a saddle point the game is strictly determined, with value 60; hence, neither company can gain any advantage by locating in town 1 or 3, unless it knows for certain that the other company intends to make one of these "suboptimal" moves.   ◆

It is worth noting that we could also have solved this game by the reduction method, as in Example 3. In general, if a matrix game reduces to a 1×1 game then it is also strictly determined, so we can find its solution directly by the minimax method.

A final observation: In Section 8.4 we encountered some decision problems which bear a superficial resemblance to matrix games. There, however, the "column-player" was nature rather than a conscious opponent, so the game-theoretic approach would not properly apply. In a matrix game we always assume that *both players are free to choose among any of their available moves,* and that each will seek to make the "best" move from his own point of view.

10.1 Exercises

1. Find the reduced form of the given matrix game.

(a) $\begin{bmatrix} -2 & 3 & 1 \\ 4 & 0 & -2 \\ 2 & -5 & -2 \end{bmatrix}$ (b) $\begin{bmatrix} 4 & 0 & 3 & -1 \\ 1 & 2 & 2 & 3 \\ -1 & 0 & 4 & 1 \end{bmatrix}$

2. Find the reduced form of the given matrix game.

(a) $\begin{bmatrix} 1 & -1 \\ 0 & 5 \\ 2 & 0 \\ -1 & 2 \end{bmatrix}$ (b) $\begin{bmatrix} 0 & 2 & 5 \\ 1 & 0 & -1 \\ 2 & 3 & 4 \\ -1 & 6 & 0 \end{bmatrix}$

In Exercises 3–10, decide whether the given matrix game is strictly determined. If so, find all minimax optimal moves and the value of the game.

3. $\begin{bmatrix} 2 & 1 \\ -1 & 0 \end{bmatrix}$ 4. $\begin{bmatrix} 1 & -2 \\ -3 & 4 \end{bmatrix}$

5. $\begin{bmatrix} 4 & -1 \\ -2 & 3 \\ 1 & 0 \end{bmatrix}$ 6. $\begin{bmatrix} -1 & -3 & 3 & 0 \\ 0 & 2 & 1 & 0 \\ -1 & 4 & -3 & -2 \end{bmatrix}$

7. $\begin{bmatrix} -1 & 0 & 4 \\ 2 & 1 & 3 \\ 5 & -1 & -2 \end{bmatrix}$ 8. $\begin{bmatrix} 1 & -1 & 2 \\ 0 & 3 & -2 \\ -1 & 2 & 0 \end{bmatrix}$

9.
$$\begin{bmatrix} -1 & 2 \\ 0 & 1 \\ 1 & 0 \\ 2 & -1 \end{bmatrix}$$

10.
$$\begin{bmatrix} 2 & 5 & 4 & 2 \\ 1 & -1 & 5 & 0 \\ -1 & 7 & -2 & 1 \\ 2 & 6 & 3 & 2 \end{bmatrix}$$

In Exercises 11–16, show that the problem can be formulated as a strictly determined matrix game. Find all minimax optimal moves and the value of the game.

11. Player R has three bills, a $1, a $10, and a $50, while player C has two bills, a $5 and a $20. At a given signal each player must put one of his bills down on a table. The player who puts down the greater bill wins the amount put down by his opponent. What should each player do?

12. Player R holds a black ace (value = 1), a black 3, and a red 5, while player C holds a red 2, a red 4, and a black 6. At a given signal each player shows one of his cards. If they are the same color the higher one wins, otherwise the lower one wins. In either case, the winner collects from the loser the value of the loser's card. What card should each player show?

13. Two competing banks, R and C, are each planning to open a branch in one of two nearby towns. Town 1 contains 60% of their total population, town 2 contains 40%. If the banks locate in different towns, the one in each town will get all that town's business. If they locate in the same town bank R will get 30% of that town's business (bank C is better known) but 50% of the other town's business (bank R provides free shuttle service). Where should each bank locate?

14. Country R must decide between two costly defense systems against a possible missile attack by country C. System 1 has a 40% chance of success against land-launched missiles, 50% against sea-launched missiles, and 90% against airborne missiles. For system 2 the corresponding figures are 80%, 60%, and 70%. Country C can afford to develop only one of the three possible delivery systems (land, sea, or air). Naturally, country C wants to minimize the chance of a successful defense by country R. Which choice is optimal for each side?

15. Candidates R and C are running against each other for a senate seat. The state's 1 million voters are bitterly divided on nuclear power, with 400,000 in favor and 600,000 opposed. Among those in favor 120,000 prefer R and 280,000 prefer C, while among those opposed 330,000 prefer R and 270,000 prefer C. However, neither candidate has taken a stand on the issue yet. If they take different stands then each candidate will capture the votes of one faction, but if they take the same stand the voters who disagree with it will boycott the election. The goal of each candidate is to maximize his share of those actually voting. What stand should each candidate take?

16. The labor union R and management C of a company have reached an impasse. The union demands a 10¢ an hour wage increase, while management claims it cannot afford any increase at all. The arbitrator is known to favor a compromise of 5¢ an hour. Hence, he has asked the union to submit to him its final demand, x cents ($5 \leq x \leq 10$) and management to submit its final offer, y cents ($0 \leq y \leq 5$). He informs them that whichever figure is closer to 5 will prevail. If they both come equally close the increase will be 5¢. What is the best strategy for each side?

17. Prove: In a matrix game, every row minimum is less than or equal to every column maximum.

18. Prove: A matrix game is strictly determined if and only if the greatest row minimum is equal to the smallest column maximum.

19. Prove: If a matrix game is strictly determined, then every minimax optimal solution occurs at a saddle point, and every saddle point is minimax optimal.

20. Prove: At any two saddle points of a matrix game the entries are the same.

10.2 Games of Strategy

In a strictly determined game, we have seen that a rational player should always make the minimax optimal move. Turning our attention to non-strictly determined games, we see that the element of surprise can play a much greater role.

Strategies   The players in a matrix game always have minimax optimal moves, but when the game lacks a saddle point they may be tempted to make other moves in order to gain an advantage.

Example 1   What strategies might the players adopt in the following matrix game?

$$
\begin{array}{c}
 \\
\text{Player R}
\end{array}
\begin{array}{cc}
 & \text{Player C} \\
 & \begin{array}{cc} 1 & 2 \end{array} \\
\begin{array}{c} 1 \\ 2 \end{array} &
\begin{bmatrix} 1 & -2 \\ -3 & 4 \end{bmatrix}
\end{array}
$$

Discussion   The reader should verify that the minimax optimal moves are row 1 for player R and column 1 for player C. However, none of the four entries is a saddle point, so the game is nonstrictly determined. Because of this, the minimax optimal solution is *unstable*.

Suppose, for example, that player R repeatedly chooses his "best move," row 1. Player C will soon notice this and switch to column 2, in order to give his opponent a 2-unit loss. If player C does this consistently, though, player R will switch to row 2 to gain a 4-unit payoff. Player C will respond by choosing column 1, whereupon player R will return to row 1 and they will end up exactly where they started!

The key to the problem is the fact that whenever one side's behavior is *predictable*, the other side can take advantage of it. This suggests that the best strategy for each player might be to choose his moves in some *random* way, so his opponent cannot anticipate what he will do next. For instance, player R might adopt the strategy

$$
P \begin{cases} \text{Choose row 1 with probability .6} \\ \text{Choose row 2 with probability .4} \end{cases}
$$

which we represent by the row vector

$$P = [.6 \quad .4].$$

This means that player R wants to mix up his two moves, playing the first row 60% of the time and the second row 40% of the time. Player C might adopt the strategy

$$Q \quad \begin{cases} \text{Choose column 1 with probability .2} \\ \text{Choose column 2 with probability .8} \end{cases}$$

which we represent by the column vector

$$Q = \begin{bmatrix} .2 \\ .8 \end{bmatrix}.$$

This means player C wants to choose the first column 20% of the time and the second column 80% of the time. There are various randomizing devices each player could use to implement these strategies. For example, player R could choose a random one-digit number from 0 to 9 before each move, playing row 1 if the number is less than 6 and row 2 otherwise. Likewise, player C could choose his own random one-digit number, playing column 1 if the number is less than 2 and column 2 otherwise. This way neither player could predict the other's next move on the basis of earlier moves.   ◇

We can generalize this notion of strategy to apply to any matrix game.

Definition

In an $m \times n$ matrix game, a **strategy for player R** is a row vector

$$P = [p_1 \quad p_2 \quad \cdots \quad p_m]$$

and a **strategy for player C** is a column vector

$$Q = \begin{bmatrix} q_1 \\ q_2 \\ \vdots \\ q_n \end{bmatrix}$$

where in each case the entries are nonnegative and sum to 1.

The idea is that player R will choose row i with probability P_i $(i = 1, \ldots, m)$, while player C will choose column j with probability q_j $(j = 1, \ldots, n)$. A player who always makes the same move is said to follow a **pure strategy.** In this case the vector P or Q will have all entries 0 except for a single entry 1. A strategy which is not pure is called **mixed.** Thus, in a 2×2 game,

$$P = [.6 \quad .4], \qquad Q = \begin{bmatrix} .2 \\ .8 \end{bmatrix}$$

are both mixed strategies, while the strategies

$$P = [.5 \quad .5], \qquad Q = \begin{bmatrix} 0 \\ 1 \end{bmatrix}$$

are mixed and pure, respectively.

Expected Payoff When a matrix game is played repeatedly, the payoff (from player C to player R) will vary from one trial to the next depending on the actual moves made. For given strategies P and Q we can calculate the *expected payoff,* which represents the average amount the row-player can expect to win per game, over the long run.

Example 1, Continued   Find the expected payoff in the game

$$A = \begin{bmatrix} 1 & -2 \\ -3 & 4 \end{bmatrix}$$

when players R and C use the respective strategies

$$P = [.6 \quad .4], \qquad Q = \begin{bmatrix} .2 \\ .8 \end{bmatrix}$$

Solution   The row-player's payoff is a random variable with four possible values: 1, -2, -3, and 4. The probability of each payoff is shown in the table below.

Payoff	Probability
1	$(.6)(.2) = .12$
-2	$(.6)(.8) = .48$
-3	$(.4)(.2) = .08$
4	$(.4)(.8) = .32$

For instance, a payoff of -2 can occur only if player R chooses row 1 (probability $= .6$) and player C chooses column 2 (probability $= .8$). Since the players' choices are made independently, we have

$$\begin{aligned} P(\text{payoff} = -2) &= P(\text{row } 1 \cap \text{column } 2) \\ &= P(\text{row } 1) \cdot P(\text{column } 2) \\ &= (.6)(.8) = .48. \end{aligned}$$

The other probabilities are calculated similarly. The expected payoff is therefore

$$\begin{aligned} E &= 1(.12) + (-2)(.48) + (-3)(.08) + 4(.32) \\ &= .12 - .96 - .24 + 1.28 \\ &= .2. \end{aligned}$$

In other words, player R stands to win .2 units per game, over the long run.   ◇

The expected payoff can be found, it turns out, simply by multiplying together the matrices P, A, and Q, in that order. First, compute

$$PA = [.6 \quad .4] \begin{bmatrix} 1 & -2 \\ -3 & 4 \end{bmatrix} = [-.6 \quad .4]$$

and then

$$PAQ = [-.6 \quad .4] \begin{bmatrix} .2 \\ .8 \end{bmatrix} = [.2].$$

Ignoring the brackets on this 1×1 matrix we recognize the expected payoff $E = .2$. We leave it for the reader to verify that this works for any 2×2 game, no matter what strategies the players use (see Exercise 23). More generally, we have the following rule:

Let A be the payoff matrix for an $m \times n$ game in which the row and column players use strategies P and Q, respectively. Then the **expected payoff** to the row-player is given by

$$E = PAQ. \tag{1}$$

Example 2 Suppose both players make their moves completely at random in the matrix game

$$\begin{array}{cc} & \text{Player C} \\ & \begin{array}{ccc} 1 & 2 & 3 \end{array} \\ \text{Player R} \begin{array}{c} 1 \\ 2 \end{array} & \begin{bmatrix} 3 & -5 & 4 \\ -2 & 2 & -3 \end{bmatrix}. \end{array}$$

Which side will have the advantage?

Solution Since player R chooses each row with probability 1/2, while player C chooses each column with probability 1/3, their respective strategies are

$$P = \begin{bmatrix} \frac{1}{2} & \frac{1}{2} \end{bmatrix}, \qquad Q = \begin{bmatrix} \frac{1}{3} \\ \frac{1}{3} \\ \frac{1}{3} \end{bmatrix}.$$

By the result stated above, the expected payoff to player R is

$$E = PAQ = \begin{bmatrix} \frac{1}{2} & \frac{1}{2} \end{bmatrix} \begin{bmatrix} 3 & -5 & 4 \\ -2 & 2 & -3 \end{bmatrix} \begin{bmatrix} \frac{1}{3} \\ \frac{1}{3} \\ \frac{1}{3} \end{bmatrix}$$

$$= \begin{bmatrix} \frac{1}{2} & -\frac{3}{2} & \frac{1}{2} \end{bmatrix} \begin{bmatrix} \frac{1}{3} \\ \frac{1}{3} \\ \frac{1}{3} \end{bmatrix} = -\frac{1}{6}.$$

Thus, player C will have the advantage, with an expectation of 1/6 unit per game over the long run.   ◆

Minimax Optimal Strategies   For each strategy P of the row-player there will be a best counterstrategy Q, the one for which the expected payoff $E = PAQ$ is least. Similarly, for each of the column-player's strategies Q there will be a best counterstrategy P, the one for which the expected payoff is greatest. A reasonable goal for player R is to try to *maximize* the expected payoff against C's best counterstrategy. Likewise, player C should try to *minimize* the expected payoff against R's best counterstrategy. Our definition of "minimax optimal moves" can therefore be generalized.

Definition

The **minimax optimal strategies** in a matrix game are determined as follows:

Row-player: For each strategy P compute the minimum value of PAQ over all strategies Q. Choose the P for which this value is greatest.

Column-player: For each strategy Q compute the maximum value of PAQ over all strategies P. Choose the Q for which this value is least.

In 1928, John von Neumann discovered a remarkable fact: *in any matrix game, the minimax optimal strategies yield a stable solution*. That is, each is the best counterstrategy to the other, and neither side can benefit by unilaterally changing to another strategy, whether pure or mixed.

To state this more precisely, let us denote the minimax optimal strategies by $P^\star$ and $Q^\star$. The statement that $P^\star$ is the best counterstrategy to $Q^\star$ means that the value of $PAQ^\star$ is greatest when $P = P^\star$. That is,

$$PAQ^\star \le P^\star AQ^\star \quad \text{for all strategies } P.$$

Similarly, the statement that $Q^\star$ is the best counterstrategy to $P^\star$ becomes

$$P^\star AQ^\star \le P^\star AQ \quad \text{for all strategies } Q.$$

Von Neumann's result may be stated as follows:

Minimax Theorem   (Theorem 1)

Let $P^\star$ and $Q^\star$ be minimax optimal strategies for players R and C in a matrix game. If R employs strategy $P^\star$, his expected payoff will be at least $P^\star AQ^\star$ no matter what strategy C adopts. Similarly, if C employs strategy $Q^\star$, his expected payoff to R will be no more than $P^\star AQ^\star$ for any strategy that R adopts.

In terms of matrices, this theorem may be stated as

$$PAQ^\star \le P^\star AQ^\star \le P^\star AQ \tag{2}$$

for all strategies P and Q. Hence, the solution $P^\star AQ^\star$ is stable: neither side can benefit by unilaterally adopting a different strategy.

The converse of the Minimax Theorem is also true (and much easier to prove): If condition (2) holds for all P and Q, then $P^\star$ and $Q^\star$ are minimax optimal strategies. The number

$$v = P^\star A Q^\star$$

is called the **value** of the game. It represents the row-player's expected payoff when both sides use their optimal strategies. If $v = 0$ the game is **fair,** since neither player has a long-run advantage.

Solution of 2 × 2 Games   How do we find minimax optimal strategies for the players in an arbitrary matrix game? Let us first consider the simplest case of a 2 × 2 game. Here the payoff matrix has the form

$$\begin{bmatrix} a & b \\ c & d \end{bmatrix}.$$

If the game is strictly determined, we can solve it by the method of Section 10.1. In this case the minimax optimal strategies $P^\star$ and $Q^\star$ are just the *pure* strategies corresponding to the minimax optimal moves. For the nonstrictly determined case we have the following theorem:

Theorem 2

Consider a 2 × 2 game

$$A = \begin{bmatrix} a & b \\ c & d \end{bmatrix}$$

which is not strictly determined. Let

$$p_1^\star = \frac{d - c}{a + d - b - c}, \qquad p_2^\star = \frac{a - b}{a + d - b - c} \tag{3a}$$

$$q_1^\star = \frac{d - b}{a + d - b - c}, \qquad q_2^\star = \frac{a - c}{a + d - b - c}. \tag{3b}$$

Then

$$P^\star = [p_1^\star \quad p_2^\star] \quad \text{and} \quad Q^\star = \begin{bmatrix} q_1^\star \\ q_2^\star \end{bmatrix}$$

are minimax optimal strategies, and the value of the game is

$$v = \frac{ad - bc}{a + d - b - c}. \tag{3c}$$

Proof   From the assumption that the game is nonstrictly determined it can be shown that the entries on one diagonal of A are both greater than those on the other: That is, either a and d are both greater than b and c, or vice versa (see Exercise 22). Hence the denominator in formulas (3a), (3b), and (3c) cannot be zero, and the given values of $p_1^\star$, $p_2^\star$, $q_1^\star$, and $q_2^\star$ are all positive.

Since also $p_1^\star + p_2^\star = 1$ and $q_1^\star + q_2^\star = 1$, the vectors $P^\star$ and $Q^\star$ are strategies. It is easily verified that for any strategies

$$P = [p_1 \quad p_2], \qquad Q = \begin{bmatrix} q_1 \\ q_2 \end{bmatrix}$$

we have $\qquad\qquad PAQ^\star = v = P^\star AQ.$

In particular, $P^\star AQ^\star = v$, and therefore inequality (2) is valid for all P and Q (since all three terms equal v). By the remark following Theorem 2 we conclude that $P^\star$ and $Q^\star$ are minimax optimal and the value of the game is v.

Example 1, Continued Solve the matrix game

Player C
  1   2

Player R $\quad \begin{matrix} 1 \\ 2 \end{matrix} \begin{bmatrix} 1 & -2 \\ -3 & 4 \end{bmatrix}$

Solution The game is nonstrictly determined, and formulas (3a) and (3b) yield

$$p_1^\star = \frac{4 - (-3)}{1 + 4 - (-2) - (-3)} = \frac{7}{10}, \qquad p_2^\star = \frac{1 - (-2)}{1 + 4 - (-2) - (-3)} = \frac{3}{10},$$

$$q_1^\star = \frac{4 - (-2)}{1 + 4 - (-2) - (-3)} = \frac{6}{10}, \qquad q_2^\star = \frac{1 - (-3)}{1 + 4 - (-2) - (-3)} = \frac{4}{10}.$$

Thus, the optimal strategies for players R and C are

$$P^\star = [.7 \quad .3] \qquad \text{and} \qquad Q^\star = \begin{bmatrix} .6 \\ .4 \end{bmatrix}$$

respectively, and the value of the game is

$$v = \frac{1 \cdot 4 - (-2)(-3)}{1 + 4 - (-2) - (-3)} = \frac{-2}{10} = -.2.$$

In other words, player C can expect to win .2 units per game from player R, over the long run, when both sides use their optimal strategies. ◆

Example 3 Solve the game of *two-finger Morra* (Section 10.1, Example 1).

Player C
  1   2

Player R $\quad \begin{matrix} 1 \\ 2 \end{matrix} \begin{bmatrix} 2 & -3 \\ -3 & 4 \end{bmatrix}$

Solution Since none of the four entries is a saddle point, this is a nonstrictly determined 2×2 game. Using Theorem 2, the optimal strategies are found to be

$$P^\star = \begin{bmatrix} \frac{7}{12} & \frac{5}{12} \end{bmatrix} \qquad \text{and} \qquad Q^\star = \begin{bmatrix} \frac{7}{12} \\ \frac{5}{12} \end{bmatrix}.$$

That is, both players should choose their first move 7/12 of the time (about 58.3%). By formula (3c), the value of the game is

$$v = \frac{2 \cdot 4 - (-3)(-3)}{2 + 4 - (-3) - (-3)} = -\frac{1}{12}$$

which shows that player C has the advantage. ◆

If a higher-dimensional matrix game can be reduced to a 2 × 2 game then Theorem 2 can be applied to find its solution.

Example 4   A smuggler can cross a mountain border through one of three passes, 1, 2, or 3. His chances of getting through undetected are 90%, 80%, and 70%, respectively. However, when the border police reinforce their patrol the smuggler's chances drop to 50% for pass 1, 70% for pass 2, and 60% for pass 3. Both the smuggler and the police know that there are only enough men to reinforce the patrol at one pass. What is the best strategy for each side?

Solution   The smuggler (player R) can attempt to cross through pass 1, 2, or 3, and the border police (player C) can reinforce the patrol at pass 1, 2, or 3. If we let the payoff be the smuggler's chance (%) of getting through, we arrive at a 3 × 3 matrix game:

Border police C

$$\begin{array}{c} & \begin{array}{ccc} 1 & 2 & 3 \end{array} \\ \text{Smuggler R} \quad \begin{array}{c} 1 \\ 2 \\ 3 \end{array} & \left[\begin{array}{ccc} 50 & 90 & 90 \\ 80 & 70 & 80 \\ 70 & 70 & 60 \end{array} \right] \end{array}$$

Comparing rows, we see that move 2 is preferable to move 3 for player R, so row 3 can be deleted:

$$\begin{array}{c} & \begin{array}{ccc} 1 & 2 & 3 \end{array} \\ \begin{array}{c} 1 \\ 2 \end{array} & \left[\begin{array}{ccc} 50 & 90 & 90 \\ 80 & 70 & 80 \\ \cdot\cdot\cdot\cdot\cdot\cdot\cdot\cdot \end{array} \right] \end{array}$$

Now move 1 (or 2) is preferable to move 3 for player C, so the game reduces to:

$$\begin{array}{c} & \begin{array}{cc} 1 & 2 \end{array} \\ \begin{array}{c} 1 \\ 2 \end{array} & \left[\begin{array}{cc} 50 & 90 \\ 80 & 70 \end{array} \right] \end{array}$$

This is a nonstrictly determined 2 × 2 game, which we can solve using Theorem 2. Formulas (3a), (3b), and (3c) yield

$$p_1^\star = 1/5, \qquad p_2^\star = 4/5$$
$$q_1^\star = 2/5, \qquad q_2^\star = 3/5$$
$$v = 74.$$

Thus, since move 3 is inferior for both sides, the optimal strategies are

$$P^\star = \begin{bmatrix} \frac{1}{5} & \frac{4}{5} & 0 \end{bmatrix} \quad \text{and} \quad Q^\star = \begin{bmatrix} \frac{2}{5} \\ \frac{3}{5} \\ 0 \end{bmatrix}$$

and the value of the game is 74, that is, the smuggler can expect to succeed 74% of the time, over the long run. ◆

In general, given a higher-dimensional game we first check whether it is strictly determined. If so, the game will have a solution in *pure* strategies, corresponding to the minimax optimal moves. If not, we can check to see whether the game reduces to a 2 × 2 game, in which case Theorem 2 applies. If the reduced game is not 2 × 2 we can solve it using linear programming, as we shall see in the next section.

10.2 Exercises

1. Which of the following are strategies for player R in a 3 × 3 matrix game? Which are pure and which are mixed?

(a) [.2   .3   .5]

(b) [.4   .2   .7]

(c) [0   1   0]

(d) $\begin{bmatrix} \frac{1}{2} & \frac{1}{2} & 0 \end{bmatrix}$

2. Which of the following are strategies for player C in a 3 × 3 matrix game? Which are pure and which are mixed?

(a) $\begin{bmatrix} .9 \\ .1 \\ 0 \end{bmatrix}$ (b) $\begin{bmatrix} 0 \\ 0 \\ 1 \end{bmatrix}$ (c) $\begin{bmatrix} \frac{1}{2} \\ \frac{1}{6} \\ \frac{1}{3} \end{bmatrix}$ (d) $\begin{bmatrix} 0 \\ \frac{1}{2} \\ \frac{1}{4} \end{bmatrix}$

3. In the matrix game

$$\begin{bmatrix} 4 & -3 & 6 \\ -1 & 7 & -4 \end{bmatrix}$$

suppose the row-player uses the strategy [.7   .3]. Which of the following strategies is better for the column-player?

(a) $\begin{bmatrix} .4 \\ .1 \\ .5 \end{bmatrix}$ (b) $\begin{bmatrix} .8 \\ .2 \\ 0 \end{bmatrix}$

4. In the matrix game of Exercise 3, suppose the column-player uses the strategy

$$\begin{bmatrix} .3 \\ .1 \\ .6 \end{bmatrix}.$$

Which of the following strategies is better for the row-player?

(a) [.8   .2]   (b) [.5   .5]

5. In the matrix game

$$\begin{bmatrix} -5 & 2 \\ 7 & -4 \end{bmatrix}$$

suppose that (i) player R rolls two fair dice, playing row 1 if the sum comes up seven and row 2 otherwise; (ii) player C tosses a fair coin, playing column 1 if it comes up heads and column 2 otherwise. Which side will have the advantage?

6. In the matrix game

$$\begin{bmatrix} 1 & -6 & 7 \\ -3 & 6 & -5 \end{bmatrix}$$

suppose that (i) player R tosses a fair coin twice, playing row 1 if both tosses come up heads and row 2 otherwise; (ii) player C chooses a card at random from a standard deck, playing column 1 if it is a spade, column 2 if it is a club, and column 3 otherwise. Which side will have the advantage?

In Exercises 7–14, solve the given matrix game by finding the optimal strategies $P^\star$ and $Q^\star$ and the value v.

7. $\begin{bmatrix} 2 & -2 \\ -1 & 3 \end{bmatrix}$

8. $\begin{bmatrix} 0 & 2 \\ 3 & 0 \end{bmatrix}$

9. $\begin{bmatrix} -2 & 3 \\ 1 & 2 \end{bmatrix}$

10. $\begin{bmatrix} -5 & 1 \\ 4 & -2 \end{bmatrix}$

11. $\begin{bmatrix} 3 & -2 & -3 \\ -2 & 3 & 2 \end{bmatrix}$

12. $\begin{bmatrix} 0 & 2 & 5 & -1 \\ 3 & -4 & -1 & 2 \end{bmatrix}$

13. $\begin{bmatrix} 2 & -3 & -5 \\ -1 & 4 & 4 \\ -3 & 2 & 3 \end{bmatrix}$

14. $\begin{bmatrix} -5 & 3 & -2 \\ 4 & 2 & 1 \\ 5 & -4 & 0 \end{bmatrix}$

15. At a given signal player C puts down either a nickel or a dime on the table and, simultaneously, player R guesses which coin was put down. If he guesses correctly player R wins the coin; otherwise player C keeps the coin and wins that amount from player R. Find the best strategy for each player. Is the game fair?

16. Player R holds a black ace (value = 1), a red 3, and a black 5, while player C holds a black 2 and a red 4. At a given signal each player shows one of his cards. If they are the same color the higher one wins, otherwise the lower one wins. In either case, the winner collects from the loser the value of the loser's card. Find the best strategy for each player. Is the game fair?

17. Competing TV stations R and C always schedule a movie or a sports even on Tuesday nights. The schedule is announced one month in advance and cannot be changed. When both stations offer the same type of program, R gets 60% of the audience and C the rest. However, when one offers a movie and the other offers

a sports event, the movie gets 55% of the audience and the sports event 45%. Find the best strategy for each station and the expected audience shares.

18. Ambassador R always travels in a convoy of two identical cars, to confuse guerrilla group C. She knows that if the lead car is attacked and she is in it she will be captured, but if she is in the second car she will have a 60% chance of escape. On the other hand, if the second car is attacked she will be captured if she is in it, but she will definitely escape if she is in the lead car. Find the best strategy for each side, and the ambassador's expected chance of escape.

19. Robin Hood and his band conduct regular monthly raids, on either village 1 or village 2. The Sheriff of Nottingham can defend only one of the two villages, as both he and Robin are well aware. The average raid on village 1 yields $900 worth of spoils, while the average raid on village 2 yields $1100 worth. However, if the band raids a village which is defended it must retreat, with no gain. Find the best strategy for each side, and the band's expected spoils.

20. On first down, the Raiders always call either a draw play or a pass play against their rivals, the Chiefs. The average gain depends on whether the Chiefs use man-to-man coverage, a zone defense, or a pass rush—a decision the Chiefs can make just before the play begins. The Raider coach knows that, on the average, a draw play will gain 4 yards against man-to-man coverage, 4 yards against a zone defense, and 7 yards against a pass rush. For a pass play, the corresponding figures are 6 yards, 5 yards, and 3 yards. Find the best strategy for each team, and the expected gain for the Raiders.

In Exercises 21–24, let

$$A = \begin{bmatrix} a & b \\ c & d \end{bmatrix}$$

be the payoff matrix for a 2 × 2 matrix game.

21. Prove: If $a = b$, the game is strictly determined.

22. Prove: If the game is nonstrictly determined, then either a and d are both greater than b and c, or b and c are both greater than a and d. (Hint: Consider whether $a > b$ or $a < b$; see Exercise 21.)

23. Prove: If the players use strategies

$$P = [p_1 \quad p_2] \quad \text{and} \quad Q = \begin{bmatrix} q_1 \\ q_2 \end{bmatrix}$$

then the expected payoff is given by $E = PAQ$.

24. Prove: If the entry a is a saddle point, then

$$P^\star = [1 \quad 0] \quad \text{and} \quad Q^\star = \begin{bmatrix} 1 \\ 0 \end{bmatrix}$$

are minimax optimal strategies. (Hint: Verify condition (2) in the text.)

Computer Projects	The GAME program simulates 2000 plays of an $m \times n$ matrix game between players using prescribed strategies P and Q. The program computes the row-player's average payoff. This program can be used to confirm the theoretical expected payoff in Exercises 3–20.

10.3 Simplex Solution of $m \times n$ Games

In this section we present a method for solving any matrix game, using linear programming.

Games with Positive Value We shall see shortly that the problem of finding a minimax optimal strategy in a matrix game can be expressed as a linear program, provided the game has positive value v. If *at least one row of the payoff matrix contains all positive entries* this will clearly be the case, since player R can choose this row repeatedly and be assured of a positive payoff no matter what player C does. If no such row exists we can simply add a suitably large constant k to each payoff entry. This increases the value of the game by k, but the optimal strategies $P^\star$ and $Q^\star$ remain the same.

Example 1 Convert the following into a game with positive value and find the optimal strategies.

$$
\begin{array}{c}
\text{Player C} \\
1 2
\end{array}
$$

$$
\text{Player R}\quad
\begin{array}{c}
1 \\ 2
\end{array}
\begin{bmatrix}
-2 & 1 \\
3 & -4
\end{bmatrix}
$$

Discussion If we add $k = 3$ to each payoff entry we obtain a matrix A with a positive row.

$$
A = \begin{bmatrix} -2+3 & 1+3 \\ 3+3 & -4+3 \end{bmatrix} = \begin{bmatrix} 1 & 4 \\ 6 & -1 \end{bmatrix}
$$

This new game A is obviously favorable to player R, since he can receive at least 1 unit playing row 1, no matter what player C does. In fact, using our 2×2 formulas we find that the original game has value $-1/2$, while the game A has value $v = 5/2 = -1/2 + 3$. The reader can verify that the optimal strategies are the same for both games, namely

$$
P^\star = \begin{bmatrix} \frac{7}{10} & \frac{3}{10} \end{bmatrix} \quad \text{and} \quad Q^\star = \begin{bmatrix} \frac{1}{2} \\ \frac{1}{2} \end{bmatrix} \quad \diamondsuit
$$

In the same way, any $m \times n$ matrix game can be converted into a game A with positive value v, in which the optimal strategies remain unchanged.

Matrix Games and Linear Programs   To explain the connection between matrix games and linear programs, let us consider the simplest case: a 2×2 game

$$A = \begin{bmatrix} a & b \\ c & d \end{bmatrix}$$

with positive value v. Suppose we analyze the game first from the *column-player's* viewpoint: For each possible strategy Q, player C is interested in the expected payoff against R's best counterstrategy, that is, the maximum value of PAQ over all strategies P. If we denote this quantity by v_Q, then player C's goal is to minimize v_Q. Now,

$$PAQ = \begin{bmatrix} p_1 & p_2 \end{bmatrix} \begin{bmatrix} a & b \\ c & d \end{bmatrix} \begin{bmatrix} q_1 \\ q_2 \end{bmatrix}$$

$$= \begin{bmatrix} p_1 & p_2 \end{bmatrix} \begin{bmatrix} aq_1 + bq_2 \\ cq_1 + dq_2 \end{bmatrix}$$

$$= p_1(aq_1 + bq_2) + p_2(cq_1 + dq_2).$$

It is not hard to show that as P ranges over all strategies, this expression takes on all values between $aq_1 + bq_2$ and $cq_1 + dq_2$ (see Exercise 19). Hence the maximum value of PAQ over all strategies P is just the larger of the numbers $aq_1 + bq_2$ and $cq_1 + dq_2$; that is,

$$v_Q = \max \{aq_1 + bq_2, cq_1 + dq_2\}.$$

When Q is minimax optimal, v_Q will assume its minimum value v. Thus, the value of the game is the least number v for which

$$aq_1 + bq_2 \leq v$$
$$cq_2 + dq_2 \leq v$$

for all strategies Q. In other words, our search for the optimal strategy Q can be stated in the form:

> Minimize v
> subject to:
> $$aq_1 + bq_2 \leq v$$
> $$cq_1 + dq_2 \leq v.$$
> $$(q_1 + q_2 = 1; q_1, q_2 \geq 0)$$

This is almost what we want, but not quite. First, we divide the inequalities by v to obtain

$$a\frac{q_1}{v} + b\frac{q_2}{v} \le 1$$

$$c\frac{q_1}{v} + d\frac{q_2}{v} \le 1.$$

(Here is where we use the assumption that $v > 0$.) Second, from the conditions $q_1 + q_2 = 1$, $q_1 \ge 0$, $q_2 \ge 0$ we see that

$$\frac{q_1}{v} + \frac{q_2}{v} = \frac{1}{v}, \qquad \frac{q_1}{v} \ge 0, \qquad \frac{q_2}{v} \ge 0.$$

If we introduce new variables

$$y_1 = \frac{q_1}{v}, \qquad y_2 = \frac{q_2}{v}, \qquad F = \frac{1}{v}$$

and observe that minimizing v is the same as maximizing $F = 1/v$, the problem becomes:

$$\text{Maximize} \quad F = y_1 + y_2$$
subject to:
$$ay_1 + by_2 \le 1 \qquad (1)$$
$$cy_1 + dy_2 \le 1.$$
$$(y_1, y_2 \ge 0)$$

This linear program can be solved for y_1, y_2 and F, and the values v, q_1, q_2 obtained using the relations

$$v = \frac{1}{F}, \qquad q_1 = vy_1, \qquad q_2 = vy_2. \qquad (2)$$

Example 1,
Continued
For the game with payoff matrix

$$A = \begin{bmatrix} 1 & 4 \\ 6 & -1 \end{bmatrix}$$

use linear programming to find the value v and the optimal strategy $Q^\star$ for player C.

Solution
Since the matrix has a positive row we know that $v > 0$. The associated linear program (1) is:

$$\text{Maximize} \quad F = y_1 + y_2$$
subject to:
$$y_1 + 4y_2 \le 1$$
$$6y_1 - y_2 \le 1.$$
$$(y_1, y_2 \ge 0)$$

Introducing slack variables s_1, $s_2 \geq 0$ we can write the program in equational form as

$$\text{Maximize} \quad F = y_1 + y_2$$
subject to:
$$y_1 + 4y_2 + s_1 \qquad = 1$$
$$6y_1 - y_2 \qquad + s_2 = 1.$$
$$(y_1, y_2, s_1, s_2 \geq 0)$$

The initial simplex tableau is the following:

y_1	y_2	s_1	s_2	F	
1	4	1	0	0	1
6	−1	0	1	0	1
−1	−1	0	0	1	0

Notice that the payoff matrix A appears in the first two rows and columns, and both decision-variable indicators are −1. Now apply the simplex algorithm (Chapter 5, Section 5.1).

y_1	y_2	s_1	s_2	F		Quotients
1	4	1	0	0	1	$1/1 = 1$
6★	−1	0	1	0	1	$1/6 = \frac{1}{6}$ ←
−1	−1	0	0	1	0	Smallest quotient

↑
Most negative (tie)

y_1	y_2	s_1	s_2	F	
0	$\frac{25}{6}$★	1	$-\frac{1}{6}$	0	$\frac{5}{6}$
1	$-\frac{1}{6}$	0	$\frac{1}{6}$	0	$\frac{1}{6}$
0	$-\frac{7}{6}$	0	$\frac{1}{6}$	1	$\frac{1}{6}$

↑

y_1	y_2	s_1	s_2	F	
0	1	$\frac{6}{25}$	$-\frac{1}{25}$	0	$\frac{1}{5}$
1	0	$\frac{1}{25}$	$\frac{4}{25}$	0	$\frac{1}{5}$
0	0	$\frac{7}{25}$	$\frac{3}{25}$	1	$\frac{2}{5}$

From this terminal tableau we read the optimal solution

$$\text{Max } F = \frac{2}{5} \quad \text{when} \quad y_1 = \frac{1}{5}, \quad y_2 = \frac{1}{5}.$$

Relations (2) now give

$$v = \frac{1}{F} = \frac{5}{2}, \qquad q_1 = vy_1 = \frac{1}{2}, \qquad q_2 = vy_2 = \frac{1}{2}.$$

That is, the value of the game is $v = 5/2$, and the column-player's optimal strategy is

$$Q^\star = \begin{bmatrix} \frac{1}{2} \\ \frac{1}{2} \end{bmatrix}$$

which agrees with the result we found earlier. ◆

We can also find the *row*-player's (Player R) optimal strategy $P = [p_1 \quad p_2]$ by solving an appropriate linear program (see Exercise 20). However, there turns out to be a much easier way.* In the terminal tableau

y_1	y_2	s_1	s_2	F	
0	1	$\frac{6}{25}$	$-\frac{1}{25}$	0	$\frac{1}{5}$
1	0	$\frac{1}{25}$	$\frac{4}{25}$	0	$\frac{1}{5}$
0	0	$\frac{7}{25}$	$\frac{3}{25}$	1	$\frac{2}{5}$
		$\uparrow$	$\uparrow$		
		x_1	x_2		

obtained above, notice that the indicators in the slack-variable columns take the values

$$x_1 = \frac{7}{25}, \quad x_2 = \frac{3}{25}$$

respectively. We call these numbers the **terminal slack indicators.** It can be shown that the row-player's strategy is optimal when $p_1 = vx_1$ and $p_2 = vx_2$, where $v = 1/F$ is the value of the game. Thus, in our example we have

$$v = \frac{1}{F} = \frac{5}{2}, \qquad p_1 = vx_1 = \frac{7}{10}, \qquad p_2 = vx_2 = \frac{3}{10}$$

so the row-player's optimal strategy is

$$P^\star = \begin{bmatrix} \frac{7}{10} & \frac{3}{10} \end{bmatrix}$$

which checks with our earlier result.

Simplex Solution of Matrix Games The method just used for the 2×2 case generalizes to any matrix game. The steps are summarized as follows:

---

*This method rests on the concept of *duality*. See for example David G. Luenberger, *Introduction to Linear and Nonlinear Programming* (Reading, Mass.: Addison-Wesley, 1973).

Simplex Solution of an $m \times n$ Matrix Game

1. Add a suitable constant k to each payoff entry (if necessary) to obtain a matrix

$$
A = \begin{bmatrix}
a_{11} & a_{12} & \cdots & a_{1n} \\
a_{21} & a_{22} & \cdots & a_{2n} \\
\vdots & \vdots & & \vdots \\
a_{m1} & a_{m2} & \cdots & a_{mn}
\end{bmatrix}
$$

in which some row contains all positive entries.

2. Apply the simplex algorithm to the linear program

$$\text{Maximize} \quad F = y_1 + y_2 + \cdots + y_n$$

subject to:

$$
\begin{aligned}
a_{11}y_1 + a_{12}y_2 + \cdots + a_{1n}y_n &\leq 1 \\
a_{21}y_1 + a_{22}y_2 + \cdots + a_{2n}y_n &\leq 1 \\
&\vdots \\
a_{m1}y_1 + a_{m2}y_2 + \cdots + a_{mn}y_n &\leq 1. \\
(y_i &\geq 0)
\end{aligned}
$$

3. Read the optimal solution $y_1, \ldots, y_n$; F and the terminal slack indicators $x_1, \ldots, x_m$.

4. Let $v = 1/F$. Then

$$
P^\star = \begin{bmatrix} vx_1 & vx_2 & \cdots & vx_m \end{bmatrix} \quad \text{and} \quad Q^\star = \begin{bmatrix} vy_1 \\ vy_2 \\ \vdots \\ vy_n \end{bmatrix}
$$

are minimax optimal strategies, and v is the value of the game A.

5. Subtract k from v to obtain the value of the original game.

This procedure works even for a strictly determined or nonreduced game. However, the reader should eliminate these cases in advance to save unnecessary work.

Example 2   Solve the matrix game

$$
\begin{array}{c}
& \text{Player C} \\
& \begin{array}{ccc} 1 & 2 & 3 \end{array} \\
\text{Player R} \quad \begin{array}{c} 1 \\ 2 \end{array} & \begin{bmatrix} 0 & 1 & 2 \\ 1 & 0 & -2 \end{bmatrix}.
\end{array}
$$

Solution   After checking that the game has no saddle point and is in reduced form, we proceed as follows:

Step 1   Since neither row is positive we add $k = 1$ to each payoff entry to obtain the game

$$A = \begin{bmatrix} 1 & 2 & 3 \\ 2 & 1 & -1 \end{bmatrix}$$

which has positive value.

Step 2   Applying the simplex algorithm to the linear program

Maximize $F = y_1 + y_2 + y_3$
subject to:

$$y_1 + 2y_2 + 3y_3 \le 1$$
$$2y_1 + y_2 - y_3 \le 1$$
$$(y_1, y_2, y_3 \ge 0)$$

we obtain the following sequence of tableaus.

y_1	y_2	y_3	s_1	s_2	F		Quotients
1	2	3	1	0	0	1	$1/1 = 1$
$2\star$	1	-1	0	1	0	1	$1/2 = \frac{1}{2}$ $\leftarrow$
-1	-1	-1	0	0	1	0	
$\uparrow$							

y_1	y_2	y_3	s_1	s_2	F	
0	$\frac{3}{2}$	$\frac{7}{2}\star$	1	$-\frac{1}{2}$	0	$\frac{1}{2}$
1	$\frac{1}{2}$	$-\frac{1}{2}$	0	$\frac{1}{2}$	0	$\frac{1}{2}$
0	$-\frac{1}{2}$	$-\frac{3}{2}$	0	$\frac{1}{2}$	1	$\frac{1}{2}$
		$\uparrow$				

y_1	y_2	y_3	s_1	s_2	F	
0	$\frac{3}{7}$	1	$\frac{2}{7}$	$-\frac{1}{7}$	0	$\frac{1}{7}$
1	$\frac{5}{7}$	0	$\frac{1}{7}$	$\frac{3}{7}$	0	$\frac{4}{7}$
0	$\frac{1}{7}$	0	$\frac{3}{7}$	$\frac{2}{7}$	1	$\frac{5}{7}$
			$\uparrow$	$\uparrow$		
			x_1	x_2		

Step 3   From the terminal tableau we read the optimal solution

$$y_1 = \frac{4}{7}, \quad y_2 = 0, \quad y_3 = \frac{1}{7}; \quad F = \frac{5}{7}$$

and the terminal slack indicators

$$x_1 = \frac{3}{7}, \quad x_2 = \frac{2}{7}.$$

Step 4 Letting $v = 1/F = 7/5$, we obtain the minimax optimal strategies

$$P^\star = [vx_1 \quad vx_2] = \begin{bmatrix} \frac{3}{5} & \frac{2}{5} \end{bmatrix}, \qquad Q^\star = \begin{bmatrix} vy_1 \\ vy_2 \\ vy_3 \end{bmatrix} = \begin{bmatrix} \frac{4}{5} \\ 0 \\ \frac{1}{5} \end{bmatrix}.$$

Step 5 The value of the original game is

$$v - k = \frac{7}{5} - 1 = \frac{2}{5}. \quad \blacklozenge$$

Nonzero-Sum and *n*-Person Games   In this chapter we have considered only zero-sum games, in which one player's loss is the other's gain. When the theory is extended to include nonzero-sum games, we find that the notion of "rational" behavior is much harder to characterize. A famous game-theoretic paradox created by A. W. Tucker illustrates.

Example 3   *Prisoner's Dilemma* Prisoners R and C are both implicated in a crime. The district attorney informs them that if neither confesses each will get a 1-year sentence on some minor charge. If one confesses but the other doesn't, the one cooperating will go free while his accomplice will get a 10-year sentence. If they both confess, they will each get a reduced sentence of 5 years. The prisoners are separated and given 24 hours to respond.

Discussion   Each of the prisoners has two moves, "confess" or "don't confess." If we interpret a sentence of *n* years as a payoff of $-n$, then for each choice of moves there will be *two* payoffs, one for player R and one for player C. These payoff-pairs are indicated in the table below.

		Prisoner C	
		Confess	*Don't confess*
Prisoner R	*Confess*	$-5, -5$	$0, -10$
	Don't confess	$-10, 0$	$-1, -1$

The first number in each pair represents R's payoff, the second C's payoff. For example, if R confesses but C does not then R receives a payoff of zero while C receives a payoff of -10.

   Now, prisoner R will no doubt observe that his payoff is uniformly greater in row 1 than in row 2 (-5 compared with -10, and 0 compared with -1). In other words—disregarding any moral qualms or fears about squealing—he will be *better off confessing, no matter what C does*. Likewise, prisoner C will be better off confessing no matter what R does, since his own payoff is uniformly greater in column 1 than in column 2. Thus, if both prisoners make the "rational" move they will each receive 5-year sentences, whereas if they both made the "irrational" move they would only receive 1-year sentences!   ◆

Of course, if the prisoners were permitted to communicate they might promise each other to remain silent, since it is in both their interests to do so. Such a pact would be inherently unstable, however, because each side could gain by double-crossing the other. Many real-life negotiations on such issues as disarmament, tariff reduction and the like present essentially the same features as this example. That is, the best solution for both sides requires a certain amount of mutual trust.

Games with more than two players also arise in applications—such as when several countries are competing for the same resources, or several corporations for the same market. The theory of n-person games is complicated by the large number of different coalitions possible among the players, and at present this theory is far less satisfactory than the theory of two-person games. Despite the early enthusiasm for game theory in the 1940s and 1950s, there still seems to be a wide gulf between what the theory can handle and what actually occurs in most conflict situations.

10.3 Exercises

In Exercises 1–4, solve the given 2×2 game using the simplex algorithm. Check your result using the 2×2 formulas in Section 10.2:

1. $\begin{bmatrix} 1 & -1 \\ -1 & 1 \end{bmatrix}$

2. $\begin{bmatrix} 1 & 0 \\ 0 & 3 \end{bmatrix}$

3. $\begin{bmatrix} 1 & -2 \\ -3 & 4 \end{bmatrix}$

4. $\begin{bmatrix} 3 & -1 \\ -3 & 1 \end{bmatrix}$

5. Solve the strictly determined game

$$\begin{bmatrix} 2 & 1 & 3 \\ 4 & 0 & 2 \end{bmatrix}$$

using the simplex algorithm. Check your result using the method of Section 10.1.

6. Solve the nonreduced game

$$\begin{bmatrix} 1 & -2 & 2 \\ -3 & 0 & 3 \end{bmatrix}$$

using the simplex algorithm. Check your result using the method of Section 10.2.

In Exercises 7–14, solve the matrix game using the simplex algorithm.

7. $\begin{bmatrix} 3 & -1 & 0 \\ -2 & 1 & -1 \end{bmatrix}$

8. $\begin{bmatrix} -3 & 1 & 3 \\ 2 & 0 & -2 \end{bmatrix}$

9. $\begin{bmatrix} 1 & -2 & 2 \\ 0 & 5 & -1 \end{bmatrix}$

10. $\begin{bmatrix} 2 & 1 & 3 \\ 1 & 5 & 0 \end{bmatrix}$

11.
$$\begin{bmatrix} 2 & -3 \\ -2 & 3 \\ -1 & 1 \end{bmatrix}$$

12.
$$\begin{bmatrix} -1 & 3 \\ 0 & -1 \\ 3 & -2 \end{bmatrix}$$

13.
$$\begin{bmatrix} 0 & 1 & 0 \\ 1 & -1 & -1 \\ -1 & -1 & 3 \end{bmatrix}$$

14.
$$\begin{bmatrix} 1 & 0 & 0 \\ 0 & -2 & 1 \\ 0 & 1 & 0 \end{bmatrix}$$

15. An inmate R breaks out regularly from the state prison. When he heads for the desert (D) his chance of escape is 30%, 20%, or 10%, depending on whether the warden sends 0, 1, or 2 search units to that area. When he heads for the mountains (M) his chance of escape is 50%, 40%, or 25%, according as 0, 1, or 2 search units are in the area. The warden has only two search units available. They can both be sent to the desert (DD), one can be sent to each area (DM), or they can both be sent to the mountains (MM). Find the best strategy for each side, and the inmate's expected chance of escape.

16. Army R plans to attack one of two outposts held by army C. Outpost A is worth 3 points and outpost B is worth 5 points. Army C can deploy both its two divisions at A (AA), one at A and one at B (AB), or both at B (BB). The commander of army R knows he can capture either objective; however, he estimates he will suffer losses worth 2 or 3 points, respectively, if the outpost he attacks is defended by 1 or 2 divisions. Find the best strategy for each side.

17. *Three-finger Morra* Two players, R and C, each simultaneously show one, two, or three fingers. If the sum of the numbers shown is even, R wins that many dollars from C. If the sum is odd, C wins that many dollars from R. Show that this game is fair, and find the best strategy for each player.

18. *Stone-scissors-paper* In a well-known children's game, each of two players simultaneously shows either a fist ("stone"), two fingers ("scissors") or a palm ("paper"). If they both make the same move the game is a tie; otherwise the winner is determined by the rule "stone breaks scissors, scissors cuts paper, paper covers stone." Suppose the loser must pay the winner a penny. Show that this game is fair, and find the best strategy for each player.

19. Prove: As $P = [p_1 \quad p_2]$ ranges over all strategies, the quantity

$$p_1 x_1 + p_2 x_2$$

takes on all values between x_1 and x_2, inclusive. (Hint: Consider whether $x_1 < x_2$, $x_1 = x_2$ or $x_1 > x_2$.)

20. Let

$$A = \begin{bmatrix} a & b \\ c & d \end{bmatrix}$$

be a 2 × 2 game with a positive value v, and let $P = [p_1 \quad p_2]$ be the row-player's optimal strategy. If

$$x_1 = \frac{p_1}{v}, x_2 = \frac{p_2}{v}, G = \frac{1}{v}$$

prove that x_1, x_2, G is the optimal solution of the linear program

$$\text{Minimize} \quad G = x_1 + x_2$$
subject to:

$$ax_1 + cx_2 \geq 1$$
$$bx_1 + dx_2 \geq 1.$$
$$(x_1, x_2 \geq 0)$$

Computer Projects   The SIMPL program computes the optimal solution to a linear program by the simplex method. This program can be used in Exercises 1–18.

Chapter 10   Summary

Important Terms

10.1 matrix game
payoff matrix
reduced form
minimax optimal moves
saddle point
strictly determined game

10.2 pure and mixed strategies
expected payoff
minimax optimal strategies

Review Exercises

In Exercises 1–4, find the reduced form of the given matrix game.

1.
$$\begin{bmatrix} 3 & -1 & 2 \\ -1 & 4 & -3 \end{bmatrix}$$

2.
$$\begin{bmatrix} -2 & 3 & 2 & 0 \\ 3 & 2 & 5 & -1 \end{bmatrix}$$

3.
$$\begin{bmatrix} 2 & -1 & 5 & 7 \\ 3 & 2 & -6 & -2 \\ 0 & 4 & 1 & 3 \\ 2 & 1 & -8 & 2 \end{bmatrix}$$

4.
$$\begin{bmatrix} 9 & -5 & 2 \\ 1 & 7 & -4 \\ 3 & 2 & 1 \\ -5 & 0 & 5 \end{bmatrix}$$

In Exercises 5–8, decide whether the matrix game is strictly determined. If so, find all minimax optimal moves and the value of the game.

5.
$$\begin{bmatrix} 3 & 2 \\ -1 & 5 \\ 4 & 0 \end{bmatrix}$$

6.
$$\begin{bmatrix} 6 & 0 & -3 \\ -2 & -1 & 7 \\ 4 & 2 & 5 \end{bmatrix}$$

7.
$$\begin{bmatrix} -1 & 4 & -3 \\ 0 & -2 & 5 \\ 1 & 3 & 2 \end{bmatrix}$$

8.
$$\begin{bmatrix} 1 & -2 & 5 & -1 \\ 2 & 4 & 3 & 2 \\ 0 & 6 & -1 & 1 \end{bmatrix}$$

9. Consider the game with the following payoff matrix:

$$\begin{bmatrix} 50 & -90 \\ -20 & 40 \end{bmatrix}$$

(a) Suppose each player tosses a fair coin to determine which of his two moves to make. Who will have the advantage?

(b) What is the optimal strategy for each player? If they use these strategies, who will have the advantage?

10. Consider the game with the following payoff matrix:

$$\begin{bmatrix} 5 & 0 & -1 \\ -3 & 4 & 1 \end{bmatrix}$$

(a) Suppose that player R chooses row 1 50% of the time and row 2 50% of the time, while player C chooses column 1 20% of the time, column 2 30% of the time, and column 3 50% of the time. Who will have the advantage?

(b) What is the optimal strategy for each player? If they use these strategies, who will have the advantage?

In Exercises 11–12, solve the matrix game using the simplex algorithm.

11. $$\begin{bmatrix} 1 & -5 & 9 \\ -1 & 0 & -2 \end{bmatrix}$$

12. $$\begin{bmatrix} 3 & -1 \\ -2 & 2 \\ -1 & 0 \end{bmatrix}$$

13. Two players, R and C, each simultaneously put one, two, or three dollars into a pot. If the total amount put in is even, R wins the amount put in by C. If it is odd, C wins the amount put in by R. Find the best strategy for each player. Is the game fair?

14. Army C must send a supply convoy from point A to point B along one of the three mountain passes shown below.

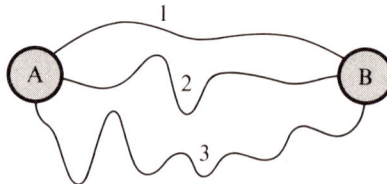

The convoy will take three days along route 1, four days along route 2, and six days along route 3. Army R can plan to ambush the convoy along any one of the three routes. If the ambush is planned along the same route taken by the convoy, army R will be able to attack the convoy during each day of travel. However, if the ambush is planned along a different route than that taken by the convoy, exactly two days of attack will be lost by army R. Find the best strategy for each side, and the expected number of days army R will be able to attack the convoy.

11 Statistics

Statistics is the branch of mathematics that classifies and interprets numerical data. Because of its many important applications this subject has developed into a specialized discipline in its own right, encompassing a vast range of different techniques. In this chapter we present a very brief survey of some topics related to our earlier study of probability theory.

11.1 Descriptive Statistics

It is common to distinguish between descriptive statistics, which deals with the analysis and description of numerical information, and inferential statistics, which concerns the use of such information to reach conclusions and make predictions. In this section we discuss the organization of data and introduce two fundamental descriptive tools; the mean, and standard deviation.

Qualitative and Quantitative Data Empirical data can be classified as *qualitative* or *quantitative,* according to the nature of the categories into which the numbers or measurements fall. The classification "countries which produce steel" contains nonnumerical categories and is qualitative. The classification "age groups of the U.S. population" contains categories which can be put in numerical order and is quantitative.

Largest steel producers* (millions of metric tons)		U.S. population by age† (millions)	
U.S.S.R.	141.3	0–20	69.2
U.S.	105.8	20–40	69.6
Japan	102.3	40–60	45.0
W. Germany	40.4	60–80	29.2
China	29.0	80–100	5.0
Table 1		**Table 2**	

The qualitative data in Table 1 can be represented by means of various graphic devices, such as the *bar graph* of Figure 1. For quantitative data like those in Table 2, we can employ a similar type of graph called a **histogram** (see Figure 2). Here the bars are constructed over the *class intervals* 0–20, 20–40, and so on, with the height of each bar representing the corresponding *class frequency;* that is, the number of people whose ages fall in the given interval. Notice that, unlike a bar graph, a histogram has no spaces between the bars, and they all have the same width.

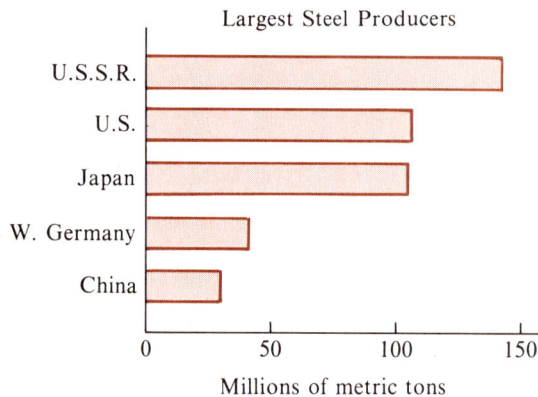

Figure 1

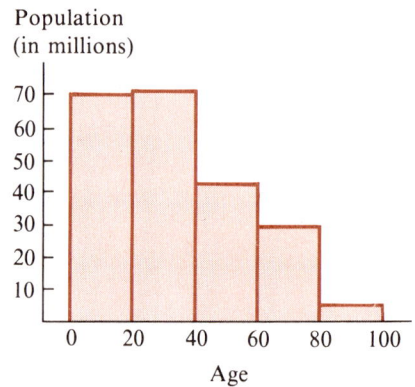

Figure 2

Table 3 illustrates two difficulties that often arise with quantitative data.

Income of U.S. families‡	
Below $5000	9.3%
$5000–10,000	18.1%
$10,000–15,000	18.5%
$15,000–25,000	31.7%
Over $25,000	22.4%
Table 3	

---

*1975 figures
†1980 figures
‡1977 figures

Notice that the last class interval ("Over $25,000") is open-ended, that is, no upper limit is specified. Moreover, the third and fourth class intervals cover unequal ranges of values ($5000 and $10,000, respectively). In general, a histogram is used only for quantitative data in which *all the class intervals are closed and cover equal ranges of values,* as in Table 2.

Grouped and Ungrouped Data The first step in organizing a set of raw data for statistical analysis is to put it in a comprehensible form. The procedure is illustrated in the following example.

Example 1 An auto plant records the following numbers of assembly-line absentees on fifty consecutive working days. How can this data be summarized usefully?

17	32	45	35	38	22	40	24	48	53
29	47	57	26	33	42	47	36	37	16
41	35	49	55	39	22	20	26	48	43
21	45	29	35	52	32	26	31	38	47
27	41	12	23	32	37	55	35	48	46

Discussion These raw numbers become more manageable when expressed as **grouped data.** To do this we first observe that the given figures vary from 12 (least) to 57 (greatest), for a range of 57 − 12 = 45 units. Thus, five class intervals each covering ten units will accommodate all the values.* Tallying the number of observations that fall in each of the intervals 10–20, 20–30, and so on, we arrive at the **frequency distribution** shown in Table 4.

Number of absentees	Tally	Frequency
10–20	||||	4
20–30	⫴⫴ |	11
30–40	⫴⫴⫴ |	16
40–50	⫴⫴ ||||	14
50–60	⫴	5
	Total	50

Table 4

Note that two of the given values (40 in row 1, and 20 in row 3) fall exactly on a class boundary. We can arbitrarily agree to put all such boundary values in the *lower* class interval; thus, we count 20 in the interval 10–20 rather than 20–30. In this way each data value is tallied exactly once, so the class frequencies add up to the total number of observations (50).

---

*The usual practice is to group raw data into at least five and at most fifteen classes.

Frequency
(Number of days)

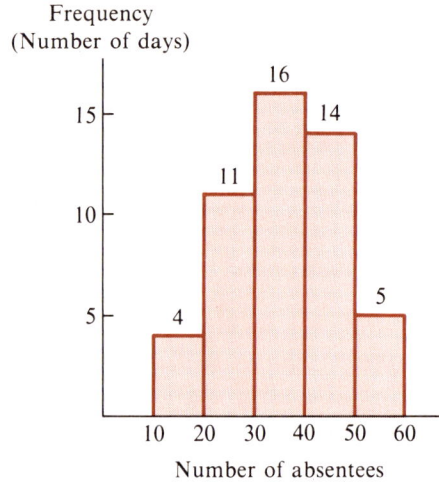

Number of absentees

Figure 3

The frequency distribution in Table 4 can be represented by a histogram, as in Figure 3. This gives us a much better feel for the data than we had before. For instance, we can see at a glance that the number of absentees is usually between 20 and 50; indeed, in our 50-day sample the number of absentees fell in this range $11 + 16 + 14 = 41$ times, or 82% of the time. The price we pay for this advantage of grouped data is the *loss of information*. For example, if we want to know how often the number of absentees was between 25 and 45, our grouped distribution is of no use; we must go back to the original raw data. ◇

Measures of Central Tendency   In describing a set of data the first task is to characterize its **central tendency;** that is, the "typical" value represented. The most widely used measure of central tendency is the *mean* (or arithmetic mean), defined as follows:

Definition

The **mean** $\bar{x}$ of a set of n numbers is their sum, divided by n.

$$\text{mean}\quad \bar{x} = \frac{x_1 + x_2 + \cdots + x_n}{n} \qquad (1a)$$

Here $x_1, x_2, \ldots,$ and x_n represent a set of n measurements or observations in ungrouped form. If we use the **summation symbol** Σ (read "sum of"), we can write (1a) in the more compact form

$$\text{mean}\quad \bar{x} = \frac{\Sigma x}{n}.$$

For grouped data the values in each class interval are approximated by the *midpoint* of the interval. Thus, if the class intervals have midpoints $x_1, \ldots, x_k$ and respective frequencies $f_1, \ldots, f_k$, the formula becomes

$$\textbf{mean} \quad \bar{x} = \frac{x_1 f_1 + x_2 f_2 + \cdots + x_k f_k}{n} = \frac{\Sigma xf}{n} \tag{1b}$$

where $n = \Sigma f$ is the total number of observations.

Example 1, Continued Find the mean for both the ungrouped and grouped data of Example 1.

Solution For the ungrouped (raw) data of Example 1, a tedious but straightforward calculation yields

$$\bar{x} = \frac{\Sigma x}{n} = \frac{17 + 32 + 45 + \cdots + 46}{50}$$

$$= \frac{1814}{50} = 36.28$$

This is the true mean, or average, of the fifty values observed. To find the mean of the grouped data we reproduce the frequency distribution of Table 4, showing the midpoint x of each class interval.

Interval	Midpoint x	Frequency f	xf
10–20	15	4	60
20–30	25	11	275
30–40	35	16	560
40–50	45	14	630
50–60	55	5	275
Totals		50	1800

The column at the right gives the product xf of each class midpoint with the corresponding class frequency. The sum of these products is shown at the bottom of the column. Formula (1b) gives the grouped mean as

$$\bar{x} = \frac{\Sigma xf}{n} = \frac{1800}{50} = 36.$$

This agrees fairly well with the true mean 36.28. The discrepancy is due to the information lost when we grouped the data, because in computing the grouped mean we essentially replace all values between 10 and 20 by the midpoint value 15, all values between 20 and 30 by 25, and so on. ◇

Another common measure of central tendency is the *median*, defined on the following page.

Definition

The **median** $\tilde{x}$ of a set of n numbers is the middle value when the numbers are arranged in increasing order (or the average of the two middle values if n is even).

For example, if we list the raw data of Example 1 in increasing order (a dreary procedure!) we obtain

12	16	17	20	21	22	22	23	24	26
26	26	27	29	29	31	32	32	32	33
35	35	35	35	36★	37★	37	38	38	39
40	41	41	42	43	45	45	46	47	47
47	48	48	48	49	52	53	55	55	57

The middle values are 36 and 37 (starred). Hence the median is

$$\tilde{x} = \frac{36 + 37}{2} = 36.5.$$

The mean is generally easier to compute than the median, since it does not require ordering the data. The median has the advantage, however, that is is less affected by a few extreme, "atypical" values (see Exercises 13–14).

Measures of Dispersion In addition to the central tendency of a set of data we are also interested in its degree of dispersion, that is, how close together or far apart the values tend to be. If most of the numbers $x_1, \ldots, x_n$ are close to the mean $\bar{x}$, then the *squared deviations*

$$(x_1 - \bar{x})^2, \quad (x_2 - \bar{x})^2, \quad \ldots, \quad (x_n - \bar{x})^2$$

will be small in magnitude, for the most part. We can therefore measure the dispersion of data by taking an average of these squared deviations.

Definition

The **variance** of a set of n numbers, denoted by s^2, is their mean squared deviation.

$$\textbf{variance} \quad s^2 = \frac{(x_1 - \bar{x})^2 + (x_2 - \bar{x})^2 + \cdots + (x_n - \bar{x})^2}{n} \tag{2a}$$

The square root of the variance is called the **standard deviation** and denoted by s.*

---

*When the data represent a sample of measurements from a larger population, the denominator in (2a) is often replaced by $n - 1$, for certain technical reasons. We then refer to the *sample variance* and *sample standard deviation*, respectively. We shall not use these notions.

Using the summation symbol, formula (2a) can be abbreviated as

$$\text{variance} \quad s^2 = \frac{\Sigma(x - \bar{x})^2}{n}$$

For grouped data we use the midpoints $x_1, \ldots, x_k$ of the class intervals to approximate the actual data, as earlier. Thus, if the class frequencies are $f_1, \ldots, f_k$, respectively, the formula becomes

$$\text{variance} \quad s^2 = \frac{(x_1 - \bar{x})^2 f_1 + \cdots + (x_k - \bar{x})^2 f_k}{n} = \frac{\Sigma(x - \bar{x})^2 f}{n} \qquad (2b)$$

where $n = \Sigma f$ is the total number of observations.

Example 1, Continued

Find the standard deviation for both the ungrouped and grouped data of Example 1.

Solution

(a) For the ungrouped data we calculated the mean $\bar{x} = 36.28$. Using formula (2a) we find, after much calculation:

$$s^2 = \frac{\Sigma(x - \bar{x})^2}{n}$$

$$= \frac{(17 - 36.28)^2 + (32 - 36.28)^2 + \cdots + (46 - 36.28)^2}{50}$$

$$\doteq \frac{371.72 + 18.32 + \cdots + 94.48}{50}$$

$$= \frac{6188.08}{50} \doteq 123.76.$$

The standard deviation is therefore

$$s = \sqrt{123.76} \doteq 11.12.$$

(b) For the grouped data the computation is much simpler. We already calculated the grouped mean $\bar{x} = 36$. The table below shows the frequency distribution of Table 4, together with the midpoint x of each class interval, the deviation $x - \bar{x}$, the squared deviation $(x - \bar{x})^2$, and the product $(x - \bar{x})^2 f$.

Interval	Midpoint x	Frequency f	$x - \bar{x}$	$(x - \bar{x})^2$	$(x - \bar{x})^2 f$
10–20	15	4	−21	441	1764
20–30	25	11	−11	121	1331
30–40	35	16	−1	1	16
40–50	45	14	9	81	1134
50–60	55	5	19	361	1805
Total		50			6050

Formula (2b) gives the variance

$$s^2 = \frac{\Sigma(x - \bar{x})^2 f}{n} = \frac{6050}{50} = 121.$$

Taking the square root, we obtain the standard deviation

$$s = \sqrt{121} = 11.$$

Note that once again, the grouped data have approximately the same statistical characteristics as the ungrouped data. ◆

Example 2   Find the mean and standard deviation for the age of an individual in the U.S. population (Table 2).

Solution   The frequency distribution in Table 2 is reproduced below, together with the midpoint of each class interval.

Interval	Midpoint x	Frequency f	xf
0–20	10	69.2	692
20–40	30	69.6	2088
40–60	50	45.0	2250
60–80	70	29.2	2044
80–100	90	5.0	450
	Total	218.0	7524

(Note: The frequency of each class is expressed in millions, as is the total number of observations $n = \Sigma f = 218.0$.) Formula (1b) for the mean of grouped data gives

$$\bar{x} = \frac{\Sigma xf}{n} = \frac{7524}{218} \doteq 34.5.$$

This is the average age of an individual in the U.S. population. To find the standard deviation we extend the above table to include columns for $x - \bar{x}$, $(x - \bar{x})^2$, and $(x - \bar{x})^2 f$, using the value of $\bar{x}$ just computed.

Midpoint x	Frequency f	$x - \bar{x}$	$(x - \bar{x})^2$	$(x - \bar{x})^2 f$
10	69.2	−24.5	600.75	41,571.90
30	69.6	−4.5	20.25	1409.40
50	45.0	15.5	240.25	10,811.25
70	29.2	35.5	1260.25	36,799.30
90	5.0	55.5	3080.25	15,401.25
Total	218.0			105,993.10

The variance is

$$s^2 = \frac{\Sigma(x - \bar{x})^2 f}{n} \doteq \frac{105{,}993.10}{218} \doteq 486.21$$

and the standard deviation is

$$s = \sqrt{486.21} \doteq 22.05. \quad \blacklozenge$$

The next example illustrates the significance of the standard deviation as a measure of dispersion.

Example 3 Find the mean and standard deviation for the following sets of ungrouped data:

(a) 1, 1, 1, 2, 2, 4, 4, 5, 5, 5.

(b) 2, 2, 2, 3, 3, 3, 3, 4, 4, 4.

Solution (a) There are $n = 10$ observations. Using formula (1a) for the mean of ungrouped data, we find that

$$\bar{x} = \frac{\Sigma x}{n} = \frac{1 + 1 + 1 + 2 + 2 + 4 + 4 + 5 + 5 + 5}{10}$$

$$= \frac{30}{10} = 3.$$

We record the data values x, the deviations $x - \bar{x}$, and the squared deviations $(x - \bar{x})^2$, as follows:

x:	1	1	1	2	2	4	4	5	5	5
$x - \bar{x}$:	-2	-2	-2	-1	-1	1	1	2	2	2
$(x - \bar{x})^2$:	4	4	4	1	1	1	1	4	4	4

Formula (2a) gives the variance:

$$s^2 = \frac{\Sigma(x - \bar{x})^2}{n} = \frac{4 + 4 + 4 + 1 + 1 + 1 + 1 + 4 + 4 + 4}{10}$$

$$= \frac{28}{10} = 2.8.$$

Thus, the standard deviation is

$$s = \sqrt{2.8} \doteq 1.67.$$

(b) As in part (a), we find that

$$\bar{x} = \frac{\Sigma x}{n} = \frac{30}{10} = 3.$$

The squared deviations for the (b) data are shown below.

$$x: \quad 2 \quad 2 \quad 2 \quad 3 \quad 3 \quad 3 \quad 3 \quad 4 \quad 4 \quad 4$$
$$x - \bar{x}: \quad -1 \quad -1 \quad -1 \quad 0 \quad 0 \quad 0 \quad 0 \quad 1 \quad 1 \quad 1$$
$$(x - \bar{x})^2: \quad 1 \quad 1 \quad 1 \quad 0 \quad 0 \quad 0 \quad 0 \quad 1 \quad 1 \quad 1$$

Hence, the variance is

$$s^2 = \frac{\Sigma(x - \bar{x})^2}{n} = \frac{1 + 1 + 1 + 0 + 0 + 0 + 0 + 1 + 1 + 1}{10}$$

$$= \frac{6}{10} = .6$$

and the standard deviation is $\quad s = \sqrt{.6} \doteq .77$.

Thus both data sets have the same mean, but (a) has a greater standard deviation than (b). This agrees with our intuitive impression that the values in (a) are more dispersed than those in (b). ◆

   Since the data sets in Example 3 involve any integer values, we can summarize them in grouped form as follows:

(a) Value	Frequency		(b) Value	Frequency
1	3		1	0
2	2		2	3
3	0		3	4
4	2		4	3
5	3		5	0

These frequency distributions are represented in the histograms of Figure 4, where we make each integer value $x = 1, 2, 3, 4, 5$ the midpoint of a "class interval" of width 1 unit. Notice that the "balance point" for both distributions is the same, namely the mean value $\bar{x} = 3$. As we see from the figure, however, distribution (a) is more widely dispersed ($s = 1.67$) while distribution (b) is concentrated near the mean ($s = .77$).

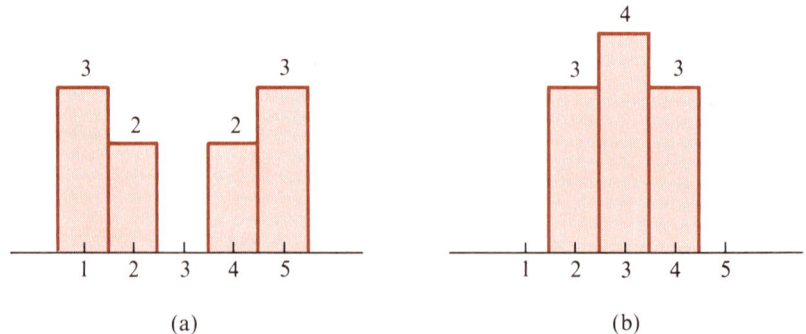

(a)   (b)

Figure 4

Another measure of dispersion is the **mean deviation,** defined by the formula

$$\text{M.D.} = \frac{\Sigma|x - \bar{x}|}{n} = \frac{|x_1 - \bar{x}| + |x_2 - \bar{x}| + \cdots + |x_n - \bar{x}|}{n}.$$

We leave it to the reader to verify that the data sets (a) and (b) of Example 3 have mean deviation 1.6 and .6 respectively. Thus, both the standard deviation and the mean deviation lead us to the conclusion that the data in (a) are more dispersed than those in (b). For mathematical purposes the standard deviation is the most useful measure of dispersion. We shall see some of its applications in the next two sections.

11.1 Exercises

1. Represent the data in the following table using a bar graph:

Advertising expenditures (billions of dollars)	
1960	11.9
1965	15.3
1970	19.6
1975	28.2

2. Represent the data in the following table using a bar graph:

Health expenses per capita, 1977	
Hospital care	$297
Physicians	$146
Nursing homes	$57
Drugs	$57
Other	$89

3. A test of the tensile strength (breaking point under stress) of a sample of 80 cables produced the distribution show below.

Tensile strength (lbs)	Number of cables
1400–1500	11
1500–1600	29
1600–1700	26
1700–1800	9
1800–1900	5

Represent this data by means of a histogram.

4. The speeds of 100 random cars on a major interstate highway were distributed as follows:

Speed (mph)	Cars
40–50	15
50–60	31
60–70	35
70–80	17
80–90	2

Represent this data by means of a histogram.

5. The daily low temperature readings for a certain area during the month of September were as follows (in degrees Fahrenheit):

41	43	39	34	38	42
44	47	48	53	58	54
51	59	53	46	47	41
36	31	36	43	43	47
49	48	49	54	57	54

Write this data in grouped form, using 30–35 as the first class interval. Represent the grouped data with a histogram.

6. The systolic blood pressures for a group of 40 young males are as follows (in millimeters of mercury):

141	98	132	135	122	117	131	116
139	118	121	128	127	113	138	115
127	142	108	147	117	118	127	138
116	131	119	122	121	94	129	118
119	129	108	123	131	124	131	102

Write this data in grouped form, using 90–100 as the first class interval. Represent the grouped data with a histogram.

7. The rates charged by ten Lompoc moving companies for local moves are as follows (in dollars per hour):

52, 57, 53, 59, 48, 50, 56, 52, 54, 49

Find the mean, median, and standard deviation for this ungrouped data.

8. Twelve randomly selected owners of Generic Motors cars were asked how many years they had been driving their present automobile. The responses were as follows:

2, 8, 5, 11, 7, 6, 1, 14, 4, 5, 3, 6

Find the mean, median, and standard deviation for this ungrouped data.

9. Find the mean and standard deviation for the grouped data in Exercise 3.

10. Find the mean and standard deviation for the grouped data in Exercise 4.

11. Find the mean and standard deviation for the grouped data in Exercise 5. How does the grouped mean compare with the true mean?

12. Find the mean and standard deviation for the grouped data in Exercise 6. How does the grouped mean compare with the true mean?

13. The salaries of ten randomly chosen major league pitchers are shown below.

$185,000 $200,000 $340,000 $405,000 $1,500,000
$360,000 $315,000 $210,000 $170,000 $205,000

Find the mean and median for this ungrouped data. If you were considering becoming a major league pitcher, which of these two measures would give you a better idea of what you could expect?

14. In ten consecutive trips to Las Vegas, a gambler had the following net gains and losses:

$1500 $1225 $1375 $2250   − $4725
$750   − $8750 $3150   − $2725 $2150

Find the mean and median for these ungrouped data. If you were considering becoming a professional gambler, which of these two measures would give you a better idea of what you could expect?

15. The tables below show the weekly sales records for two realtors, A and B, during the year just ended.

| Realtor A | | Realtor B | |
Number of houses sold	Frequency	Number of houses sold	Frequency
0	4	0	10
1	25	1	11
2	17	2	12
3	5	3	9
4	1	4	8
5	0	5	2

Thus, there were 4 weeks in which realtor A sold no houses, 25 weeks in which he sold exactly one house, and so on.

(a) Find the mean and standard deviation in weekly sales, for each realtor.

(b) Which realtor had the better sales record? Which was more consistent?

16. The lifetimes of two brands of light bulbs were compared by randomly sampling 100 bulbs from each company, with the following results:

| Brand A | | Brand B | |
Lifetime (hours)	Frequency	Lifetime (hours)	Frequency
500–600	12	500–600	5
600–700	23	600–700	30
700–800	26	700–800	47
800–900	21	800–900	16
900–1000	18	900–1000	2

Thus, 12 of the brand A bulbs lasted 500–600 hours, 23 lasted 600–700 hours, and so on.

(a) Find the mean and standard deviation in bulb life, for each brand.

(b) Which brand lasts longer, on the average? Which is more dependable?

17. Compare the means and standard deviations for the grouped data sets whose histograms are shown below.

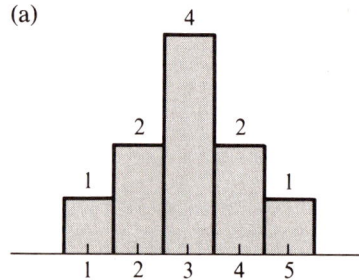

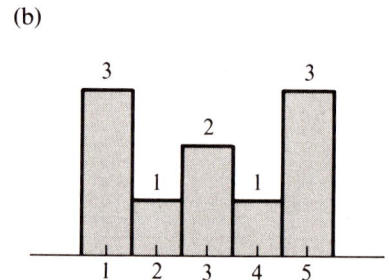

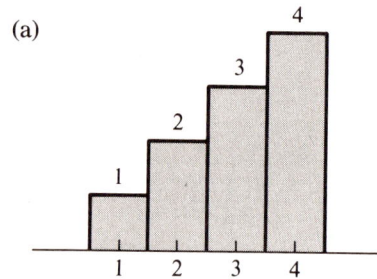

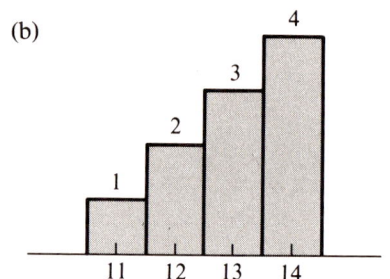

(a)

(b)

18. Compare the means and standard deviations for the grouped data sets whose histograms are shown below.

(a)

(b)

19. Prove: Let $\bar{x}$ and s be the mean and standard deviation for the ungrouped data $x_1, \ldots, x_n$. If each number is multiplied by $k > 0$, the new data will have mean $k\bar{x}$ and standard deviation ks.

20. Prove: The variance s^2 is equal to *the mean square of the data, minus the squared mean.*

$$\text{ungrouped:} \quad s^2 = \frac{\Sigma x^2}{n} - \left(\frac{\Sigma x}{n}\right)^2$$

$$\text{grouped:} \quad s^2 = \frac{\Sigma x^2 f}{n} - \left(\frac{\Sigma xf}{n}\right)^2$$

These formulas simplify the computation of s.

| Computer Projects | A. The DATA program computes the mean, median, and standard deviation for a given set of raw data and groups the data into prescribed class intervals. Use this program to confirm your answer in Exercises 5–8 and 11–14. |
| | B. The STAT program computes the mean and standard deviation for a given set of grouped data. Use this program to confirm your answer in Exercises 9–12 and 15–18. |

11.2 Probability Distributions

The notion of a frequency distribution for a set of data is connected with the notion of a probability distribution, which we encountered in Section 8.3. In this section we explore this connection and see how to extend the definition of mean, variance, and standard deviation to arbitrary probability distributions.

Random Variables   Quantitative data always represent measurements of some quantity X over a certain population, such as the ages of individuals in the U.S., the number of auto plant absentees on various days, or the number of hours a light bulb will last before burning out. Since the value of X depends on which member of the population we choose, we can view it as a **random variable** in the sense of Section 8.3, that is, a numerical quantity whose value x depends on the outcome of an experiment.

Example 1   An auto towing service keeps a record of the number of service requests received each day over a 60-day period, with the following results. Chart the probability of a certain number of requests on a given day.

Number of requests	Frequency
0	6
1	15
2	18
3	12
4	6
5	3
Total	60

Discussion   The table above shows the frequency distribution for the random variable

$$X = \text{number of daily requests}.$$

If we divide each frequency by the total number of observations (60) we obtain the corresponding *relative* frequency, or empirical probability. This leads to the *probability distribution* shown in Table 1 on the next page.

Number of requests	Probability
0	.10
1	.25
2	.30
3	.20
4	.10
5	.05
Total	1.00

Table 1

We can represent this distribution by a histogram as in Figure 1, where each value x is the midpoint of a "class interval" extending one-half unit to the left and right. The height of each bar represents the probability of the corresponding value (frequency divided by 60). Thus, the probability histogram has the same shape as the frequency histogram, the only difference being the units of measurement on the vertical axis.

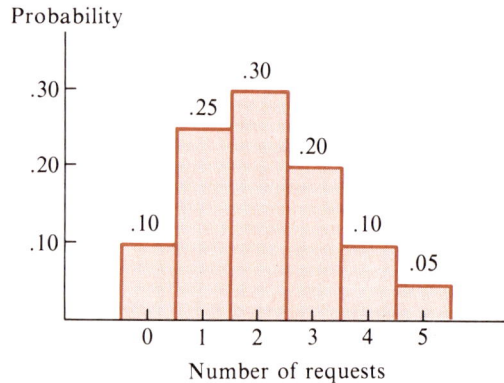

Figure 1

Example 1, Continued   Find the probability that the number of service requests received by the towing service on a given day will be between 1 and 3, inclusive.

Solution   From Table 1 we see that

$$P(1 \le X \le 3) = P(X = 1 \cup X = 2 \cup X = 3)$$
$$= P(X = 1) + P(X = 2) + P(X = 3)$$
$$= .25 + .30 + .20$$
$$= .75.$$

This is illustrated in Figure 2, where the part of the histogram corresponding to the values $x = 1, 2, 3$ is shaded. Since each bar of the histogram has unit width, the total area of the shaded portion is

$$(.25) \cdot 1 + (.30) \cdot 1 + (.20) \cdot 1 = .75$$

which is just the probability in question.

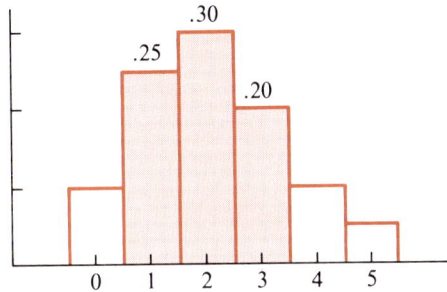

Figure 2

◇

Example 1 illustrates the following theorem:

Theorem 1

Let X be a random variable with consecutive integer values. The probability that X will assume a value between a and b (inclusive) is equal to the area under the probability histogram for X, between $a - 1/2$ and $b + 1/2$.

Thus, we see from Figure 1 that the probability of receiving between 1 and 4 service requests is

$$P(1 \le X \le 4) = \text{area between 0.5 and 4.5} = .85$$

while the probability of receiving at least 3 requests is

$$P(X \ge 3) = \text{area to the right of 2.5} = .35.$$

The corresponding regions in the histogram are shaded in Figure 3. Notice in particular that the *total* area under a probability histogram is always 1.

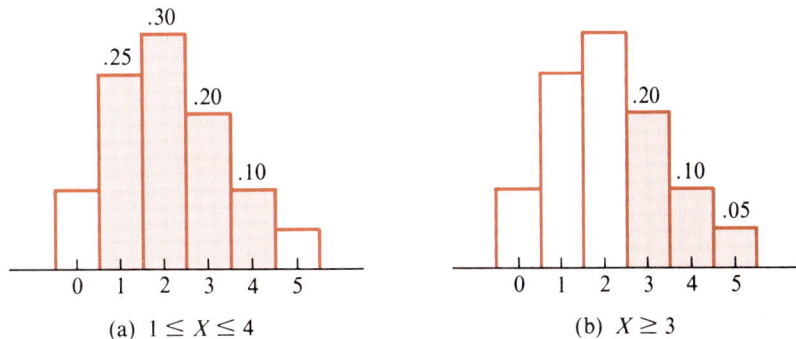

(a) $1 \le X \le 4$

(b) $X \ge 3$

Figure 3

For grouped data we take the values of the random variable X to be the midpoints $x_1, \ldots, x_k$ of the class intervals. If these values occur with respective frequencies $f_1, \ldots, f_k$, then the frequency and probability distributions for X take the form

Value	Frequency	Probability
x_1	f_1	p_1
x_2	f_2	p_2
.	.	.
.	.	.
.	.	.
x_k	f_k	p_k
Totals	n	1

Table 2

where $n = \Sigma f$ is the total number of observations, and $p_i = f_i/n$ is the empirical probability of the value x_i. The probability histogram is constructed as in Example 1, with the bars centered over the values $x_1, \ldots, x_k$.*

Mean and Variance   As we have seen, an empirical frequency distribution for quantitative data gives rise to a probability distribution when we replace frequencies by *relative* frequencies. From Table 2 above we can calculate the grouped mean and variance of the data using the formulas given in Section 11.1.

$$\bar{x} = \frac{\Sigma xf}{n} = \frac{x_1 f_1 + x_2 f_2 + \cdots + x_k f_k}{n}$$

$$= x_1 \frac{f_1}{n} + x_2 \frac{f_2}{n} + \cdots + x_k \frac{f_k}{n}$$

$$= x_1 p_1 + x_2 p_2 + \cdots + x_k p_k$$

$$s^2 = \frac{\Sigma (x - \bar{x})^2 f}{n} = \frac{(x_1 - \bar{x})^2 f_1 + \cdots + (x_k - \bar{x})^2 f_k}{n}$$

$$= (x_1 - \bar{x})^2 \frac{f_1}{n} + \cdots + (x_k - \bar{x})^2 \frac{f_k}{n}$$

$$= (x_1 - \bar{x})^2 p_1 + \cdots + (x_k - \bar{x})^2 p_k$$

From these formulas we see that both the mean $\bar{x}$ and variance s^2 can be expressed in terms of the probabilities $p_i = f_i/n$, instead of the frequencies f_i. Hence we can generalize the notion of mean and variance to apply to any random variable.

---

*Theorem 1 must be modified slightly if the values $x_1, \ldots, x_k$ are not consecutive integers, because the bars in the histogram will not have unit width. This problem does not arise in our later examples.

Definition

Given a random variable X which takes distinct values $x_1, \ldots, x_k$ with respective probabilities $p_1, \ldots, p_k$, the **mean** of X is defined by

$$\mu = \Sigma xp = x_1 p_1 + \cdots + x_k p_k \tag{1}$$

and the **variance** of X is defined by

$$\sigma^2 = \Sigma(x - \mu)^2 p = (x_1 - \mu)^2 p_1 + \cdots + (x_k - \mu)^2 p_k \tag{2}$$

The square root of the variance is called the **standard deviation,** and denoted by σ.*

The mean of X as defined by (1) is the same as the expected value $E(X)$, which was defined in Section 8.3. It predicts the average value X will take over many trials. The variance σ^2 represents the **expected squared deviation** of X, that is, the expected value of $(X - \mu)^2$.

Example 1, Continued

Find the mean and standard deviation for the number of daily requests received by the towing service.

Solution

The probability distribution of Table 1 is reproduced below, together with some auxiliary columns.

Requests	Probability				
x	p	xp	$x - \mu$	$(x - \mu)^2$	$(x - \mu)^2 p$
0	.10	0	-2.1	4.41	.4410
1	.25	.25	-1.1	1.21	.3025
2	.30	.60	-0.1	.01	.0030
3	.20	.60	0.9	.81	.1620
4	.10	.40	1.9	3.61	.3610
5	.05	.25	2.9	8.41	.4205
Total	1.00	2.10			1.6900

Adding the products xp in column 3 we obtain

$$\mu = \Sigma xp = 2.10.$$

This is the mean (or expected) number of daily requests. Using this value of μ we complete the remaining columns of the table. The last column gives us the variance

$$\sigma^2 = \Sigma(x - \mu)^2 p = 1.69.$$

---

*The Greek letters μ(mu) and σ(sigma) are used here instead of $\bar{x}$ and s to emphasize the fact that definitions (1) and (2) apply to any random variable, whether the probability distribution is empirical or theoretical.

Taking the square root, we obtain the standard deviation

$$\sigma = \sqrt{1.69} = 1.3.$$

The reader should check that these results agree with the mean $\bar{x}$ and standard deviation s for the data in Example 1.   ◆

Example 2   Find the mean and standard deviation for the sum obtained when two fair dice are rolled.

Solution   The sum X when two fair dice are rolled has the *theoretical* probability distribution shown below (see Section 7.2, Example 8).

Sum	Probability
2	1/36
3	2/36
4	3/36
5	4/36
6	5/36
7	6/36
8	5/36
9	4/36
10	3/36
11	2/36
12	1/36
Total	1

Adding a column for the product xp, as in Example 1, we find that

$$\mu = \Sigma xp = 2 \cdot \frac{1}{36} + 3 \cdot \frac{2}{36} + 4 \cdot \frac{3}{36} + \cdots + 12 \cdot \frac{1}{36}$$

$$= \frac{252}{36} = 7.$$

That is, the mean (or expected) sum is 7. Computing columns for $x - \mu$, $(x - \mu)^2$, and $(x - \mu)^2 p$ and adding them, we obtain the variance

$$\sigma^2 = \Sigma(x - \mu)^2 p$$

$$= (2 - 7)^2 \cdot \frac{1}{36} + (3 - 7)^2 \cdot \frac{2}{36} + \cdots + (12 - 7)^2 \cdot \frac{1}{36}$$

$$= \frac{210}{36} \doteq 5.83$$

so the standard deviation is given by

$$\sigma = \sqrt{5.83} \doteq 2.42.$$

The histogram for the distribution is shown in Figure 4. As always, the total area under the histogram is 1.

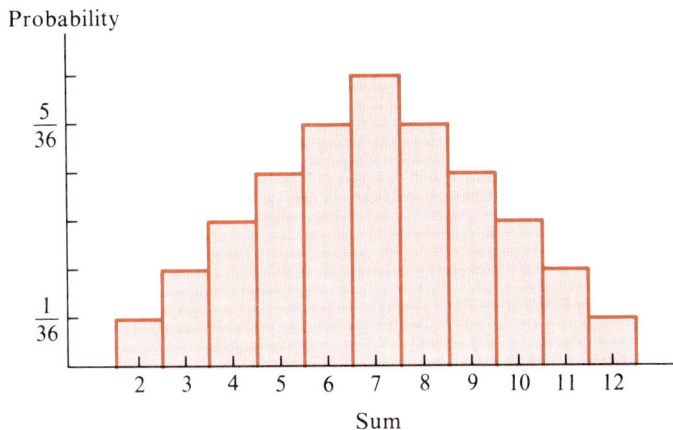

Figure 4 ◆

Binomial Distributions Recall that in Section 8.2 we defined a *Bernoulli process* to be a sequence of *n* trials of an experiment, where

1. each trial has only two possible outcomes, denoted *S* (success) and *F* (failure);
2. the trials are independent; and
3. the probability of success is the same for each trial.

If *X* is the number of successes in a sequence of *n* Bernoulli trials, then *X* is a random variable which takes the possible values $0, 1, 2, \ldots, n$. The distribution of *X* is given by Bernoulli's Formula

$$P(k \text{ successes in } n \text{ trials}) = C(n, k)p^k q^{n-k} \quad (k = 0, 1, \ldots, n)$$

where *p* and *q* are the single-trial probabilities of success and failure, respectively. This distribution is called a **binomial distribution.**[*]

Example 3 Find the probability distribution for the number of heads obtained when a fair coin is tossed ten times.

Solution If we define heads as a "success" and tails as a "failure," then we are interested in the distribution of the random variable

$$X = \text{number of successes in } n \text{ trials}$$

---

[*]For the connection between Bernoulli's Formula and the binomial expansion, see Exercise 26 in Section 8.2.

where $n = 10$ and the single-trial probabilities of success and failure are $p = 1/2$ and $q = 1/2$ respectively. Using Bernoulli's Formula, we arrive at the binomial probability distribution shown in the table below.

Number of heads	Probability
0	.001
1	.010
2	.044
3	.117
4	.205
5	.246
6	.205
7	.117
8	.044
9	.010
10	.001
Total	1.000

For instance, the probability of obtaining exactly 3 heads is computed as follows:

$$P(3 \text{ successes in 10 trials}) = C(10, 3)p^3 q^7$$

$$= 120\left(\frac{1}{2}\right)^3\left(\frac{1}{2}\right)^7$$

$$= \frac{120}{1024} \doteq .117$$

and similarly for the other values. The histogram for this distribution is shown in Figure 5. As we see from the figure, the *most* likely number of

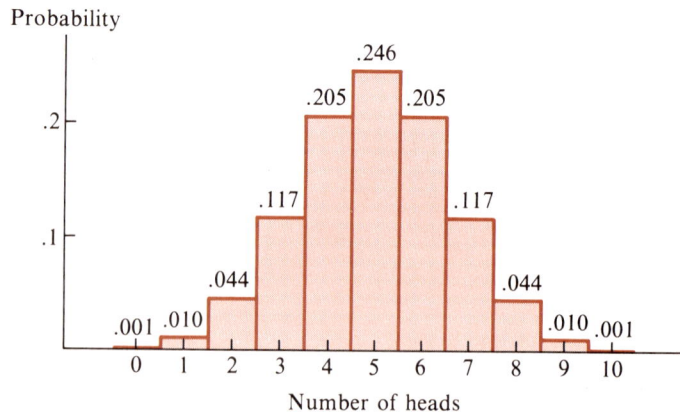

Figure 5

heads is 5, which has probability .246. Using Theorem 1, we find that the probability of tossing *either 4, 5, or 6* heads is given by

$$P(4 \leq X \leq 6) = \text{area between 3.5 and 6.5} = .656$$

while the probability of tossing between 3 and 7 heads is

$$P(3 \leq X \leq 7) = \text{area between 2.5 and 7.5} = .890. \quad \diamond$$

Using formulas (1) and (2) we can calculate the mean and standard deviation for the distribution in the table. Fortunately, however, we have a general result about binomial distributions which makes this computation unnecessary.

Theorem 2

Let X be the number of successes in n Bernoulli trials, where p and q represent the single-trial probabilities of success and failure. Then X has mean

$$\mu = np \tag{3}$$

and standard deviation

$$\sigma = \sqrt{npq}. \tag{4}$$

Example 3, Continued

Find the mean and standard deviation for the number of heads obtained when a fair coin is tossed ten times.

Solution

Here $n = 10$ and $p = q = 1/2$. By Theorem 2, the mean (or expected) number of heads is

$$\mu = np = 10\left(\frac{1}{2}\right) = 5$$

and the standard deviation is

$$\sigma = \sqrt{npq} = \sqrt{10\left(\frac{1}{2}\right)\left(\frac{1}{2}\right)} \doteq 1.58.$$

Formula (3) makes intuitive sense: If the single-trial probability of success is p, then in n trials we would "expect" np successes. Note though that if we toss a fair coin 100 times, the probability of obtaining *exactly* 50 heads is only

$$C(100, 50)\left(\frac{1}{2}\right)^{50}\left(\frac{1}{2}\right)^{50} \doteq .0796$$

even though $np = 50$ is the expected (and indeed, the most likely) value! ◆

Binomial distributions arise in a great many applications. The next example illustrates its use in testing the efficacy of drugs.

Example 4   It is known that 40% of those having a certain disease recover spontaneously, without any treatment. Suppose eight individuals with the disease are given an experimental drug, and five of them recover. What is the probability that this is due merely to chance?

Solution   In other words: Under the hypothesis that the drug has no effect, what is the probability that among eight patients, five or more would recover? To answer this, we view the experiment as a Bernoulli process with $n = 8$ trials, in which the single-trial probability of success (recovery) is $p = .4$. The number X of successes is binomially distributed, with mean

$$\mu = np = 8(.4) = 3.2$$

and standard deviation

$$\sigma = \sqrt{npq} = \sqrt{8(.4)(.6)} \doteq 1.39.$$

The distribution of X is shown in the following table.

Number of successes	Probability
0	.017
1	.090
2	.209
3	.279
4	.232
5	.124
6	.041
7	.008
8	.001

For instance, the probability that exactly three of the eight patients will recover is

$$P(3 \text{ successes in 8 trials}) = C(8, 3)p^3q^5$$
$$= 56(.4)^3(.6)^5$$
$$\doteq .279.$$

From the histogram for this distribution in Figure 6 we see that the probability of 5 or more successes in 8 trials is given by

$$P(X \geq 5) = \text{area to the right of } 4.5$$
$$= .124 + .041 + .008 + .001$$
$$= .174.$$

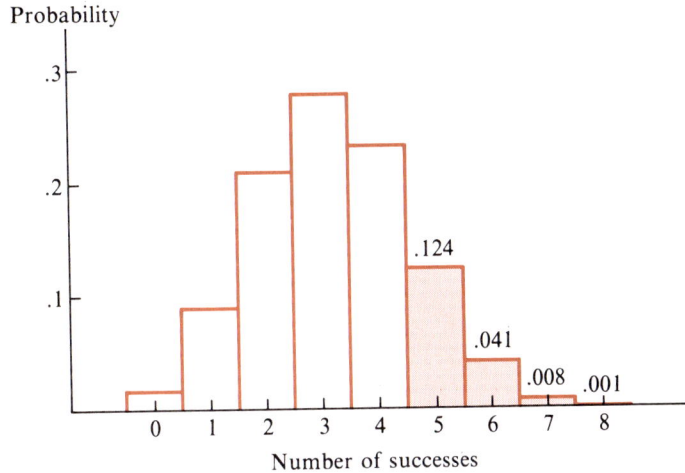

Figure 6

Thus, there is a fairly good chance (17.4%) that five or more patients would have recovered anyway, even without having taken the drug. Few statisticians would regard a success rate of five out of eight as evidence of the drug's efficacy, because of the high probability that it could occur by chance alone. On the other hand, the probability of *six* or more successes out of eight is only

$$P(X \geq 6) = \text{area to the right of } 5.5 = .05.$$

Hence, if six of the patients recovered we would say that the test result was *statistically significant at the 5% level.* ◆

When the number *n* of trials is large the binomial distribution becomes unwieldy. In such cases we can approximate the binomial probability histogram by a smooth curve called the *normal* curve. In the next section we shall see how this is done.

11.2 Exercises

1. An insurance company checked the records of 80 random motorists to determine how many moving violations each had been cited for within the last five years. The results are shown below.

Violations	Frequency
0	34
1	28
2	12
3	4
4	2
Total	80

Find the empirical probability distribution for the number of violations. Represent this distribution by a histogram.

2. A Labor Department statistician interviewed 120 unemployed persons in an inner-city area, asking each one how many years of high school he or she had completed. The responses are shown below.

Number of years	Frequency
0	6
1	24
2	39
3	30
4	21
Total	120

Find the empirical probability distribution for the number of years completed. Represent this distribution by a histogram.

3. A baker finds that the number of chocolate cakes requested on any given day is a random variable with the following distribution:

Requests	Probability
0	.11
1	.20
2	.32
3	.19
4	.12
5	.05
6	.01

(a) Represent this distribution by a histogram.

(b) Using areas under the histogram, find the probability that on a given day the number of requests will be (i) exactly two; (ii) at most two; (iii) at least four.

4. A stockbroker observes that the delivery time for her in-city mailings is a random variable with the following distribution:

Delivery time (working days)	Probability
1	.218
2	.318
3	.272
4	.130
5	.062

(a) Represent this distribution by a histogram.

(b) Using areas under the histogram, find the probability that an item mailed on Monday will arrive (i) on Wednesday, Thursday or Friday; (ii) by Thursday at the latest.

5. A random variable X takes the values $x = 0, 1, 2, 3, 4$, with the probabilities indicated in the histogram below.

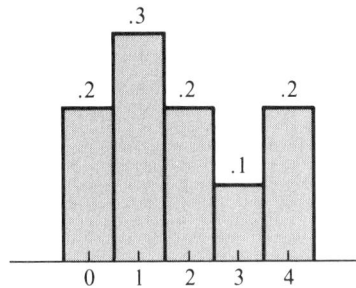

Find the probability that X will assume a value
(a) between 1 and 3, inclusive; (b) greater than or equal to 2; (c) less than 3.

6. A random variable X takes the values $x = 0, 1, \ldots, 10$, with the probabilities indicated in the histogram below.

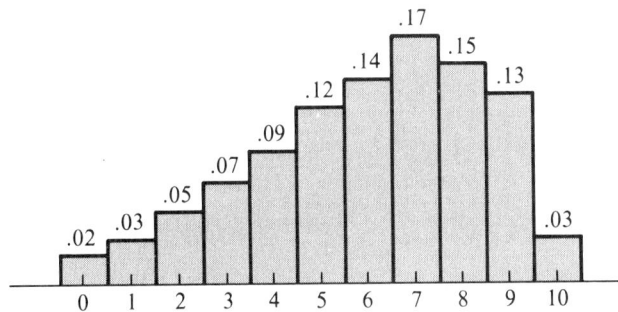

Find the probability that X will assume a value
(a) between 4 and 8, inclusive; (b) greater than 5; (c) less than or equal to 6.

7. In Exercise 1, find the mean and standard deviation for the number of moving violations, using

   (a) the frequency distribution

   (b) the probability distribution.

8. In Exercise 2, find the mean and standard deviation for the number of years of high school completed, using

   (a) the frequency distribution

   (b) the probability distribution.

9. In Exercise 3, find the mean and standard deviation for the number of cakes requested per day.

10. In Exercise 4, find the mean and standard deviation for the delivery time of an in-city mailing.

11. Find the mean and standard deviation of the random variable in Exercise 5.

12. Find the mean and standard deviation of the random variable in Exercise 6.

13. An urn contains five marbles, of which two are red and three are white. A player draws one marble after another out of the urn at random until finally a red marble is obtained. Let X be the number of marbles drawn.

    (a) Find the probability distribution for X.

    (b) Find the mean and standard deviation of X.

†14. Let X be the number of clubs obtained in a random three-card hand.

    (a) Find the probability distribution for X.

    (b) Find the mean and standard deviation of X.

15. Let X be the number of successes in $n = 3$ Bernoulli trials, where $p = q = 1/2$.

    (a) Find the probability distribution for X.

    (b) Represent this distribution by a histogram.

    (c) Find the mean and standard deviation of X.

16. Let X be the number of successes in $n = 5$ Bernoulli trials, where $p = .25$ and $q = .75$

    (a) Find the probability distribution for X.

    (b) Represent this distribution by a histogram.

    (c) Find the mean and standard deviation of X.

17. Let X be the number of times a sum of seven occurs when two fair dice are rolled 90 times. Find the mean and standard deviation of X.

18. Let X be the number of times an 80% shooter hits the target, out of 50 shots taken. Find the mean and standard deviation of X.

19. In American roulette, 18 of the 38 possible outcomes are favorable for red. Suppose a player bets on red 50 times in a row, and let X be the number of times she wins. Find the mean and standard deviation of X.

20. A mail solicitation is sent to 10,000 individuals, each of whom has a 9% chance of responding. Let X be the number of responses. Find the mean and standard deviation of X.

21. A machine produces 0.5% defectives. Let X be the number of defectives in a day's production of 8000 items. Find the mean and standard deviation of X.

22. A true-false exam contains 100 questions, each worth one point. Each question has two possible answers, one of which is correct. Let X be the score of a student who randomly guesses the answer to each question. Find the mean and standard deviation of X.

23. An airline has found that there is a 10% chance that any given person with a plane reservation will fail to show up. Suppose that 350 individuals have been booked for a certain flight, and let X be the number who show up. Find the mean and standard deviation of X.

---

†This exercise requires counting techniques discussed in Section 6.3.

24. Each infant in a pediatric ward needs a nurse's attention 15% of the time. Suppose there are 120 infants in the ward, and let X be the number who need attention at any given moment. Find the mean and standard deviation of X.

25. Seventy-five percent of the voters in a certain district favor a tax cut. Suppose 200 voters are sampled at random, and let X be the number in the sample who favor a tax cut. Find the mean and standard deviation of X.

26. The recovery rate for a certain disease is 60%. A pharmaceutical house claims that its new drug increases the chance of recovery to 80%. Suppose the drug is given to 50 patients with the disease, and let X be the number who recover.

   (a) Find the mean and standard deviation of X, under the assumption that the manufacturer's claim is true.

   (b) Find the mean and standard deviation of X, under the assumption that the drug is in fact worthless.

In Exercises 27 and 28 we denote the mean μ and variance σ^2 of a random variable X by $E(X)$ and $V(X)$, respectively.

27. Prove: For any random variable X and constant k,

   (a) $E(kX) = kE(X)$ and $V(kX) = k^2V(X)$

   (b) $E(X + k) = E(X) + k$ and $V(X + k) = V(X)$.

28. Prove: For any random variable X,

$$V(X) = E(X^2) - E(X)^2 = \Sigma x^2 p - (\Sigma xp)^2$$

11.3 Normal Distributions

To this point we have dealt only with random variables which have finitely many possible values. Probability distributions can also be defined for **continuous** random variables, those whose values are measured on a continuous scale. In this section we discuss the most important case, namely random variables which are normally distributed.

Continuous Distributions When we measure the distance of a javelin throw, the weight of a newborn infant, or the amount of cholesterol in a person's blood, we are dealing with continuous random variables. While we can approximate such measurements by rounding them off to a finite number of places, the fact remains that—in theory, at least—the possible values cover a continuous interval of real numbers.

   To obtain the probability distribution for a continuous random variable X, we imagine dividing the interval of possible values into many small subintervals, as in Figure 1. Over each subinterval (5.3 to 5.4 in the figure) we construct a bar whose *area* represents the probability that X will assume a value in that subinterval. Using smaller and smaller subintervals, we find that the resulting histogram approaches a smooth curve (Figure 2). This curve is called the **probability distribution** (or **probability density curve**) for X. If we think of it as a histogram with "infinitely thin" bars, we see that it has the following property:

The area under the curve between any two values a and b represents the probability that X will assume a value in this interval.

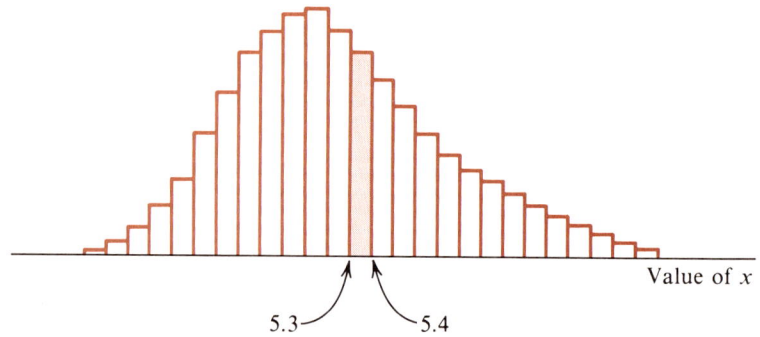

Figure 1

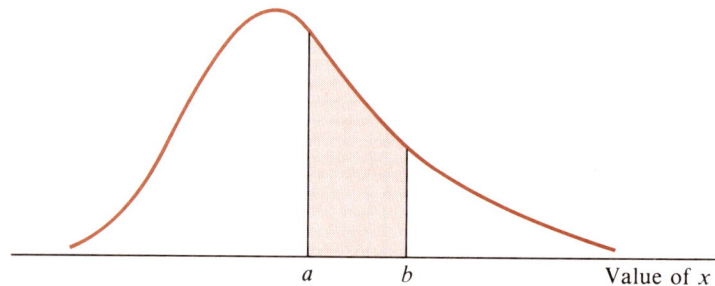

Figure 2

In particular, the *total* area under the curve is always 1. Note that when $b = a$ in Figure 2, the area between them is zero. Hence, *the probability that a continuous random variable will take any particular value a is zero*. This is not as unreasonable as it sounds at first. For instance, it should seem highly unlikely—impossible, for all practical purposes—that the next throw of a javelin will result in a distance of *exactly* 71.6325 meters. In any case, this points up an important difference between finite and continuous random variables. For the former we consider the probability of a given *value*, but for the latter we consider the probability of a given *interval*.

Using calculus it is possible to define the mean and variance of a continuous random variable X. These numbers have the same significance as in the finite cases, namely, the mean μ represents the expected value of X, while the variance σ^2 represents the expected squared deviation of X; that is, the expected value of $(X - \mu)^2$. Thus, μ measures the central tendency or "middle" of the distribution, while σ measures its dispersion or "spread."

Normal Distributions   In his study of random experimental errors, the French mathematician Abraham de Moivre (1667–1754) discovered that repeated measurements of a physical quantity tend to be distributed according to a precise mathematical pattern known as a **normal distribution.*** Figure 3 shows a typical normal distribution, with mean μ and standard deviation σ. There is a unique normal curve for each value of μ and σ, and it can be shown (using calculus) that every normal curve has the following properties:

Properties of a Normal Curve

1. The curve is bell-shaped, with its highest point at $x = \mu$.
2. The curve extends infinitely in both directions, and comes arbitrarily close to the x-axis.
3. The curve is symmetric with respect to the vertical line $x = \mu$.
4. The total area under the curve is 1.

Figure 3 also shows the points $x = \mu \pm \sigma$ located at a distance σ on either side of the mean. It can be shown that a normal curve twists or *inflects* at these points: the curve is concave downward (like an inverted bowl) between $\mu - \sigma$ and $\mu + \sigma$, and concave upward outside this interval.

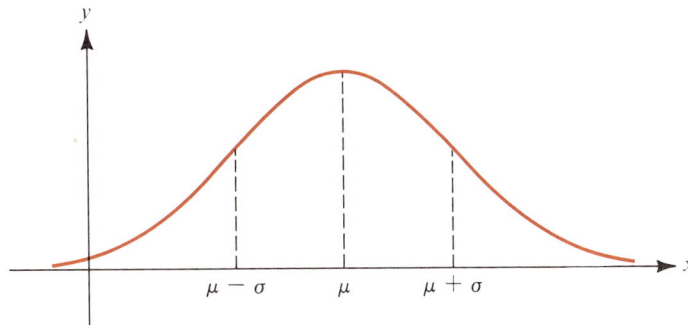

Figure 3

The mean μ determines the position of a normal curve, while the standard deviation σ determines its shape. This is illustrated in Figure 4, which shows three normal curves having the same mean $\mu = 5$. Note that a small

---

*Or **Gaussian** distribution.

value of σ gives a tall, narrow curve, indicating that the values are concentrated near the mean; while a large value of σ gives a lower, flatter curve, indicating a greater dispersion or spread of the values.

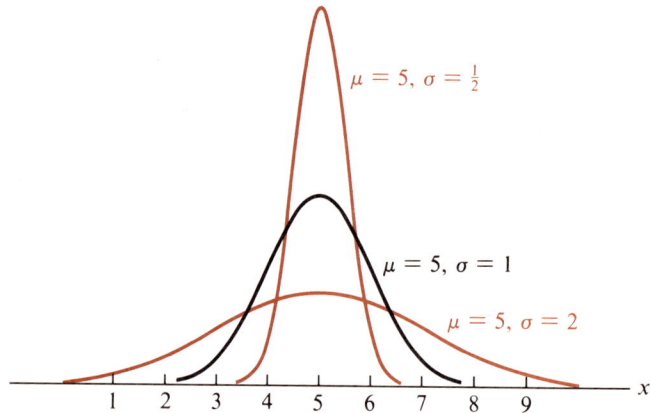

Figure 4

Standard Units   It can be shown that for any normal distribution, approximately 68.26% of the area under the curve will lie within one standard deviation of the mean, and approximately 95.44% of the area will lie within two standard deviations of the mean (see Figure 5). We can convert any x-value into **standard units** (or **z-units**) by the formula

$$z = \frac{x - \mu}{\sigma}. \tag{1}$$

This just amounts to a change of scale as illustrated in Figure 5, with z measuring the number of deviations above or below the mean for a given value of x. In terms of standard units, the figure shows that 68.26% of the area under a

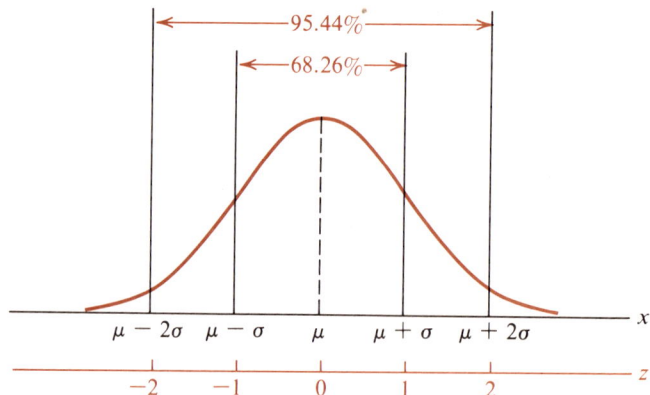

Figure 5

normal curve lies between $z = -1$ and $z = 1$, while 95.44% of the area lies between $z = -2$ and $z = 2$.

The area under a normal curve between any two z-values is a constant, independent of the mean μ or standard deviation σ of the distribution.

Table 1 in the Appendix gives *the area $A(z)$ under a normal curve to the left of z,* as illustrated in Figure 6. Using this table we can find the area under a normal curve between any two z-values, by subtraction.

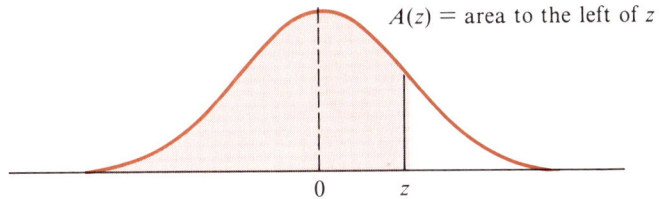

$A(z)$ = area to the left of z

Figure 6

Example 1 A normal distribution has mean 20 and standard deviation 4. Find the area under the curve between 17 and 22.

Solution The region is shaded in Figure 7. Using formula (1), we convert $x = 17$ and $x = 22$ to standard units.

$$x = 17 \qquad z = \frac{17 - 20}{4} = -.75$$

$$x = 22 \qquad z = \frac{22 - 20}{4} = .50$$

That is, $x = 17$ lies .75 standard deviations below the mean, while $x = 22$ lies .50 standard deviations above the mean. From Table 1 in the Appendix we find that $A(-.75) = .2266$ and $A(.50) = .6915$. Hence, the area between

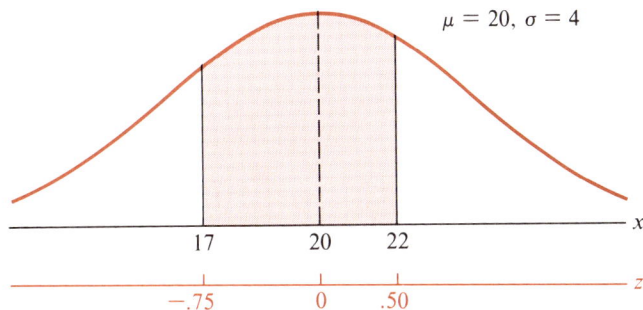

$\mu = 20, \sigma = 4$

Figure 7

$x = 17$ and $x = 22$ is given by

$$A(.50) - A(-.75) = .6915 - .2266$$
$$= .4649.$$

That is, 46.49% of the total area under the curve lies between these two values.   ◇

Example 1,   For the distribution with mean 20 and standard deviation 4, find the area un-
Continued   der the curve to the right of 24.

Solution   In standard units, $x = 24$ becomes $z = \dfrac{24 - 20}{4} = 1.00.$

That is, $x = 24$ lies one standard deviation above the mean (see Figure 8). From Table 1, the area to the *left* of $x = 24$ is $A(1.00) = .8413$. Since the total area under the curve is 1, the area to the *right* of $x = 24$ is given by

$$1 - A(1.00) = 1 - .8413$$
$$= .1587.$$

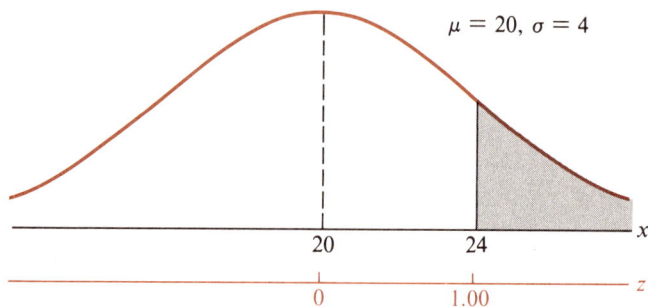

Figure 8

Applications   Many of the random variables that arise in statistical applications turn out to be normally distributed. In such cases we can calculate probabilities by finding areas under a normal curve.

Example 2   The annual rainfall in a certain city follows a normal distribution, with mean 48.5 inches and standard deviation 6.2 inches. Find the probability that the rainfall in a given year will be less than 42 inches.

Solution   The amount X of rainfall per year (in inches) has the normal distribution shown in Figure 9. In standard units, $x = 42$ becomes $z = -1.05$. Hence,

$$P(X < 42) = \text{area to the left of 42}$$
$$= A(-1.05)$$
$$= .1469.   \text{(from Table 1)}$$

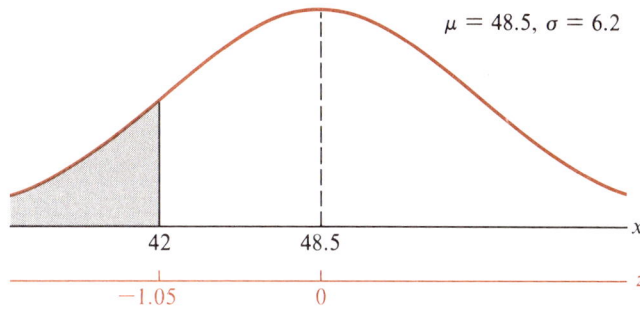

$\mu = 48.5, \sigma = 6.2$

Figure 9

Example 3 The mean systolic blood pressure of adult males is 138 mm Hg (millimeters of mercury), with a standard deviation of 9.7. Assuming a normal distribution, what percentage of adult males have blood pressure between 145 and 150 mm Hg?

Solution Let X be the systolic blood pressure of a random adult male. Then X has the normal distribution shown in Figure 10, with mean $\mu = 138$ and standard deviation $\sigma = 9.7$. In standard units, $x = 145$ and $x = 150$ become $z = .72$ and $z = 1.24$, respectively. Hence,

$$
\begin{aligned}
P(145 \le X \le 150) &= A(1.24) - A(.72) \\
&= .8925 - .7642 \qquad \text{(from Table 1)} \\
&= .1283.
\end{aligned}
$$

That is, approximately 12.83% of all adult males will have blood pressure in this range.

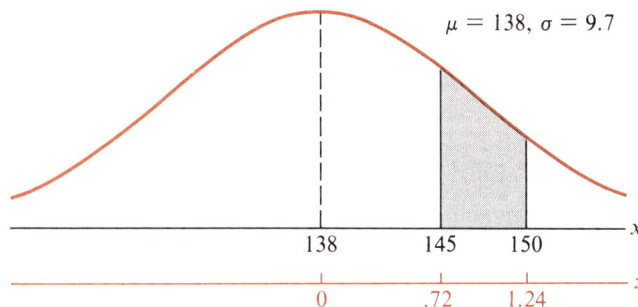

$\mu = 138, \sigma = 9.7$

Figure 10

Normal Approximation to the Binomial Distribution One of the most important applications of normal distributions rests on the fact that they give good approximations to binomial histograms.

Example 4   Let X be the number of heads obtained when a fair coin is tossed ten times.

Discussion   We saw in Section 11.2, X is binomially distributed, with mean

$$\mu = np = 5$$

and standard deviation

$$\sigma = \sqrt{npq} \doteq 1.58.$$

The histogram for this distribution is reproduced in Figure 11, together with the normal curve having the same mean and standard deviation ($\mu = 5$, $\sigma = 1.58$). As we see from the figure, the normal curve gives a remarkably good "smooth" approximation to the binomial histogram. Consequently, we can use areas under the normal curve to approximate areas under the histogram.

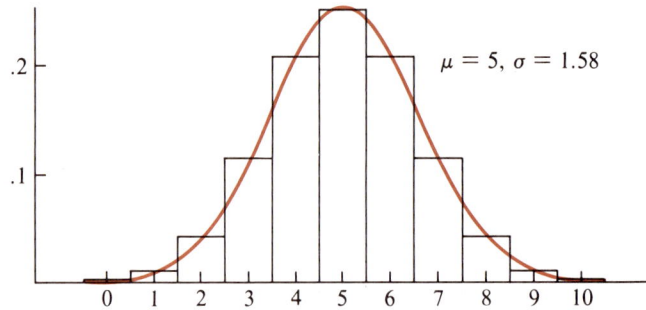

Figure 11

Example 4,   Use the normal approximation to find the probability of tossing
Continued   (a) exactly five heads;
   (b) either four, five, or six heads.

Solution   (a) By Theorem 1 of Section 11.2, the probability of tossing exactly five heads is equal to the area under the histogram between 4.5 and 5.5. This is approximately equal to *the area under the normal curve* between 4.5 and 5.5. Converting these values to standard units, we obtain

$$x = 4.5 \qquad z = \frac{4.5 - 5}{1.58} \doteq -.32$$

$$x = 5.5 \qquad z = \frac{5.5 - 5}{1.58} \doteq .32.$$

Hence,

$$P(X = 5) \doteq A(.32) - A(-.32)$$
$$= .6255 - .3745 \qquad \text{(from Table 1)}$$
$$= .2510.$$

The true value (to three places) was found to be .246 in Section 11.2.

(b) The probability of tossing either four, five, or six heads is equal to the area under the histogram between 3.5 and 6.5. Since these values become $-.95$ and $.95$ in standard units, the normal approximation gives

$$P(4 \leq X \leq 6) \doteq A(.95) - A(-.95)$$
$$= .8289 - .1711 \qquad \text{(from Table 1)}$$
$$= .6578.$$

This is quite close to the true value .656, which we calculated in Section 11.2. ◆

In general, the **normal approximation to a binomial distribution** is more accurate when it is used over a larger interval of values, as we see from the example above. Also, the approximation improves as the number n of trials increases. A precise statement of this result, known as the **De Moivre-Laplace Limit Theorem,** is beyond the scope of this text. However, we can summarize it roughly as follows.

Let X be the number of successes in n Bernoulli trials, where p and q represent the single-trial probabilities of success and failure. If n is large and neither p nor q is close to 0, the probability histogram for X can be approximated by the normal curve with mean

$$\mu = np$$

and standard deviation

$$\sigma = \sqrt{npq}.$$

As a practical rule of thumb, *both np and nq should be greater than 5* when we use a normal curve to approximate a binomial distribution.

Example 5 It is known that 40% of those having a certain disease recover spontaneously, without any treatment. Suppose 20 individuals with the disease are given an experimental drug, and 13 of them recover. What is the probability that this is due merely to chance?

Solution If we assume the drug had no effect, then the 20 patients represent $n = 20$ Bernoulli trials, with probability $p = .4$ of success (recovery) on each trial. The number X of successes is binomially distributed, with mean

$$\mu = np = 20(.4) = 8$$

and standard deviation

$$\sigma = \sqrt{npq} = \sqrt{20(.4)(.6)} \doteq 2.19.$$

Since $np = 8$ and $nq = 12$ both exceed 5, we can approximate this distribution by a normal curve having the same mean and standard deviation (see

Figure 12). The probability of 13 or more successes in 20 trials is approximately equal to the area under the curve to the right of 12.5. Since 12.5 becomes 2.05 in standard units, the normal approximation gives

$$P(X \geq 13) \doteq 1 - A(2.05)$$
$$= 1 - .9798 \qquad \text{(from Table 1)}$$
$$= .0202.$$

That is, there is about a 2% probability that the observed number of recoveries would have occurred by chance alone. Since this value is less than 5% but greater than 1%, we could say that the experimental evidence of the drug's effectiveness is significant at the 5% level, but not at the 1% level.

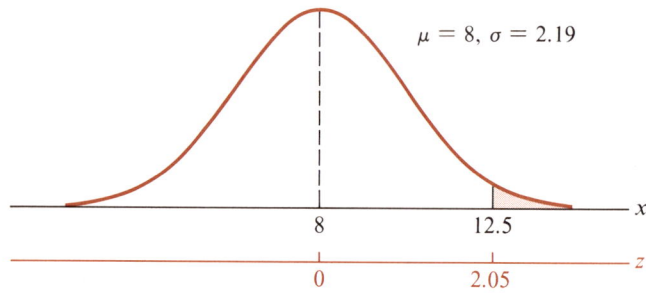

$\mu = 8, \sigma = 2.19$

Figure 12

Example 6   Actuarial tables indicate an annual accidental death rate in the U.S. of 56 persons per 100,000. If an insurance company covers 500,000 individuals for accidental death benefits, what is the probability that the number of claims in a given year will be between 260 and 300 (inclusive)?

Solution   The number X of claims among the policyholders is binomially distributed, with $n = 500,000$, $p = .00056$, and $q = .99944$. The mean or expected number of claims is

$$\mu = np = 280$$

and the standard deviation is

$$\sigma = \sqrt{npq} \doteq 16.73.$$

Since np and nq both exceed 5, we can use a normal approximation (Figure 13). The probability that the number of claims will be between 260 and 300 is represented by the area under the normal curve between 259.5 and 300.5. Converting these values to standard units, we obtain

$$P(260 \leq X \leq 300) \doteq A(1.23) - A(-1.23)$$
$$= .8907 - .1093 \qquad \text{(from Table 1)}$$
$$= .7814.$$

That is, the chances are about 78% that the number of claims will fall in this range.

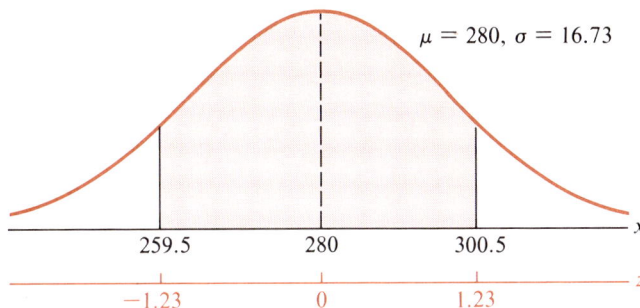

Figure 13

---

1. Find the area under a normal curve, over the given z-interval:

   (a) to the left of $z = .58$

   (b) between $z = -1.25$ and $z = 1.25$

   (c) to the right of $z = -.02$.

2. Find the area under a normal curve, over the given z-interval:

   (a) to the left of $z = -1.04$

   (b) between $z = .47$ and $z = 1.63$

   (c) to the right of $z = 2.81$.

3. Find the value of z (standard units) for which the shaded region under a normal curve has the indicated area.

   (a) Area .1635

   (b) Area .1020

   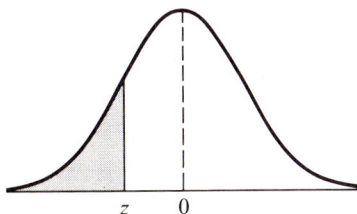
   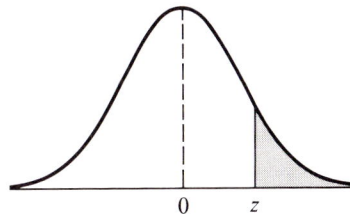

4. Find the value of z (standard units) for which the shaded region under a normal curve has the indicated area.

   (a) Area .3708

   (b) Area .6266

   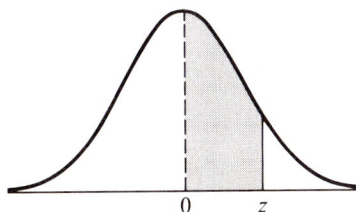
   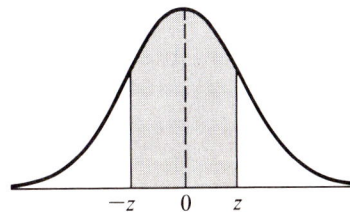

5. A normal distribution has mean 120 and standard deviation 12. Find the area under the curve

    (a) to the left of 100   (b) between 110 and 130   (c) to the right of 150.

6. A normal distribution has mean 8.75 and standard deviation .05. Find the area under the curve

    (a) to the left of 8.77   (b) between 8.62 and 8.68   (c) to the right of 8.72.

7. Find the x-value for which the shaded region under the given normal curve has the indicated area.

    (a)   Area .6915   (b)   Area .0668

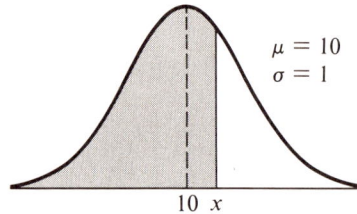 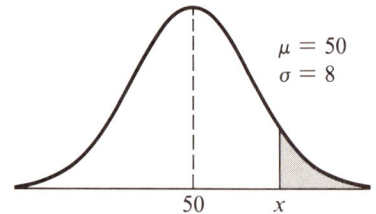

8. Find the x-value for which the shaded region under the given normal curve has the indicated area.

    (a)   Area .2734   (b)   Area .8664

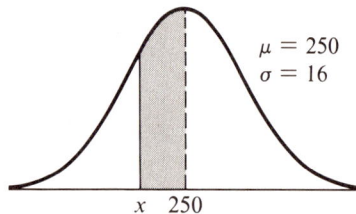 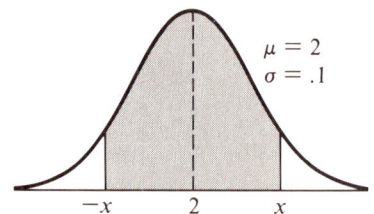

9. A skier finds that her time on a certain ski course follows a normal distribution, with mean 4 minutes, 15 seconds and standard deviation 12 seconds. What is the probability that her time will be under 4 minutes on a given run?

10. In order to function properly a certain machine part must have diameter 12 cm (centimeters), to within a tolerance of .01 cm. If the diameters of these parts are normally distributed with mean 12 cm and standard deviation .004 cm, what percentage of them will fail to function?

11. A certain brand of light bulb has a mean lifetime of 742 hours, with a standard deviation of 105 hours. Assuming a normal distribution, what proportion of the bulbs will last at least 700 hours?

12. The patients in a doctor's office wait an average of 21 minutes, with a standard deviation of 7 minutes. Assuming a normal distribution, what percentage of the patients wait between 15 minutes and a half hour?

13. The weekly demand for poultry at a certain market is normally distributed, with a mean of 2700 lb and a standard deviation of 240 lb.

    (a) If the market stocks 2500 lb per week, what proportion of the time will it all be sold?

(b) Approximately how much should the market stock per week in order to be 90% sure of selling it all?

14. Scholastic Aptitude Test scores are normally distributed, with mean 500 and standard deviation 100.

   (a) What percentage of the scores are above 550?

   (b) Approximately what score puts someone in the top 10%?

15. The heights of adult males are normally distributed, with mean 69 inches and standard deviation 2.5 inches. Approximately what height range contains the middle 50% of adult males?

16. The mean annual salary of workers in a certain profession is $28,500, with a standard deviation of $2300. Assuming the salaries are normally distributed, approximately what salary range contains the middle 75% of the workers?

In Exercises 17–26 use a normal approximation to a binomial distribution. See the corresponding exercises in Section 11.2.

17. If two fair dice are rolled 90 times, what is the probability that a sum of seven will occur between 10 and 20 times (inclusive)?

18. What is the probability that an 80% shooter will hit a target between 35 and 45 times (inclusive), out of 50 shots taken?

19. Suppose a player repeatedly bets $1 on red in roulette. What is the probability that she will be at least $10 ahead after 50 rolls—in other words, that red will come up at least 30 times?

20. A mail solicitation is sent to 10,000 individuals, each of whom has a 9% chance of responding. What is the probability that at least 850 individuals will respond?

21. A machine produces 0.5% defectives. What is the probability that a day's production of 8000 items will contain at most 50 defectives?

22. A true-false exam contains 100 questions, each worth one point. If a student guesses the answer to each question at random, what is the probability that he will score at most 60?

23. An airline has found that there is a 10% chance that any given person with a plane reservation will fail to show up. Suppose 350 individuals have been booked for a flight which has only 320 seats. What is the probability that everyone who shows up will be seated?

24. Each infant in a pediatric ward needs a nurse's attention 15% of the time. Suppose there are 120 infants in the ward and 20 nurses on call. What percentage of the time will some infant go unattended?

25. Seventy-five percent of the voters in a certain district favor a tax cut. Suppose 200 voters are sampled at random. What is the probability that between 70% and 80% of those sampled (inclusive) will favor a tax cut?

26. The recovery rate for a certain disease is 60%. A pharmaceutical house claims that its new drug increases the chance of recovery to 80%. Suppose the drug is given to 50 patients with the disease, and 35 of them recover.

   (a) What is the probability of a result this *low*, if the maker's claim is true?
   (b) What is the probability of a result this *high*, if in fact the drug is worthless?

Computer Projects

A. The NORMAL program computes areas under a normal curve. This program can be used in Exercises 1–2, 5–6, 9–14, and 17–26.

B. For a Bernoulli process with given n, p, and q, the BINOM program computes the theoretical probability of obtaining between k_1 and k_2 successes, inclusive. This program can be used to compare the normal and binomial distributions in Exercises 17–26.

Chapter 11 Summary

Important Terms

11.1 histogram
grouped data
mean $\bar{x}$
median $\tilde{x}$
variance s^2
standard deviation s

11.2 random variable
mean μ

variance σ^2
standard deviation σ
binomial distribution

11.3 normal distribution
standard units
normal approximation to the
binomial distribution

Review Exercises

1. The typing speeds (in words per minute) of 80 secretaries were distributed as follows:

Speed (wpm)	Secretaries
40–50	8
50–60	12
60–70	31
70–80	22
80–90	7

(a) Represent these data by means of a histogram.

(b) Find the mean and standard deviation for these grouped data.

2. A group of 25 high school seniors had the following scores on a scholastic aptitude test:

672	419	523	614	531
724	367	431	602	612
569	472	763	525	643
675	798	664	372	599
518	349	617	558	593

(a) Write these data in grouped form, using 300–400 as the first class interval. Represent the grouped data with a histogram.

(b) Find the mean and standard deviation for these grouped data. How does the grouped mean compare with the true mean?

3. Five independent gem experts were asked to estimate the value of a diamond ring. Their appraisals were as follows:

$5000 $7000 $6500 $8000 $7500

Find the mean, median, and standard deviation for these ungrouped data.

4. The percentage price increases in a certain year for ten selected consumer goods are shown below:

7%   12%   8%   10%   11%

9%   13%   126%   14%   10%

Find the mean and median for these ungrouped data. Which of these two measures would give you a better idea of the "typical" increase in consumer prices for the year?

5. A stereo equipment salesman finds that the number of tape decks sold in a given day is a random variable with the following distribution:

Units sold	Probability
0	.19
1	.35
2	.24
3	.12
4	.09
5	.01

Find the mean and standard deviation for the number of units sold per day.

6. An urn contains 60 marbles, of which 10 are red, 20 are white, and 30 are blue. A player chooses a single marble at random from the urn. If it is red he wins $4 and if it is white he wins $1, but if it is blue he loses $2. Find the mean and standard deviation for the player's net gain.

7. Let X be the number of successes in $n = 15$ Bernoulli trials, where $p = 1/3$ and $q = 2/3$. Find the mean and standard deviation of X.

8. Thirty percent of the registered voters in a certain area are senior citizens. Suppose that 50 voters are chosen at random for jury duty, and let X be the number of senior citizens chosen. Find the mean and standard deviation of X.

9. Observing many trees of a certain species, a botanical research team finds that their heights are normally distributed, with mean 8.44 meters and standard deviation 2.4 meters. What percentage of the trees are more than 10 meters tall?

10. The speeds of cars on a certain interstate highway are found to be normally distributed, with mean 61.3 mph and standard deviation 8.4 mph. What percentage of the cars are observing the speed limit of 55 mph?

11. If a fair coin is tossed 100 times, what is the probability that heads will come up between 45 and 55 times (inclusive)?

12. It is known that 25% of those having a certain disease recover spontaneously. Suppose 20 individuals with the disease are given an experimental drug, and exactly half of them recover. What is the probability that a recovery rate this high is due merely to chance?

12 Mathematics of Finance

Everyone who has a savings account or makes installment purchases has had some contact with the world of finance. In this chapter we investigate the mathematics involved in the use of money, employing as our central tool the notion of a *difference equation*. As we shall see, difference equations can be used to analyze simple and compound interest, mortgages and other installment loans, annuities, various kinds of depreciation, and other processes involving periodic growth or decline.*

12.1 Simple and Compound Interest

Interest is money paid for the use of someone else's money over a period of time. In this section we develop the basic equations governing simple and compound interest, which enter into virtually all financial calculations.

Difference Equations   An infinite list of numbers

$$x_0, \quad x_1, \quad x_2, \quad \cdots$$

is called a **sequence** or **progression**. In this section and the next we shall deal with sequences defined by conditions of the form

$$x_{n+1} = ax_n + b \qquad\qquad (\star)$$

where a and b are constants. An equation of this type is called a **first-order linear difference equation.** Given the **initial value** x_0, we can use ($\star$) to successively calculate all the terms in the sequence.

---

*We assume the reader has a calculator with an exponential function y^x.

Example 1 Find the first five terms of the sequence generated by the difference equation

$$x_{n+1} = 2x_n + 3$$

with initial value $x_0 = 1$.

Discussion Letting $n = 0, 1, 2, 3, 4$ in the above equation, we successively compute

$$
\begin{aligned}
x_0 &= 1 \\
x_1 &= 2x_0 + 3 = 5 \\
x_2 &= 2x_1 + 3 = 13 \\
x_3 &= 2x_2 + 3 = 29 \\
x_4 &= 2x_3 + 3 = 61
\end{aligned}
$$

Thus, the given equation generates the sequence

$$1, \quad 5, \quad 13, \quad 29, \quad 61, \quad \dots \quad \blacklozenge$$

Solving a difference equation means finding an explicit formula for x_n in terms of n. We shall see in Section 12.2 that the above equation has the solution

$$x_n = 4 \cdot 2^n - 3$$

which the reader can check yields the first five values calculated above. In this section we shall need only the following two special cases of equation ($\star$).

Case 1: $a = 1$ In this case the difference equation ($\star$) becomes

$$x_{n+1} = x_n + b. \tag{$\star\star$}$$

That is, each term differs from the preceding term by a fixed amount b. Such a sequence is called an **arithmetic progression.** Given the initial value x_0, equation ($\star\star$) successively generates the terms

$$
\begin{aligned}
x_0 & \\
x_1 &= x_0 + b \\
x_2 &= x_1 + b = x_0 + 2b \\
x_3 &= x_2 + b = x_0 + 3b \\
x_4 &= x_3 + b = x_0 + 4b
\end{aligned}
$$

with the general term given by

$$x_n = x_0 + nb. \tag{1}$$

That is, formula (1) is the explicit solution of equation ($\star\star$).

Case 2: $b = 0$ In this case, equation ($\star$) becomes

$$x_{n+1} = ax_n. \tag{$\star\star\star$}$$

Here the *ratio* of successive terms is constant; that is, $x_{n+1}/x_n = a$. Such a sequence is called a **geometric progression.** From the first few values

$$x_0$$
$$x_1 = ax_0$$
$$x_2 = ax_1 = a^2x_0$$
$$x_3 = ax_2 = a^3x_0$$
$$x_4 = ax_3 = a^4x_0$$

we see that the general term is given by

$$x_n = a^n x_0. \tag{2}$$

Example 2   Solve the difference equations

(a) $x_{n+1} = x_n + 5$, $x_0 = -8$;   (b) $x_{n+1} = 3x_n$, $x_0 = 7$.

Solution   (a) The first few terms in the sequence are

$$-8, \quad -3, \quad 2, \quad 7, \quad 12, \quad 17, \quad \ldots .$$

This is an arithmetic progression, with difference 5 between successive terms. By formula (1), the general term is

$$x_n = -8 + 5n.$$

Thus, for example, we find that $x_{10} = 42$, $x_{50} = 242$, and $x_{100} = 492$.
(b) Here the sequence starts out

$$7, \quad 21, \quad 63, \quad 189, \quad 567, \quad \ldots .$$

This is a geometric progression, in which successive terms have the constant ratio 3. Formula (2) gives the solution

$$x_n = 3^n \cdot 7.$$

The terms in this progression grow very rapidly. For instance, x_{10} is almost a half million, x_{15} is over 100 million, and x_{50} exceeds the number of grains of sand on all the beaches in the world!   ◆

Arithmetic and geometric progressions arise quite naturally in the mathematics of finance. When money is invested in a savings account or lent to a commercial borrower, the initial amount is called the **principal** and the cost of using it is called **interest.** The sum of principal and interest is the **accumulated sum** (or **balance**) on the investment or loan.

Simple Interest   There are several different ways of charging interest. **Simple interest** is computed as a certain fraction or percentage of the principal, per unit time.

Example 3 An individual borrows $5000 at 12% per annum simple interest. What is the total debt at the end of three years?

Solution The interest each year is 12% of the principal.

$$(.12)(5000) = 600 \quad \text{(dollars)}$$

Thus, the total debt will be

$$\$5000 + \$600 = \$5600 \quad \text{after one year,}$$
$$\$5600 + \$600 = \$6200 \quad \text{after two years,}$$
$$\$6200 + \$600 = \$6800 \quad \text{after three years.}$$

That is, the borrower must repay $6800 at the end of three years, of which $5000 represents repayment of the principal and $1800 represents interest. ◇

In this example the time unit, or **interest period,** is one year and the **interest rate** is .12 (12%) per period. More generally, suppose a principal P earns simple interest at a rate r per period, and let S_n denote the accumulated sum, or balance, after n periods. Since the interest for each period is rP, we have the relation

$$S_{n+1} = S_n + rP. \tag{3}$$

That is, the balance after $n + 1$ periods is the balance after n periods, plus the interest for one period. This is a difference equation of type (★★), with initial value $S_0 = P$. By formula (1), its solution is

$$S_n = S_0 + n(rP)$$
$$= P + nrP$$
$$= (1 + nr)P.$$

We therefore have the following formula:

Suppose a principal P earns simple interest at a rate r per period. Then the accumulated sum after n periods is

$$S = (1 + nr)P. \tag{4}$$

Thus in Example 3, with $P = 5000$, $r = .12$ and $n = 3$, we obtain

$$S = (1 + 3(.12))(5000)$$
$$= (1.36)(5000)$$
$$= 6800 \quad \text{(dollars)}$$

which agrees with our earlier result. Formula (4) is used for simple interest calculations even when n is not a whole number, as in the next problem.

Example 4   A manufacturer borrows $250,000 for 9 months at 15% per annum simple interest. How much will he owe when the loan is due?

Solution   Since the interest period is one year, the term of the loan is $n = 9/12 = 3/4$ periods. Thus, the total debt will be

$$S = (1 + nr)P$$
$$= \left(1 + \frac{3}{4}(.15)\right)(250,000)$$
$$= (1.1125)(250,000)$$
$$= 278,125 \quad \text{(dollars)}.$$

That is, the manufacturer must repay the $250,000 principal, plus an additional $28,125 in interest. ◆

Compound Interest   **Compound interest** is computed as a certain fraction or percentage of the *accumulated sum,* instead of the principal. In other words, interest is charged on accumulated interest as well as the original investment or loan.

Example 3, Revisited   An individual borrows $5000 at 12% per annum *compound* interest. What is the total debt at the end of three years?

Solution   The interest for the first year is 12% of the principal, or $600. Thus, at the end of the first year the borrower owes

$$\$5000 + \$600 = \$5600.$$

The lender now treats this amount as the new principal: That is, for the second year of the loan the interest charge is 12% of $5600, or $672. Hence, at the end of two years the borrower owes

$$\$5600 + \$672 = \$6272.$$

For the third year the interest is 12% of $6272, or $752.64. Thus, the total debt at the end of three years is

$$\$6272 + \$752.64 = \$7024.64.$$

Note that this is significantly more than the corresponding debt under simple interest. In fact, a 12% per annum compound interest rate over a three-year period is equivalent to about a 13.5% simple interest rate.* ◆

To develop a general formula, let us assume that a principal P earns compound interest at a rate r per period, and let S_n denote the accumulated

---

*This is assuming that no payments are made until the end of the third year, when the loan is due. The case of paying off a loan in installments is discussed in Section 12.2.

sum, or balance, after n periods. Then the interest for the next period (period $n + 1$) will be rS_n, so the balance after $n + 1$ periods is given by

$$S_{n+1} = S_n + rS_n = (1 + r)S_n. \tag{5}$$

This is a difference equation of type (★★★), with initial value $S_0 = P$. By formula (2), its solution is

$$\begin{aligned} S_n &= (1 + r)^n S_0 \\ &= (1 + r)^n P. \end{aligned}$$

In other words:

Suppose a principal P earns compound interest at a rate r per period. Then the **accumulated sum** after n periods is

$$S = (1 + r)^n P. \tag{6}$$

Thus, in our example where $5000 is borrowed at 12% per annum compound interest, the total debt after three years will be

$$\begin{aligned} S &= (1 + .12)^3(5000) \\ &= (1.12)^3(5000) \\ &= (1.404928)(5000) \\ &= 7024.64 \quad \text{(dollars)} \end{aligned}$$

as we already found out the long way.

The compound interest rate on savings accounts is usually stated as a certain percentage per annum (year). Often, though, there are several partial interest payments during the year, which result in a somewhat higher growth rate.

Example 5   Compare the accumulated sum when $1000 is deposited for five years at 9% per annum interest, compounded
(a) annually   (b) quarterly   (c) monthly.

Solution   (a) With annual compounding the interest period is one year, and the interest rate is $r = .09$ per period. Hence, after $n = 5$ years we have

$$\begin{aligned} S &= (1 + r)^n p \\ &= (1.09)^5(1000) \\ &\doteq 1538.62 \quad \text{(dollars)}. \end{aligned}$$

(b) With quarterly compounding the interest period is three months rather than a year, and the periodic interest rate is $.09/4 = .0225$. That is, at the end of each quarter the balance is increased by 2.25%, instead of increasing it by 9% at the end of each year. We therefore have a principal $P = \$1000$ earning compound interest at a rate $r = .0225$ per period, for

$n = 5(4) = 20$ periods (5 years). By formula (7), the accumulated sum is

$$S = (1.0225)^{20}(1000)$$
$$\doteq 1560.51 \quad \text{(dollars)}.$$

(c) Compounding monthly we have $n = 5(12) = 60$ periods, with an interest rate $r = .09/12 = .0075$ per period. Hence

$$S = (1.0075)^{60}(1000)$$
$$\doteq 1565.68 \quad \text{(dollars)}. \quad \blacklozenge$$

As the frequency of compounding increases so does the sum S, but with a steadily diminishing gain. Daily compounding in Example 5, for instance, gives an accumulated sum \$1568.22, and *continuous* compounding—every millionth of a second, say—gives \$1568.31.

Nominal and Effective Rates   Since the rate of growth of a savings deposit depends on both the annual interest rate and the frequency of compounding, it is sometimes useful to convert the stated or **nominal** interest rate into an equivalent simple rate of interest, called the **effective rate.**

Example 6   What is the effective interest rate corresponding to 8% per annum compounded quarterly?

Solution   We choose an arbitrary principal, say $P = \$100$, and see what happens to it in one year. The nominal interest rate of 8% per annum actually represents a rate of $r = .08/4 = .02$ per quarter. Hence in $n = 4$ quarters (1 year), the principal will grow to

$$S = (1.02)^4(100)$$
$$\doteq 108.24 \quad \text{(dollars)}.$$

This is an increase of 8.24%. In other words, the nominal rate of 8%, compounded quarterly, yields an effective rate of 8.24% per annum. $\blacklozenge$

Future and Present Value   The simple and compound interest formulas give the accumulated sum S produced by a principal P after n interest periods. We say that S is the **future value** of P, and P is the **present value** of S. To compute future values we use formula (5) or (7), as in the preceding examples. To find present values we simply turn these formulas around, solving for P in terms of S.

Simple Interest

$$\text{present value } P = \frac{S}{1 + nr}$$

Compound Interest

$$\text{present value } P = \frac{S}{(1 + r)^n}$$

Example 7 How much money, invested now, will grow to $12,000 in 15 years at the following rates?
(a) 18% per annum simple interest
(b) 10% per annum interest, compounded quarterly

Solution (a) We want to find the present value of $12,000 in 15 years, at 18% per annum simple interest. Using the simple interest formula, we obtain

$$P = \frac{S}{1 + nr}$$

$$= \frac{12,000}{1 + 15(.18)}$$

$$= \frac{12,000}{3.7}$$

$$\doteq 3243.24 \quad \text{(dollars)}.$$

(b) With quarterly compounding, 15 years represent $n = 15(4) = 60$ periods, at an interest rate $r = .10/4 = .025$ per period. By the compound interest formula, the present value of $12,000 in 15 years is therefore

$$P = \frac{S}{(1 + r)^n}$$

$$= \frac{12,000}{(1.025)^{60}}$$

$$\doteq \frac{12,000}{4.3998} = 2727.40 \quad \text{(dollars)}. \quad \blacklozenge$$

Depreciation Difference equations can also be used to model the decline in value of a depreciable asset. If a building, machine or other tangible asset has original cost C and scrap value S, the difference $W = C - S$ is called the **wearing value** of the asset. For tax purposes, this amount can be depreciated in several ways over the useful life of the asset. Under **linear** or **straight-line** depreciation, the book value of the asset decreases each year by a constant amount until it reaches the scrap value. Under **declining-balance** depreciation, the book value decreases each year by a constant percentage until it reaches the scrap value.

Example 8 A machine costs $150,000 initially and has scrap value $35,000 at the end of 10 years. Find a formula for the book value of the machine after n years, using
(a) straight-line depreciation
(b) declining-balance depreciation at 20% a year.

Solution (a) Let V_n be the book value of the machine after n years. Using the straight-line method, this value decreases by a fixed amount A each year. Hence

$$V_{n+1} = V_n - A.$$

The initial book value is the cost of the machine, so $V_0 = 150,000$. The difference equation above is of type (★★). By formula (1) its solution is

$$V_n = V_0 + n(-A) = 150,000 - nA.$$

Since the machine has scrap value $35,000 after 10 years, we have

$$V_{10} = 150,000 - 10A = 35,000$$

from which we find that $A = \$11,500$. Thus, the book value of the machine after n years is given by

$$V_n = 150,000 - 11,500n.$$

(b) Using declining-balance depreciation at 20% a year, the difference equation becomes

$$V_{n+1} = V_n - .2V_n = .8V_n$$

which is of type (★★★). The solution, by formula (2), is

$$V_n = (.8)^n V_0 = (.8)^n(150,000).$$

Table 1 below shows the book value at the end of each year $n = 0$, 1, . . ., 10, for both methods of depreciation.

Year	Book value Straight-line	Declining-balance
0	$150,000	$150,000
1	138,500	120,000
2	127,000	96,000
3	115,500	76,800
4	104,000	61,440
5	92,500	49,152
6	81,000	39,322
7	69,500	31,457
8	58,000	25,166
9	46,500	20,133
10	35,000	16,106

Table 1

Note that under the declining-balance method the machine depreciates rapidly in the first few years, then more slowly as time goes on. Since depreciation is a deductible business expense, such accelerated write-offs can have important tax advantages. * ◆

---

*In actual practice the declining-balance method could only be used through the sixth year, since the book value cannot go below the scrap value of the asset. The remaining book value is then depreciated by the straight-line method.

The examples given in this section involve only arithmetic and geometric progressions. In the next section we consider the general first-order difference equation (★), which also has important applications in finance.

12.1 Exercises

1. For each of the following difference equations, generate the first five terms of the sequence. Determine whether the progression is arithmetic, geometric, or neither.

   (a) $x_{n+1} = 5x_n - 1$, $x_0 = 3$

   (b) $x_{n+1} = x_n + \frac{3}{2}$, $x_0 = 0$

   (c) $x_{n+1} = \frac{1}{2}x_n$, $x_0 = 1$

   (d) $x_{n+1} = -.2x_n + 40$, $x_0 = 5$

2. For each of the following difference equations, generate the first five terms of the sequence. Determine whether the progression is arithmetic, geometric, or neither.

   (a) $x_{n+1} = -1.8x_n$, $x_0 = 625$

   (b) $x_{n+1} = 2x_n - 7$, $x_0 = 6$

   (c) $x_{n+1} = 1.1x_n + 100$, $x_0 = 5000$

   (d) $x_{n+1} = x_n - 3.2$, $x_0 = 24.8$

3. Solve each of the following difference equations. Use your solution to calculate x_{10} and x_{20}.

   (a) $x_{n+1} = x_n + 6$, $x_0 = 2$

   (b) $x_{n+1} = 2x_n$, $x_0 = 3$

   (c) $x_{n+1} = x_n - \frac{5}{2}$, $x_0 = 35$

   (d) $x_{n+1} = -\frac{2}{3}x_n$, $x_0 = 1$

4. Solve each of the following difference equations. Use your solution to calculate x_{10} and x_{20}.

   (a) $x_{n+1} = 1.5x_n$, $x_0 = 1$

   (b) $x_{n+1} = x_n + 12$, $x_0 = -32$

   (c) $x_{n+1} = .9x_n$, $x_0 = 100$

   (d) $x_{n+1} = x_n - .17$, $x_0 = 5$

5. An individual borrows $6800 for 18 months at 14% annum simple interest. How much must she repay when the loan is due? How much of this is interest?

6. A corporation borrows $2 million for $3\frac{1}{2}$ years at 14% per annum simple interest. How much must be repaid when the loan is due? How much of this is interest?

7. A lender offers $750 on condition that the borrower repay $900 in 12 months. What simple interest rate does this represent?

8. A buyer pays $4000 for a bond which is redeemable at its face value of $5000 in another 15 months. What simple interest rate does this represent?

9. A consumer makes a $250 purchase using her credit card, which charges 18% per annum interest, compounded monthly. If she makes no payments for three months, how much will she owe?

10. A savings and loan pays 12% per annum interest, compounded quarterly, on its long-term accounts (30 months or more). If $3200 is deposited initially, what will the accumulated sum be after 30 months?

11. Compare the accumulated sum when $475 is deposited for three years at 13% per annum interest, compounded

   (a) annually   (b) semiannually   (c) quarterly.

12. Compare the accumulated sum when $1500 is deposited for two years at 10% per annum interest, compounded

    (a) quarterly   (b) monthly   (c) daily.

13. What is the effective interest rate corresponding to

    (a) 8.4% per annum compounded monthly?

    (b) 8.5% per annum compounded quarterly?

    Which return is better for the investor?

14. What is the effective interest rate corresponding to

    (a) 15% per annum compounded semiannually?

    (b) 14.6% per annum compounded daily?

    Which return is better for the investor?

15. A commodities speculator needs cash immediately. The sale of his property will be final in 15 months, giving him $180,000 with which to repay a loan. How much can he afford to borrow now

    (a) at 16% per annum simple interest?

    (b) at 14% per annum interest compounded quarterly?

16. A company will have to replace a piece of equipment in nine years at a cost of $2.6 million. It has been decided to invest a certain amount now so that the accumulated sum in nine years will pay for the new equipment. How much must be invested, if the money will earn

    (a) 12% per annum simple interest?

    (b) 10% per annum interest compounded semiannually?

17. If money can be invested at 8% per annum compounded quarterly, which is worth more: $5000 immediately, or $5200 in six months?

18. If money can be invested at 9% per annum compounded monthly, how much should you be willing to pay for

    (a) a note redeemable for $7500 in one year?

    (b) a note redeemable for $8000 in two years?

19. An apartment building costs $356,000. Assume the building has a useful life of 40 years and no scrap value. Find a formula for the book value of the building after n years, using

    (a) straight-line depreciation

    (b) declining-balance depreciation at 5% a year.

    Compare the book value under both methods at the end of years 10, 20, and 30.

20. A small business computer costs $15,600. The company intends to use it for eight years, at which time its estimated scrap value will be $1200. Find a formula for the book value of the computer after n years, using

    (a) straight-line depreciation

    (b) declining-balance depreciation at 25% a year.

    Compare the book value under both methods at the end of years 1, 2, 3, and 4.

21. Food prices have been increasing by 8% a year. Let C_n be the cost of a standard shopping list of groceries n years from now, assuming the present price trend continues.

    (a) Express C_{n+1} in terms of C_n using a difference equation.

    (b) Solve this equation.

    (c) If the current cost of the items is $100, what will it be in ten years? What percentage increase is this?

22. The population of a certain nation has been increasing at an annual rate of 2.8%. Let P_n be the population n years from now, assuming this growth rate continues.

    (a) Express p_{n+1} in terms of p_n using a difference equation.

    (b) Solve this equation.

    (c) If the current population is 12.4 million, what will it be in twenty years? What percentage increase is this?

23. One method of dating archaeological and geological samples uses the radioactive isotope carbon-14. The amount of this substance in a given sample decreases by about 1.2% every 100 years due to radioactive decay. Let C_n be the amount left in the sample after n centuries.

    (a) Express C_{n+1} in terms of C_n using a difference equation.

    (b) Solve this equation.

    (c) If the sample contained 1 unit of carbon-14 initially, how much of it will remain after 5000 years? What percentage decrease is this?

24. A certain drug, when injected intravenously, is eliminated exclusively through the kidneys at the rate of 7% per hour; that is, in each one-hour period the amount of drug still in the body decreases by 7%. Let A_n be the amount remaining after n hours.

    (a) Express A_{n+1} in terms of A_n using a difference equation.

    (b) Solve this equation.

    (c) If a patient receives 300 mg of the drug at 8 am, what amount will still be in his body at 5 pm?

Computer Projects   Suppose $100 is deposited at 8% per annum interest, compounded quarterly. Write a BASIC program to show the accumulated sum at the end of each quarter for ten years. Approximately how long will it take for the original amount to double?

12.2 Installment Loans and Annuities

Loans for the purchase of a home or car are usually paid off in monthly installments. In this section we use difference equations to analyze such installment loans, as well as some other topics in finance which involve regular periodic payments.

The General First-Order Difference Equation   In the preceding section we saw how to solve the first-order linear difference equation

$$x_{n+1} = ax_n + b \qquad (\star)$$

for the special cases $a = 1$ and $b = 0$. Suppose now we are given an equation of the form $(\star)$ where $a \neq 1$ and $b \neq 0$. By means of a simple substitution, we can reduce it to the case $b = 0$, that is, to an equation defining a geometric progression. Specifically, we let

$$y_n = x_n - c \qquad (n = 0, 1, 2, \ldots)$$

where c is a constant whose value we leave temporarily unspecified. Since $x_n = y_n + c$ and $x_{n+1} = y_{n+1} + c$, equation $(\star)$ becomes

$$y_{n+1} + c = a(y_n + c) + b$$

or

$$y_{n+1} + c = ay_n + (ac + b).$$

If we now choose c so that $c = ac + b$, then this last equation will have the same constant term on both sides. That is, if

$$c = \frac{b}{1 - a}$$

the equation simplifies to $y_{n+1} = ay_n$.

This equation has the solution $y_n = a^n y_0$

as we saw in Section 12.1. Substituting $x_n - c$ for y_n and $x_0 - c$ for y_0 brings us to

$$x_n - c = a^n(x_0 - c)$$

or

$$x_n = a^n(x_0 - c) + c.$$

This result is summarized below.

For $a \neq 1$, the difference equation

$$x_{n+1} = ax_n + b \qquad (\star)$$

has solution

$$x_n = a^n(x_0 - c) + c \qquad (1)$$

where $c = \dfrac{b}{1 - a}$.

Example 1   Solve the difference equation

$$x_{n+1} = 2x_n + 3$$

with initial value $x_0 = 1$.

Solution In Example 1 of Section 12.1 we saw that the above equation generates the sequence

$$1, \quad 5, \quad 13, \quad 29, \quad 61, \quad \ldots$$

Since $a = 2 \neq 1$, we let

$$c = \frac{b}{1 - a} = \frac{3}{1 - 2} = -3$$

in formula (1). The solution is

$$\begin{aligned}
x_n &= a^n(x_0 - c) + c \\
&= 2^n(1 - (-3)) + (-3) \\
&= 4 \cdot 2^n - 3.
\end{aligned}$$

We see from this formula that the terms x_n increase steadily without bound. ◆

Example 2 A lumber company owns a stand of 8000 redwoods. The company plans to harvest 10% of its trees each year and plant 650 new ones.
(a) How many trees will there be after 10 years? After 20 years?
(b) How many trees will there be in the long run?

Solution (a) We let x_n be the number of trees in the stand after n years. In year $n + 1$ the number of trees harvested will be 10% of those standing, or $.1x_n$. Since 650 additional trees will be planted, the number at the end of $n + 1$ years will be

$$\begin{aligned}
x_{n+1} &= x_n - .1x_n + 650 \\
x_{n+1} &= .9x_n + 650.
\end{aligned}$$

This is a difference equation of type ($\star$), with initial value $x_0 = 8000$. To solve the equation we use formula (1) with

$$c = \frac{b}{1 - a} = \frac{650}{1 - .9} = 6500$$

to obtain

$$\begin{aligned}
x_n &= a^n(x_0 - c) + c \\
&= (.9)^n(8000 - 6500) + 6500 \\
&= 1500(.9)^n + 6500.
\end{aligned} \qquad (2)$$

Thus, after 10 years the number of trees will be

$$x_{10} = 1500(.9)^{10} + 6500 \doteq 7023$$

and after 20 years

$$x_{20} = 1500(.9)^{20} + 6500 \doteq 6682.$$

(b) As n gets larger, the quantity $(.9)^n$ in formula (2) will steadily decrease, getting closer and closer to zero.* Hence we see from the formula that

---

*In general, a^n tends to zero as n increases, whenever $|a| < 1$.

x_n approaches the limiting value 6500 as n increases; thus, the number of trees will stabilize at 6500 over the long run. (Note that the number of trees harvested will then exactly equal the number replaced.) ◆

Installment Loans   Home mortgages and many other loans are repaid in **installments:** That is, the borrower makes regular payments of a certain amount, part of which pays the interest on the loan and part of which reduces the principal. The loan may be completely repaid in this way, or there may be a final "balloon payment" covering the outstanding balance when the loan is due.

Example 3   An individual borrows $2000 for 12 months at 15% per annum interest. The loan is to be repaid in monthly installments of $150, with a final balloon payment at the end of the year. Find the amount of the final payment.

Discussion   The interest period for an installment loan is always the same as the payment period. Thus, in our example it is understood that the interest is compounded monthly on the outstanding balance, at the rate of $.15/12 = .0125$ per month. For the first month the interest is 1.25% of $2000, or $25. Hence, at the end of the first month we have

$$\text{new balance} = \text{old balance} + \text{interest} - \text{payment}$$
$$= 2000 + 25 - 150$$
$$= \$1875.$$

The second month's interest is 1.25% of $1875, or approximately $23.44. Hence, at the end of two months we have

$$\text{new balance} = \text{old balance} + \text{interest} - \text{payment}$$
$$= 1875 + 23.44 - 150$$
$$= \$1748.44.$$

Continuing in this way we obtain the table shown below.

Month	Old balance	Interest	Payment	New balance
1	$2000.00	$25.00	$150	$1875.00
2	1875.00	23.44	150	1748.44
3	1748.44	21.86	150	1620.29
4	1620.29	20.25	150	1490.55
5	1490.55	18.63	150	1359.18
6	1359.18	16.99	150	1226.17
7	1226.17	15.33	150	1091.50
8	1091.50	13.64	150	955.14
9	955.14	11.94	150	817.08
10	817.08	10.21	150	677.29
11	677.29	8.47	150	535.76
12	535.76	6.70	150	392.45

Thus, in addition to the final installment of $150 at the end of month 12 the borrower must make a balloon payment of $392.45 to retire the debt. ◆

Note that the monthly interest decreases as payments continue, and therefore more of each payment goes toward reducing the principal. We can calculate the total interest paid during any period by adding the corresponding figures in the "Interest" column. An easier method, however, is to use the relation

$$\text{interest} = \text{payments} - \text{reduction of principal.}$$

For instance, after 12 months the principal has been reduced by

$$2000 - 392.45 = \$1607.55.$$

Since the total payments were $12(150) = \$1800$, the difference

$$1800 - 1607.55 = \$192.45$$

represents the interest paid during the year.

The step-by-step method used above can be very tedious for long-term installment loans such as home mortgages, so it is natural to look for a short-cut. Suppose we let L be the original amount of the loan, p the periodic payment, and r the periodic interest rate (.0125, in our example). We would like a formula for B_n, the balance owed after n periods. Since the interest for period $n + 1$ is rB_n, the relation

$$\text{new balance} = \text{old balance} + \text{interest} - \text{payment}$$

leads to the equation

$$B_{n+1} = B_n + rB_n - p$$

or

$$B_{n+1} = (1 + r)B_n - p.$$

This is a difference equation of the form ($\star$) with $a = 1 + r$ and $b = -p$. If we let

$$c = \frac{b}{1 - a} = \frac{-p}{1 - (1 + r)} = \frac{p}{r}$$

then formula (1) gives the solution

$$B_n = a^n(B_0 - c) + c$$

$$= (1 + r)^n\left(B_0 - \frac{p}{r}\right) + \frac{p}{r}.$$

Substituting the initial balance $B_0 = L$, we arrive at the desired formula.

Let L be the original amount of an installment loan, p the periodic payment, and r the periodic interest rate. Then the balance owed after n periods is

$$B_n = (1 + r)^n\left(L - \frac{p}{r}\right) + \frac{p}{r}. \tag{3}$$

Thus, in Example 3 where $L = 2000$, $p = 150$ and $r = .15/12 = .0125$, the balance after n months is given by

$$B_n = (1.0125)^n\left(2000 - \frac{150}{.0125}\right) + \frac{150}{.0125}$$
$$= (1.0125)^n(2000 - 12{,}000) + 12{,}000$$
$$= 12{,}000 - 10{,}000(1.0125)^n$$

For instance, after 12 months the balance is

$$B_{12} = 12{,}000 - 10{,}000(1.0125)^{12}$$
$$\doteq 12{,}000 - 11{,}607.55$$
$$= \$392.45$$

as we found out using the step-by-step approach.

Example 4   A borrower pays \$240 a month on a \$20,000 loan at 12% per annum interest. How much will she owe after 5 years? 10 years? 15 years?

Solution   The monthly interest rate is $.12/12 = .01$. By formula (3), the balance owed after n months is

$$B_n = (1.01)^n\left(20{,}000 - \frac{240}{.01}\right) + \frac{240}{.01}$$
$$= (1.01)^n(20{,}000 - 24{,}000) + 24{,}000$$
$$= 24{,}000 - 4000(1.01)^n.$$

Hence, after 5 years (60 months) she will owe

$$B_{60} = 24{,}000 - 4000(1.01)^{60} \doteq \$16{,}733.21;$$

after 10 years,

$$B_{120} = 24{,}000 - 4000(1.01)^{120} \doteq \$10{,}798.45;$$

and after 15 years,

$$B_{180} = 24{,}000 - 4000(1.01)^{180} \doteq \$16.79.$$

Thus, for all practical purposes the debt will be retired after 15 years. Note that here as in Example 3 the balance decreases slowly at first, because most of the early payments go toward interest rather than reduction of principal.   ◆

Amortization   We say that the loan in Example 4 is **amortized** over a 15-year term. That is, the loan will be completely paid off (except for a negligible end balance) after $n = 180$ payments. If a loan is amortized over n periods, the balance B_n will be zero. By formula (3), this gives us the equation

$$(1 + r)^n\left(L - \frac{p}{r}\right) + \frac{p}{r} = 0$$

or equivalently,

$$(1 + r)^n L = \frac{p}{r}[(1 + r)^n - 1].$$

Solving this equation for p, we obtain the following formula:

If a loan L is amortized over n periods at a periodic interest rate r, then the periodic payment is

$$p = \frac{rL}{1 - (1 + r)^{-n}} \qquad (4)$$

This is the formula used to compute the monthly payment on a home mortgage or other amortized loan.

Example 5 A buyer puts $20,000 down on a $100,000 home and takes a 30-year mortgage for the balance. If the interest rate is $13\frac{1}{2}\%$ per annum (fixed), what will the monthly payment be?

Solution The amount of the loan is $80,000, and the monthly interest rate is $.135/12 = .01125$. Since the loan is amortized over $30(12) = 360$ months, formula (4) gives a monthly payment of

$$p = \frac{.01125(80,000)}{1 - (1.01125)^{-360}}$$
$$\doteq \$916.33. \qquad \blacklozenge$$

If we solve formula (4) for L in terms of p, we obtain

$$L = p\left[\frac{1 - (1 + r)^{-n}}{r}\right] \qquad (5)$$

The quantity in brackets is sometimes abbreviated by the symbol $a_{\overline{n}|r}$, read "a angle n at r." Formula (5) can be used to determine how much a person can afford to borrow.

Example 6 A car dealer will finance new car purchases at 18% per annum, amortized over 48 months. If a buyer can afford to pay $125 a month, what is the maximum amount he can borrow?

Solution Substituting $r = .18/12 = .015$, $n = 48$, and $p = 125$ in formula (5), we obtain

$$L = 125\left[\frac{1 - (1.015)^{-48}}{.015}\right]$$
$$\doteq \$4255.32.$$

Note that the buyer will pay a total of 48(125) = $6000 over the term of the loan. Thus, when the car is finally paid off he will have paid 6000 − 4255.32 = $1744.68 in interest.   ◆

Annuities An **annuity** is an interest-bearing fund, or account, into which equal payments are made at regular intervals. The accumulated sum can eventually be used for retirement or children's college expenses, for example. In the case of an **ordinary annuity** a certain amount is paid into the fund at the end of each payment period, and interest on the previous balance is added. In other words,

$$\text{new balance} = \text{old balance} + \text{interest} + \text{payment}.$$

If we let p be the periodic payment, r the periodic interest rate, and S_n the accumulated sum, or balance, after n periods, then

$$S_{n+1} = S_n + rS_n + p = (1 + r)S_n + p.$$

This equation is a difference equation of form $(\star)$, with initial value $S_0 = 0$ (since we assume the fund is empty to begin with). By Formula (1), the solution to this equation is

$$S_n = (1 + r)^n\left(S_0 + \frac{p}{r}\right) - \frac{p}{r}.$$

Substituting $S_0 = 0$ and simplifying, we obtain the following formula:

For an annuity with periodic payment p and periodic interest rate r, the accumulated sum after n periods is given by

$$S = p\left[\frac{(1 + r)^n - 1}{r}\right] \qquad (6)$$

The quantity in brackets is sometimes abbreviated $s_{\overline{n}|r}$, read "s angle n at r." The payment p is sometimes called the **periodic rent,** and denoted by R.

Example 7 An individual pays $100 quarterly into a retirement fund which earns 9% per annum interest, compounded quarterly. How much will the fund contain after 30 years?

Solution The periodic (quarterly) interest rate is .09/4 = .0225. By formula (6), the accumulated sum after 30(4) = 120 quarters will be

$$S = 100\left[\frac{(1.0225)^{120} - 1}{.0225}\right]$$

$$\doteq \$59{,}737.89. \qquad ◆$$

Sinking Funds   An annuity which is established for the purpose of generating a specific amount at some future time is called a **sinking fund.** If the goal is to accumulate a sum S after n periods, then the periodic payment can be found by solving equation (6) for p.

$$p = \frac{rS}{(1 + r)^n - 1} \tag{7}$$

Example 8   Under a payroll savings plan, an employee can set aside a certain fixed amount of his salary each month. If the money earns $10\frac{1}{2}\%$ per annum interest, compounded monthly, what should the monthly deduction be in order to accumulate $5000 in two years?

Solution   The monthly interest rate is $.105/12 = .00875$. To accumulate a sum of $5000 after $2(12) = 24$ months, the monthly payment must be

$$p = \frac{.00875(5000)}{(1.00875)^{24} - 1}$$
$$\doteq \$188.13. \quad \blacklozenge$$

Under the **sinking-fund depreciation** method, a fixed amount p is paid each year into an interest-bearing account, the proceeds of which will cover the replacement cost of a depreciable asset at the end of its useful life.

Example 9   A machine costs $150,000 initially and has scrap value $35,000 at the end of 10 years. How much should be paid each year into a depreciation fund earning $8\frac{1}{2}\%$ per annum interest, in order to pay for replacement of the machine at the end of year 10?

Solution   The wearing value of the machine is

$$\text{cost} - \text{scrap value} = 150{,}000 - 35{,}000 = \$115{,}000.$$

We want the annuity to generate an accumulated sum of $115,000 in 10 years. Since the yearly interest rate is $.085$, the yearly payment is given by formula (7).

$$p = \frac{.085(115{,}000)}{(1.085)^{10} - 1}$$
$$\doteq \$7751.89 \quad \blacklozenge$$

Future and Present Value   Suppose some benefactor offers you an annuity as a gift. That is, a fund is to be established in your name, into which an amount p will be paid periodically. It is stipulated that the fund will earn interest at a rate r per period, and that payments will continue for n periods.

The **future value** of such an annuity is just the accumulated sum at the end of its term, which is given by formula (6).

$$S = p\left[\frac{(1 + r)^n - 1}{r}\right]$$

We may also want to know the **present value** of the annuity. That is, what amount P, invested now at the same compound interest rate, would produce the same accumulated sum after n periods? From Section 12.1 we know that the present value of an amount S in n periods is given by

$$P = \frac{S}{(1 + r)^n}.$$

Thus, for an annuity we have

present value $P = p\left[\dfrac{1 - (1 + r)^{-n}}{r}\right].$ (8)

Example 10   Assuming that money can be invested at an effective rate of 10% per annum, which is worth more: an outright gift of \$35,000 now, or an installment gift of \$5000 a year for 10 years?

Solution   The installment gift is an annuity with a term of $n = 10$ periods, and periodic (annual) interest rate $r = .1$. By formula (8), the present value of this annuity is

$$P = 5000\left[\frac{1 - (1.1)^{-10}}{.1}\right]$$
$$\doteq \$30,722.84.$$

Thus, the outright gift of \$35,000 is worth more.   ◆

There is another way to interpret the present value of an annuity. Given the periodic interest rate r, we can say that *a present amount P is equivalent to a series of n periodic payments p*. Hence, if an initial principal P is invested in an account earning interest at a rate r per period, the accountholder will be able to withdraw an amount p each period for n periods, before the account is exhausted.

Example 11   A bank account earns 9% per annum interest, compounded monthly. How much must you deposit initially if you plan to withdraw \$200 a month for 15 years?

Solution   Putting the question in other words: At an interest rate of $.09/12 = .0075$ per month, what is the present value of an annuity of \$200 a month for $15(12) = 180$ months? Using formula (8), we obtain

$$P = 200 \left[\frac{1 - (1.0075)^{-180}}{.0075} \right]$$

$$\doteq \$19,718.68.$$

With an initial deposit of this amount, you could withdraw $200 a month for 15 years, at the end of which time the account would be empty.   ◆

It is no coincidence that formula (8) for the present value of an annuity is the same as formula (5) for the amount of a fully amortized loan. In effect, a lender gives a borrower an amount L now, in return for a series of payments p over n periods. That is, the lender "buys an annuity" from the borrower. Many texts in fact treat installment loans as a special case of annuities, deriving the amortization formulas from the formula for present value. Since installment loans are more familiar to most readers than annuities, we have chosen to deal with these topics separately.

12.2 Exercises

1. Solve each of the following difference equations. Use your solution to calculate x_5 and x_{10}.

   (a) $x_{n+1} = 3x_n + 4,\quad x_0 = 5$

   (b) $x_{n+1} = \frac{1}{2}x_n + 1,\quad x_0 = 34$

   (c) $x_{n+1} = 1.5x_n - 3,\quad x_0 = 8$

   (d) $x_{n+1} = -.6x_n + 24,\quad x_0 = 10$

2. Solve each of the following difference equations. Use your solution to calculate x_5 and x_{10}:

   (a) $x_{n+1} = 2x_n - 5,\quad x_0 = 6$

   (b) $x_{n+1} = 1.01x_n + 25,\quad x_0 = 100$

   (c) $x_{n+1} = -\frac{2}{3}x_n + 5, x_0 = 4$

   (d) $x_{n+1} = 1.2x_n - 2.4,\quad x_0 = 12$

3. Suppose $500 is deposited initially in an account which earns 8% per annum interest, compounded quarterly, and at the end of each quarter an additional $50 is deposited. Let S_n be the accumulated sum, or balance, after n quarters.

   (a) Express S_{n+1} in terms of S_n using a difference equation.

   (b) Solve this equation.

   (c) How much will be in the account at the end of five years?

4. A university has raised $1.5 million for a scholarship endowment fund. The money is invested at 10% per annum interest, compounded annually, and at the end of each year $120,000 is withdrawn from the fund for scholarships. Let B_n be the balance in the account after n years.

   (a) Express B_{n+1} in terms of B_n using a difference equation.

   (b) Solve this equation.

   (c) How much will the fund contain at the end of ten years?

5. An individual borrows $10,000 for two years at 16% per annum interest. The loan is to be repaid in quarterly installments of $1200, with a final balloon payment at the end of the second year.

   (a) How much will the balloon payment be?

   (b) How much interest will be paid during the term of the loan?

6. A $90,000 home mortgage at 12% per annum interest is paid off in monthly installments of $925.75.

(a) How much will be owed after 15 years? 30 years?

(b) How much interest will be paid during the first 15 years?

7. A couple puts 20% down on $1800 worth of furniture and pays off the balance in equal installments over 24 months. If the interest rate is 18% per annum, what will the monthly payment be?

8. What is the monthly payment on a mortgage of $75,000, amortized over 20 years at $14\frac{1}{4}$% per annum interest?

9. The Small Business Administration offers business loans at $16\frac{1}{2}$% per annum interest, amortized monthly over 7 years. If the owner of a restaurant can afford monthly payments of $800, what is the maximum amount he can borrow?

10. A woman pays $180 a month on her car loan, which is amortized over 36 months at 15% per annum interest. Of the total amount paid, how much goes toward the car and how much goes toward interest?

11. If monthly deposits of $75 are made to an account paying 9% per annum interest, compounded monthly, how much will the account contain after ten years?

12. A man pays $85 quarterly for theft insurance on his oriental rugs, which are worth $7500. Show that if he invested the money instead at 10% per annum interest, compounded quarterly, he would have enough to cover the loss of the rugs after 12 years.

13. To prepare for their son's college education Ramon's parents have decided to make a deposit each year, starting on his first birthday, to an account earning an effective rate of $8\frac{1}{2}$% per annum. How much should the yearly deposit be in order to accumulate $20,000 by the time their son turns eighteen?

14. A homeowner decides to set aside a certain amount each month to cover his $1300 yearly property tax bill, which is payable on December 31. If money can be invested at $7\frac{1}{2}$% per annum, compounded monthly, how much should he set aside each month?

15. A municipality must retire a $1,250,000 airport bond issue in ten years. To meet this future obligation a sinking fund is set up, into which semiannual payments are made. If the fund earns 7% per annum interest, compounded semiannually, how much must the payments be?

16. An airline estimates that its present fleet of 707s will have to be replaced in 12 years at a cost of $130 million. If the scrap value of the planes is $15 million, how much should be paid each quarter into a depreciation fund earning 9% per annum interest, compounded quarterly, in order to pay for replacement of the fleet at the end of year 12?

17. A rich uncle stipulated in his will that his niece is to receive $10,000 a year for 25 years, the first payment to be made one year after his death. Assuming that money can be invested at an effective rate of 12% per annum, what is the equivalent cash value of this inheritance?

18. A bank holds a 30-year mortage with payments of $560 a month. At the end of year 10 (with 20 years of payments remaining) the bank wishes to sell the mort-

gage to another bank. How much is it worth, if money can be invested at 15% per annum, compounded monthly?

19. A medical institute is attempting to raise endowment funds to support an annual $25,000 research fellowship, to be awarded each year for the next 20 years. How much money must be raised for this project, if the institute can invest its money at an effective rate of 11% per annum?

20. A professional athlete estimates that his playing career will last another five years. He decides to put a certain amount of his salary at the end of each of these years into a retirement fund earning 10% per annum interest, compounded annually. When he retires, he wants to be able to draw $30,000 a year from this fund for 35 years. How much should each of his five payments be?

21. A recent medical manpower study concluded that Xanadu's current supply of 4700 physicians is inadequate for the population's future needs. Moreover, the study found that 5% of the country's doctors emigrate or retire from practice each year, while 375 new ones graduate from the medical school at Pakkadur. If these trends continue,

(a) how many doctors will there be in 10 years? In 20 years?

(b) how many doctors will there be in the long run?

22. A government agency received initial funding of $15 million. Each year the agency requests $5 million more than it received the previous year, but the final approved budget is always 10% less than the amount requested.

(a) How much will the agency receive after 10 years? After 20 years?

(b) How much will the agency receive in the long run?

23. Suppose the sequence $x_0, x_1, \ldots$ satisfies a difference equation

$$x_{n+1} = ax_n + b$$

where $a \neq 1$, and let $c = b/(1 - a)$. Prove:

(a) If $x_0 = c$ then x_n is constant.

(b) If $|a| < 1$ then x_n approaches the limiting value c as n increases.

24. Let L be the original amount of an installment loan, r the periodic interest rate.

(a) Show that if the loan is amortized over n periods, then the balance after k periods is

$$B_k = \frac{(1 + r)^n - (1 + r)^k}{(1 + r)^n - 1} L$$

(b) Using the formula in (a), find the balance owed after 15 years on a 30-year mortgage for $80,000, paid off in monthly installments at $13\frac{1}{2}$% per annum interest.

Computer Projects

A. Suppose an individual borrows $15,000 to be completely repaid in equal monthly installments over 5 years at 15% per annum interest. Write a BASIC program to show the balance owed and the interest paid for each month the loan is outstanding. Approximately how long will it take to reduce the debt by half?

B. Suppose $100 is deposited at the end of each quarter into an annuity account which earns 8% per annum interest, compounded quarterly. Write a BASIC program to show the accumulated sum at the end of each quarter for ten years. Approximately how long will it take for the account to reach $5000?

Chapter 12   Summary

Important Terms

12.1 difference equation
arithmetic progression
geometric progression
simple interest
compound interest
accumulated sum
effective rate of interest
present value

12.2 installment loan
amortization
ordinary annuity
sinking fund
future value of an annuity
present value of an annuity

Review Exercises

1. Solve each of the following difference equations:

   (a) $x_{n+1} = x_n + 2$, $x_0 = -1$

   (b) $x_{n+1} = .8x_n$, $x_0 = 125$

   (c) $x_{n+1} = 4x_n + 9$, $x_0 = 0$

   (d) $x_{n+1} = \frac{1}{2}x_n + 1$, $x_0 = 2$

2. Solve each of the following difference equations:

   (a) $x_{n+1} = 1.02x_n - 50$, $x_0 = 1000$

   (b) $x_{n+1} = -2x_n$, $x_0 = 3$

   (c) $x_{n+1} = -\frac{2}{3}x_n + \frac{5}{3}$, $x_0 = 8$

   (d) $x_{n+1} = x_n + .75$, $x_0 = .5$

3. A small business borrows $75,000 for 6 months at 16% per annum simple interest. How much must be repaid when the loan is due? How much of this is interest?

4. Suppose that $6000 deposited in a monetary fund increases to $6810 in 9 months. What simple interest rate does this represent?

5. Compare the accumulated sum when $2500 is deposited for two years at $13\frac{1}{2}\%$ per annum interest, compounded

   (a) annually   (b) quarterly   (c) monthly.

6. What is the effective rate of interest corresponding to 15% per annum compounded monthly?

7. How much invested now will grow to $30,000 in 7 years, at 12% per annum interest compounded quarterly?

8. Suppose you hold a note which promises to pay you $15,000 in three years. If money can be invested at $16\frac{1}{2}\%$ per annum compounded monthly, what is the present value of this note?

9. Oscar has just received a $25,000 inheritance from his uncle. He plans to deposit the money in an account which earns 15% per annum interest, compounded monthly. At the end of each month he will withdraw $500 to cover his living expenses while he is away at college. Let B_n be the balance in the account after n months.

   (a) Express B_{n+1} in terms of B_n using a difference equation.

   (b) Solve this equation.

   (c) How much will the account contain in four years, when Oscar graduates?

10. The owner of an apartment building borrows $12,000 for 12 months at 18% per annum interest. The loan is to be repaid in monthly installments of $1000, with a final balloon payment at the end of the year. How much will this balloon payment be?

11. A buyer puts $1500 down on a new car which costs $7200. The balance is paid off in equal monthly installments over 3 years, at 18% per annum interest. What will the monthly payment be?

12. A savings and loan offers 30-year home mortgages at 15% per annum interest. If a buyer can afford monthly payments of $625, what is the maximum amount he can borrow toward the purchase of a home?

13. At the end of each quarter, Hilda deposits $250 in an account which earns 12% per annum interest, compounded quarterly. If the first payment was made on March 31, 1980, how much will the account contain on January 1, 1985?

14. A manufacturer has determined that his plant's present equipment will have to be replaced in 15 years at an estimated cost of $750,000. In order to finance this future obligation he has decided to make quarterly payments to a depreciation fund earning 8% per annum interest. How much will the quarterly payment be?

15. A power company finds that its monthly fuel cost is $1.5 million. Assuming that money can be invested at $16\frac{1}{2}\%$ per annum interest, compounded monthly, how much does the company need now to pay for its fuel for the coming year?

16. A bank account earns 14% per annum interest, compounded quarterly. How much must you deposit initially if you plan to withdraw $1500 at the end of each quarter for the next 10 years?

17. Each year the population of a certain country increases by 3% due to natural factors, and 450,000 new immigrants are absorbed. If the current population is 84.2 million, what will it be in 10 years?

18. A patient is given a loading dose of 1000 mg of a certain antibiotic at time 0. Every six hours thereafter an additional 300 mg booster dose is administered. During each six-hour period between doses, 60% of the drug present in the body is eliminated.

   (a) How much will be present in the body immediately after the fourth booster dose? Immediately before it?

   (b) How much will be present in the body immediately after the nth booster dose when n is large? Immediately before it?

Appendix

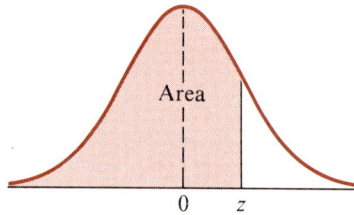

Area

0   z

Table 1   Area Under a Normal Curve to the Left of z, Where $z = \dfrac{x - \mu}{\sigma}$

z	0.00	0.01	0.02	0.03	0.04	0.05	0.06	0.07	0.08	0.09
−3.4	0.0003	0.0003	0.0003	0.0003	0.0003	0.0003	0.0003	0.0003	0.0003	0.0002
−3.3	0.0005	0.0005	0.0005	0.0004	0.0004	0.0004	0.0004	0.0004	0.0004	0.0003
−3.2	0.0007	0.0007	0.0006	0.0006	0.0006	0.0006	0.0006	0.0005	0.0005	0.0005
−3.1	0.0010	0.0009	0.0009	0.0009	0.0008	0.0008	0.0008	0.0008	0.0007	0.0007
−3.0	0.0013	0.0013	0.0013	0.0012	0.0012	0.0011	0.0011	0.0011	0.0010	0.0010
−2.9	0.0019	0.0018	0.0017	0.0017	0.0016	0.0016	0.0015	0.0015	0.0014	0.0014
−2.8	0.0026	0.0025	0.0024	0.0023	0.0023	0.0022	0.0021	0.0021	0.0020	0.0019
−2.7	0.0035	0.0034	0.0033	0.0032	0.0031	0.0030	0.0029	0.0028	0.0027	0.0026
−2.6	0.0047	0.0045	0.0044	0.0043	0.0041	0.0040	0.0039	0.0038	0.0037	0.0036
−2.5	0.0062	0.0060	0.0059	0.0057	0.0055	0.0054	0.0052	0.0051	0.0049	0.0048
−2.4	0.0082	0.0080	0.0078	0.0075	0.0073	0.0071	0.0069	0.0068	0.0066	0.0064
−2.3	0.0107	0.0104	0.0102	0.0099	0.0096	0.0094	0.0091	0.0089	0.0087	0.0084
−2.2	0.0139	0.0136	0.0132	0.0129	0.0125	0.0122	0.0119	0.0116	0.0113	0.0110
−2.1	0.0179	0.0174	0.0170	0.0166	0.0162	0.0158	0.0154	0.0150	0.0146	0.0143
−2.0	0.0228	0.0222	0.0217	0.0212	0.0207	0.0202	0.0197	0.0192	0.0188	0.0183
−1.9	0.0287	0.0281	0.0274	0.0268	0.0262	0.0256	0.0250	0.0244	0.0239	0.0233
−1.8	0.0359	0.0352	0.0344	0.0336	0.0329	0.0322	0.0314	0.0307	0.0301	0.0294
−1.7	0.0446	0.0436	0.0427	0.0418	0.0409	0.0401	0.0392	0.0384	0.0375	0.0367
−1.6	0.0548	0.0537	0.0526	0.0516	0.0505	0.0495	0.0485	0.0475	0.0465	0.0455
−1.5	0.0668	0.0655	0.0643	0.0630	0.0618	0.0606	0.0594	0.0582	0.0571	0.0559
−1.4	0.0808	0.0793	0.0778	0.0764	0.0749	0.0735	0.0722	0.0708	0.0694	0.0681
−1.3	0.0968	0.0951	0.0934	0.0918	0.0901	0.0885	0.0869	0.0853	0.0838	0.0823
−1.2	0.1151	0.1131	0.1112	0.1093	0.1075	0.1056	0.1038	0.1020	0.1003	0.0985
−1.1	0.1357	0.1335	0.1314	0.1292	0.1271	0.1251	0.1230	0.1210	0.1190	0.1170
−1.0	0.1587	0.1562	0.1539	0.1515	0.1492	0.1469	0.1446	0.1423	0.1401	0.1379

Reprinted with permission of Macmillan Publishing Co., Inc. from PROBABILITY AND STATISTICS FOR ENGINEERS AND SCIENTISTS, 2nd ed., by Ronald E. Walpole and Raymond H. Myers. Copyright © 1972, 1978 by Macmillan Publishing Co., Inc.

Table 1   (Continued)

z	0.00	0.01	0.02	0.03	0.04	0.05	0.06	0.07	0.08	0.09
−0.9	0.1841	0.1814	0.1788	0.1762	0.1736	0.1711	0.1685	0.1660	0.1635	0.1611
−0.8	0.2119	0.2090	0.2061	0.2033	0.2005	0.1977	0.1949	0.1922	0.1894	0.1867
−0.7	0.2420	0.2389	0.2358	0.2327	0.2296	0.2266	0.2236	0.2206	0.2177	0.2148
−0.6	0.2743	0.2709	0.2676	0.2643	0.2611	0.2578	0.2546	0.2514	0.2483	0.2451
−0.5	0.3085	0.3050	0.3015	0.2981	0.2946	0.2912	0.2877	0.2843	0.2810	0.2776
−0.4	0.3446	0.3409	0.3372	0.3336	0.3300	0.3264	0.3228	0.3192	0.3156	0.3121
−0.3	0.3821	0.3783	0.3745	0.3707	0.3669	0.3632	0.3594	0.3557	0.3520	0.3483
−0.2	0.4207	0.4168	0.4129	0.4090	0.4052	0.4013	0.3974	0.3936	0.3897	0.3859
−0.1	0.4602	0.4562	0.4522	0.4483	0.4443	0.4404	0.4364	0.4325	0.4286	0.4247
−0.0	0.5000	0.4960	0.4920	0.4880	0.4840	0.4801	0.4761	0.4721	0.4681	0.4641
0.0	0.5000	0.5040	0.5080	0.5120	0.5160	0.5199	0.5239	0.5279	0.5319	0.5359
0.1	0.5398	0.5438	0.5478	0.5517	0.5557	0.5596	0.5636	0.5675	0.5714	0.5753
0.2	0.5793	0.5832	0.5871	0.5910	0.5948	0.5987	0.6026	0.6064	0.6103	0.6141
0.3	0.6179	0.6217	0.6255	0.6293	0.6331	0.6368	0.6406	0.6443	0.6480	0.6517
0.4	0.6554	0.6591	0.6628	0.6664	0.6700	0.6736	0.6772	0.6808	0.6844	0.6879
0.5	0.6915	0.6950	0.6985	0.7019	0.7054	0.7088	0.7123	0.7157	0.7190	0.7224
0.6	0.7257	0.7291	0.7324	0.7357	0.7389	0.7422	0.7454	0.7486	0.7517	0.7549
0.7	0.7580	0.7611	0.7642	0.7673	0.7704	0.7734	0.7764	0.7794	0.7823	0.7852
0.8	0.7881	0.7910	0.7939	0.7967	0.7995	0.8023	0.8051	0.8078	0.8106	0.8133
0.9	0.8159	0.8186	0.8212	0.8238	0.8264	0.8289	0.8315	0.8340	0.8365	0.8389
1.0	0.8413	0.8438	0.8461	0.8485	0.8508	0.8531	0.8554	0.8577	0.8599	0.8621
1.1	0.8643	0.8665	0.8686	0.8708	0.8729	0.8749	0.8770	0.8790	0.8810	0.8830
1.2	0.8849	0.8869	0.8888	0.8907	0.8925	0.8944	0.8962	0.8980	0.8997	0.9015
1.3	0.9032	0.9049	0.9066	0.9082	0.9099	0.9115	0.9131	0.9147	0.9162	0.9177
1.4	0.9192	0.9207	0.9222	0.9236	0.9251	0.9265	0.9278	0.9292	0.9306	0.9319
1.5	0.9332	0.9345	0.9357	0.9370	0.9382	0.9394	0.9406	0.9418	0.9429	0.9441
1.6	0.9452	0.9463	0.9474	0.9484	0.9495	0.9505	0.9515	0.9525	0.9535	0.9545
1.7	0.9554	0.9564	0.9573	0.9582	0.9591	0.9599	0.9608	0.9616	0.9625	0.9633
1.8	0.9641	0.9649	0.9656	0.9664	0.9671	0.9678	0.9686	0.9693	0.9699	0.9706
1.9	0.9713	0.9719	0.9726	0.9732	0.9738	0.9744	0.9750	0.9756	0.9761	0.9767
2.0	0.9772	0.9778	0.9783	0.9788	0.9793	0.9798	0.9803	0.9808	0.9812	0.9817
2.1	0.9821	0.9826	0.9830	0.9834	0.9838	0.9842	0.9846	0.9850	0.9854	0.9857
2.2	0.9861	0.9864	0.9868	0.9871	0.9875	0.9878	0.9881	0.9884	0.9887	0.9890
2.3	0.9893	0.9896	0.9898	0.9901	0.9904	0.9906	0.9909	0.9911	0.9913	0.9916
2.4	0.9918	0.9920	0.9922	0.9925	0.9927	0.9929	0.9931	0.9932	0.9934	0.9936
2.5	0.9938	0.9940	0.9941	0.9943	0.9945	0.9946	0.9948	0.9949	0.9951	0.9952
2.6	0.9953	0.9955	0.9956	0.9957	0.9959	0.9960	0.9961	0.9962	0.9963	0.9964
2.7	0.9965	0.9966	0.9967	0.9968	0.9969	0.9970	0.9971	0.9972	0.9973	0.9974
2.8	0.9974	0.9975	0.9976	0.9977	0.9977	0.9978	0.9979	0.9979	0.9980	0.9981
2.9	0.9981	0.9982	0.9982	0.9983	0.9984	0.9984	0.9985	0.9985	0.9986	0.9986
3.0	0.9987	0.9987	0.9987	0.9988	0.9988	0.9989	0.9989	0.9989	0.9990	0.9990
3.1	0.9990	0.9991	0.9991	0.9991	0.9992	0.9992	0.9992	0.9992	0.9993	0.9993
3.2	0.9993	0.9993	0.9994	0.9994	0.9994	0.9994	0.9994	0.9995	0.9995	0.9995
3.3	0.9995	0.9995	0.9995	0.9996	0.9996	0.9996	0.9996	0.9996	0.9996	0.9997
3.4	0.9997	0.9997	0.9997	0.9997	0.9997	0.9997	0.9997	0.9997	0.9997	0.9998

Table 2   250 Random 5-Digit Binary Numbers

00100	10100	01001	10001	11001	01001	01001	11001	01010	10111
01011	00000	01111	01010	01010	10010	10001	11000	11110	00101
00110	00000	10001	01101	11100	10110	00111	00000	00101	00110
01001	10111	11101	10101	10110	01101	00001	11111	00001	01100
10001	11111	00111	01100	00101	11111	11110	11110	10100	11100
01100	00100	11001	01101	01011	00001	01101	01001	11011	10000
10100	01110	11101	10111	10010	10100	00000	00111	01010	00001
00011	10000	10001	00011	00010	01011	10100	11110	11100	00101
01010	10010	01001	01010	11100	01110	00011	11001	01010	01110
11110	00110	11010	11010	01101	10111	01110	11000	01011	01101
00001	11001	01110	11011	01000	10111	11001	11010	11101	10101
11011	10001	10100	11110	10000	10100	11100	00000	11010	11000
10001	01110	01000	10111	10100	11101	10010	11100	10001	00010
00101	00000	01011	01100	00011	11100	00111	10000	01000	00110
01101	11011	01111	11011	01110	01100	01011	11000	01101	11001
00010	10101	01011	11010	11110	10111	10101	10101	00010	00100
01000	11101	11110	10111	00001	10111	01101	00001	10010	10010
00100	10111	10001	11001	10010	11011	01111	00010	10110	01111
11111	00110	01010	00111	01011	11011	10101	11010	11101	00101
10110	01101	10000	11000	10100	00001	11110	11011	00010	00011
10011	11100	00000	10000	01010	01110	10010	11001	10001	00001
01110	00101	10100	11010	11101	01101	10101	01100	00111	00010
01111	10001	00110	01011	10101	00100	11001	10000	10000	11000
01110	10101	01101	00101	10101	00010	10100	00000	11101	01000
01100	00101	10110	00111	10010	11101	10001	01011	01100	11000

Answers to Selected Exercises

Chapter 1

Section 1.1, page 7

1.

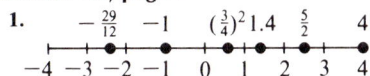

-4 -3 -2 -1 0 1 2 3 4 with $-\frac{29}{12}$ -1 $(\frac{3}{4})^2$ 1.4 $\frac{5}{2}$ 4

3.

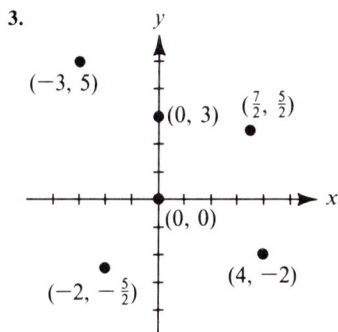

$(-3, 5)$ $(0, 3)$ $(\frac{7}{2}, \frac{5}{2})$ $(0, 0)$ $(-2, -\frac{5}{2})$ $(4, -2)$

5. $2x + y = 1$

7. $x - y = 0$

9. $3x + 5y = 0$

11. $5x - 12y = 30$

13.

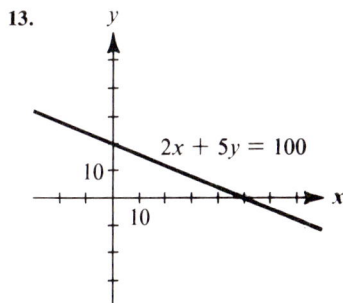

$2x + 5y = 100$

15.

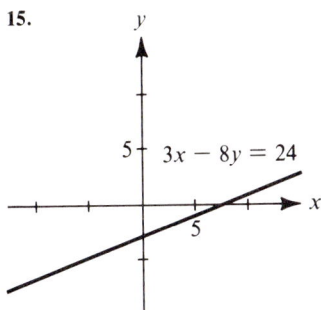

$3x - 8y = 24$

17.

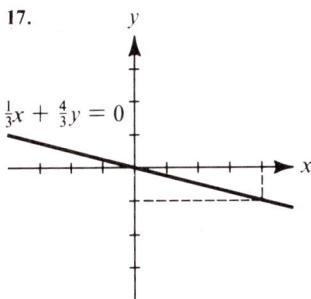

$\frac{1}{3}x + \frac{4}{3}y = 0$

19.

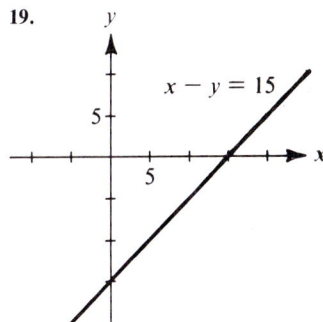

$x - y = 15$

21.

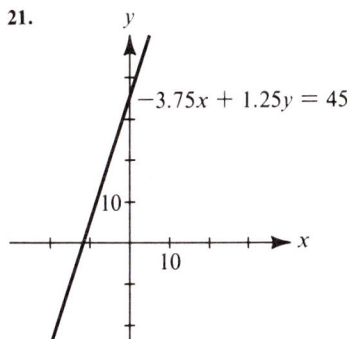

$-3.75x + 1.25y = 45$

23.

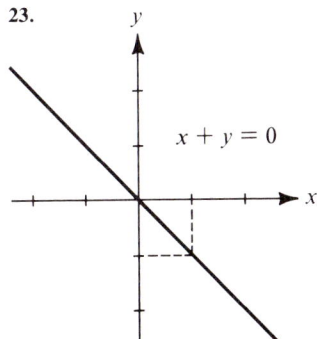

$x + y = 0$

25.

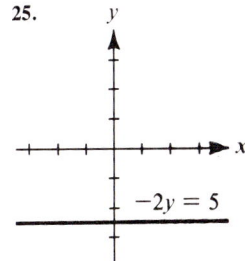

$-2y = 5$

493

27.

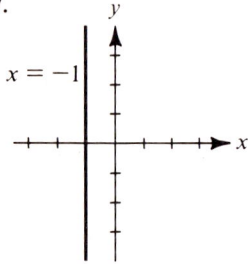

$x = -1$

29.

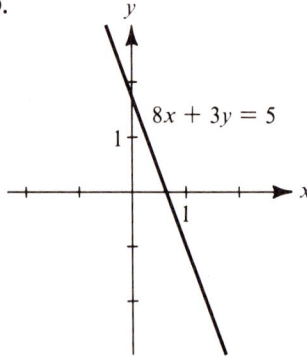

$8x + 3y = 5$

31.

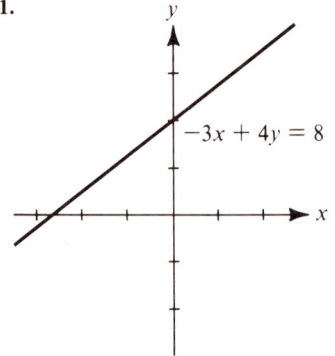

$-3x + 4y = 8$

Section 1.2, page 16

1. (2, 1)   **3.** (6, −3)   **5.** (3, 4)   **7.** (64, 18)   **9.** (11/2, 5/2)   **11.** Parallel   **13.** (30, 5)
15. 20 oz Food A, 5 oz Food B.   **17.** (a) 1440 lb Blend A, 1260 lb Blend B.   (b) not possible (−160 lb Blend A, 4560 lb Blend B).   **19.** 80 liters Solution A, 20 liters Solution B.   **21.** Arnold $600, Berenice $300.
23. Investment 1: $40,000; Investment 2: $30,000.

Section 1.3, page 26

1. slope −2/5, y-intercept (0, 2)   **3.** slope 3, y-intercept (0, −4)   **5.** no slope or yintercept   **7.** slope 1/5,
y-intercept (0, −4/5)   **9.** slope −1, y-intercept (0, 0)   **11.** $3x + 4y = 16$ **13.** $y = 3$
15. $-2x + 3y = 14$ **17.** $x = -1$ **19.** $x + y = 1$ **21.** $9x - y = 35$ **23.** $4x - 9y = 0$
25. $3x + 4y = 2$ **27.** $2x - y = 0$ **29.** $x + y = 4$ **31.** $3x - 2y = 0$ **33.** (a) $5450; $5900; $6350
(b) $S = 450t + 5000$ (c) 5 yrs, 4 mos.   **35.** (a) $F = \frac{9}{5}C + 32$ (b) 68°F

Section 1.4, page 35

1. Profit $P = .75x - 8400$
   Break-even point: $x = 11,200$

3. x = number of chips per day
   Profit $P = .80x - 1200$
   Break-even point: $x = 1500$

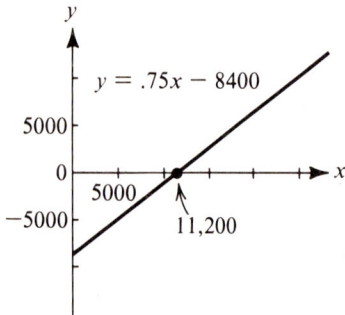

$y = .75x - 8400$

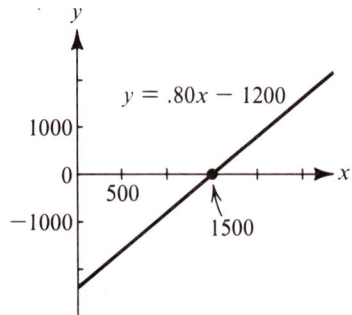

$y = .80x - 1200$

5. (a) 19,200   (b) 16,800   (c) 33,600   **7.** (a) 5250   (b) 4000   (c) 5000

9. (a) Method 1: $x = 11{,}200$; Method 2: $x = 12{,}000$

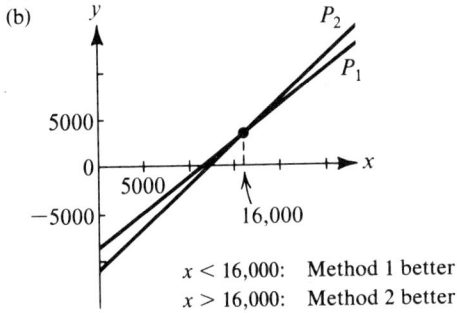

(b)

$x < 16{,}000$: Method 1 better
$x > 16{,}000$: Method 2 better

11. x = number of cars produced per day

Present system: Profit $P_1 = 1250x - 60{,}000$
Computerized system: Profit $P_2 = 1850x - 120{,}000$
(a) Present: $x = 48$; Computerized: $x = 64.9$ (approx.)
(b)

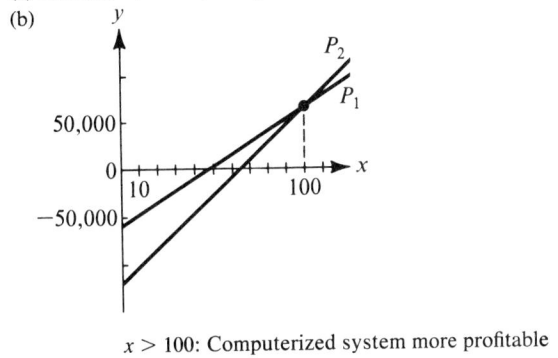

$x > 100$: Computerized system more profitable

13. x = number of copies printed

Printer 1: Cost $C_1 = 2.80x + 1000$
Printer 2: Cost $C_2 = 1.80x + 4000$

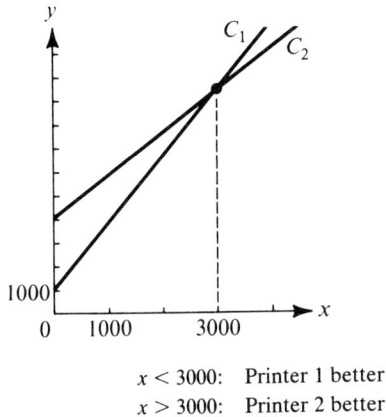

$x < 3000$: Printer 1 better
$x > 3000$: Printer 2 better

15. Red Cab: Cost $C_R = .20x + .65$
White Cab: Cost $C_W = .15x + .85$
Blue Cab: Cost $C_B = .12x + 1.15$

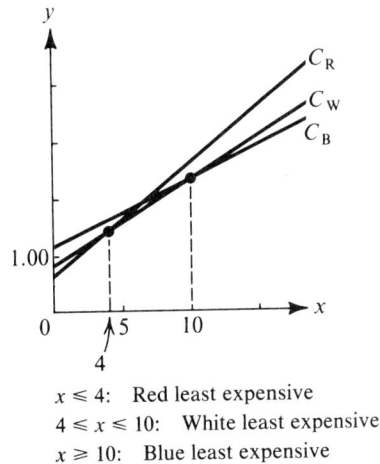

$x \leq 4$: Red least expensive
$4 \leq x \leq 10$: White least expensive
$x \geq 10$: Blue least expensive

17. Company 1: $P_1 = 1{,}000{,}000$
Company 2: $P_2 = .20x + 100{,}000$
Company 3: $P_3 = .10x + 250{,}000$

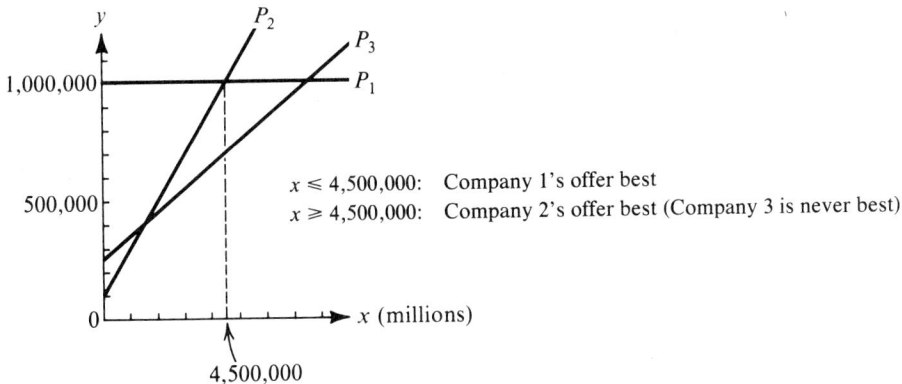

$x \leq 4{,}500{,}000$: Company 1's offer best
$x \geq 4{,}500{,}000$: Company 2's offer best (Company 3 is never best)

Review Exercises, page 38

1.

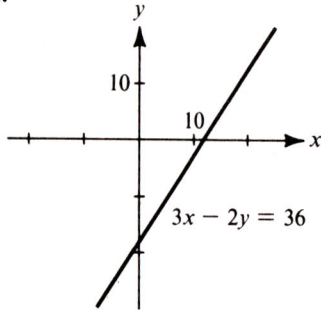

$3x - 2y = 36$

3. (15, 4)   **5.** 750 type A fish, 250 type B fish.
7. slope 5/3, y-intercept (, -4)   **9.** \$164,000; \$136,000
11. (a) 859,375   (b) 1,328,125   (c) 1,250,000   (d) 1,100,000
13. (a) 1834;2110   (b) If company can produce (and sell)
more than 6250 cube per week.

Chapter 2

Section 2.1, page 46

1. (a) (10,20) feasible, $C = 400$
(b) (25, 2) not feasible
(c) (15, 15) feasible, $C = 375$
(d) (35, 0) feasible, $C = 350$

3. Maximize $R = 16x + 12y$
subject to:
$$8x + 3y \leqslant 480$$
$$4x + 9y \leqslant 720$$
$$x \geqslant 27$$
$$x + y \leqslant 80$$
$$x \geqslant 0$$
$$y \geqslant 0$$

5. x = number of lbs of Mixture A
y = number of lbs of Mixture B

Maximize $R = 3.5x + 4y$
subject to:
$$\tfrac{1}{4}x + \tfrac{3}{5}y \leqslant 300$$
$$\tfrac{1}{2}x + \tfrac{1}{5}y \leqslant 250$$
$$\tfrac{1}{4}x + \tfrac{1}{5}y \leqslant 150$$
$$x \geqslant 0$$
$$y \geqslant 0$$

7. x = number of acres alfalfa
y = number of acres beets
Maximize $P = 30x + 40y$
subject to:
$$20x + 30y \leqslant 45,000$$
$$25x + 75y \leqslant 90,000$$
$$x + y \leqslant 2000$$
$$x \geqslant 0$$
$$y \geqslant 0$$

9. x = number of type A units
y = number of type B units
Maximize $P = 600x + 1000y$
subject to:
$$10,000x + 15,000y \leqslant 300,000$$
$$x + 2y \leqslant 36$$
$$x + y \leqslant 28$$
$$x \geqslant 0$$
$$y \geqslant 0$$

11. x = number of type A rolls
y = number of type B rolls
Minimize $C = 100x + 300y$
subject to:
$$2x + 5y \geqslant 1000$$
$$35x + 25y \leqslant 10,500$$
$$x \geqslant 0$$
$$y \geqslant 0$$

13. x = number of units policy A
y = number of units policy B
Minimize $C = 500x + 450y$
subject to:
$$10,000x + 10,000y \geqslant 400,000$$
$$5,000x + 10,000y \geqslant 400,000$$
$$20,000x + 10,000y \geqslant 1,000,000$$
$$x \geqslant 0$$
$$y \geqslant 0$$

Section 2.2, page 55

1. A, B, E satisfy   **3.** C, F satisfy   **5.** E satisfies   **7.** None satisfy

9.

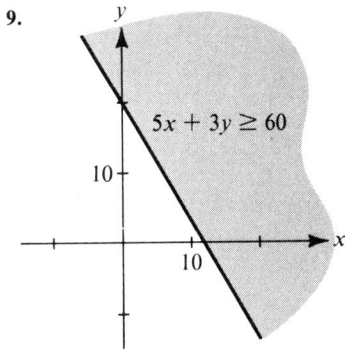

$5x + 3y \geq 60$

11.

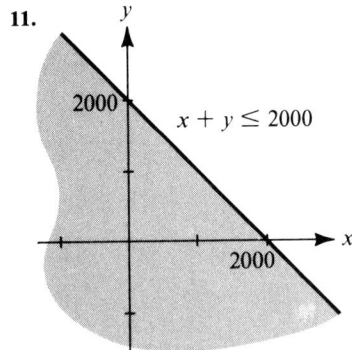

$x + y \leq 2000$

13.

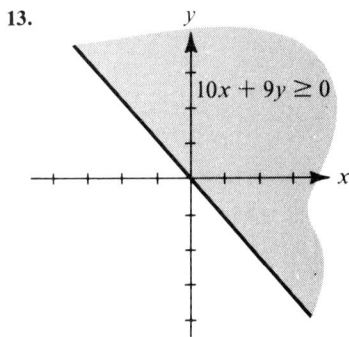

$10x + 9y \geq 0$

15.

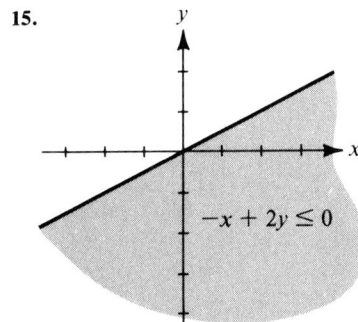

$-x + 2y \leq 0$

17.

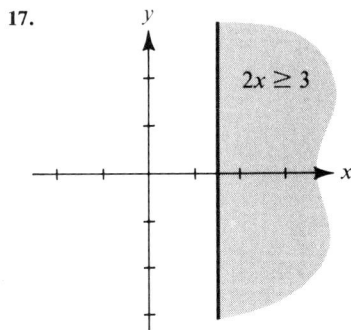

$2x \geq 3$

19.

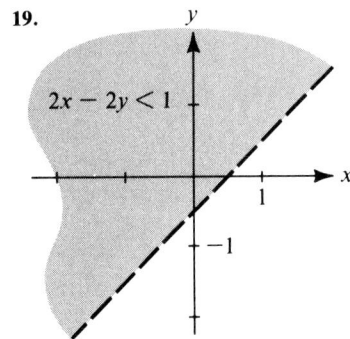

$2x - 2y < 1$

21.

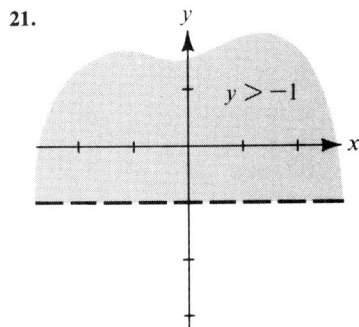

$y > -1$

23. $x - y \leq 1$ **25.** $3x + 5y < 0$ **27.** $15x + 16y < 6000$ **29.** $-x + 2y \geq 100$ **31.** (a) $5x - 6y \leq 0$

(b) $4x - 3y \geq 0$ (c) $x - y \geq 0$

Section 2.3, page 64

1.

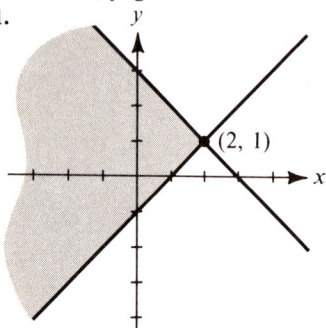

(2, 1)

3.

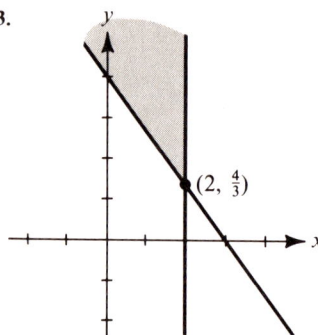

$(2, \frac{4}{3})$

5.

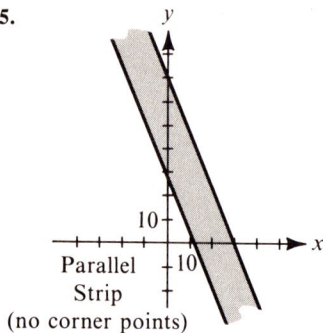

Parallel
Strip
(no corner points)

7.

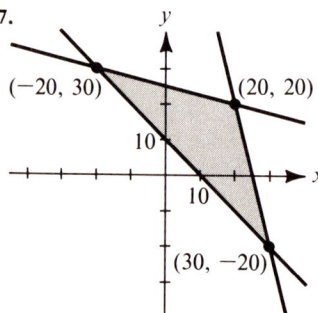

$(-20, 30)$ $(20, 20)$

$(30, -20)$

9.

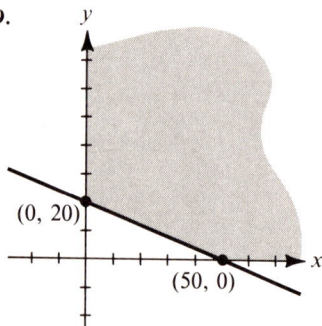

$(0, 20)$

$(50, 0)$

11.

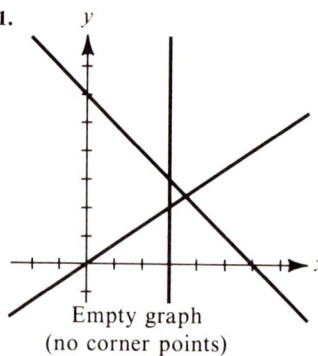

Empty graph
(no corner points)

13.

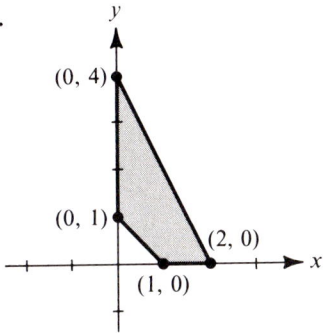

15.

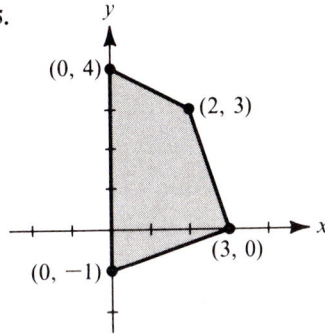

17.

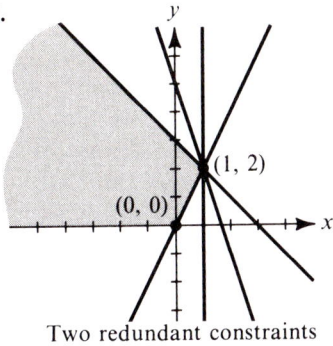

Two redundant constraints

19.

Bounding lines	Basic point	Feasible?
L_1, L_2	(1, 0)	Yes
L_1, L_3	(0, 1)	Yes
L_2, L_3	(2, 1)	Yes

21.

Bounding lines	Basic point	Feasible?
L_1, L_2	(3, 3)	Yes
L_1, L_3	(1, 5)	No
L_1, L_4	(6, 0)	Yes
L_2, L_3	(1, 1)	Yes
L_2, L_4	(0, 0)	No
L_3, L_4	(1, 0)	Yes

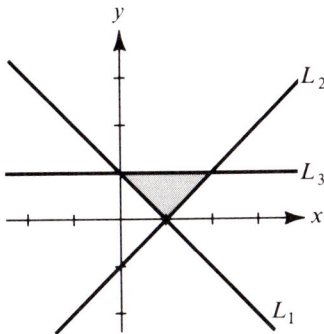

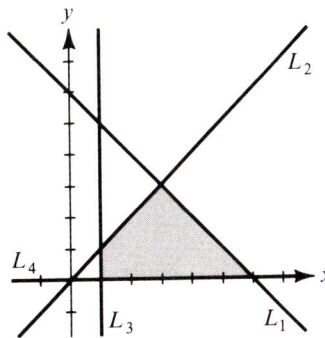

23.

Bounding lines	Basic point	Feasible?
L_1, L_2	(8, 8)	No
L_1, L_3	(6, 9)	Yes
L_1, L_4	(0, 12)	Yes
L_1, L_5	(24, 0)	No
L_2, L_3	(9, 6)	Yes
L_2, L_4	(0, 24)	No
L_2, L_5	(12, 0)	Yes
L_3, L_4	(0, 15)	No
L_3, L_5	(15, 0)	No
L_4, L_5	(0, 0)	Yes

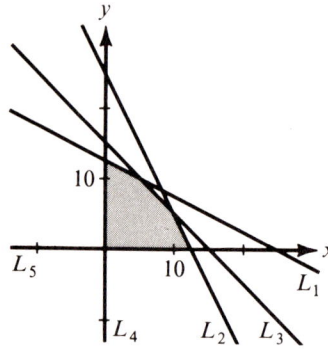

25. $2x - y \leq 2$
$-x + 2y \leq 2$
$x + y \geq 1$

27. $-x + y \leq 1$
$x - y \leq 1$
$x + y \leq 3$
$x + y \geq 1$

Section 2.4, page 71

1. Minimum $C = \$3.10$ when $x = 5$ and $y = 20$. **3.** Maximum $R = 1152$ when $x = 48$ and $y = 32$. **5.** Maximum $R = 2550$ when $x = 300$ and $y = 375$. **7.** Maximum $P = 65,000$ when $x = 1500$ and $y = 500$. **9.** Maximum $P = 19,200$ when $x = 12$ and $y = 12$. **11.** Minimum $C = 55,600$ when $x = 220$ and $y = 112$. **13.** Minimum $C = 29,000$ when $x = 40$ and $y = 20$. **15.** Maximum $F = 100$ when $x = 20$ and $y = 20$. **17.** Minimum $G = 2$ when (i) $x = 0$ and $y = 1$, or (ii) $x = 1$ and $y = 0$. Since both these corner points are optimal, so is any point on the segment joining them. **19.** Maximum $F = 9$ when $x = 1$ and $y = 2$.

Section 2.5, page 79

1.

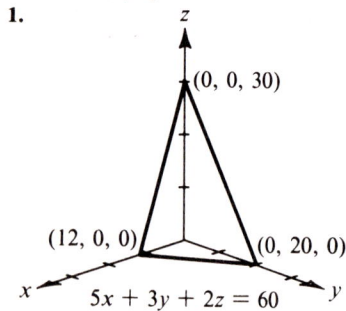

3.

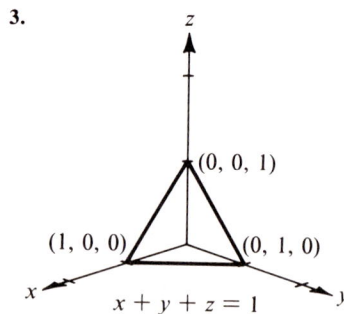

5.

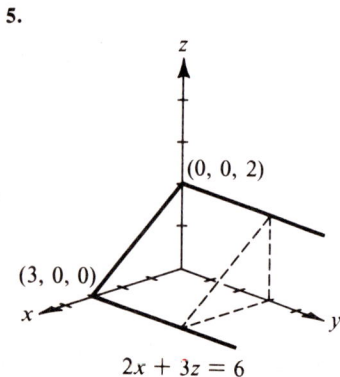

7.

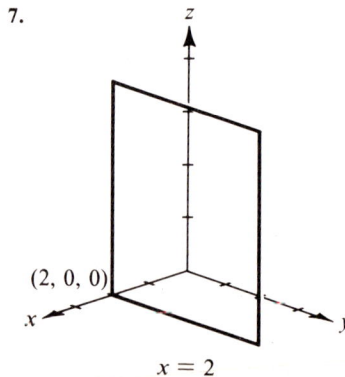

9.

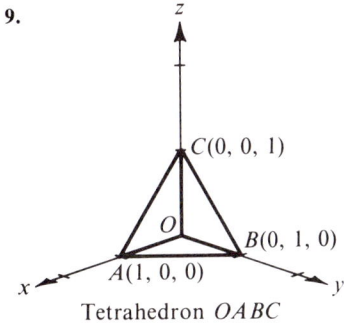

Tetrahedron $OABC$

11.

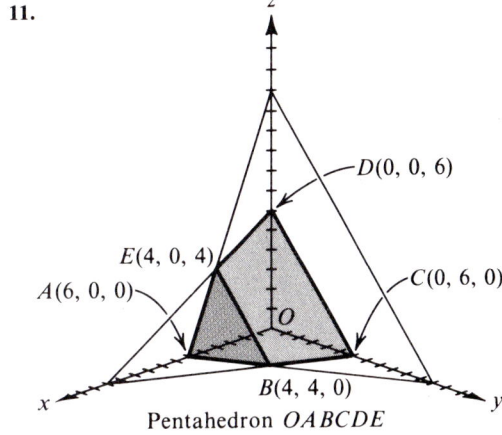

Pentahedron $OABCDE$

13. Maximum $R = \$1380$ when $x = 0$ cans Blend A, $y = 40$ cans Blend B, and $z = 60$ cans Blend C.

15. Minimum $C = \$66,000$ when $x = 6$ type A planes, $y = 12$ type B planes, and $z = 0$ type C planes.

Review Exercises, page 81

1.

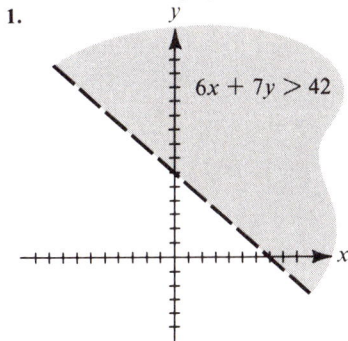

$6x + 7y > 42$

3. $3x - 5y \le 0$

5.

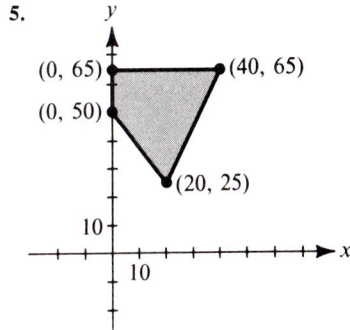

7. Maximum $F = 106$ when $x = 30$ and $y = 4$.

9. Maximum $P = 240$ projects, when $x = 0$ type A teams, $y = 30$ type B teams.

11. Maximum $P = \$510,000$ when $x = 0$ single-family homes, $y = 30$ condos.

Chapter 3

Section 3.1, page 91

1. $\begin{array}{cc|c} 2 & 3 & 12 \\ 1 & -\frac{1}{4} & \frac{3}{4} \end{array}$

3. $\begin{array}{ccc|c} 1 & -2 & \frac{1}{2} & 4 \\ 0 & \frac{1}{2} & -1 & -5 \\ 0 & 4 & -\frac{5}{3} & -2 \end{array}$

5. $\begin{array}{cccc|c} 2 & -1 & 1 & 0 & 6 \\ 0 & 1 & \frac{5}{2} & -2 & 4 \\ \frac{3}{4} & -1 & 1 & 2 & 2 \\ -1 & 0 & 1 & 1 & -5 \end{array}$

7. $\begin{array}{cc|c} 14 & 0 & 21 \\ -4 & 1 & -3 \end{array}$

9. $\begin{array}{ccc|c} 2 & -4 & 1 & 8 \\ 2 & -\frac{7}{2} & 0 & 3 \\ \frac{10}{3} & -\frac{8}{3} & 0 & \frac{34}{3} \end{array}$

11. $\begin{array}{cccc|c} 2 & -1 & 1 & 0 & 6 \\ \frac{3}{8} & 0 & \frac{7}{4} & 0 & 3 \\ \frac{3}{8} & -\frac{1}{2} & \frac{1}{2} & 1 & 1 \\ -\frac{11}{8} & \frac{1}{2} & \frac{1}{2} & 0 & -6 \end{array}$

13. $R_3 = (-1)R_3$ **15.** $R_2 = \frac{2}{3}R_2$ **17.** $R_2 = R_2 + \left(-\frac{2}{3}\right)R_3$ **19.** $x = 8, y = 5$ **21.** $x = 2, y = 1, z = 4$
23. $x = 0, y = 1, z = 0$ **25.** $x = 1, y = 3, z = 2$ **27.** $x = -1, y = 0, z = 1, w = 2$ **29.** $R_i = \frac{1}{a}R_i$
31. $R_i = R_i + R_j$
    $R_j = R_j + (-1)R_i$
    $R_i = R_i + R_j$
    $R_j = (-1)R_i$

Section 3.2, page 102

1. No solutions   **3.** Infinitely many solutions (1 free variable)   **5.** No solutions   **7.** No solutions
9. Infinitely many solutions (1 free variable)   **11.** Infinitely many solutions (1 free variable)
13. Infinitely many solutions (2 free variables)   **15.** No solutions
17. (a) General solution: $x = 4 + z, y = -1 - 2z$
    Basic solution: $(4, -1, 0)$
  (b) General solution: $x = \frac{7}{2} - \frac{1}{2}y, z = -\frac{1}{2} - \frac{1}{2}y$
    Basic solution: $\left(\frac{7}{2}, 0, -\frac{1}{2}\right)$
  (c) General solution: $y = 7 - 2x, z = -4 + x$
    Basic solution: $(0, 7, -4)$
19. (a) General: $x = 15 - u - \frac{1}{2}v, y = 20 - \frac{1}{2}u - v, z = 25 - \frac{1}{2}u - \frac{1}{2}v$
    Basic: $(15, 20, 25, 0, 0)$
  (b) General: $x = -35 + 2z + \frac{1}{2}v, y = -5 + z - \frac{1}{2}v, u = 50 - 2z - v$
    Basic: $(-35, -5, 0, 50, 0)$
  (c) General: $y = -40 - x + 3z, u = -20 - 2x + 2z, v = 70 + 2x - 4z$
    Basic: $(0, -40, 0, -20, 70)$

Section 3.3, page 111

1. 12 days mine 1, 16 days mine 2, 3 days mine 3.   **3.** 450 lb Type A, 100 lb Type B, 50 lb Type C.
5. (a) $-5x \quad\quad + 15z + \quad 25w = 120,000$ (b) $x = -24,000 + 3z + 5w, y = 30,000 - 3z - 5w$
    $25x + 50y + 75z + 125w = 900,000$ (c) For feasible solution must have $24,000 \le 3z + 5w \le 30,000$
7. $y = x^2 - 5x + 8$ **9.** $y = 2x^3 + x^2 - x + 1$ **11.** $x^2 + y^2 - 2x - 4 = 0$ **13.** $y = \frac{1}{2}x + \frac{3}{2}$

Review Exercises, page 114

1. $x = 7, y = -6$ **3.** $x = 5, y = 2, z = -1$ **5.** Infinitely many solutions: $x = 3 - 3z - w, y = 1 - z - w$
7. $x = -3, y = 1, z = 2$
9. General solution: $z = 5 + x - 2y, u = -8 - x + y, v = 3 - 2x - 4y$
    Basic solution: $(0, 0, 5, -8, 3)$
11. \$24 million to City A, \$48 million to City B, \$48 million to City C.

Chapter 4

Section 4.1, page 123

1. 2×2 **3.** 2×1; column vector   **5.** 2×3 **7.** (a) $a_{21} = 7, a_{22} = -4, a_{23} = 1$
(b) $a_{13} = 9, a_{23} = 1, a_{33} = 6$ **9.** $x = 3, y = 4, z = 1$

11. (a) $\begin{bmatrix} 4 & 4 & 0 \\ 9 & 8 & 2 \end{bmatrix}$
(b) $\begin{bmatrix} 0 & 2 & 2 \\ -1 & 2 & 2 \end{bmatrix}$
(c) $\begin{bmatrix} 2 & 4 \\ 3 & 5 \\ 1 & 2 \end{bmatrix}$

(d) $\begin{bmatrix} 3/2 & 2 \\ 5/2 & 1/2 \\ -1 & 1 \end{bmatrix}$
(e) $\begin{bmatrix} 2 & 7 & 5 \\ 2 & 9 & 6 \end{bmatrix}$
(f) $\begin{bmatrix} 7 & 14 \\ 7 & 7 \\ -4 & 2 \end{bmatrix}$

13. (a) $D_1 = \begin{array}{c} A \\ B \end{array}\begin{matrix} E & M & W & S \\ \begin{bmatrix} 25{,}000 & 20{,}000 & 30{,}000 & 15{,}000 \\ 40{,}000 & 15{,}000 & 35{,}000 & 10{,}000 \end{bmatrix} \end{matrix}$

15. $A + B = \begin{bmatrix} 2+1 & 4+(-2) \\ -3+2 & 5+3 \end{bmatrix} = \begin{bmatrix} 3 & 2 \\ -1 & 8 \end{bmatrix}$

(b) $D_2 = 1.15\,D_1 = \begin{bmatrix} 28{,}750 & 23{,}000 & 34{,}500 & 17{,}250 \\ 46{,}000 & 17{,}250 & 40{,}250 & 11{,}500 \end{bmatrix}$

$B + A = \begin{bmatrix} 1+2 & -2+4 \\ 2+(-3) & 3+5 \end{bmatrix} = \begin{bmatrix} 3 & 2 \\ -1 & 8 \end{bmatrix}$

(c) $D = D_1 + D_2 = \begin{bmatrix} 53{,}750 & 43{,}000 & 64{,}500 & 32{,}250 \\ 86{,}000 & 32{,}250 & 75{,}250 & 21{,}500 \end{bmatrix}$

17. $A + O = \begin{bmatrix} 2 & 4 \\ -3 & 5 \end{bmatrix} + \begin{bmatrix} 0 & 0 \\ 0 & 0 \end{bmatrix} = \begin{bmatrix} 2 & 4 \\ -3 & 5 \end{bmatrix} = A$

19. $c(A + B) = c\begin{bmatrix} a_{11}+b_{11} & a_{12}+b_{12} \\ a_{21}+b_{21} & a_{22}+b_{22} \end{bmatrix} = \begin{bmatrix} ca_{11}+cb_{11} & ca_{12}+cb_{12} \\ ca_{21}+cb_{21} & ca_{22}+cb_{22} \end{bmatrix}$

$cA + cB = \begin{bmatrix} ca_{11} & ca_{12} \\ ca_{21} & ca_{22} \end{bmatrix} + \begin{bmatrix} cb_{11} & cb_{12} \\ cb_{21} & cb_{22} \end{bmatrix} = \begin{bmatrix} ca_{11}+cb_{11} & ca_{12}+cb_{12} \\ ca_{21}+cb_{21} & ca_{22}+cb_{22} \end{bmatrix}$

21. $c(dA) = c\begin{bmatrix} da_{11} & da_{12} \\ da_{21} & da_{22} \end{bmatrix} = \begin{bmatrix} cda_{11} & cda_{12} \\ cda_{21} & cda_{22} \end{bmatrix}$

23. $X = B - \frac{7}{2}A = \begin{bmatrix} 1 & -4 \\ 3 & 5 \\ 8 & -9 \end{bmatrix}$

$(cd)A = (cd)\begin{bmatrix} a_{11} & a_{12} \\ a_{21} & a_{22} \end{bmatrix} = \begin{bmatrix} cda_{11} & cda_{12} \\ cda_{21} & cda_{22} \end{bmatrix}$

25. (a) $(A + B)^t = \begin{bmatrix} a_{11}+b_{11} & a_{12}+b_{12} \\ a_{21}+b_{21} & a_{22}+b_{22} \end{bmatrix}^t = \begin{bmatrix} a_{11}+b_{11} & a_{21}+b_{21} \\ a_{12}+b_{12} & a_{22}+b_{22} \end{bmatrix}$

$A^t + B^t = \begin{bmatrix} a_{11} & a_{21} \\ a_{12} & a_{22} \end{bmatrix} + \begin{bmatrix} b_{11} & b_{21} \\ b_{12} & b_{22} \end{bmatrix} = \begin{bmatrix} a_{11}+b_{11} & a_{21}+b_{21} \\ a_{12}+b_{12} & a_{22}+b_{22} \end{bmatrix}$

Section 4.2, page 132

1. 23   **3.** 1   **5.** 16   **7.** (a) 2×2 (b) 2×4 (c) undefined   (d) undefined   (e) 3×4 (f) 2×2

9. $AB = \begin{bmatrix} 8 & 5 \\ 20 & 13 \end{bmatrix}, \qquad BA = \begin{bmatrix} 13 & 20 \\ 5 & 8 \end{bmatrix}$

11. $AB\,[16], \; BA = \begin{bmatrix} 6 & 15 \\ 4 & 10 \end{bmatrix}$

13. $LR = \begin{array}{c} A \\ B \\ C \end{array}\begin{matrix} X & Y \\ \begin{bmatrix} 39 & 44 \\ 61 & 70 \\ 22 & 26 \end{bmatrix} \end{matrix} = $ labor cost (dollars/chair)

15. $X^2 - X - 2I = \begin{bmatrix} 6 & -5 \\ 2 & -1 \end{bmatrix} - \begin{bmatrix} 4 & -5 \\ 2 & -3 \end{bmatrix} - \begin{bmatrix} 2 & 0 \\ 0 & 2 \end{bmatrix} = \begin{bmatrix} 0 & 0 \\ 0 & 0 \end{bmatrix} = 0$

17. E.g., let $A = \begin{bmatrix} 1 & 2 \\ 3 & 4 \end{bmatrix}, B = \begin{bmatrix} 4 & 3 \\ 2 & 1 \end{bmatrix}.$

19. $A(BC) = \begin{bmatrix} 2 & 4 \\ -3 & 5 \end{bmatrix}\begin{bmatrix} 2 & 4 \\ -3 & 8 \end{bmatrix} = \begin{bmatrix} -8 & 40 \\ -21 & 28 \end{bmatrix}$

Then $(A + B)^2 = \begin{bmatrix} 50 & 50 \\ 50 & 50 \end{bmatrix}, A^2 + 2AB + B^2 = \begin{bmatrix} 45 & 35 \\ 65 & 55 \end{bmatrix}.$

$(AB)C = \begin{bmatrix} 10 & 8 \\ 7 & 21 \end{bmatrix}\begin{bmatrix} 0 & 4 \\ -1 & 0 \end{bmatrix} = \begin{bmatrix} -8 & 40 \\ -21 & 28 \end{bmatrix}$

21. $(A + B)C = \begin{bmatrix} 3 & 2 \\ -1 & 8 \end{bmatrix}\begin{bmatrix} 0 & 4 \\ -1 & 0 \end{bmatrix} = \begin{bmatrix} -2 & 12 \\ -8 & -4 \end{bmatrix}$

$AC + BC = \begin{bmatrix} -4 & 8 \\ -5 & -12 \end{bmatrix} + \begin{bmatrix} 2 & 4 \\ -3 & 8 \end{bmatrix} = \begin{bmatrix} -2 & 12 \\ -8 & -4 \end{bmatrix}$

23. Let $A = \begin{bmatrix} a_{11} & a_{12} \\ a_{21} & a_{22} \end{bmatrix}$, $B = \begin{bmatrix} b_{11} & b_{12} \\ b_{21} & b_{22} \end{bmatrix}$

$(AB)^t = \begin{bmatrix} a_{11}b_{11} + a_{12}b_{21} & a_{11}b_{12} + a_{12}b_{22} \\ a_{21}b_{11} + a_{22}b_{21} & a_{21}b_{12} + a_{22}b_{22} \end{bmatrix}^t = \begin{bmatrix} a_{11}b_{11} + a_{12}b_{21} & a_{21}b_{11} + a_{22}b_{21} \\ a_{11}b_{12} + a_{12}b_{22} & a_{21}b_{12} + a_{22}b_{22} \end{bmatrix}$

$B^tA^t = \begin{bmatrix} b_{11} & b_{21} \\ b_{12} & b_{22} \end{bmatrix}\begin{bmatrix} a_{11} & a_{21} \\ a_{12} & a_{22} \end{bmatrix} = \begin{bmatrix} b_{11}a_{11} + b_{21}a_{12} & b_{11}a_{21} + b_{21}a_{22} \\ b_{12}a_{11} + b_{22}a_{12} & b_{12}a_{21} + b_{22}a_{22} \end{bmatrix}$

Hence $(AB)^t = B^tA^t$

25. $AB = \begin{bmatrix} 2 & 1 \\ 6 & 3 \end{bmatrix}\begin{bmatrix} -1 & 2 \\ 2 & -4 \end{bmatrix} = \begin{bmatrix} 0 & 0 \\ 0 & 0 \end{bmatrix} = 0$, but $A \neq 0, B \neq 0$.

Section 4.3, page 144

1. $\begin{bmatrix} 1 & 1 \\ 1 & -1 \end{bmatrix}\begin{bmatrix} x \\ y \end{bmatrix} = \begin{bmatrix} 4 \\ 2 \end{bmatrix}$

3. $\begin{bmatrix} 2 & 2 & 1 & 1 \\ 1 & 2 & 2 & 1 \\ 1 & 1 & 2 & 2 \end{bmatrix}\begin{bmatrix} x \\ y \\ z \\ w \end{bmatrix} = \begin{bmatrix} -1 \\ 0 \\ 1 \end{bmatrix}$

5. $X = \begin{bmatrix} -4 \\ 5 \end{bmatrix}$

7. $X = \begin{bmatrix} 10 \\ 20 \\ 30 \end{bmatrix}$

9. $\begin{bmatrix} 2 & -5 \\ -1 & 3 \end{bmatrix}$

11. $\begin{bmatrix} 15 & -11 & -8 \\ -4 & 3 & 2 \\ -1 & 1 & 1 \end{bmatrix}$

13. noninvertible

15. $\begin{bmatrix} 0 & 1 & 0 & -1 \\ 1 & 0 & 0 & 0 \\ 0 & 0 & 0 & 1 \\ -1 & 0 & 1 & 0 \end{bmatrix}$

17. $A^{-1} = \begin{bmatrix} 6/5 & 4/15 \\ 2/5 & 6/5 \end{bmatrix}$, $X = \begin{bmatrix} 80 \\ 110 \end{bmatrix}$

19. $A^{-1} = \begin{bmatrix} -7 & 5 \\ 3 & -2 \end{bmatrix}$; (a) $X = \begin{bmatrix} -2 \\ 1 \end{bmatrix}$ (b) $X = \begin{bmatrix} -7 \\ 3 \end{bmatrix}$ (c) $X = \begin{bmatrix} 5 \\ -2 \end{bmatrix}$

21. (a) $\begin{bmatrix} 2 & 4 \\ 2 & 2 \end{bmatrix}\begin{bmatrix} x \\ y \end{bmatrix} = \begin{bmatrix} b_1 \\ b_2 \end{bmatrix}$ (b) $x = -\frac{1}{2}b_1 + b_2, y = \frac{1}{2}b_1 - \frac{1}{2}b_2$

23. Assume $AX = I = XA$ and $AY = I = YA$. By M9 and M13: $X = XI = X(AY) = (XA)Y = IY = Y$.

25. Assume A is invertible and $AB = O$. Then: $B = IB = (A^{-1}A)B = A^{-1}(AB) = A^{-1}O = O$.

27. $(A^{-1})^{-1} = \begin{bmatrix} 1 & 1 \\ -1 & -2 \end{bmatrix}^{-1} = \begin{bmatrix} 2 & 1 \\ -1 & -1 \end{bmatrix} = A$

29. $(A^t)^{-1} = \begin{bmatrix} 2 & -1 \\ 1 & -1 \end{bmatrix}^{-1} = \begin{bmatrix} 1 & -1 \\ 1 & -2 \end{bmatrix}$

$(A^{-1})^t = \begin{bmatrix} 1 & 1 \\ -1 & -2 \end{bmatrix}^t = \begin{bmatrix} 1 & -1 \\ 1 & -2 \end{bmatrix}$

31. $\begin{bmatrix} 2 & -5 \\ -1 & 3 \end{bmatrix}$

Section 4.4, page 155

1. nonmetals; services   **3.** metals: $300,000 worth; nonmetals: $1,650,000; energy: $225,000; services: $825,000.

5. agriculture; machinery   **7.** agriculture: $5 billion worth; textiles: $0; chemicals: $400 million; machinery: $1 billion; transportation: $1.4 billion.

9. $\begin{bmatrix} .3 & .2 \\ .4 & .1 \end{bmatrix}$ **11. (a)** $\begin{bmatrix} 1250 \\ 1800 \end{bmatrix}$ **(b)** $\begin{bmatrix} 169 \\ 204 \end{bmatrix}$ **(c)** $\begin{bmatrix} 1081 \\ 1596 \end{bmatrix}$

(d)

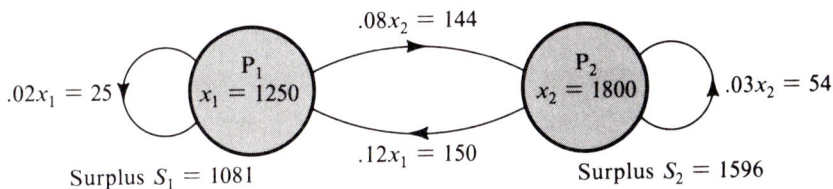

13. (a) $\begin{bmatrix} 106 \\ 116 \end{bmatrix}$ **(b)** $\begin{bmatrix} 340 \\ 240 \end{bmatrix}$

15. (b) $\begin{bmatrix} 4100 \\ 6290 \\ 5610 \end{bmatrix}$ **(c)**

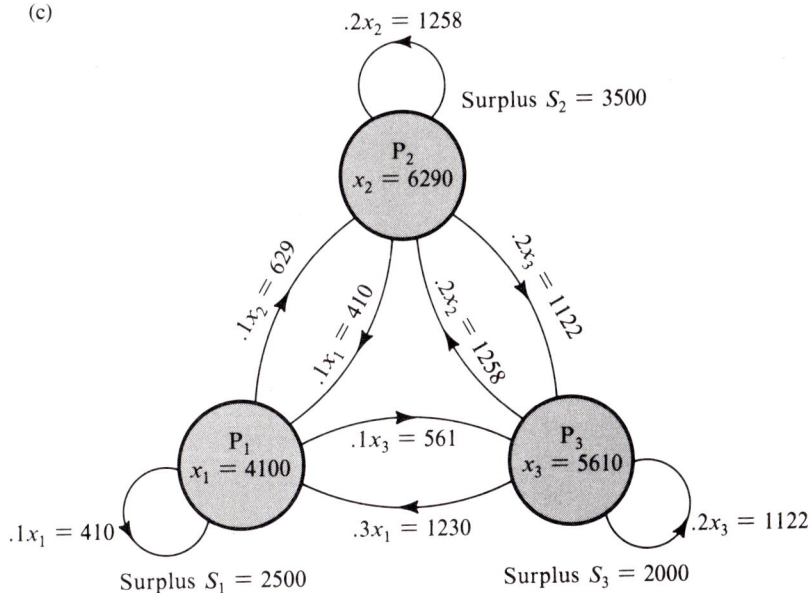

17. If $(I - A)^{-1} \geq O$ and $D \geq O$ then $X = (I - A)^{-1}D \geq O.$

19. (a) Not Leontief: $(I - A)$ is noninvertible   **(c)** Leontief: $(I - A)^{-1} = \begin{bmatrix} 1.75 & .025 \\ 5 & 1.5 \end{bmatrix} \geq O$

(b) Leontief: $(I - A)^{-1} = \begin{bmatrix} 1.75 & .75 \\ .75 & 1.75 \end{bmatrix} \geq O$ **(d)** Not Leontief: $(I - A)^{-1} = \begin{bmatrix} -14 & -18 \\ -12 & -14 \end{bmatrix} \not\geq O$

Review Exercises, page 159

1. undefined

3. $\begin{bmatrix} 12 & -5 & 2 \\ 9 & 19 & -4 \end{bmatrix}$

5. $\begin{bmatrix} 5 & -9 & 4 \\ 13 & 13 & -12 \end{bmatrix}$

7. $\begin{bmatrix} 14 \\ -13 \end{bmatrix}$

9. $X = 3A - 2B = \begin{bmatrix} 2 & -3 \\ 9 & 1 \end{bmatrix}$

11. $X = A^{-1} = \begin{bmatrix} 3 & -1 \\ -5 & 2 \end{bmatrix}$

13. $x = 7b_1 - 3b_2,\ y = -2b_1 + b_2$

15. $x_1 = 9568$ units goods, $x_2 = 8480$ units services

Chapter 5

Section 5.1, page 169

1.

x	y	u	v	w	F	
1	1	1	0	0	0	10
2	3	0	1	0	0	24
0	1	0	0	1	0	6
-3	-7	0	0	0	1	0

3. Maximum $F = 51$ when $x = 3$ and $y = 6$.

5. Maximum $F = 75$ when $x = 15$ and $y = 0$.
Edge-path: $(0, 0) \rightarrow (15, 0)$.

7. Maximum $F = 40$ when $x = 3$ and $y = 2$.
Edge-path: $(0, 0) \rightarrow (3, 0) \rightarrow (3, 2)$.

9. Maximum $F = 40$ when $x = 20$ and $y = 10$.
Edge-path: $(0, 0) \rightarrow (10, 0) \rightarrow (20, 10)$.

11. Maximum $F = 120$ when $x = 60$, $y = 0$, $z = 0$.

13. Maximum $F = 7$ when $x = 3$, $y = 1$, $z = 0$.

15. Maximum $F = 10$ when $x = 2$, $y = 4$, $z = 0$, $w = 0$.

17. Maximum $R = \$2750$ when $x = 0$ lb mixture A, $y = 125$ lb mixture B, $z = 0$ lb mixture C, $w = 450$ lb mixture D

19. Maximum $P = \$225,000$ when $x = 500$ acres asparagus, $y = 500$ acres brussels sprouts, $z = 0$ acres cauliflower.

21. Edge-path: $(0, 0) \rightarrow (30, 0) \rightarrow (30, 10)$.

23. (a) Maximum $F = 600$ when $x = 0$ and $y = 24$.
(b) Maximum $F = 600$ when $x = 15$ and $y = 15$.

Section 5.2, page 178

1.

x	y	u	v	w	-G	
-1	1	1	0	0	0	5
-1	-2	0	1	0	0	-20
-1	-1	0	0	1	0	-15
5	4	0	0	0	1	0

3. Minimum $G = 65$ when $x = 5$ and $y = 10$.

5. Maximum $F = 125$ when $x = 5$ and $y = 15$.
Phase 1: Pivot on 1 in x-column
Path: $(0, 0) \rightarrow (20, 0) \rightarrow (5, 15)$.

6. Phase 1: Pivot on -1 in x-column

7. Minimum $G = 70$ when $x = 5$ and $y = 20$.
Phase 1: Pivot on -1 in second row, y-column
Path: $(0, 0) \rightarrow (0, 40) \rightarrow (5, 20)$.

8. Phase 1: Pivot on -2 in y-column

9. Minimum $G = 85$ when $x = 5$ and $y = 15$.
Phase 1: Pivot on -1 in first row, y-column, then on -2 in x-column.
Path: $(0, 0) \rightarrow (0, 20) \rightarrow (5, 15)$.

10. Phase 1: Pivot on 1 in x-column, then on -1 in y-column.

11. Minimum $G = 60$ when $x = 0$, $y = 0$, $z = 60$.
Phase 1: Pivot on -1 in z-column.

12. Phase 1: Pivot on -1 in x-column.

13. Maximum $F = 22$ when $x = 3$, $y = 0$, $z = 5$.
Phase 1: Pivot on 1 in z-column.

14. Phase 1: Pivot on -1 in first row, z-column.

15. Minimum $G = 110$ when $x = 10$, $y = 40$, $z = 50$, $w = 0$.

Phase 1: Pivot on -1 in x-column, then on -1 in y-column.

16. Phase 1: Pivot on -1 in x-column, then on 1 in z-column

17. Minimum $C = \$3450$ when $x = 5$ tons oil, $y = 5$ tons coal.

Phase 1: Pivot on -40 in x-column.

18. Phase 1: Pivot on -2 in x-column.

19. Minimum $C = \$66,000$ when $x = 6$ type A planes, $y = 12$ type B planes, $z = 0$ type C planes.

Phase 1: Pivot on -60 in y-column.

20. Phase 1: Pivot on -1 in y-column.

Section 5.3, page 188

1. Minimum $C = \$650$ when $x = 50$ lb mixture A, $y = 75$ lb mixture B.   **3.** Minimum $C = \$1300$ when $x = 1000$ gal grade A, $y = 1000$ gal grade B, $z = 0$ gal grade C.

5. Minimum $M = 6750$ miles, when $x_{11} = 5$, $x_{12} = 0$, $x_{13} = 25$, $x_{21} = 10$, $x_{22} = 20$, $x_{23} = 0$.

(x_{ij} = number of cars driven from city A_i to city B_j).

7. Minimum $C = \$24,600$ when $x_{1A} = 0$, $x_{1B} = 24$, $x_{2A} = 30$, $x_{2B} = 9$.

(x_{ij} = number of type i trucks sent to city j).

9. Minimum $C = \$900$ when $x_1 = 0$ projectors rented for day 1, $x_2 = 5$ rented for day 2, $x_3 = 0$ rented for day 3, $x_4 = 5$ rented for days 1–2, $x_5 = 10$ rented for days 2–3.   **11.** Minimum $C = \$1,500,000$ when $x_1 = 0$ tankers leased for Jan, $x_2 = 4$ leased for Feb, $x_3 = 0$ leased for Mar, $x_4 = 0$ leased for Jan–Feb, $x_5 = 0$ leased for Feb–Mar, $x_6 = 10$ leased for Jan–Mar.

13. Minimum trim-waste $W = 800$ roll-ft, when either

(i) $x_1 = 250$, $x_2 = 0$, $x_3 = 0$, $x_4 = 150$, $x_5 = 0$, or

(ii) $x_1 = 200$, $x_2 = 0$, $x_3 = 0$, $x_4 = 200$, $x_5 = 0$.

(x_i = number of standard rolls cut according to pattern i).

Review Exercises, page 191

1. Maximum $F = 44$ when $x = 4$ and $y = 12$.

3. Minimum $G = 130$ when $x = 30$ and $y = 20$.

Phase 1: Pivot on -1 in x-column

5. Maximum $F = 32$ when $x = 0$, $y = 4$, $z = 8$.   **7.** Maximum $T = 13$ tons when $x = 100$ tons ore from Mine A, $y = 0$ tons from Mine B, $z = 400$ tons from Mine C.

9. Minimum $C = \$23,100$ when $x_{11} = 8$, $x_{12} = 0$, $x_{21} = 4$, $x_{22} = 7$.

(x_{ij} = number of carloads bought from farm F_i and sent to distributor D_j).

Chapter 6

Section 6.1, page 200

1. $\{3, 4, 5, 6, 7, 8\}$ **3.** $\{1, 2, 5\}$ **5.** $\{1, 2, 5, 10\}$ **7.** $\varnothing$ **9.** (a) $\{a\}$, $\{b\}$, $\{c\}$, $\{d\}$ (b) $\{a, b\}$, $\{a, c\}$, $\{a, d\}$, $\{b, c\}$, $\{b, d\}$, $\{c, d\}$ (c) $\{a, b, c\}$, $\{a, b, d\}$, $\{a, c, d\}$, $\{b, c, d\}$ **11.** (a) $\{y, u\}$ (b) $\{x, y, z, u, v, w\}$ (c) $\varnothing$

(d) $\{x, z, w\}$ (e) $\{z\}$ (f) $\{x, v, w\}$ (g) $\{x, y, u, v, w\}$ (h) $\{x\}$

13. (a)

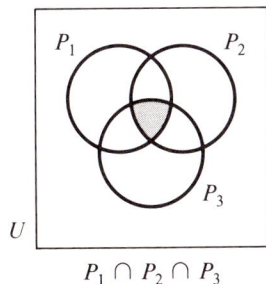

$P_1 \cap P_2 \cap P_3$

(b)

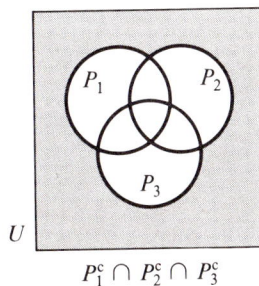

$P_1^c \cap P_2^c \cap P_3^c$

(c)

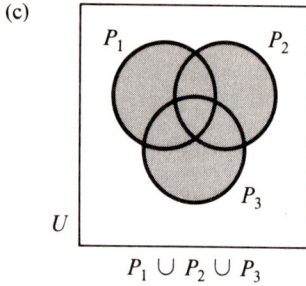

$$P_1 \cup P_2 \cup P_3$$

(d)

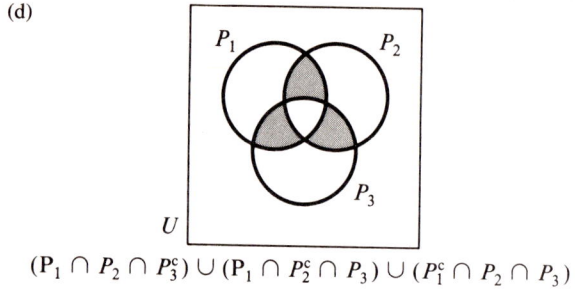

$$(P_1 \cap P_2 \cap P_3^c) \cup (P_1 \cap P_2^c \cap P_3) \cup (P_1^c \cap P_2 \cap P_3)$$

15.

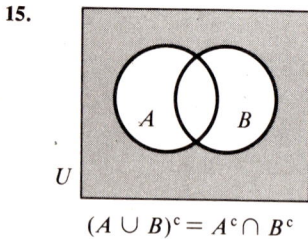

$$(A \cup B)^c = A^c \cap B^c$$

17.

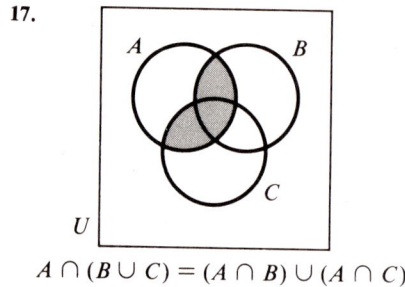

$$A \cap (B \cup C) = (A \cap B) \cup (A \cap C)$$

19. Yes; the bill will get at least 53 votes.   **21.** 38

23. (a) A^+: 285 B^+: 60 AB^+: 30 O^+: 270
    A^-: 45 B^-: 15 AB^-: 0 O^-: 45
(b) 74%   (c) 80%

25. 430 husbands and 215 wives were satisfied.

Section 6.2, page 208

1. 16   **3.** 192   **5.** 13,800   **7.** 9,000,000   **9.** 362,880; 5040   **11.** 60,466,176   **13.** (a) 11,881,376
(b) 7,893,600   **15.** 380   **17.** (a) 1024   (b) 59,049   **19.** (a) 2^n (b) 2^n **21.** (a) 90,000   (b) 3125
(c) 20,000   (d) 37,512

23. Let $A = \{a_1, a_2, \ldots, a_n\}$. For a subset $X \subset A$ as follows:
  C_1: Decide whether or not $a_1 \in X$: 2 ways.
  C_2: Decide whether or not $a_2 \in X$: 2 ways.

  $\vdots$

  C_n: Decide whether or not $a_n \in X$: 2 ways
  By the Counting Principle, these decisions can be made in $2 \cdot 2 \cdots \cdots 2 = 2^n$ ways.

Section 6.3, page 218

1. (a) 2520   (b) 36   (c) 720   (d) 8   (e) 3360   (f) 252

3. (a) ab   ba   ca   da   ea
      ac   bc   cb   db   eb
      ad   bd   cd   dc   ec
      ae   be   ce   de   ed
(b) {a, b}, {a, c}, {a, d}, {a, e}, {b, c}, {b, d}, {b, e}, {c, d}, {c, e}, {d, e}

5. 2730   **7.** (a) 1,860,480   (b) 15,504   **9.** 40,320   **11.** 46,080   **13.** (a) 45   (b) 120   **15.** (a) 126

(b) 84   (c) 256   **17.** (a) 38,760   (b) 10,010   **19.** 411,600   **21.** 11   **23.** (a) 212,520   (b) 78,624

25. (a) 1728   (b) 576   **27.** (a) 3,628,800   (b) 2184   (c) 8   (d) 220

29. $P(n, k) = n(n - 1) \cdots (n - k + 1)$

$$= \frac{n(n - 1) \cdots (n - k + 1)(n - k) \cdots 3 \cdot 2 \cdot 1}{(n - k)(n - k - 1) \cdots 3 \cdot 2 \cdot 1}$$

$$= \frac{n!}{(n - k)!}$$

$$C(n, k) = \frac{P(n, k)}{P(k, k)} = \frac{n!}{k!(n - k)!}$$

31. $C(n, 0) = \dfrac{n!}{0!(n - 0)!} = \dfrac{n!}{1 \cdot n!} = 1$

$C(n, n) = \dfrac{n!}{n!(n - n)!} = \dfrac{n!}{n!0!} = 1$

There is exactly one combination of n objects taken 0 at a time (the empty set).

There is exactly one combination of n objects taken n at a time (the whole set).

Section 6.4, page 228

1. (a) 20   (b) 126   (c) 455   (d) 60   (e) 4950   (f) 1120

3. $x^7 + 7x^6y + 21x^5y^2 + 35x^4y^3 + 35x^3y^4 + 21x^2y^5 + 7xy^6 + y^7$ **5.** (a) 1   (b) 91   (c) 3432

7. Row $n = 10$: 1, 10, 45, 120, 210, 252, 210, 120, 45, 10, 1.

9. (a) $n = 0$: $1 = 2^0$

$n = 1$: $1 + 1 = 2 = 2^1$

$n = 2$: $1 + 2 + 1 = 4 = 2^2$, and so on.

(b) Let $x = y = 1$ in Theorem 1.   (c) By Exercise 23 in Section 6.2, a set with n elements has 2^n subsets. The

number of subsets containing exactly k elements ($k = 0, 1, \ldots, n$) is $C(n, k) = \dbinom{n}{k}$. Hence:

$$2^n = \binom{n}{0} + \binom{n}{1} + \cdots + \binom{n}{n}.$$

11. $\dbinom{n}{k} = \dfrac{n!}{k!(n - k)!} = \dfrac{n - k + 1}{k} \cdot \dfrac{n!}{(k - 1)!(n - k + 1)!} = \dfrac{n - k + 1}{k} \cdot \dbinom{n}{k - 1}$

13. $x^3 + y^3 + z^3 + 3x^2y + 3xy^2 + 3x^2z + 3xz^2 + 3y^2z + 3yz^2 + 6xyz$ **15.** (a) 1   (b) 15   (c) 180

17. 2520   **19.** (a) $52!/(13!)^4$ (b) $52!/(5!)^4 \cdot 32!$ **21.** (a) 420   (b) 60   (c) 37,800   **23.** 92,400

25. (a) Represent each route by a word containing four E's and three N's.

(b) Number of routes = number of words = $\dbinom{7}{4, 3} = 35$

(c) Number of routes = $\dbinom{m + n}{m, n} = \dfrac{(m + n)!}{m!n!}$

Review Exercises, page 232

1. True   **3.** False   **5.** False

7.

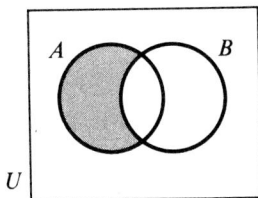

9. 138   **11.** 2880   **13.** (a) 270,725   (b) 3,628,800   (c) 4896   (d) 2184   **15.** 24   **17.** 280

Chapter 7

Note Answers with a   +   or   −   are rounded to three significant figures; for example, $1/7 = .143^-$ and $4/7 = .571^+$.

Section 7.1, page 244

1. $S = \{0, 1, 2, \ldots, 15\}$ **3.** $S = \{HH, HT, TH, TT\}$
   (a) $\{0\}$ (b) $\{0, 1, 2\}$ (c) $\{8, 9, \ldots, 15\}$ (a) $\{HH, TH\}$ (b) $\{HH, HT, TH\}$ (c) $\emptyset$
5. Yes   **7.** No   **9.** (i) 3/8   (ii) 2/5   (iii) .45   **11.** (i) 1   (ii) 1   (iii) 1   **13.** $.158^+$
15. (a) Number of

requests	Probability
0	.11
1	.20
2	.32
3	.19
4	.12
5	.05
6	.01

   (b) (i) .32   (ii) .63   (iii) .18   (c) 2
17. .28   **19.** $.444^+$ **21.** (a), (b) relative frequency   (c) comparison class   (d) neither   **23.** (a) .875
(b) .5   (c) $.455^-$ (d) .4

Section 7.2, page 255

1. (a) .5   (b) $.333^+$ (c) $.167^-$ (d) 0   **3.** $.0849^+$; $.255^-$ **5.** (a) .47   (b) .34   (c) .25   **7.** (a) .18
(b) .70   (c) .72   **9.** (a) $.417^-$ (b) $.444^+$ (c) $.167^-$ **11.** (a) .125   (b) .375   (c) .5
13. (a) $.0833^+$ (b) $.233^+$ **15.** (a) $.0478^-$ (b) $.0476^+$ **17.** $.145^+$ **19.** .1   **21.** (a) $.0518^-$
(b) $.00235^+$ (c) $.169^+$ **23.** (a) $.0167^-$ (b) .5

Section 7.3, page 268

1. (a)

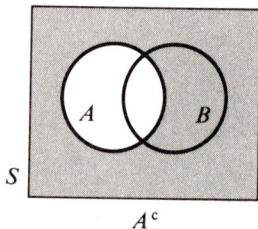

A^c

(b)

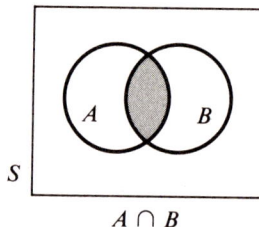

$A \cap B$

(c)

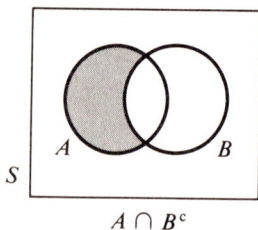

$A \cap B^c$

(d)

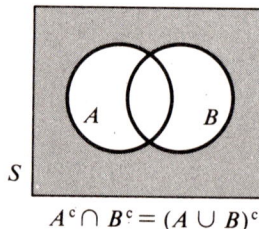

$A^c \cap B^c = (A \cup B)^c$

3. (a) Card is red and higher than five; $.308^-$ (b) Card is red or higher than five; $.808^-$ (c) Card is red but not higher than five; $.192^+$ (d) Card is neither red nor higher than five; $.192^+$ **5.** (a) .47   (b) .77   (c) .23
7. .875   **9.** $.656^+$ **11.** $.601^-$ **13.** (a) .95   (b) .05   **15.** $.632^-$

17. *N:* Norman likes Sylvia

S: Sylvia likes Norman

$$1 \geqslant P(N \cup S) = P(N) + P(S) - P(N \cap S) = \frac{11}{12} + \frac{7}{12} - P(N \cap S)$$

$$\therefore P(N \cap S) \geqslant \frac{11}{12} + \frac{7}{12} - 1 = \frac{6}{12} = .5$$

19. .45   **21.** (a) .46   (b) .90   (c) .92   **23.** $.738^+$ **25.** (a) Points in $A \cap B^c \cap C$ are counted once as elements of A and once as elements of C. Points in $A \cap B \cap C$ are counted once as elements of A, once as elements of B, and once as elements of C.   **27.** (a) .32   (b) .41

Section 7.4, page 278

1. (a) $.667^-$ (b) $.906^-$ **3.** (a) $.417^-$ (b) $.636^+$ (c) .5   **5.** (a) .875   (b) .5   (c) .75

7. $\frac{1}{13} = \frac{4}{52} \doteq .0769$ **9.** (a) $.610^-$ (b) $.790^+$ **11.** $.471^-$ **13.** $.848^+$ **15.** $.170^+$ **17.** $.125^-$

19. By Exercise 26 in Section 7.3:

$$P(E \cap A^c) = P(E - A) = P(E) - P(E \cap A)$$

Dividing by $P(E)$:

$$P(A^c|E) = \frac{P(E \cap A^c)}{P(E)} = 1 - \frac{P(E \cap A)}{P(E)} = 1 - P(A|E)$$

21. Given only the circumstantial evidence that Snavely's car has properties (a) and (b), the conditional probability that he is the murderer is only $\frac{1}{5}$, not $1 - 5/735,000$ as the prosecutor is trying to imply.

Section 7.5, page 290

1. .18   **3.** .06   **5.** (a) .72   (b) .94   **7.** (a) .15   (b) .25   (c) .4   **9.** $.559^-$ **11.** (a) .8125

(b) $.913^-$ **13.** Independent   **15.** Independent   **17.** (a) $P(A \cap B) = 0$, $P(A \cup B) = .95$

(b) $P(A \cap B) = .195$, $P(A \cup B) = .755$ **19.** .828   **21.** .835   **23.** (a) .018   (b) .43

25. $0 = P(\varnothing) = P(A \cap B) = P(A) \cdot P(B)$, so $P(A) = 0$ or $P(B) = 0$.

27. (a) Player's point   Probability of winning

4	3/9
5	4/10
6	5/11
8	5/11
9	4/10
10	3/9

(b) $\frac{244}{495} \doteq .493$

29. $.664^+$

Review Exercises, page 294

1. (a) .12   (b) .33   (c) .67   (d) .91   **3.** .72   **5.** $.143^-$ **7.** $.332^+$ **9.** (a) .6   (b) .4   (c) .5

11. .58   **13.** $.191^+$ **15.** .18   **17.** .44

Chapter 8

Note Answers with a $+$ or $-$ are rounded to three significant figures; for example, $1/7 = .143^-$ and $4/7 = .571^+$.

Section 8.1, page 302

1. .4   **3.** .7   **5.** $.533^+$ **7.** .9775   **9.** approx. 73.7%   **11.** $.667^-$ **13.** .8   **15.** 0.143%

17. (a) $.0531^+$ (b) $.0326^-$

Section 8.2, page 311

1. (a) $.185^+$ (b) $.420^+$ (c) $.118^-$ (d) $.882^+$ **3.** (a) $.579^-$ (b) $.347^+$ (c) $.069^+$ (d) $.005^-$
5. (a) $.0503^+$ (b) $.274^+$ **7.** $.258^+$ **9.** $.914^-$ **11.** $.316^+$ **13.** $.665^+$ **15.** $P(\text{A wins}) = .665^+$;
$P(\text{B wins}) = .619^-$; $P(\text{C wins}) = .597^+$ **17.** $.167^+$ **19.** 7 **21.** $.156^+$ **23.** (a) $.0702^-$ (b) $.0547^-$
25. The outcome of a Bernoulli process can be represented by an n-letter "word" containing only the letters S and F.
The number of words containing exactly k S's is $C(n, k)$ and each of these outcomes has probability $p^k q^{n-k}$. Hence
P (k successes in n trials) $= C(n, k)p^k q^{n-k}$.

Section 8.3, page 322

1. 2.2 **3.** $1.60 **5.** $2.5 million **7.** $106 **9.** 1.66 **11.** $-7.7¢$ (approx.) **13.** player
(approx. 33¢) **15.** $-5.3¢$ (approx.); house percentage is 5.3% **17.** $-29.7¢$ (approx.); house percentage 24.7%
19. $-50¢$; $-75¢$ **21.** $-10.5¢$ (approx.) **23.** $2p - 1$; .45 **25.** 3.5

Section 8.4, page 332

1. Maximin: a_3; maximax: a_3; Bayes: a_3. **3.** (a) set up the equipment (b) 70% **5.** 2 papers ($E = 64¢$)
7. $E = 58$ **9.** Hire the lobbyist ($E = \$650,000$). **11.** Look for a space on the street. If found, put money in
the meter ($E = -79¢$). **13.** File the suit. If the company offers to settle, accept ($E = \$1950$). **15.** 3 cameras
(Exp. loss $= \$24,500$). **17.** 13 (Exp. net gain $= \$6.50$). **19.** 10 **21.** $3900

Review Exercise, page 336

1. $.296^+$ **3.** (a) .375 (b) .875 **5.** (a) $.817^+$ (b) $.984^-$ **7.** $90 **9.** $-13.9¢$ (approx.)
11. 10,000 copies ($E = \$101,000$) **13.** Yes ($E = \$13,000$)

Chapter 9

Note Answers with a $+$ or $-$ are rounded to three significant figures; for example, $1/7 = .143^-$ and
$4/7 = .571^+$.

Section 9.1, page 347

1. (a)

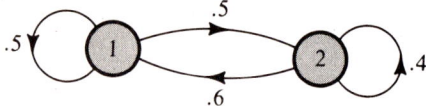

(b)

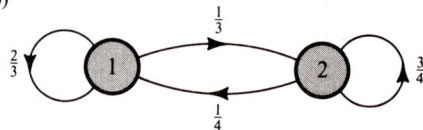

(c)

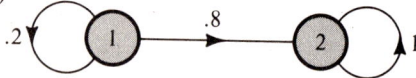

3. (a) $\begin{bmatrix} .9 & .1 \\ .1 & .9 \end{bmatrix}$ (b) $\begin{bmatrix} .3 & .7 \\ 1 & 0 \end{bmatrix}$
5. (a) $[.3 \quad .7]$ (b) $[.2375 \quad .7625]$ **7.** .8752 **9.** 42.92% **11.** 3 years (50.29%) **13.** .375
15. 10.3% upper, 61.6% middle, 28.1% lower
17.

		0	1	2	3	4
	0	.2	.6	.2	0	0
	1	.2	.6	.2	0	0
Current	2	0	.2	.6	.2	0
state	3	0	0	.2	.6	.2
	4	0	0	0	.2	.8

Next state

Section 9.2, page 358

1. (a) regular   (b) not regular   3. (a) $[1 \quad 0], [0 \quad 1], [1 \quad 0], [0 \quad 1]$ (b) $[0 \quad 1], [1 \quad 0], [0 \quad 1], [1 \quad 0]$ (c) No.

5. $[.2 \quad .8]$ 7. 87.5%   9. 25%   11. 2/3   13. 1/3   15. 12.5% upper, 62.5% middle, 25% lower.

17. 2.45

19. (a)

$$\begin{array}{c} \text{Current} \\ \text{state} \end{array} \quad \begin{array}{c} \quad\; \text{Next state} \\ \quad\; A \quad\; D \\ \begin{array}{c} A \\ D \end{array} \begin{bmatrix} .6 & .4 \\ .1 & .9 \end{bmatrix} \end{array}$$

(b) Equilibrium vector $X = [.2 \quad .8]$

Section 9.3, page 368

1. (a) absorbing   (b) not absorbing   3. (a) 3 steps; 2.5 steps   (b) $p = 1$ in both cases   5. 13 years (make W an absorbing state)   7. (a) 3 weeks   (b) .8   9. (a) 5 ballots   (b) .25   11. 6 tosses   13. (a) 3 games   (b) .75   (c) 4 games; .5

Section 9.4, page 376

1. (a) $AB = 1 \cdot 3 = 3$ (b) $\dfrac{B}{A + B} = \dfrac{3}{1 + 3} = .75$ (c) $AB = 2 \cdot 2 = 4$; $\dfrac{B}{A + B} = \dfrac{2}{2 + 2} = .5$ 3. (a) .4

(b) 150   5. .884$^-$ 7. .018$^+$; .413$^-$ 9. 100 tosses

11. $P(\text{player ruined}) = \dfrac{1 - r^B}{1 - r^{A+B}} \geqslant 1 - r^B$ since $r = \dfrac{p}{q} < 1$.

13. $p_x = \dfrac{N - x}{N}$: Equation (2) becomes

$$\frac{N - x}{N} = \frac{1}{2} \cdot \frac{N - (x + 1)}{N} + \frac{1}{2} \cdot \frac{N - (x - 1)}{N}$$

which is verified by algebraic simplification. Since

$$p_0 = \frac{N - 0}{N} = 1, p_N = \frac{N - N}{N} = 0$$

the boundary conditions (1) are satisfied.

Section 9.5, page 384

1. (a) .25   (b) .5   3. .8   7. First generation: 64% AA, 32% Aa, 4% aa
5. (a)   Second generation: Same.

9. 32%

$$\begin{array}{c} \text{Current} \\ \text{genotype} \end{array} \quad \begin{array}{c} \quad\;\; \text{Next genotype} \\ \quad\; AA \quad Aa \quad aa \\ \begin{array}{c} AA \\ Aa \\ aa \end{array} \begin{bmatrix} 1/2 & 1/2 & 0 \\ 1/4 & 1/2 & 1/4 \\ 0 & 1/2 & 1/2 \end{bmatrix} = P \end{array}$$

(b) .225; .5; .275   (c) .25 red, .5 pink, .25 white

11. Let $u = p + \frac{1}{2}q$, $v = \frac{1}{2}q + r$. Then genotype proportions in next generation are $p_1 = u^2$, $q_1 = 2uv$, $r_1 = v^2$.
   (i) $p_1 = p, q_1 = q, r_1 = r$ implies $q^2 = q_1^2 = 4u^2v^2 = 4pr$.
   (ii) $q^2 = 4pr$ implies $p_1 = u^2 = p^2 + pq + \frac{1}{4}q^2 = p(p + q + r) = p$,   and so on.

13. (a) 1; 1/3   (b) 1/12

Review Exercises, page 386

1. $\begin{bmatrix} .1 & .9 \\ 1 & 0 \end{bmatrix}$ regular

3. $\begin{bmatrix} 0 & .5 & .5 \\ 0 & 0 & 1 \\ 0 & 1 & 0 \end{bmatrix}$ neither regular nor absorbing

5. (a) $[.26 \quad .18 \quad .56]$ (b) $[.3 \quad .2. \quad 5]$ **7.** 4%   **9.** (a) $.615^+$ (b) 1000   **11.** (a) $.422^-$ (b) $.0156^+$

Chapter 10

Section 10.1, page 396

1. (a) $\begin{bmatrix} -2 & \vdots & 1 \\ 4 & \vdots & -2 \\ \cdots & \cdots & \cdots \end{bmatrix}$

(b) $\begin{bmatrix} 4 & 0 & \vdots & -1 \\ 1 & 2 & \vdots & 3 \\ \cdots & \cdots & \cdots \end{bmatrix}$

3. Strictly determined; row 1, column 2; value 1   **5.** Not strictly determined.   **7.** Strictly determined; row 2, column 2; value 1   **9.** Not strictly determined   **11.** R put down $50, C put down $5; value $5   **13.** R in town 2, C in town 1; value = 40% of total population.   **15.** Both candidates should oppose nuclear power; value = 55% of those voting.

17. Let $A = [a_{ij}]$ be the payoff matrix. Let r be the index of an arbitrary fixed row and s the index of an arbitrary fixed column. Then

$$\min_j a_{rj} \le a_{rs} \le \max_i a_{is}.$$

(Note: "$\min_j a_{rj}$" means the minimum of the set $\{a_{r1}, a_{r2}, \ldots, a_{rn}\}$; that is, the smallest entry in the rth row. Also,

"$\max_i a_{is}$" means the maximum of the set $\{a_{1s}, a_{2s}, \ldots, a_{ms}\}$; that is, the largest entry in the sth column.)

19. If the game $A = [a_{ij}]$ is strictly determined, then by Exercise 18

$$\max_i \min_j a_{ij} = \min_j \max_i a_{ij}.$$

(Note: "$\max_i \min_j a_{ij}$" means the number found by taking the minimum entry in each row, putting these minimum entries together in a set, and then taking the maximum of this set. The number "$\min_j \max_i a_{ij}$" is found by taking the maximum entry in each column, putting these maximum entries together in a set, and then taking the minimum of this set.) If row r and column s are minimax optimal, then

$$\min_j a_{rj} = \max_i \min_j a_{ij} = \min_j \max_i a_{ij} = \max_i a_{is}.$$

Since $\min_j a_{rj} \le a_{rs} \le \max_i a_{is}$, it follows that $\min_j a_{rj} = a_{rs} = \max_i a_{is}$ so a_{rs} is a saddle point. The converse uses Exercise 17.

Section 10.2, page 406

1. (a), (c), (d) strategies; (c) pure, (a) and (d) mixed.   **3.** (a) $E = 2.5$ (b) $E = 2$; strategy   (b) better.
5. Player R $(E = 1)$ **7.** $P^\star = [1/2 \quad 1/2]$, $Q^\star = [5/8 \quad 3/8]^t$; $v = 1/2$ **9.** $P^\star = [0 \quad 1]$, $Q^\star = [1 \quad 0]^t$; $v = 1$

11. $P^\star = [2/5 \quad 3/5]$, $Q^\star = [1/2 \quad 0 \quad 1/2]^t$; $v = 0$ **13.** $P^\star = [5/12 \quad 7/12 \quad 0]$, $Q^\star = [3/4 \quad 0 \quad 1/4]^t$; $v = 1/4$

15. R: Nickel 1/2, dime 1/2

   C: Nickel 2/3, dime 1/3. Game is fair.

17. R: movie 3/4, sports 1/4

   C: movie 1/4, sports 3/4

   Exp. audience shares: 56.25% R, 43.75% C

19. Robin Hood: Raid village 1 55%, village 2 45%

   Sheriff: Defend village 1 45%, village 2 55%

   Exp. spoils: $495

21. Suppose $a = b$. If $a \geq c$ then a is a saddle point. If $b \geq d$ then b is a saddle point. Otherwise $a < c$ and $b < d$. But then smaller of c and d is a saddle point.

23. $E = ap_1q_1 + bp_1q_2 + cp_2q_1 + dp_2q_2$

$$PAQ = [p_1 \quad p_2]\begin{bmatrix} a & b \\ c & d \end{bmatrix}\begin{bmatrix} q_1 \\ q_2 \end{bmatrix}$$

$$= [ap_1 + cp_2 \quad bp_1 + dp_2]\begin{bmatrix} q_1 \\ q_2 \end{bmatrix}$$

$$= ap_1q_1 + bp_1q_2 + cp_2q_1 + dp_2q_2$$

Section 10.3, page 417

1. $P^\star = [1/2 \quad 1/2]$, $Q^\star = [1/2 \quad 1/2]^t$; value 0   **3.** $P^\star = [7/10 \quad 3/10]$, $Q^\star = [3/5 \quad 2/5]^t$; value $-1/5$

5. $P^\star = [1 \quad 0]$, $Q^\star = [0 \quad 1 \quad 0]^t$; value 1   **7.** $P^\star = [2/3 \quad 1/3)]$, $Q^\star = [0 \quad 1/3 \quad 2/3]^t$; value $-1/3$

9. $P^\star = [5/8 \quad 3/8]$, $Q^\star = [7/8 \quad 1/8 \quad 0]^t$; value 5/8   **11.** $P^\star = [1/2 \quad 1/2 \quad 0]$, $Q^\star = [3/5 \quad 2/5]^t$; value 0

13. $P^\star = [4/7 \quad 2/7 \quad 1/7]$, $Q^\star = [4/7 \quad 1/7 \quad 2/7]^t$; value 1/7

15. Inmate: Desert 5/9, mountains 4/9

   Warden: Both units to desert 1/9, both to mountains 8/9.

   Exp. chance of escape: 27.78% (approx.)

17. R: Show 1 finger 1/4, 2 fingers 1/2, 3 fingers 1/4.

   C: Same; value 0.

19. If $x_1 = x_2 = x$, then $p_1x_1 + p_2x_2 = (p_1 + p_2)x = x$ for all P, so can assume, say, $x_1 < x_2$: For any $P = [p_1 \quad p_2]$,

$$x_1 = (p_1 + p_2)x_1 = p_1x_1 + p_2x_1 \leq p_1x_1 + p_2x_2$$
$$x_2 = (p_1 + p_2)x_2 = p_1x_2 + p_2x_2 \geq p_1x_1 + p_2x_2$$

So $x_1 \leq p_1x_1 + p_2x_2 \leq x_2$. Moreover, if $x_1 \leq x \leq x_2$, then

$$p_1 = \frac{x_2 - x}{x_2 - x_1}, p_2 = \frac{x - x_1}{x_2 - x_1}$$

gives a strategy $P = [p_1 \quad p_2]$ with $p_1x_1 + p_2x_2 = x$.

Review Exercises, page 419

1. $\begin{bmatrix} \vdots & -1 & 2 \\ \vdots & & \\ \vdots & 4 & -3 \end{bmatrix}$ **3.** $\begin{bmatrix} 2 & -1 & 5 & \vdots \\ 3 & 2 & -6 & \vdots \\ 0 & 4 & 1 & \vdots \\ \cdots & \cdots & \cdots & \vdots \end{bmatrix}$

5. Not strictly determined   **7.** Strictly determined; row 3, column 1; value 1   **9.** (a) Player C $(E = -5)$

(b) $P^\star = [.3 \quad .7]$, $Q^\star = [.65 \quad .35]^t$; $v = 1$ (advantage R)   **11.** $P^\star = [1/7 \quad 6/7]$, $Q^\star = [5/7 \quad 2/7 \quad 0]^t$; value $-5/7$

13. R: Put \$1 with probability 2/3, \$2 with probability 1/3
   C: Put \$1 with probability 1/2, \$2 with probability 1/2
   Value 0 (fair game)

Chapter 11

Section 11.1, page 431

1.

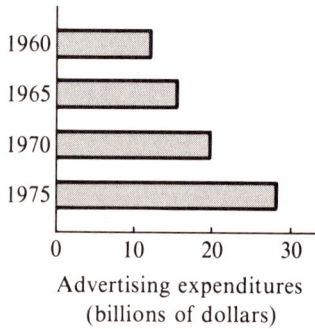

Advertising expenditures
(billions of dollars)

3.

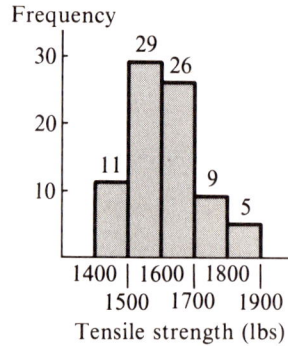

Tensile strength (lbs)

5.

Interval	Frequency
30–35	2
35–40	4
40–45	7
45–50	8
50–55	6
55–60	3
Total	30

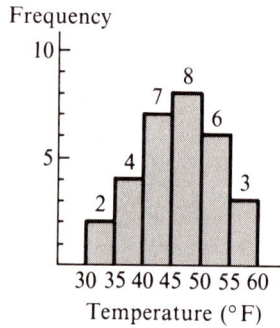

Temperature (°F)

7. $\bar{x} = 53$, $\tilde{x} = 52.5$, $s \doteq 3.38$.   **9.** $\bar{x} = 1610$, $s \doteq 105.59$.   **11.** Grouped $\bar{x} = 46$, $s \doteq 6.85$; true $\bar{x} \doteq 46.17$
13. $\bar{x} = 389{,}000$; $\tilde{x} = 262{,}500$. Median gives better measure of central tendency since one value is "atypical" and distorts the mean.   **15.** (a) A: $\bar{x} = 1.5$, $s \doteq .84$; B: $\bar{x} = 2$, $s \doteq 1.45$.   (b) B had better record but A was more consistent.   **17.** (a) $\bar{x} = 3$, $s \doteq 1.10$ (b) $\bar{x} = 3$, $s \doteq 1.61$.

19. Mean of new data $= \dfrac{kx_1 + \cdots + kx_n}{n} = k\left(\dfrac{x_1 + \cdots + x_n}{n}\right) = k\bar{x}.$

Variance of new data $= \dfrac{(kx_1 - k\bar{x})^2 + \cdots + (kx_n - k\bar{x})^2}{n}$

$= \dfrac{k^2(x_1 - \bar{x})^2 + \cdots + k^2(x_n - \bar{x})^2}{n}$

$= k^2\left(\dfrac{(x_1 - \bar{x})^2 + \cdots + (x_n - \bar{x})^2}{n}\right) = k^2 s^2$

$\therefore$ S.d. of new data $= \sqrt{k^2 s^2} = ks$

Section 11.2, page 445

1.

Violations	Probability
0	.425
1	.350
2	.150
3	.050
4	.025
Total	1.000

Probability

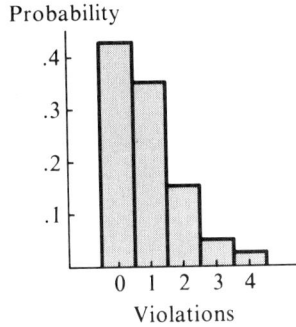

Violations

3. (a) Probability

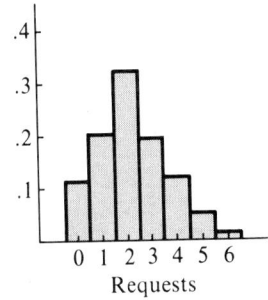

Requests

(b) (i) .32   (ii) .63   (iii) .18

5. (i) .6   (ii) .5   (iii) .7   **7.** (a) $\bar{x} = .9, s \doteq .995$ (b) $\mu = .9, \sigma \doteq .995$.   **9.** $\mu = 2.2, \sigma \doteq 1.37$.

11. $\mu = 1.8, \sigma = 1.4$.

13. (a)

Number drawn	Probability
1	.4
2	.3
3	.2
4	.1

(b) $\mu = 2, \sigma = 1$.

15. (a)

No. of successes	Probability
0	.125
1	.375
2	.375
3	.125

(b)

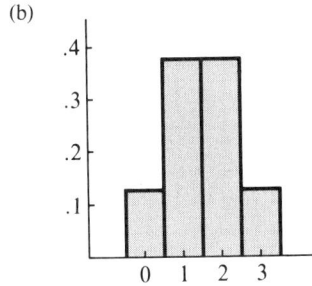

(c) $\mu = 1.5, \sigma \doteq .866$

17. $\mu = 15, \sigma \doteq 3.54$ **19.** $\mu \doteq 23.68, \sigma \doteq 3.53$ **21.** $\mu = 40, \sigma \doteq 6.31$ **23.** $\mu = 315, \sigma \doteq 5.61$

25. $\mu = 150, \sigma \doteq 6.12$

27. (a) $E(kX) = \Sigma(kx)p = k\Sigma xp = kE(X) = k\mu$

$V(kX) = \Sigma(kx - k\mu)^2p = k^2\Sigma(x - \mu)^2p = k^2V(X)$.

(b) $E(X + k) = \Sigma(x + k)p = \Sigma xp + \Sigma kp = E(X) + k$

$V(X + k) = \Sigma[(x + k) - (\mu + k)]^2p = \Sigma(x - \mu)^2p = V(X)$.

Section 11.3, page 459

1. (a) .7190   (b) .7888   (c) .5080   **3.** (a) $-.98$ (b) 1.27   **5.** (a) .0475   (b) .5934   (c) .0062

7. (a) 10.5   (b) 62   **9.** .1056   **11.** 65.54%   **13.** (a) 79.67%   (b) 2393 lb.   **15.** 67.3 to 70.7

17. .8788   **19.** .0495   **21.** .9515   **23.** .8365   **25.** .9146

Review Exercises, page 462

1. (a)

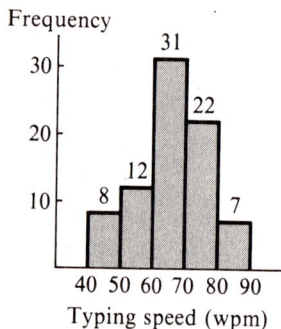

(b) $\bar{x} = 66$, $s \doteq 10.79$

3. $\tilde{x} = 6800$, $\bar{x} = 7000$, $s \doteq 1029.56$ **5.** $\mu = 1.6$, $\sigma \doteq 1.23$ **7.** $\mu = 5$, $\sigma \doteq 1.83$ **9.** 25.78%   **11.** .7286

Chapter 12

Section 12.1, page 473

1. (a) 3, 14, 69, 344, 1719, . . .; neither   (b) 0, 3/2, 3, 9/2, 6, . . .; arithmetic   (c) 1, 1/2, 1/4, 1/8, 1/16, . . .;
geometric   (d) 5, 39, 32.2, 33.56, 33.288, . . .; neither   **3.** (a) $x_n = 2 + 6n$; $x_{10} = 62$, $x_{20} = 122$
(b) $x_n = 3 \cdot 2^n$; $x_{10} = 3072$, $x_{20} = 3{,}145{,}728$ (c) $x_n = 35 - \frac{5}{2}n$; $x_{10} = 10$, $x_{20} = -15$ (d) $x_n = \left(-\frac{2}{3}\right)^n$;
$x_{10} \doteq .0173$, $x_{20} \doteq .000301$ **5.** $8228; $1428 interest   **7.** 20% per annum   **9.** $261.42   **11.** (a) $685.38
(b) $693.09   (c) $697.23   **13.** (a) 8.73%   (b) 8.77%   **15.** (a) $150,000   (b) $151,555.17   **17.** $5000
immediately (present value of $5200 in six months = $4998.08).   **19.** (a) $V_n = 356{,}000 - 8900n$
(b) $V_n = 356{,}000 (.95)^n$

Year	Straight-line Book value	Declining balance
10	$267,000	$213,150
20	178,000	127,621
30	89,000	76,411

21. (a) $C_{n+1} = 1.08\, C_n$ (b) $C_n = (1.08)^n C_0$ (c) $215.89 (116%)   **23.** (a) $C_{n+1} = .988\, C_n$
(b) $C_n = (.988)^n C_0$ (c) .547 units (45.3%)

Section 12.2, page 485

1. (a) $x_n = 7 \cdot 3^n - 2$; $x_5 = 1699$, $x_{10} = 413{,}341$ (b) $x_n = 32 \left(\frac{1}{2}\right)^n + 2$; $x_5 = 3$, $x_{10} = 2.03125$
(c) $x_n = 2(1.5)^n + 6$; $x_5 = 21.1875$, $x_{10} = 121.33$ (d) $x_n = (-5)(-.6)^n + 15$; $x_5 = 15.3888$, $x_{10} \doteq 14.970$
3. (a) $S_{n+1} = 1.02\, S_n + 50$, $S_0 = 500$ (b) $S_n = 3000(1.02)^n - 2500$ (c) $S_{20} \doteq $1957.84   **5.** (a) $2628.62
(b) $2228.62   **7.** $71.89   **9.** $39,706.39   **11.** $14,513.57   **13.** $508.61   **15.** $44,201.35
17. $78,431.39   **19.** $199,083.20   **21.** (a) 5824; 6496   (b) 7500   **23.** By Theorem 1, $x_n = a^n(x_0 - c) + c$
(a) If $x_0 = c$ then $x_n = c$ for all n.   (b) If $|a| < 1$ then $a^n \to 0$ as $n \to \infty$, so $x_n \to c$.

Review Exercises, page 488

1. (a) $x_n = -1 + 2n$ (b) $x_n = 125(.8)^n$ (c) $x_n = 3 \cdot 4^n - 3$ (d) $x_n = 2$ **3.** $81,000; $6000 interest
5. (a) $3220.56   (b) $3260.35   (c) $3269.98   **7.** $13,112.30   **9.** (a) $B_{n+1} = 1.0125\, B_n - 500$, $B_0 = 25{,}000$
(b) $B_n = -15{,}000(1.0125)^n + 40{,}000$ (c) $B_{48} = $12,769.68   **11.** $206.07   **13.** $6717.59
15. $16,489,380.75   **17.** 118.32 million

Index

A

Absorbing chain, 360–61
  and gambler's ruin, 371–72
Absorbing state, 360–61
Absorption times, 364
Accumulated sum, 468–69
Algebra
  of linear programming, 160–92
  matrix, 115–59
Alleles, 377
Alternatives
  complete set of, 297
Amortization, 480
  payment formula, 481
Annuity, 482
  present and future value of, 484
  sinking fund, 483
Artificial variables, 176
Augmented matrix, 136

B

Balance formula, 479
Balloon payment, 478
Basic point, 61–62, 78
Basic solution
  defined, 98
Basic variable, 97–98
Bayes, Thomas, n 298
Bayes' criterion, 326–27, 328

Bayes' formula, 296
  application, testing, 301–2
  defined, 297
Bayesian analysis, 330–31
Bernoulli, Jacob, n 305, 371
Bernoulli process, 305–6
  and absorbing chains, 367, 371–72
  applications, 308–11
  defined, 305, 441
  Formula, 307, 336
  See also Binomial distribution
Binomial coefficients, 221, 222
Binomial distribution, 441
  application of, 444–45
  normal approximation to, 455–58
  See also Bernoulli process
Binomial theorem, 221
Bounding lines, 57–59
Break-even analysis, 27
Break-even point, 27

C

Canonical transition matrix P^*, 363–64
Central tendency, 424–26
Certain event
  defined, 235
Chance, games of
  See Games of chance
Chance nodes, 329–31
Choice-nodes, 329–31

Classical formula, 249
Coefficients
  matrix of, 136
Column vector, 116, 118
Combinations, 213–14
  defined, 213
  and factorials, 216–17
  formula, 214, 217
  See also Counting
Comparison class, 243
Complement A^c, 196
Complement Rule, 261
Complete set of alternatives, 297
Compound interest, 468
  present value formula, 470
Conditional probability, 272–78
  applications, 276–78
  $P(A|E)$, defined, 274
Constant(s)
  negative, 173
  vector of right hand, 136
Constant linear function, 24
Constraints, 42
  equality, 187
  functional, 161
  of $\geq$ type, 173
  inequality sign in, 44
  joint, 57
  nonnegativity, 161
  right-hand, 161
  See also Inequality

Constraint sets, 57
  closed, 61
  corner points of, 61–62
  isolines and, 66–67
Continuous distributions, 449–50
Convex, 61
Convex polygonal sets, 61
Convex polyhedral sets, 77
Coordinate, 1, 3
  rectangular, 2
Coordinate axes, 2–3
Coordinate planes, 74
Coordinate system, 1–2
  xy, 2
Corner point, 58, 61–62
  in nonstandard programs, 174–76
  theorem, 67–68, 77–78
Cost
  fixed, 27
  and matrix formula, 122–23
  total, 27
  variable, 27
Cost and revenue lines, 28
Counting
  and combinations, 213–14, 216–17
  factorial notation, 216–17
  and games of chance, 252–53
  multinomial coefficients and, 225–27
  and permutations, 210–12, 213, 214, 216–17
  principle, 204–5
  principle, examples, 205–8
  and sampling, 254–55
  scheme, 203
  techniques of, 203–7
  tree diagrams and, 203–4
  Venn diagrams and, 199–200
  See also Set(s)
Cubic functions, 108
Curve
  fitting, 108–9
  normal, 451–52, 453

D

Data
  grouped and ungrouped, 423
  qualitative and quantitative, 421–22

Decision making
  value of perfect information for, 335–36
Decision theory, 325
  applications, 325–28, 329–31, 331–32
  under uncertainty, 325
Decision trees, 329–31
Decision variables, 42, 161
Declining-balance depreciation, 471
Decreasing linear function, 24
Degenerate standard maximum program, 168
Degree of the polynomial, 108
Demand D, 151
De Méré, Chevalier, 284, 308, 309
De Moivre-Laplace Limit Theorem, 457
Density curve, 450
Dependent variable, 23
Depreciation, 471
  declining-balance, 471
  sinking fund, 483
Descartes, René, 1
Deviation
  expected squared, 439
  squared, 426
  standard, 426, 439, 443
Diagrams
  stochastic, 282–83, 299, 300, 301, 341
  transition, 340
  tree, 203–4, 236–37
  Venn, 196–200
Difference equations, 464
Dimension $m \times n$, 115
Disjoint sets, 196
Dispersion
  measures of, 426–27
Distribution
  binomial, 441
  continuous, 449
  normal, 451–52
Dominant gene, 377
Dot product $A \cdot 20 \cdot B$
  defined, 126
  examples, 126–27

E

Economy
  n-sector theorem of, 152
  See also Input-output theory
Effective rate, 470
Elementary operations, 84
Elements, set, 193
Elimination
  Gauss-Jordan, 83–93
  Gauss-Jordan, $m \times n$, 99
  Gauss-Jordan, $n \times n$, 83, 89
  method, 10–12, 14–15
Empty set, 194
End-nodes, 329–31
Entries, 115
  scalar, 116
Equality $A = B$, 195
Equality constraints, 187
Equality of matrices, 117
Equations
  difference, 464
  first-order linear difference, 464
  general first-order difference, 476
  linear, 3–4, 5, 7
  linear, in the unknowns, 84
  linear, in x, y, and z, 74–75
  normal, 111
  objective function, 161, 162
  point-slope form, 21–23
  systems of, 9–15
  systems of linear, 83–114
  systems of scalar, 117
  See also Formula(s)
Equilibrium vector, 351–52, 353–54
Event E
  defined, 235
Events
  as sets, 259
  independent, 285–86
  mutually exclusive, 266
  operations on, 260
Expected gain, 321
Expected payoff, n 326, 400–402
Expected squared deviation, 439
Expected value, 314–21
  applications, 316–21
  defined, 315
  formula, 315

Experiment
  defined, 234
  repeatable actions within, 304–6, 307–11
  stochastic process, 281–82
  sums of random variables within, 324

F

Factorial notation, 216
Feasibility, 42, 73
  standard maximum program, 161–62
Feasible point, 57
Fixed costs, 27
Formula(s)
  accumulated sum, 469
  amortization, 480
  annuity, 482
  balance, 479
  Bayes', 296–302
  Bernoulli's, 307, 336
  classical, 249
  counting, 211, 212, 214, 217
  difference, $m \times n$ matrix, 119
  expected payoff, 401
  expected value, 315
  future value, 484
  general linear equation, 3, 38
  laws of matrix algebra, 120–21, 130–31, 158
  mean, 424–24, 438, 439
  mean deviation, 431
  permutation, of n objects, 211–12
  point-slope form, 21, 38
  present value, 484
  probability $P(E)$, 237, 239
  relative frequency, 238
  simple interest, 467
  slope, 18, 38
  slope-intercept form, 20, 38
  sums of random variables, 324
  variance, 427, 439
  See also Equations
Free variable, 97–98
Frequency distribution, 423
Function, 17
  linear, 65

  objective, 161
  revenue, 66
Functional constraints, 161
Fundamental matrix, 363
Future value, 470, 483–84

G

Gambler's ruin, 371–73
  applications, 374–75
Games, 388–420
  duration of play, 375–76
  with positive value, 409
  value of, 403
  See also Games of chance
Games of chance
  expected value and, 318–21
  gambler's ruin and, 371–73
  matrix, 388–420
  nonzero-sum and n-person, 416–17
  and probability, 250–54
  simplex solution of $m \times n$, 409–16
  of strategy, 398–406
  strictly determined, 388–96
  value of defined, 319
  See also Games
Gauss, Carl F., n 83
Gauss-Jordan elimination
  mastery techniques, 91
  $m \times n$ system, examples, 99–102
  $m \times n$ system solution, 99
  $n \times n$ system, 83–93
  $n \times n$ system, examples, 84–91
  $n \times n$ system, procedure, 89, 136–37
  and product-mix problems, 104–8
  See also Matrix, inversion
Gene, 377
General first-order difference equation, 476
General solution, 108
Genetics, 377–79
Genotypes, 377
Geometric progression, 465
Geometry of the invisible, 78–79
Graph(s)
  cost and revenue line, 28
  histogram, 422

  by intercepts, 4–5
  isoline, 66
  linear equation, 3–4, 5–7
  linear equation, defined, 3
  linear equation in x, y, and z, 75
  linear inequality, 51–55
  of polynomial functions, 108
  profit line, 29
  system of joint constraints, 57–61
Gross production, 150
Grouped data, 423

H

Half-plane, 51
Hardy-Weinberg Principle, 380–82
Histogram, 422
House percentage, 319
Huygens, 371
Hybrid genotype, 377

I

Identity matrix I
  defined, 131
  examples, 132
Impossible event
  defined, 235
Inclusion $A \supset B$, 195
Inconsistent system, 12, 93–94
Increasing linear function, 24
Independent events, 285–86
  defined, 286
Independent variable, 23
Indicators, 162
Inequality
  laws, 48–49
  laws, examples of, 49–55, 81
  linear, 48–55
  in linear programs, 41–45
  signs, weak and strict, 48
Initial simplex tableau formation, 161–62
Initial value x_0, 464
Input-output
  matrix A, 148–50
  ratio a_{ij}, 148–49
  theory, 147–55

Installment loans, 478–80
Intensity, 149–50
Intensity vector X, 150
Intercepts, 4–5
Interest
    compound, 468
    nominal rate, 470
    period, 467
    rate, 467
    simple, 466
Internal consumption vector U, 150
Intersecting lines, 9–15
Intersection $A \cap B$, 196
Inverse of a matrix A^{-1}, 139
Isolines, 66

J

Joint constraints, 57
    basic point for, 61–62
graphs of, 57–61

K

k-termed vector, 116

L

Least squares
    approximation, 113
    criterion, 111
Leibniz, Gottfried, n 224
Leontief, Wassily, 147
Leontief input-output model, 146–55
Linear
    depreciation, 471
    function, 65
Linear equation(s), 3–4, 5, 83–111
    forms, 7
    point-slope form, 21
    slope-intercept form, 20–21
    two, in two unknowns, 9–15
    in the unknowns $x_1, \ldots, x_n$, 84
    in x, y, and z, 74–75
Linear functions, 23–24
    applications of, 27–34
    first-degree polynomials as, 108

Linear inequalities, 48–55
    explained, 50
    graphing of, 51–55
    system of, 55, 57–60
Linear program, 40–42, 65, 160
    Corner Point Theorem of, 68
    examples, 42–45
    formation of, 72–78, 187
    geometric solution of, 65–71
    geometry of, 40–82
    matrix game using, 409–17
    models, using simplex algorithms, 181–87
    optimal solution, 43, 171
    solution of, 75–78
    standard maximum, defined, 161
    See also Simplex algorithm
Linear systems
    analysis of, 93–102
    applications of, 104–11
    matrix solutions of, 135–44
    $m \times n$, 84, 99
    $n \times n$, 89, 95–96, 97, 148
Long-run stability, 362–63

M

Markov, A. A., n 339
Markov chains, 339–47
    absorbing, 360–61, 367, 371–73
    applications, 346–47
    conditions of, 339
    and genetics, 379
    regular, 352–54
    Theorem 1, 353
    Theorem 2, 364
    transient states, 361
Mathematical modeling, 41
Mathematical expectation of X, n 315
Matrix
    of absorption probabilities A, 364
    algebra, 115–59
    applications, 121–23
    augmented, 136
    of coefficients, 135–36
    defined, 115
    dot product $A \cdot B$, 126
    equality of, 117
    form, linear systems in, 135–36

fundamental, 363
identity, 131
input-output, defined, 148
input-output, example, 148–50
inversion, 138–39
labor, 127–28
laws of, algebra, 120–21, 130–31, 132, 158
multiplication, 125–35
multiplication, defined, 128
noninvertible, 142–43
operations on, 118–21
payoff, 325–27, 389
solution of $AX = B$, 140–41
square, 115
transition, 341
    See also Matrix game(s)
Matrix game(s), 388–89
    applications, 394–96
    and linear programs, 410–11
    reduced form of, 390–91
    of strategy, 398–406
    strictly determined, 393
    value, 393
    See also Games; Matrix
Matrix product AB, 128
    example, 128–30
Maximax criterion, 326
Maximin criterion, 326
Mean, 438
    defined, 424
    formula, 439
    Theorem 2 on, 443
Mean deviation formula, 431
Mean value of X, n 315
Median, 425–26
Members, set, 193
Mendel, Gregor, 377
Mendelian Theory, 377–79
Midpoint, 425
Minimax optimal moves, 392, 393
Minimax optimal strategies, 402
Minimax Theorem, 402–3
Minimum program, 44
Moves, 388
Multinomial coefficients, 224
    examples, 225–27
Multinomial theorem, 224
Mutually exclusive events, 266

$m \times n$ linear system, 84
  in matrix form, 136
$m \times n$ matrix, 115
  difference formula, 119
  form, 116
  See also Matrix game(s)

N

Negative indicators, 171
Negative of a matrix $-A$, 118–19
Net production vector S, 150
Nodes, 329–31
  value of, 330
Nominal rates, 470
Noninvertible matrices, 142–43
Nonnegativity constraints, 161
Nonstandard program, 172
  negative constants in, 173
Nonzero-Sum games, 416–17
Normal curve, 451–52
  area under, 453
Normal distributions, 449–59
n-person games, 416–17
Null set, 194
n-variable program
  geometric solution of, 79
$n \times n$ linear system
  cases, review, 97
  inconsistency, 93–94
  redundancy, 95–96
  solution of, 89
$n \times n$ matrix A
  inversion of, 139

O

Objective function, 42, 161
Octants, 74
Odds
  defined, 243
Operations
  algebraic, on matrices, 118–19
  elementary types, 84, 86
  on events, 260
  matrix multiplication, 125–35
  pivot transformation, 88

Optimal solution(s), 43, 164–65
  multiple, determination of, 171
Optimization, 40
Ordered pair, 2
Ordered partition, 226
Origin, 1–3
Overdetermined, 101–2

P

Parabola, 109
Parallel strip, 59
Parameter, 97
Pascal, Blaise, n 222, 233, 250, 284, 308
Pascal's Triangle, 222, 223
Payoff, 388
  expected, 400–402
  matrix, 325–27, 389
Periodic rent, 482
Permutation, 210–11
  defined, 210
  examples, 210–12
  factorials and, 216–17
  formula, 211, 212, 217, 231
  See also Counting
Phase 1 pivots, 174–76
Pivot transformation, 88–89
  on simplex algorithm, 162–65
Plane(s)
  coordinate, 74
  half-, 51
  linear equations as, 74–75
Players, 388
  row and column, 392, 402
Point-slope form, 21–23
Polynomial functions, 108
Population shifts, 346–47
Positive direction, 1
Predictive value, n 301
Present value, 470, 483–84
Primary input, n 151
Prime numbers, 194
Principal, 466
Probability
  basic laws of, 261, 263–64, 266, 282, 287, 294
  and Bayes' formula, 298–99

Bernoulli process and, 304–6, 307, 311, 336
  classical formula, 249
  conditional, 272–78
  conditional, applications, 276–78
  conditional, $P(A|E)$, defined, 274
  and decision making, 325–34
  defined, 233
  defined empirically, 237
  density curve, 450
  and diagnostic testing, 301–2
  distribution. *See* Probability distribution
  of event E, defined, 239–40
  expected value of, 314–15
  of gambler's ruin, 371–73
  and games of chance, 250–54
  and genetics, 377–79
  and Markov chain, 339–47
  and odds, 243–44
  $P(E)$, 237
  space, 237, 248
  space, defined, 239
  and stochastic processes, 281–90
  Theorem 1, gambler's ruin, 373
  theory, 233–44
  transition, 340–41
  and uniform space, 248
  See also Matrix game(s)
Probability distributions, 435–45
  normal, 449–59
  Theorem 1 and, 437
  for X, 315
  See also Probability
Producing sector, 147
Product-mix problems, 104
Profit, 29
  functions, 28–30
  functions, comparison of, 30–34
Progression, 464
  geometric, 465
Pure dominant genotype, 377
Pure recessive genotype, 377

Q

Quadrants, 3
Quadratic functions, 108

R

Raiffa, Howard, 321
Random processes, 233
Random variable, 314, 435
Rate of change, 23, 25
Rates of convergence, 357
Real number line, 1–2
Recessive gene, 377
Rectangular coordinates, 1, 2
Reduced form, 390–91
Reduced sample spaces, 272–73
Redundant system, 13, 95–96
Relative frequency, 237, 238
Revenue, 27
  lines, 28
Revenue function, 66
Rise over the run, 18
Risk aversion, 328–29
Row vector, 116, 118, 343

S

Saddle points, 393
Sample points, 234–35
Sample space(s)
  defined, 234
  reduced, 272–73
Sampling, 254–55
Scalar, 116
  system of equations, 117
Scalar product cA, 118–19
Segregation, law of, 378
Sequence, 464
Set
  -builder $\{x|P(x)\}$, 193–94
  complete, of alternatives, 297
  defined, 193
  disjoint, 196
  empty or null, 194
  operations on, 195–96
  ordered partition of , 226
  theory, elements of, 193–200
  universal, 194
  Venn diagrams, 196–200
  See also Counting; Probability
Simple interest, 466
  formula, 467
  present value formula, 470

Simplex algorithm, 79
  minimum programs, 172
  nonstandard programs, 172–78
  nonstandard programs, summary, 178
  sequence for, 163
  standard maximum programs, 160–68, 172
Simplex pivot, 162–65
  meaning, 166–68
Simplex solution
  of $m \times n$ games, 409–17
  procedure, 414
Simulation, 258
Sinking fund, 483
Slack variable, 161
Slope, 17–19
  defined, 18
  -intercept form, 20–21
Special Union Rule, 266
Squared deviations, 426, 439
Square matrix, 115
Standard deviation, 426, n 426
  Theorem 2 on, 443
Standard maximum program, 160
  defined, 161
  degenerate, 168
  optimal solution for, 168
Standard units, 452–53
State, 340
State vector $x^{(n)}$, 342–44
Statistical sampling
  Bernoulli's formula in, 309
Statistics, 421–59
  descriptive, 421–31
  normal distributions, 449–59
  probability distributions, 435–45
Stochastic diagrams, 282–83, 299, 300, 301, 341
Stochastic processes, 281–90
Straight-line depreciation, 471
Strategy(ies), 398
  defined, 399
  games of, 398–406
  minimax optimal, 402
  pure and mixed, 399–400
Strictly determined game, 393
Subscripts, matrix, 116
Subset, 195
Substitution method, 9

Success runs, 367
Sum $A + B$, 118–19
Summation symbol, 424
Surplus, 149–50
Surplus vector S, 150

T

Tableau
  form, 86
  initial, simplex, formation, 161–62
  terminal, 164
Theorem(s)
  absorbing Markov chain, 364
  binomial, 221
  $AX = B$, 137
  Corner Point, 67–68, 77–78
  De Moivre-Laplace Limit, 457
  gambler's ruin, 373
  Markov chain, 353
  mean and standard deviation in
    Bernoulli trials, 443
  Minimax, 402–3
  multinomial, 224
  n-sector economy, 152
  probability distribution, 437
Three-dimensional space, 73–74
Total cost, 27
Transient states, 361
Transition diagram, 340
Transition matrix
  canonical form, 363
  defined, 341
  examples, 341–42
  properties of, 341
Transition probabilities, 340
Transpose A^t, 118
Tree diagrams, 203–4
  and experiments, 236–37
Tucker, A. W., 416
Two-person zero-sum games, 388–89

U

Underdetermined, 99–101
Ungrouped data, 423
Uniform spaces, 248
Union $A \cup B$, 196

Union Rule, 263–64
  Special, 266
Unit length, 1, 2
Universal set, 194
Utility, 321, 328

V

Value
  future and present, 470, 483–84
  of games, 403
  initial, x_0, 464
  of perfect information, 335–36
  wearing, 471
Variable(s), 17
  artificial, 176
  basic and free, 97–98
  decision, 42, 161
  random, 314, 435

  slack, 161
  sums of random, 324
Variable costs, 27
Variance, 426–27, 438
  defined, 426, 439
Vector, 116
  of absorption times T, 364
  demand, 151
  equilibrium, 351–52, 353–54
  intensity X, 150
  internal consumption U, 150
  of right hand constants, 136
  stable state, 352
  state, 342–43
  surplus or production, 150
  of unknowns, 135–36
Venn diagrams, 196–200
  counting applications, 199–200
Venn, John, n 196
Von Neumann, John, 388, n 388

W

Wearing value, 471

X

x-axis, 2
x-intercept, 4–7, 75

Y

y-axis, 2
y-intercept, 4–7

Z

Zero matrix, 120
z-units, 452–53

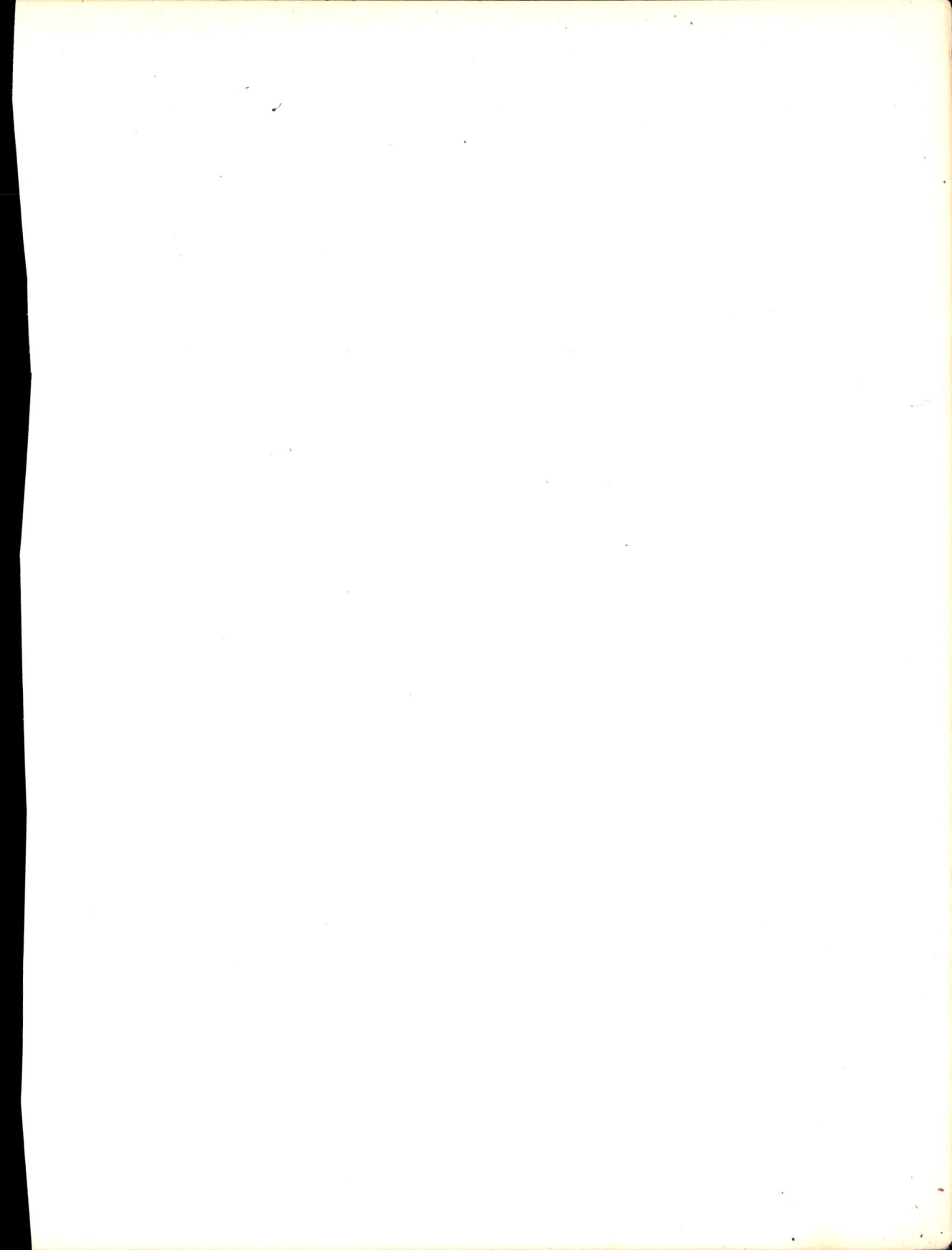

arbitration of wage dispute, 397
ambassador's convoy vs. guerrillas, 408
escaped inmate vs. search parties, 418
army strategies, 418, 420
average age of U.S. population, **428**
moving violations of motorists, 445, 447
educational attainment of unemployed, 446, 447
true-false examination, 448, 461
voters favoring tax cut, 449, 461
scholastic aptitude test scores, 461, 462
voters who are senior citizens, 463
scholarship endowment fund, 485
savings account for college education, 486
medical institute endowment for research, 487

Life Sciences, Health, Agriculture, Environment

nutrition, 16, **43**, 46, **68**, 111, 145, 180
refining oil, 16, 47
mixing chemical solution, 17
nutrients for fish, 17, 38
exporting decreasing oil reserves, **25**
Celsius temperature, 27
oxidant pollution controls, 27
ecology group's fund-raising dinner, 36
combining grains in silo, **54, 181**
blending grades of gasoline, 188
growing vegetable crops, 46, 56, 170
emission control in producing chemicals, 48
mixtures of bronze alloys, 56, 112, 188
blending grades of crude oil, 56, 112
mining grades of ore, 80, 111, **141**, 180
reducing pollutants from steel plant, 111
mixing types of fertilizer, 111
mining ore yielding titanium, 191
vitamins taken by athletes, **198, 199**
tests for diabetes, 202
blood types of hospital patients, 202
strands of DNA molecules, 230
astronaut candidates, **234**, 244
Nobel Prize finalists in biology, **241**
oil yield of drill sites, 246, 294, 322, 335
heart bypass successful, 247
pollution will melt polar ice caps, 247
drilling for natural gas, **265**
landing of Venus probe, 270, **288**
headache drug research, **274**
Framingham cardiovascular research study, 279
geneticist's observations of traits of plants, 291

women getting breast cancer, 292
sickle cell anemia, **301**, 312, 385
new oil prediction method, 302
blood bank screening for hepatitis, 303
physician's diagnosis, 304
congenital disease of island inhabitants, 304
home pregnancy test, 304
recovery rate for disease, 312
test whether drug cures disease, 313
association of blood antigen and disease, 336
sex of newborn children, 336
farmer deciding on cloud-seeding operation, 338
parasitic disease model, **341, 345, 354,** 360
patient's state of health, **357**
childhood disease states, 370
albinism, carriers, extinction of, **379**, 387
carrier of fibrocystic disease, **380**, 384
carriers of phenylketonuria, **383**
mating of gene-types, 384, 385
genotype for snapdragon color, 384
rabbit carriers of albinism, 385
genotype distribution, 385, 387
recessive gene carriers in family tree, 385
children of doctors, 387
per capita health expenses, 431
temperature reading in September, 432
systolic blood pressures, 432, 433, **455**
recovery with drug, **444**, 449, **457**, 461, 463
infants' needs in pediatric ward, 449, 461
distribution of rainfall in a city, **454**
heights of adult males, 461
heights of trees in botanic research, 463
rate of population increase, 475, 489
carbon-14 decrease over centuries, 475
elimination of drug through kidneys, 475
harvesting of redwoods by lumber company, **477**
medical personnel of Xanadu, 487
doses of antibiotic to patient, 489

General Interest, Sports

smuggling jewelry, 17
buying nails for home repair, 82
dogfood nutrition, 82
tickets for class at amusement park, 192
everyday combinations, **205**, 208, **214**, 218, 219
music groups' billing in ad, **205**
coin toss, **205, 241, 308, 367, 441, 443, 456**
Scrabble, 206